Chinese and Indian Warfare – From the Classical Age to 1870

This book examines the differences and similarities between warfare in China and India before 1870, both conceptually and on the battlefield. By focusing on Chinese and Indian warfare, the book breaks the intellectual paradigm requiring non-Western histories and cultures to be compared to the West, and allows scholarship on two of the oldest civilizations to be brought together. An international group of scholars compare and contrast the modes and conceptions of warfare in China and India, providing important original contributions to the growing study of Asian military history.

Kaushik Roy is Guru Nanak Professor in the Department of History at Jadavpur University, India and Global Fellow at PRIO, Norway.

Peter Lorge is Assistant Professor in Chinese History at Vanderbilt University, USA.

Asian states and empires

Edited by Peter Lorge
Vanderbilt University

The importance of Asia will continue to grow in the twenty-first century, but remarkably little is available in English on the history of the polities that constitute this critical area. Most current work on Asia is hindered by the extremely limited state of knowledge of the Asian past in general, and the history of Asian states and empires in particular. *Asian States and Empires* is a book series that will provide detailed accounts of the history of states and empires across Asia from earliest times until the present. It aims to explain and describe the formation, maintenance and collapse of Asian states and empires, and the means by which this was accomplished, making available the history of more than half the world's population at a level of detail comparable to the history of Western polities. In so doing, it will demonstrate that Asian peoples and civilizations had their own histories apart from the West, and provide the basis for understanding contemporary Asia in terms of its actual histories, rather than broad generalizations informed by Western categories of knowledge.

1 **The Third Chinese Revolutionary Civil War, 1945–49**
An analysis of Communist strategy and leadership
Christopher R. Lew

2 **China's Southern Tang Dynasty, 937–976**
Johannes L. Kurz

3 **War, Culture and Society in Early Modern South Asia, 1740–1849**
Kaushik Roy

4 **The Military Collapse of China's Ming Dynasty, 1618–44**
Kenneth M. Swope

5 **China's Second Capital – Nanjing under the Ming, 1368–1644**
Jun Fang

6 **Rethinking the Decline of China's Qing Dynasty**
Imperial activism and borderland management at the turn of the nineteenth century
Daniel McMahon

7 **Civil–Military Relations in Chinese History**
From ancient China to the communist takeover
Edited by Kai Filipiak

8 **Chinese and Indian Warfare – From the Classical Age to 1870**
Edited by Kaushik Roy and Peter Lorge

9 **The East Asian War, 1592–1598**
International relations, violence, and memory
Edited by James B. Lewis

10 **Warfare in Pre-British India – 1500 BCE to CE 1740**
Kaushik Roy

Chinese and Indian Warfare – From the Classical Age to 1870

Edited by Kaushik Roy and Peter Lorge

Routledge
Taylor & Francis Group
LONDON AND NEW YORK

First published 2015
by Routledge
2 Park Square, Milton Park, Abingdon, Oxon OX14 4RN

and by Routledge
711 Third Avenue, New York, NY 10017

Routledge is an imprint of the Taylor & Francis Group, an informa business

British Library Cataloguing in Publication Data
A catalogue record for this book is available from the British Library

Library of Congress Cataloging in Publication Data
A catalog record for this book has been requested

ISBN: 978-0-415-50244-3 (hbk)
ISBN: 978-1-315-74276-2 (ebk)

Typeset in Times New Roman
by Wearset Ltd, Boldon, Tyne and Wear

Printed and bound in the United States of America by Publishers Graphics, LLC on sustainably sourced paper.

The editors respectfully dedicate this volume to Professor Jeremy Black, who has done so much to support and promote the study of non-Western military history.

Contents

Notes on contributors xii

Introduction 1
KAUSHIK ROY AND PETER LORGE

PART I
Military theory/theories of warfare in China and India 15

1 **Opportune moments in early Chinese military thought: the concept of *ji* 機 in the Warring States period manuscript *Cao Mie's Battle Array*** 17
EARNEST CALDWELL

2 ***Yuddha* and *Vijaya*: concepts of war and conquest in ancient and early medieval India (up to CE 1300)** 32
KRISHNENDU RAY

PART II
Technology, geography and warfare 55

3 **Chinese border garrisons in an international context: Liaodong under the early Ming dynasty** 57
DAVID M. ROBINSON

4 **Elephants in pre-modern India** 74
DEBRAJ CHAKRABORTY

5 **British-India and Afghanistan: 1707–1842** 91
KAUSHIK ROY

6 **Technological advance in the war against the Mongols** 121
KAI FILIPIAK

7 **Bringing in the big guns: on the use of artillery in the Ming–Manchu war** 134
KENNETH M. SWOPE

8 **Battles, boats and bridges: Mughal amphibious warfare, 1571–1612** 146
PRATYAY NATH

9 **Indo-Portuguese naval battles in the Indian Ocean during the early sixteenth century** 166
K.S. MATHEW

10 **The politics of military control in the west coast: Marathas, Mughals and the Europeans, 1650–1730** 181
AMARENDRA KUMAR

PART III
Military culture, state and society 201

11 **Command, control and castration: eunuch supervisors in the armies of the Tang dynasty** 203
DAVID A. GRAFF

12 **The cultural construction of war in Tang frontier poetry** 214
MING JIANG

13 **Martial values in painting: Chinese bannerman painters at the Qing Court** 224
NIXI CURA

14 **Disorder in the general staff: a corruption case during the First Jinchuan War (1747–1749)** 234
ULRICH THEOBALD

15 **Fortified walls and social ordering in China's late eighteenth-century revolts** 246
DANIEL MCMAHON

16 **Total war: military supply and civilian resources during China's era of rebellions** 257
ELISABETH KASKE

17 **European military experience in South Asia: the Dutch and British armies in Sri Lanka in the eighteenth century** 289
CHANNA WICKREMESEKERA

18 **Military revolution and state formation reconsidered: Mir Qasim, Haider Ali and transition to colonial rule in the 1760s** 302
DOUGLAS M. PEERS

19 **Contrasting Indian and British concepts of race and authority in the East India Company armies** 324
MICHAEL H. FISHER

20 **The East India Company, the Indian Army and the China Wars, 1839–1860** 347
ROB JOHNSON

Index 368

Contributors

Earnest Caldwell is a lecturer in Chinese Law at the School of Oriental and African Studies, specializing in the study of Chinese law, legal linguistics, socio-legal studies and comparative constitutional law. His research on traditional and contemporary Chinese legal and political thought has been published in several journals including the *Law and History Review*, *Chicago-Kent Law Review* and *Early China*.

Debraj Chakraborty has recently completed his doctoral research on environmental and medical history of pre-colonial India from the Department of History, Visva-Bharati (Santiniketan, India). He had been Junior Research Fellow of the Indian Council of Historical Research, New Delhi.

Nixi Cura is Programme Director, Arts of China, at Christie's Education London/University of Glasgow and serves concurrently as Honorary Research Fellow at Glasgow. Her research interests include art of the Qing (1644–1911), especially during the Qianlong reign (1736–1795), collecting and antiquarian practices in the Qing, Republican (1911–1949) and Manchukuo (1932–1945) periods, and contemporary Chinese visual culture. In 2002 she co-founded and remains active on the board of the Arts of China Consortium.

Kai Filipiak is Privatdozent and teaches Chinese history at the University of Leipzig. He is author of two books: on the history of Chinese martial arts (2001) and the impact of military conflicts on the Chinese state during the Ming dynasty (2008). He is currently preparing a conference volume entitled *Civil–Military Relations in Chinese History: From Ancient China to the Communist Takeover*.

Michael H. Fisher holds the Danforth Chair in History at Oberlin College, USA. He has published various books on interactions between Asians and Europeans in Asia and Europe including: *Inordinately Strange Life of Dyce Sombre: Victorian Anglo Indian M.P. and Chancery 'Lunatic'* (London: Hurst; New York: Columbia University Press; Delhi: Foundation Books, 2010); *Counterflows to Colonialism: Indian Travellers and Settlers in Britain, 1600–1857* (Delhi: Permanent Black, 2004); and *Travels of Dean Mahomet* (Berkeley, CA: University of California Press, 1997).

David A. Graff is Associate Professor of History at Kansas State University. He is the author of *Medieval Chinese Warfare, 300–900* (London: Routledge, 2002) and co-editor of the *Journal of Chinese Military History*.

Ming Jiang is Professor of Chinese and Asian Studies in the Department of Languages and Cultures and the Asian Studies Program at the William Paterson University of New Jersey. He has published in the areas of Chinese literature, Chinese cinema, German literature and Chinese translations of Max Weber and Alexander Gottlieb Baumgarten.

Rob Johnson is Lecturer in the History of War and Deputy Director of the Changing Character of War Programme at University of Oxford, UK. His latest publication is *The Afghan Way of War* (London: Hurst, 2011).

Elisabeth Kaske is a historian of nineteenth- and early twentieth-century China. She holds a PhD from Heidelberg University in Germany and published two monographs, *Bismarcks Missionäre: Deutsche Militärinstrukteure in China (1884–1890)* and *The Politics of Language in Chinese Education (1895–1919)*, before turning to the study of the political economy of selling public offices and rank during the last decades of the Qing Dynasty. She is currently Associate Professor of Chinese Studies at Carnegie Mellon University in Pittsburgh.

Amarendra Kumar has a PhD on Maratha Navy and is Assistant Professor at the History Department of Viva-Bharati University, Santiniketan, India.

Peter Lorge teaches Chinese and military history at Vanderbilt University, specializing in tenth- and eleventh-century Chinese military history and thought. His recent publications include *The Asian Military Revolution: From Gunpowder to the Bomb* (2008) and *Chinese Martial Arts: From Antiquity to the Twenty-First Century* (2012). He is currently working on a history of Chinese military thought from the third to the thirteenth centuries.

K.S. Mathew is now the Honourable Director of the Institute for Research in Social Sciences and Humanities at Nirmalagiri, Kannur. He is an expert on the Portuguese interaction with India. He has published 21 books and 105 research articles.

Daniel McMahon is a professor in the Department of History at Fu Jen Catholic University, Taiwan. He specializes in the study of Qing China's borderland administration. His most recent publication is *Rethinking the Decline of China's Qing Dynasty: Imperial Activism and Borderland Management at the Turn of the Nineteenth Century* (Routledge, 2014).

Pratyay Nath is completing his PhD on Mughal Warfare from the Centre for Historical Studies, School of Social Sciences, at Jawaharlal Nehru University, New Delhi.

Douglas M. Peers is Professor of History and Dean of Arts at the University of Waterloo. He is co-editor of *India and the British Empire* for the Oxford

History of the British Empire, as well as other books and more than 20 scholarly articles addressing the relationship between imperialism and the military.

Krishnendu Ray is Reader in the Department of Ancient Indian History and Culture at the University of Calcutta, India. He has published in the *Journal of Epigraphical Society of India.*

David M. Robinson teaches East Asian history at Colgate University, specializing in early modern China. His recent publications include *Culture, Courtiers, and Competition: The Ming Court* (2008), *Empire's Twilight: Northeast Asia under the Mongols* (2009) and *Martial Spectacles of the Ming Court* (2013). His current work focuses on China and Korea in the wake of the Mongol empire's collapse in the fourteenth and fifteenth centuries.

Kaushik Roy is Global Fellow at the Peace Research Institute Oslo (PRIO), Norway, and Guru Nanak Chair Professor, Department of History, Jadavpur University, Kolkata, India. He has edited and published 22 books from various publishers like Cambridge University Press, Oxford University Press, Routledge, E.J. Brill, Ashgate, Bloomsbury, and has written more than 50 articles in various edited volumes and peer reviewed journals like the *Journal of Global History*, *Modern Asian Studies*, *War in History*, *Journal of Military History*, *First World War Studies* and *Journal of Military Ethics.*

Kenneth M. Swope is Professor of History and Fellow of the Dale Center for the Study of War and Society at the University of Southern Mississippi (USA). He is the author of *The Military Collapse of China's Ming Dynasty, 1618–1644* (Routledge, 2014), and *A Dragon's Head and A Serpent's Tail: Ming China and the First Great East Asian War, 1592–1598* (Oklahoma, 2009), as well as numerous articles and book chapters about the military history of late imperial China. He has lived and conducted research throughout Asia and is currently working on a book on the Ming–Qing transition in south-west China.

Ulrich Theobald is Senior Lecturer for Chinese Studies at Tübingen University. He has published on military labour, war finance and logistics, border peoples, Chinese empresses and monetary issues. He currently works on the changes in the appointment of local officials in nineteenth-century China.

Channa Wickremesekera obtained his PhD from Monash University, Melbourne, Australia. He is based in Melbourne and is working on the history of separatist conflict in Sri Lanka. His latest publication is *Kandy at War* (New Delhi: Manohar).

Introduction

Kaushik Roy and Peter Lorge

Military history remains a Eurocentric enterprise, with Britain, the United States and Western Europe garnering the majority of scholarly attention. Until recently, the military history of Asia was barely studied even in Asia. A number of scholars have begun to publish world military histories that engage a much broader palette of cultures, times and practices and try to break out of the pre-existing Western paradigms of warfare. Jeremy Black has been a pioneer in this, actively reading and including in his publications every scrap of scholarship on non-Western warfare in Western languages humanly possible.[1] Geoffrey Parker made some effort to include non-Western warfare in his *The Military Revolution*, but his model remained Eurocentric.[2] Wayne E. Lee has attempted to study the variety of military cultures of different regions of the world[3] and Stephen Morillo made a valuable comparison of the impact of guns and government of Japan and the West. And mention may be made of Peter Lorge's (one of the editors of this volume) book that traces the origin and spread of gunpowder technology in Asia.[4]

It may not be possible completely to explain why the overwhelming majority of history written in English focuses on modern times, roughly the sixteenth century and later. Why this is the case is less important than its effect upon our understanding of history. Western military history concentrates on the period after gunpowder and gunpowder weapons reached Europe, and stresses the dramatic increase in military power of Western Europe.[5] Two of the most important paradigms in the field of military history have been the Military Revolution thesis and the Western Way in Warfare thesis. The Military Revolution thesis, first proposed by Michael Roberts, and later taken up, altered and broadened, by Geoffrey Parker, turns on the notion that the introduction of gunpowder weapons stimulated a dramatic series of military changes that, in turn, brought about similarly dramatic political and social changes in Europe.[6] The Western Way in Warfare thesis passionately argued by Victor Davis Hanson, a classicist, presented a manner of fighting in Ancient Greece, the direct clash of equally armed infantrymen on an open field without subterfuge or manoeuvre, as the most effective way to fight. This Ancient Greek Way of Warfare, Hanson believes, underlies all subsequent Western approaches to battle, and explains why the West has bested the rest, on balance, over the course of history, up to and including the present.[7]

Both of these ideas have crumbled in recent years, but in truth neither, particularly the Western Way in Warfare, was ever fully accepted. Despite their obvious shortcomings, they have had very positive effects on the field of military history, stimulating at least a generation of scholarship. Their decline among professional military historians has been much more rapid than in popular military history. While Victor Davis Hanson failed to connect Ancient Greek warfare with subsequent battles in 'the West' beyond asserting it to be so and cherry-picking a few battles here and there to support his view,[8] the Military Revolution interlocutors generally neglected pre-modern warfare entirely. These discussions excluded a long period of history stretching from Ancient Greece to sixteenth-century Europe, and a broad swathe of the globe that included most of Eurasia, the southern hemisphere, Africa, the Middle East and the Americas. Despite such lacunae, if one can even construe gaps in a scholarly record that is mostly empty as such, many scholars have felt comfortable making broad generalizations about the nature of warfare as a universal phenomenon based solely on what was available in Western languages. Since all that was available was written in Western languages, mostly about Europeans (Ancient Greeks and Romans fitting into this category) and Americans, then all warfare could be seen from this Western perspective. Any non-Western warfare would surely fit into the Western conception; if it did not, then it could be thrown into the undifferentiated, inferior and unimportant category of 'non-Western'. There was the West and the rest.

China and India: juxtaposing and interlinking military history

Jeremy Black rightly says that the twenty-first century's military history ought to pivot around Asia.[9] But, why China and India? Some justifications are necessary. China and India, the two rising economic and military giants of twenty-first-century Asia have a lot of commonalities in their history. Both had been gigantic agrarian bureaucratic empires that were threatened by the steppe nomads for nearly two millennia until the eighteenth century. Neither China nor India were able to breed good war horses which were essential in dealing with the speed and mobility of the mounted steppe nomadic archers. And before the colonial era, both China and India were substantial military powers. The different dynasties in both China and India maintained million strong armies long before post-Roman Empire Europe could think of that. The pre-modern regimes in both China and India had to devote massive military assets to countering peasant rebellions (counter-insurgency or COIN of modern terminology). And during the eighteenth century, the imperial regimes of both China and India fell prey to internal decay and were threatened by the Western maritime colonial powers.

This is not to say that the history of China and India were always similar. Confucianism was the dominant ideology among the ruling elites in pre-modern China, Brahmanism in the early medieval era and then Islam through the eighteenth century were the dominant ideologies of the ruling classes on the subcontinent. And, unlike India, China was never totally conquered by any Western colonial power. China and India sometimes followed similar trajectories of military development and sometimes did not. The two cultures often arrived at a different solution to the same problem.

There has been some work in the field of interlinked pre-modern Chinese and South Asian histories but nothing on the military.[10] This study is the first sustained discussion of Indian and Chinese military history. It brings together 20 scholars from around the world to present their individual snapshots of the military history of China or India. Consequently, it is very much a first attempt at bringing not only two different military histories, but also two different historiographical traditions under one roof. We have not imposed a master plan on these contributions because to do so would be to obscure the separate historiographies in order to invent a common military history. The resulting *bricolage* (in Joe L. Kincheloe, rather than Claude Lévi-Strauss, or Jacques Derrida's use of the term) begins to paint a picture whose outline and ultimate form is still uncertain. In many ways the two historiographies talk past each other, one emphasizing questions and problems ignored by the other. Yet, there are also many cases of similarity. These should not be seen as dichotomies. We seek to juxtapose, rather than compare, two histories.

The concept of comparison is freighted with hegemonic meaning because of its history of use as a method intellectually to subordinate one civilization to another. Bringing together scholarship on the military history of two great, ancient civilizations implies that a certain amount of comparison will take place as the contributors wrestle with the analytical problem of which civilization did better than the other in the military sphere. This comparison of cultures has most often been done between the West, however that is conceived, and someone else. Frequently that someone else is China or India, with the a priori assumption that the West was and is, of course, superior in all things. The discussion is then one of why the other was not as good as the West, or failed to respond effectively to the arrival of the West by adopting Western culture and technology. The West versus the rest paradigm has hitherto dominated much of the military history scholarship in English that touches upon the non-West.

Our goal here is thus emphatically not comparative. We do not seek to compare Chinese and Indian military history, nor are we trying to replace one failed, Eurocentric model with another Sino-Indian one, or even a broader global paradigm. There are several reasons for this, apart from the intellectual inadequacies of the previous approach. First, both Indian and Chinese military histories are young fields of research, China more so than India, and the available scholarship is not yet adequate to make meaningful generalizations. Second, it is not clear that a set of culture specific paradigms relating China to India is useful. These two civilizations had a long history of contact, but until very recently their militaries did not interact. Third, there is no particular reason why there should be inherent commonalities between India and China just because they are both Asian cultures. There is no 'oriental' mind shared by these two cultures. In other words, the chapters collectively argue that despite the presence of some commonalities at certain moments of time, there has been no typical Indian or Chinese Way of Warfare throughout history.

In comparing warfare in China and India we are faced with several immediate questions, and several ways of comparing their military histories. The most obvious question is whether there is any value in comparing the warfare of these cultures? Assuming comparative history is valuable, why compare these cultures' warfare and not others? Are they in any way comparable? If so, how? When comparing the history of any two cultures there are a number of problems in the very process, which related directly to these larger questions. Since, by definition, we must compare the histories of these cultures, there is a basic issue of different understandings of history writing in both the primary and secondary sources. Thus it is not only what people in the past chose to write down and all the attendant problems with that, but also the shape of the modern scholarly field.

It is natural for these disparate histories and scholarships to lack coherence, but they are fundamentally united by the universal human phenomena of war. With respect to intellectual category, all of these contributions fall within the realm of military history. Juxtaposing these histories argues that they are all equally valid in the search for knowledge. Given the continuing resistance in Western academia, particularly in America, to military history and even, it must still be admitted, to Asian history, it is no small issue to create the seed of an Asian military history field. We do not claim here to have done more than begin what we see as an important conversation that must take place in Asian, military and world history. We are not so ignorant, to paraphrase Martin Luther, that we do not know what ignorance is.

Having yielded the field in overall coherence, however, we have held our ground on quality. It is always worthwhile to draw together excellent scholarship from around the world. We have grouped the contributions together around certain aspects that several seem to have in common without regard for India–China balance, chronology or evenness of coverage. We have divided the chapters into three broad categories: theories of warfare; armed forces, technology and warfare; and military culture, control and construction of armies. While Earnest Caldwell focuses on early Chinese military thought, Krishnendu Ray takes on the ancient and early medieval Indian concepts of war. Kaushik Roy, David M. Robinson, Kai Filipiak, Kenneth M. Swope, Pratyay Nath, Debraj Chakraborty, K.S. Mathew and Amarendra Kumar turn the focus on the interconnections between military technology, geography and the changing dynamics of warfare. And Michael H. Fisher, Douglas M. Peers, Channa Wickremesekera, Rob Johnson, Nixi Cura, Ming Jiang, Ulrich Theobald, Daniel McMahon, David A. Graff and Elisabeth Kaske chart the intricate relationship between the broader military culture and command culture which in turns relates with the bigger issue of the nature of the armies and the objectives of the state regarding it.

Theories of warfare

Both ancient China and India evolved complicated secular theories of warfare. The most important ancient Chinese theoreticians are Sunzi (Sun Tzu) and Wuzi. For ancient India, the most important military theoreticians are Kautilya and

Kamandaka. Several scholars agree that Sunzi's *Art of War* and Kautilya's *Arthasastra* continue to influence modern day military behaviour of Chinese and Indian armed forces, respectively.[11] Two chapters in this collection focus on ancient Chinese and Indian military thought. While Caldwell turns the limelight on a Chinese treatise about the Warring States (475–221 BCE) era, Ray turns the limelight on *yuddha* and *vijaya* tradition in ancient India. Caldwell analyses the newly discovered bamboo slips dealing with Cao Mie's *Battle Array*. The ancient Indian military theoreticians also devoted considerable time to formulating various *vyuhas* (battle arrays) that would be appropriate to the terrain and size of force deployed. Cao Mie's *Battle Array* concerns itself with the opportune moment when an action (or non-action) could tip the scales towards victory or defeat. Seizing the opportune moment to gain victory requires a general with wisdom.

As a point of comparison, not only Kautilya, but also Western military theoreticians like Machiavelli and Clausewitz emphasized the decisiveness of acting on the 'right moment' to launch a counter-attack. So, despite cultural differences, there existed some similarities in trans-continental military thought. These two chapterss challenge the prevailing view that only the West produced non-religious military theories that were crucial constituents of the Western Way of Warfare. By contrast, while pre-modern Western military theorists focused mostly on conventional war, Kautilya and his followers gave a lot of attention to the issue of internal security.

Armed forces, technology and warfare

Although technology is a major factor in world historical development, we do not want to fall into the trap of technological determinism. And at its core, the Military Revolution thesis is technology driven. The usefulness (or lack of it) of a particular piece of technology depends on the political objectives, social and cultural conditions and finally the physical geography of the theatre concerned. The sedentary agrarian bureaucratic empires of both China and India were subject to attacks by mounted steppe archers until the eighteenth century. At different moments, strong rulers of the empires of China and India followed an aggressive policy of launching military raids deep into nomadic territory. Very often the results of such military expeditions were not conclusive. Besides being costly in financial and personnel terms, often these raids resulted in disaster. One such example was the defeat of the Chinese at Tumu at the hands of Essen's nomadic cavalry. And in the case of India, the Mughal Empire's futile campaigns in the arid waste of Balkh and Badakshan in the reign of Shah Jahan, are worth remembering. Another strategy followed by the agrarian bureaucratic polities was to buy off the nomads by offering them titles and subsidies. The Mughals paid money to the chieftains along their north-western frontier in order to keep the strategic passes open. And the Chinese regimes often resorted to 'bribery' through opening markets on their northern border and providing the nomads with both tea and silk along with imperial titles to keep them in line.

However, appeasement policy did not work always. At times, subsidies merely whetted the 'hunger' of ambitious nomadic chiefs for more. In terms of military operations, both the Chinese and Indian sedentary regimes faced a problem in acquiring good horses. Good horses could not be bred in the hotter and more humid climates of China and India. Due to intense agricultural activities in the fertile plain especially in the river valleys of China and India, the arid pasture lands required for breeding horses were just not available. Both China and India were dependent on the nomads of the Eurasian steppes for good warhorses, and the nomads could create problems for the sedentary regimes by cutting off supplies of horses. In terms of horsemanship and skill, the nomadic archers were considerably better than the soldiers of the agrarian societies. Besides inferior mounts, the troopers of the sedentary societies just lacked the skill in horsemanship and archery to tackle the dispersed high-speed cavalry warfare practised by the nomads.

David M. Robinson's chapter notes the importance of the steppe frontier in the early Ming Empire. Liaodong (southern Manchuria) was important both for the rise and ultimately the fall of the Ming Empire. China's northern steppe frontier like India's northwest frontier was porous. Warhorses and military manpower moved back and forth through the borders. The regimes in Beijing and Delhi had to manipulate the flow of animals and military manpower, as well as intelligence, in order to maintain themselves. Robinson shows that the successful use of animal and demographic resources from this region enabled the Mings to consolidate power at the cost of the Mongolian Yuan Dynasty. However, the very necessity of depending on the steppe frontier forced the central government at Beijing to devolve power to the border officials and the garrison commanders of this region. In the long run, this policy enabled the rise of Jurchen power in Manchuria which in the first half of the seventeenth century overthrew the Ming and resulted in the establishment of the Manchu (Qing) Dynasty. As far as India was concerned, the piece by Kaushik Roy (one of the editors) shows that the British during the first half of the nineteenth century tried to demilitarize the northwest frontier of India and reduced their dependence on the Central Asian horses. However, an attempt to chastise the tribal chieftains and the 'restless' nomads resulted in a punitive invasion of Afghanistan. The net result was an Indian 'Tumu'. The British-Indian invading force was wiped out when it was retreating from Kabul to Jalalabad. The steppe nomadic frontier of British-Indian Empire remained unsettled.

Gunpowder weapons offered another possible response to the nomads threatening China. Kenneth Chase in his broad ranging overview about the spread of firearms in Eurasia asserts that gunpowder weapons were not that useful against the high-speed cavalry warfare of the nomads.[12] Kai Filipiak, in his thought provoking contribution to this volume, argues that the early Ming used gunpowder weapons (both hand-held firearms and field artillery) successfully against the nomads. In fact, by the mid-sixteenth century, the nomads were afraid of Chinese gunpowder weapons. Filipiak goes on to argue that the construction of gunpowder forts was another feature of the Ming defensive policy against the nomads. In contrast to West Europe, where siege artillery, as the advocates of the Military Revolution thesis propound, drove the development of military architecture, in Ming China, the driver for fortress development was completely different: response to the threat posed by the nomads who lacked gunpowder weapons but were equipped with cold steel.

Kenneth M. Swope's chapter shows that the Ming were very much interested in using gunpowder weapons both in field warfare and in siege warfare. Not only were the Ming interested in learning from the field guns used by the Portuguese but they also innovated and came up new designs. Swope asserts that, in fact, the iron guns cast by the Manchus were in many respects more advanced than the bronze cannons used by the Portuguese. In his other works, Swope has demonstrated in a masterly manner the continuous evolution of gunpowder weapon designs by the Ming political–military establishment in the last decade of the sixteenth century during the Japanese invasion of Korea. The field guns were not, however, very useful against the dispersed, high-speed cavalry warfare practised by the Jurchens.

Still, the Ming did make effective use of heavy guns in defending their fortified cities. The use of heavy artillery prevented Nurhaci from completing the destruction of the Ming regime. An action–reaction dialectic set in. Realizing the importance of gunpowder and gunpowder weapons, Nurhaci's son and successor Huang Taiji adopted gunpowder weapons in his steppe nomadic force. The Jurchens used Korean experts to manufacture gunpowder weapons. Even then, firepower superiority enabled the Ming to hold on. By the 1640s continuous and massive insurgencies and political treachery enabled the Manchus to overwhelm the Ming.

The regimes of China and India faced a similar problem: how to tackle mounted archers when good mounts were not locally available. India unlike China followed a different trajectory in its response to steppe nomadic invasions. Indian rulers used war elephants as pre-modern main battle tanks. Elephants were used in warfare both in ancient and in medieval China. But elephants became extinct in China.[13] However, elephants both for combat purpose and for logistical tasks continued to be used in the Indian subcontinent until the early nineteenth century. Debraj Chakraborty's chapter shows that superficially it seemed that elephantry had no chance against mounted archery. But in certain geographical locales elephants had their utility. So much so, that the Mughals, even though they were Chaghtai Turks from Central Asia, absorbed elephants within their burgeoning military machine.

Pratyay Nath's chapter shows the complex inter-linkages between force structure, geographical requirements and war objectives. The mounted archers who were suitable for conducting high-speed warfare in the north Indian plains were ineffective in the marshy paddy culture that dominated the high rainfall zones of east and north-east India. Disease and rain killed many Turki sowars and their mounts. So, the Mughals had to innovate. In the monsoon-fed rivers of east India, sea-going battleships were unusable. River crafts loaded with cannons and soldiers were required. The Song Navy had such a riverine fleet. And in the second half of the sixteenth century, the Mughals built such a fleet. Although battleships with high velocity guns might be useful on the Atlantic or Indian Oceans, they were useless for the geo-strategic requirements of the Mughals. Mud was plentiful in east India and the polities could mobilize lots of labour power to construct earthen forts and mud embankments. The point to be noted is that even in the first decade of the nineteenth century, the wide mud walls of Indian forts were able to resist the largest siege artillery that the British East India Company (henceforth EIC) could deploy. Like battleships, *trace italienne* or *alla moderna* fortifications should not be taken as the universal index of military modernization.

K.S. Mathew's piece shows that the hardy Portuguese ships with their bronze and iron cannon had a definite edge over the ships of the Indian powers which were constructed with timber planks and coir (but without iron nails) and lacked gunpowder weapons. As a result, during the first half of the sixteenth century the Portuguese fleet established sea control over the Indian Ocean. It seems at least that the naval component of the Military Revolution thesis of Parker seems to hold sway. The scenario was different in the coastal water, however, with recent studies showing that the large battleships of the West European maritime powers were not that effective in the narrow shoals and the winding coastal creeks of India's West (Malabar) Coast.

Amarendra Kumar notes that the marine forts established by the Marathas were very effective in blunting Portuguese efforts to control of India's western coastline. The inability of the Portuguese Army to project power ashore further diminished Portuguese hegemony. After the decline of Portuguese sea power, even the British EIC could not make much headway against the chain of interlocking Maratha coastal forts in the Malabar Coast. Only internecine rivalry among the indigenous powers and political fragmentation of the Maratha Empire allowed the British to intrude into India's West Coast. In other words, politics undid what the effective military opposition of the 'native' powers had achieved. And this brings us to the important issue of political cum military culture and the conduct of warfare.

Military culture, control and construction of the armies

The culture of command and control rather than technology seemed to be at times the decisive factor shaping defeats and victories. Douglas M. Peers compares the military resistance to the British EIC offered by two post-Mughal polities: Mysore and Bengal *Nawabi*. In pre-British India, the dominant military culture in South Asia according to Peers was individualistic heroic one to one cavalry combat. However, the introduction of a Western style infantry-artillery army by the EIC transformed indigenous military culture. Indian rulers like Haider Ali and Tipu Sultan of Mysore and *Nawab* Mir Qasim of Bengal eagerly emulated Western technology. During the second half of the eighteenth century, there was no significant technological gap between the armies of the Indian potentates and the force raised by the EIC. Besides the same types of arms and equipment, the personnel of the opposing armies were also similar. Both the Indian rulers and the EIC recruited from the South Asian military labour market. Moreover, the EIC's army also had to adapt to Indian conditions by adopting certain indigenous elements and ethos. Light cavalry, for example, was important for conducting warfare in the plains of the Indian subcontinent. However, in the field or financial organization and command culture (a clear cut hierarchical chain of command), the EIC had a definite edge over its Indian opponents. To use modern terminology, the EIC triumphed not in the sphere of technology (as the proponents of Military Revolution thesis argue) but in C3I (Command, Control, Communications and Intelligence).

Superior management capabilities on the part of the British enabled the EIC not only to recruit Indian soldiers to construct the colonial army which fought against other Indian powers, but also to use the colonial army as an instrument of power projection in order to subdue other countries in an attempt to transform them into colonies or quasi-colonies. Channa Wickremesekera and Rob Johnson focus on this issue. Wickremesekera writes that from the last decade of the eighteenth century, the Dutch and the British became interested in Kandy (the central highlands of Sri Lanka, the island at the tip of south India). The EIC and Kandy represented two differing military cultures: a high technology professional standing army geared for decisive battles versus a people's militia designed for guerrilla warfare. Taking advantage of terrain and climate, the guerrillas of Kandy obstructed smooth operations by the highly professional sepoys and the Malay soldiers in the pay of the EIC. Kandy did not have to win. If the militia was able to impose a high rate of casualties on the invading colonial army of the metropolitan powers like the Portuguese, Dutch and lastly the British, then Kandy was not worth conquering. Disease and casualties deterred the Portuguese and the Dutch from annexing Kandy. However, in the first decade of the nineteenth century, politics undid what the military had achieved. A political upheaval in Kandy resulted in the nobles surrendering the country to the British.

The British officered Indian Army or the colonial army of the British Empire in India was used not only to subjugate the subcontinent but also to project British power in other regions of Asia. The British could do this thanks to the organizational format imported into Asia from West Europe. Rob Johnson focuses on the use of the British-Indian Army in the two wars (known as the First and Second Anglo-Chinese Wars) fought against the Manchu Empire between the 1840s and the 1860s. These two wars represented a clash of cultures between the British and the Chinese. Geopolitics determined the force structure and military doctrine and thus military culture of a polity. The absence of a maritime threat by a neighbouring polity before the intrusion of the West Europeans in the South China Sea, and presence of Japanese pirates, resulted in the Manchus possessing not a battleship centric navy but an anti-piracy fleet.

The Manchu Army was geared to check internal uprisings through the display of numerical superiority. Set-piece battles with high technology weapons geared for destruction of the enemy force initially came as a shock to the Manchus. Similarly, the Manchu military culture of conducting harassing attacks with cavalry instead of fighting to the finish and use of bounty hunters by the Manchu regime also shocked the British officers steeped in regimental spirit. In the end it was not so much a technology gap but cultural difference that functioned as the principal determinant of events. A hierarchical command chain, professional officer corps and other elements that constituted the 'command climate' were something that both the Indian potentates and the Manchus lacked. Political infighting and bad leadership further hampered the Manchu military opposition to the British. In the end, we must not overestimate the threat that the British posed to the Manchu regime. True, the British traders hampered the Manchu economy and the import of opium also caused degeneration of the health of the inhabitants of China. But the real threat to the Manchus came from elsewhere. During the First Anglo-Chinese War in the 1840s, the best Manchu units were stationed in north China guarding the traditional steppe frontier. During the Second Anglo-China War, the Taiping Rebellion threatened the Manchu heartland. Nevertheless, British (or Western) military culture was not invincible. In 1857, mutiny in the Bengal Army by the sepoys and sowars forced the British to move military assets from China to India in order to suppress the mutineers.

Pre-modern China like the West European polities faced civil–military troubles. Large empires had to maintain large armies. Large standing armies under professional generals posed a threat to the emperor and his court. Military takeover by the generals was quite common in history. In the Roman Empire, several generals crowned themselves emperors. In the case of ancient India, Pushyamitra Sunga, the Commander-in-Chief of the Maurya Empire executed the last Maurya Emperor Brihadratha during a military review and established the Sunga Dynasty soon afterwards.

In early medieval China, the Tang Dynasty came up with a unique solution to resolve civil–military tensions. David A. Graff's chapter shows that Tang emperors appointed eunuchs in command of the palace standing armies. Eunuchs were sent as military supervisors during campaigns in an attempt to restore control of the centre over the field armies. Eunuch overseers were also attached to the provincial armies of the governors. The Confucian scholar-officials who generated history of pre-modern China hated the eunuchs. However, as Graff shows, eunuch supervision of the Tang armies was not a completely dysfunctional mechanism.

The basic dilemma facing the rulers of the large empires was how to inculcate aggressiveness in the armies under their control to protect the polities from internal and external enemies and simultaneously, while also taming them. Militarism needed to be both fostered and controlled by the courts. This requirement gave rise to complex cultural norms and courtly values with associated elaborate rituals. Court paintings, such as the Mughal and Rajput paintings of medieval India, and central government sponsored plays and acts provide an eye opening perspective on this. To use a post-modern terminology, reading between the lines of the visual texts generated by the members associated with the courts also provide a sort of nuanced criticism of courts' policies as regards governance of the states. The tension between *wen* (civil) and *wu* (military) was a consistent feature of Chinese political culture. Ming Jiang's chapter shows that in the early Tang period, several scholar officials tried to bridge this tension by fusing these two opposing qualities. The poems written by such scholar officials noted that a truly virtuous person requires both *wen* as well as *wu* (horse riding skill). Such a synthesis was considered necessary due to the rising importance of the army in Tang statecraft and also because military service became the stepping stone for many of the poor literati. They commemorated war in their poetry, especially combat against non-Chinese 'barbarians', as a sort of righteous war (just war). Here, one is reminded of the close parallel of *kutayuddha* and *dharmayuddha* of ancient Indian tradition.

Nixi Cura's piece in this collection turns the spotlight on the inherent tension between *wu* and *wen* in the Qing Court (1644–1911). The Qing Dynasty tried to resolve the tension between the civil culture of the Han Chinese (*wen*) with the militaristic culture of the quasi-nomadic Jurchens (*wu*). The Qing Empire established by the Jurchens from Manchuria could be categorized as a sort of 'colonial enterprise' in China. Just like the British officered Indian Army of the British Empire in South Asia, the Qing Empire had to recruit Chinese manpower into the Banner Armies, as well as taking control over veteran turncoat Ming armies. Both Jurchen and British military manpower were inadequate to conquer and then police the large indigenous populations of China and India, respectively. Sheer necessity forced these conquest regimes to use indigenous military manpower. In fact, it would not be far from the truth to argue that the Jurchen conquest of China and the British conquest of India were made possible by the Chinese and Indian soldiers who operated under Jurchen and British officers. Both the Qing Dynasty and the British Raj had to develop the ideological construct of a harmonious multi-ethnic military system. Cura makes a case study of a Chinese Bannerman, the painter Zhu Lunhan who lived from 1680 to 1760. His poetry and painting throw light on the hybrid military system of the Qing that had Jurchen organizational format (just like the British regimental format) but Chinese manpower.

The problem of ensuring proper military control especially over the commanders commanding large detachments far away from the centre continued to dog the Qing court. During the First Jinchuan War (1747–1749), Ulrich Theobald's chapter shows, the Qing central government had problems in maintaining obedience from the gifted commander Zhang Guangsi. Zhang was charged with showing disrespect to the 'Son of Heaven', but he argued that the emperor was only listening to his wicked cronies and not attending to the real situation. The court brought a false charge of corruption against Zhang. Such a practice remains common even in the modern era all over the world. The court got rid of Zhang by bringing the charges of insubordination, failure to execute proper tactics and the inability to initiate proper counter-espionage measures.

The conquest dynasties controlling sedentary societies faced frequent rebellions (modern insurgencies). The imperial response involved an amalgam of coercion and persuasion or carrot and stick policy. In modern terminology it is known as the use of minimum necessary force and winning the hearts and minds policy. Total destruction of the internal rebels was not attempted because it would destroy the legitimacy of the regime and devastate the agrarian base of the polity. Moreover, the soldiery engaged in counter-insurgency (COIN) duties was actually drawn from the peasant stratum of the indigenous societies, part of which was in rebellion against the governments concerned. Daniel McMahon turns our attention towards the armed rebellions that rocked the Qing Empire during the last decade of the eighteenth century and first two decades of the nineteenth century. While the Miao uprising by the highlanders against the territorial encroachment by the plainspeople occurred in west Hunan, the White Lotus rebellion was the product of grievances of the plainsmen.

The Qing government turned to the traditional defensive strategy of building walls with the aid of local labour in order to protect the cities, towns and villages from the 'bandits'. These walls not only protected the 'good insiders' against the 'bad outsiders' but also aided regimentation and surveillance of the insiders. In a way, these walls functioned as a moral as well as a physical barrier against the uncivilized outlaws. It is interesting to note that Arthur Waldron also speculated that the Great Wall of China functioned as more of a cultural psychological barrier (of sedentary cultured Chinese society against the pastoral nomadic barbaric steppe society) than as a physical defensive measure. Besides hunting down the rebels with military units, the static local defensive measures involved the mobilization and exploitation of the local communities. This involved bringing the locals within the security orbit and also arming them. Arming the locals to fight the 'outlaws' was always a dangerous strategy. It could prove a double-edged sword. After many debates and discussions, the Qing court finally agreed to follow this strategy. McMahon concludes that this resulted in penetration of military values into civilian society.

The culture of civil domination of the military was well entrenched in premodern China, but did not always result in military effectiveness. Elisabeth Kaske's chapter shows that during the 1850s and the 1860s when the Qing Empire was beset by numerous rebellions both in the core and periphery of the empire, continuous tussles between the civilian branch of the imperial administration and military leaders hampered combat effectiveness of the armies engaged in COIN duties against various types of rebels like Miao, Muslims etc. High-ranking civilian administrators were in charge of collecting and disbursing funds. In wartime the size of the armies expanded several fold and revenues declined. Various interlocking agencies for collecting revenues and disbursing them as well as lack of unity of command among different types of government security forces operating under separate commanders with their own separate lines of communications (an example of Chinese divide and rule policy to check the rise of the generals) further exacerbated the situation.

In the case of poorer provinces like Gansu, which was dependent on financial aid from richer neighbouring provinces, the situation went from bad to worse. Instead of sending financial aid, the comparatively well off richer provinces sent their mercenary bands as 'guest armies' to suppress the peasant rebellion. Gansu, which was a poor province, had to take responsibility for feeding these extra 'guest armies'. The situation was unmanageable for the civilian administrators. Frequent failures by the civilians to pay the soldiers regularly and substantially (especially in a time of rapid inflation) resulted in numerous mutinies. And the uncontrolled rapaciousness of the various types of mercenary bands and irregular soldiers deployed by the government became an equal if not greater danger for public security.

Michael H. Fisher traces the use of race and ethnicity in the construction of armies from Mughal to British era. Both the Mughal regime as well as the British colonial state were concerned with racial balance within their armies in order to ensure the central government's command and control. Both the Mughal and the British governments were afraid that the predominance of a particular race or a community from a particular geographical zone might result in mutiny in the army. And that was exactly what happened in 1857 when the *Purbiyas* (i.e. Brahmins and Rajputs from Bihar and Awadh who were dominant in the Bengal Army of the EIC) rebelled. The British absorbed, expanded and modified the ideas about race prevalent in pre-British India with their own assumptions and concepts of racial superiority of the Britons. And race as an organizing principle of the armies continued to circulate in the South Asian political landscape even in the late nineteenth century.

Notes

1 Among Jeremy Black's numerous publications mention can be made of *War in the World: A Comparative History, 1450–1600* (Basingstoke, Hampshire: Palgrave Macmillan, 2011).
2 Geoffrey Parker, *The Military Revolution: Military Innovation and the Rise of the West, 1500–1800* (Cambridge: Cambridge University Press, 1988).
3 See for instance his edited *Warfare and Culture in World History* (New York and London: New York University Press, 2011).
4 Peter A. Lorge, *The Asian Military Revolution: From Gunpowder to the Bomb* (Cambridge: Cambridge University Press, 2008).
5 The latest in the genre of world military history writing is the big book by Azar Gat entitled *War in Human Civilization* (Oxford: Oxford University Press, 2006). While trying to integrate sex with organized violence in an innovative way, Gat for the modern era focuses mostly on West Europe and USA.
6 For the historiography of the Military Revolution debate see Clifford J. Rogers (ed.), *The Military Revolution Debate: Readings on the Military Transformation of Early Modern Europe* (Boulder, CO and San Francisco: Westview, 1995).
7 Hanson believes that the ancient Greeks invented the concept of a decisive battle. The objective was to destroy the hostile army in one single afternoon. Hanson and later Geoffrey Parker conclude that decisive battle, technological superiority, a penchant for military theory and civic virtues constitute the chief characteristics of the much vaunted Western Way of Warfare. Victor Davis Hanson, *Warfare and Agriculture in Classical Greece* (Berkeley and Los Angeles, CA: University of California Press, 1998), and *The Western Way of War: Infantry Battles in Classical Greece* (1994, reprint, Berkeley and Los Angeles, CA: University of California Press, 2009); Geoffrey Parker (ed.), *The Cambridge Illustrated History of Warfare: The Triumph of the West* (Cambridge: Cambridge University Press, 1995). Seventeen essays along with the powerful Introduction by Geoffrey Parker trace a linear teleological connection between the hoplites and the American GIs.

8 See Victor Davis Hanson's *Carnage and Culture: Landmark Battles in the Rise of Western Power* (New York: Doubleday, 2001). Hanson's selection of nine battles (three of them naval) which resulted in the rise of the West from 480 BCE to 1968 goes back to Major-General J.F.C. Fuller's massive three volume study and far earlier to Edward Creasy's study of battles which resulted in the rise of the West against the 'Rest'.

9 J. Black, *War in the World*, p. 219.

10 Xinru Liu (ed.), *India and Central Asia: A Reader* (New Delhi: Permanent Black, 2012).

11 Thomas M. Kane, *Ancient China on Postmodern War: Enduring Ideas from the Chinese Strategic Tradition* (London and New York: Routledge, 2007); Kaushik Roy, *Hinduism and the Ethics of Warfare in South Asia: From Antiquity to the Present* (Cambridge: Cambridge University Press, 2012).

12 Kenneth Chase, *Firearms: A Global History to 1700* (Cambridge: Cambridge University Press, 2003).

13 Edward H. Schafer, 'War Elephants in Ancient and Medieval China', *Oriens*, vol. 10, no. 2 (1957), pp. 289–291.

Part I

Military theory/theories of warfare in China and India

1 Opportune moments in early Chinese military thought

The concept of *ji* 機 in the Warring States period manuscript *Cao Mie's Battle Array*

Earnest Caldwell

Introduction

The era commonly referred to as the Warring States period (戰國 *c*.475–221 BCE) was a time of great turmoil, with various states increasingly engaging in large scale military activities in order to gain control of neighbouring territories. As warfare became increasingly ubiquitous and violent there arose a concurrent interest, particularly among rulers of smaller states, in employing technical specialists of military strategy to provide strategic and tactical advice on how to bolster defences and project some deterrent form of military power. Not surprisingly, this led to the proliferation during the late spring and autumn (春秋, *c*.771–476 BCE) and Warring States periods of many of the great military masterpieces. Outside the well-known *Sunzi bingfa* 孫子兵法, the military texts (*bingshu* 兵書) section of the *Han shu* 漢書 'bibliographic treatise' 藝文志 lists some 53 schools of military thought (*jia* 家) and a corpus consisting of 790 fascicles (*pian* 篇) of text and 43 rolls (*juan* 卷) of maps and illustrations, many of which were purportedly composed during the Warring States period.[1] Many of these texts emphasize the necessity for small states to cultivate awareness of 'opportune moments', both on and off the battlefield, that could offer crucial advantages in warding off threats from stronger neighbours. Unfortunately the occasional burnings of imperial libraries and a host of other historical incidents allowed for only a few of these texts to survive down to the present. Although these surviving texts provide ample references to opportune moments in battle, they also leave many questions unanswered. For example, most references are rarely organized collectively into a systematic framework, making it difficult to fully contextualize opportune moments within the progression of a battle scene; nor do these texts give reasons for *why* a particular moment is deemed to be strategically opportune. The recent discovery an ancient bamboo manuscript of a heretofore unknown Chinese military treatise, entitled *Cao Mie's Battle Array* 曹蔑之陳, provides some 'new' material to begin answering some of these questions.[2]

Dated to the fourth century BCE, the bamboo manuscript *Cao Mie's Battle Array* is currently the oldest extant source on Chinese military thought. On a total of 65 bamboo slips, the text comprises two dialogues between Duke Zhuang of Lu (魯莊公, r. 693–662 BCE) and Cao Mie, a notable statesman and military theoretician in early China. As the duke is keenly interested in protecting his small state of Lu from invasion by its much larger and more powerful neighbours, the discussions between these two figures cover a range of issues including statecraft, law, tactical decision-making, and military reengagement after suffering a defeat. Unlike other Chinese military texts, however, an entire section of *Cao Mie's Battle Array* is dedicated to a systematic explication of opportune moments.[3] This systematic approach allows for a matching of specific indicators of tactical advantage to actions taken by an enemy at specific points during battle. In this chapter, I will first provide a translation of the section of the *Cao Mie's Battle Array* manuscript dealing with 'opportune moments', and then contextualize Cao Mie's theory of opportune moments within the broader discourse of other early Chinese military texts. In so doing, I will offer a more comprehensive view of a key concept in early Chinese military thought, the opportune moment.

'Opportune moments'

As in many other Warring States period texts (e.g. *Mengzi*, *Xunzi* and *Han Feizi*), the question-and-answer based dialogic framework utilized in *Cao Mie's Battle Array* evinces a certain didactic quality embedded within the text. Thus, while the text concerns a historical event some several centuries prior to the actual writing of this particular manuscript, its overall textual content would retain a pedagogical relevance to later generations, particularly those people possessing high levels of political or military authority. This explains the overt emphasis running throughout the text on the necessity of a perspicacious leader capable of both ascertaining one's own strengths and weaknesses, as well as those of an adversary, and of executing appropriate actions so as to exploit an advantage or buttress a deficiency. It is in the section of *Cao Mie's Battle Array*, that I call 'Opportune Moments' that such leadership attributes are explicitly associated with military context. Here Duke Zhuang inquires about observable signs which indicate opportune moments for carrying out actions at four different stages of battle.

Translation of 'Opportune Moments'[4]

Slip 40 partial

> 莊公曰既成教矣出師有幾(機)虖答曰有臣聞之三軍出
> Duke Zhuang said: '[I] have already completed instruction.[5] When deploying forces are there opportune moments?' [Cao Mie] replied: 'There are. Your servant has heard that when the three armies are deployed…

Slip 42

亓將卑父兄不廌(存)由邦御之此出師之幾(機)莊公或問曰三軍散裹有幾(機)虖答曰有臣聞

...if their general is inferior; [his] father and elder brother are not living;[6] or the ruler manipulates him, these are the opportune moments for the deployment of forces.' Duke Zhuang then inquired: 'When the three armies are scattered or congregating are there opportune moments?' [Cao Mie] replied: 'There are. Your servant has heard...

Slip 43

之三軍未成陳未豫行阪濟障此散裹之幾(機)莊公或問曰戰有幾(機)虖答曰有亓去之

...this: "When the three armies have yet to complete the battle array; have yet to encamp; or are traversing a ridge or fording some [water] obstacle, these are opportune moments when scattered and congregating."' Duke Zhuang then asked: 'During battle are there opportune moments?' [Cao Mie] replied: 'There are. If when withdrawing they are...

Slip 44

不速亓就之不附亓啟節不疾此戰之幾(機)是故疑陳敗疑戰死莊公或問曰既戰有幾(機)虖

...not rapid; when advancing they do not draw near; and when their launch measure is not fierce,[7] these are the opportune moments during battle. Therefore, doubt when arraying for battle will bring defeat; doubt during battle will bring demise.' Duke Zhuang then inquired: 'After having battled are there opportune moments?'

Slip 45

答曰有亓賞淺且不中亓誅厚不察死者弗收傷者弗問既戰而有怠=(怠心)此既戰之幾(機)

[Cao Mie] replied: 'There are. If their rewards are light and not centred (=fair); their punishments are heavy and not scrutinized (=arbitrary); the dead are not reclaimed and the wounded are not inquired about; then having already battled and having weary hearts, these are opportune moments after having battled.

忌 *or* 機 *and the question of perspective*

In the above translation the graph [] is transcribed as *ji* [幾] (OCM *kəi) and I have chosen to follow Chen Jian's reading of it as the word *ji* {機} (OCM *kəi), 'opportune moment.' In his initial study of the manuscript, Li Ling properly transcribed the graph as *ji* [幾], but then reads the graph as a phonetic loan for the word *ji* {忌} (OCM *gəh), 'taboo' or 'something to be avoided'.[8] Following such a reading, Duke Zhuang's question in slip 40 would read, 'Are there things to be avoided when deploying the troops? 出師有忌乎'. Thus, according to Li Ling each of Cao Mie's responses must be understood from a defensive perspective as warnings against faulty decision-making associated with battle. A failure by Duke Zhuang to assess qualified generals, to avoid the hazardous situations or to maintain post-combat morale would place his armies at a serious disadvantage and guarantee a defeat. While the phonology of the two words allows for the possibility of such a loan, some scholars, such as Chen Jian and Gao Youren, have identified certain lexical, grammatical and contextual concerns that make Li Ling's interpretation problematic.[9]

First, while it is phonologically possible to loan *ji* [幾] for *ji* {忌}, other excavated materials dating to the Warring States period frequently represent Li Ling's sense of 'taboo' or 'something to avoid' with the graph-word *hui* {諱} (OCM *hwəih).[10] To my knowledge there are no other extant passages in excavated materials where *ji* [幾] is interpreted to mean *ji* {忌}. This does not necessarily mean that such a loan is impossible, but the other two concerns illustrated by Chen Jian and Gao Youren show the weakness of such a reading.

Second, Chen Jian has pointed to the presence of the graph [亓]=*qi* [其] (OCM *gə) at the beginning of slip 42, and notes that as the possessive pronoun 'their', it refers to the enemy forces. Following this reading, in the line 三軍出亓將卑 ... (slips 40 and 42), the 三軍出 would refer to the deployment of the enemy troops and 亓將卑 would refer to the fact that *their* general is inferior. Furthermore, in the same section of text the graph *qi* [其] is also found in similar contexts in slips 43, 44 and 45, and in each instance Chen Jian believes that it is referring to the enemy forces. He supports this by referring to the similar usage of *qi* [其] in slips 17 and 18 of the same manuscript. The following is an excerpt from advice given by Cao Mie concerning methods of bribing enemy officials.

> 毋愛貨資子女以事其便嬖
> [One] ought not to be stingy with material goods or boys or girls. Employ [them] in servicing *their* [i.e. the enemy's] favoured ministers.

Finally, a brief survey of early Chinese military texts provides ample evidence that the word *ji* {機}, meaning an 'opportune moment' or a 'crucial factor', was commonly used in military contexts. Clear examples of the usage of *ji* [機] in contexts similar to that of *ji* [幾] in the *Cao Mie* manuscript can be found in several passages from the *Wei Liaozi* 尉繚子 and *Wuzi* 吳子.

> 機在於應事.
> Opportunities are found through (proper) responses to events.[11]

> 吳起儒服以兵機見魏文侯.
> Wu Qi donning Confucian robes met with the Marquis Wen of Wei to discuss military opportunities.[12]

When considering the content of the above passages, and of the *Cao Mie* manuscript as a whole, reading the graph *ji* [幾] as *ji* {機}, 'opportune moment' would seem suitable. Such arguments, however, completely change the perspective of the questions posed by Duke Zhuang and Cao Mie's responses from Li Ling's defensive avoidance to the exploitation of an offensive advantage.

Finally, when translating any text, one often struggles to maintain as accurate a reading as possible in an attempt to stay 'true' to the author's original meaning. Thus, the outcome of arguments over *ji* 忌 and *ji* 機 have consequences not just for the content of this particular text, but also for the way in which we understand Cao Mie's own thought. For each of Cao Mie's discussions, I have read them as positive statements regarding tactical advantages for Duke Zhuang's military forces. In the context of their conversations, it seems that this is the best reading. That being said, however, someone reading the text as a source of military philosophy would likely find the basic meaning of this particular passage to be that specific actions by rulers and generals create opportune moments for force A against force B, and that one ideally wants their force to be in the position of force A and not force B. So, when Cao Mie advises the duke that an enemy army led by an inferior general is an advantage, one could also understand the statement in the negative which would imply that rulers should never send out their forces under the command of an inferior general. Thus, while we translate the text to read *ji* [幾] explicitly as *ji* {機}, 'opportune moment', through an expanded interpretive meaning of passage one could infer Li Ling's *ji* {忌}, 'something to be avoided', as an implied didactic element.

Opportune moments for deployment of the forces (出師之機)

Before committing large quantities of armed forces and state resources to carry out a military offensive against an opponent, it is imperative that a ruler or commander has adequate information related to the strengths and weaknesses of the enemy in order to assess the probability of success or failure. Military writers in early China were particularly interested in evaluating the leadership qualities of the enemy's command staff. The *Wuzi*, for example, claims that before battle it is 'essential first to divine the enemy's general and ascertain his abilities [必先占其將而察其才]'.[13] In his response to Duke Zhuang's initial question about signs of opportune moments for deploying forces, Cao Mie evinces a similar pattern of strategic thought. In keeping with the overall theme of the manuscript, the response is concerned primarily with assessing the weaknesses of the enemy's leadership and on the basis of such information deciding whether or not to go to battle.

First, in terms of decision-making skills, an inferior general will naturally prove less formidable than a superior general. In various military texts, there are myriad qualities associated with inferior generals ranging from avarice and inexperience to indecision and indolence. A general possessing such qualities will likely calculate battle-related decisions according to the potential for personal gain or selfishness, and thus be impaired from making proper judgments on the battlefield or in commanding the respect of his officers and soldiers. Indeed, the *Wuzi* claims that armies led by inferior generals and careless officers may be attacked without recourse to pre-battle divination.[14]

The second phrase, 父兄不膺(存), concerns a detailed assessment of the personal life of the enemy general in order to ascertain his true commitment to a potential battle. As mentioned above (see Note 6), many scholars have read this phrase as 父兄不膺(薦) 'father and elder brother do not recommend [him]', and then read it in conjunction with the third phrase, with both serving as supporting clauses that emphasize the extent of the general's inferior qualities mentioned in the first phrase. Though phonologically plausible, I believe an alternative (perhaps, more creative) reading in which the graph is read as *cun* 存 'exist', or 'to be alive', might be better.[15] In saying this, I understand the whole sentence to consist of three individual phrases. Each is linked to the common subject of the general, but not necessarily *causally* linked as other interpretations have assumed. By reading the phrase 父兄不存 as '[his] father and elder brothers are not living', I consider the advantage (or opportunity) that Cao Mie is expounding to be a potential psychological conflict within the general between his ancestral-familial responsibilities and a commitment to battle. A general whose father and elder brother are dead, thus making him the eldest male descendant, would have a far greater range of familial responsibilities (carrying out worship of the ancestors, continuing the family line, etc.), and this might condition his judgment on the battlefield, particularly in the areas of risk assessment. Such a general would likely be more wary of taking risks which may threaten his life (and by extension his lineage); and, therefore, he may hesitate to enter battle or hastily call for a retreat. There are several references within early Chinese military texts, particularly in the *Wuzi*, that warn of defeat if one were to go into battle with generals, officers or common soldiers who have only a desire to live and are not fully committed to dying on the battlefield in order to secure victory.[16]

Last, Cao Mie points out that failure by the ruler to invest in his general sole control over the forces once dispatched will inevitably lead to defeat. The government, or more specifically the ruler, should not interfere once a charge of command has been awarded. For Cao Mie, if a general's commands are constantly undermined by subsequent orders issued from the capital, then this represents an ideal, opportune moment to attack.

Cao Mie is not alone in pointing this out, for similar statements can be found in other early Chinese military texts. The *Sun Bin bingfa* 孙膑兵法, excavated from a Han dynasty tomb at Yinqueshan, states, 'There are five situations in which one will never be victorious: If [the ruler] manipulates the general, one will not be victorious [恆不勝有五御將不勝]'.[17] This is further elaborated on a partial slip belonging to another section entitled, Virtues of the General 將德. It states, '[missing section] … and not manipulate the general. The ruler's orders do not enter into the army's gates. This is a constant principle of leading an army [而不御將君令不入軍門將軍之恆也]'.[18] And perhaps the most explicit reference to the principles underlying this theory is found in a section of the *Liu Tao* that details the appointment of generals. It describes the following scene:

> 將已受命，拜而報君曰：臣聞國不可從外治，軍不可從中御二心不可以事君，疑志不可以應敵. 臣既受命, 專斧鉞之威. 臣不敢生還，愿君亦垂一言之命于臣 君不許臣，臣不敢將 . . . 軍中之事，不聞君命，皆由將出。臨敵決戰，無有二心
>
> Having received the mandate, the general makes obeisance and states to the lord, 'Your humble servant has heard that a state cannot be administered from the outside, and an army cannot be manipulated from the center. One with two minds cannot be used to serve the lord; one with doubt cannot respond to the enemy. Your humble servant having already received the mandate has sole possession of the might of the *fu* and *yue*. Your humble servant dares not return alive. I request that milord also bestow upon me complete command. If milord does not so permit your humble servant, then I dare not serve as general.'... Thus, in military affairs one does not consider the lord's commands; all [orders] emanate from the general. When approaching the enemy and determining whether to do battle, they are not of two minds.[19]

As the above quotations demonstrate, there was clearly a great concern among early Chinese military theoreticians over the potential intermeddling of politics and military affairs. One particular fear concerned the ruler issuing orders after the command had been assigned to a general. It is the general who is out in the battlefield and is conscious of the multifaceted nature of the military situation unfolding before him. Therefore, the decisions of a ruler could certainly endanger the military campaign.

From this section, it is clear that according to Cao Mie the greatest opportunity for a smaller state to send out its military forces would be when the enemy forces are led by inferior generals. Other military texts affirm that without proper leadership, any force, regardless of size and resources, will certainly be susceptible to defeat by smaller forces. For a small state like Lu, ascertaining such opportunities *well before* sending out its own troops was necessary for survival.

Opportune moments when scattered or congregating (散裹之機)

Most modern scholars typically follow Li Ling's understanding of the graphs [散裹] to mean 'dispelling encirclement'. However, I read these as two separate adverbial forms depicting 'opportune moments' based on the organizational situation of the enemy forces, either scattered (散) or congregating (裹). Cao Mie's reply to the duke supports this reading as he offers two different, but interrelated, situations in which an enemy in motion will be at a serious disadvantage and open to repulsion.

The first situation concerns enemy forces that have yet to form their battle arrays or properly set up their camp. At these moments such forces are disorganized and unprepared for combat, increasing the probability of a successful assault against them. The second situation concerns enemy forces that are traversing various types of difficult terrain: water obstacles, steep inclines and narrow pathways necessitate alternative configurations of troop organization and also impede swift movement. Each of these factors creates negative effects on the battle-readiness of the enemy soldiers and decreases the commander's ability to control his troops effectively.

In reading Cao Mie's brief reply, one is immediately reminded of the state of Song's 宋 disastrous battle with Chu 楚 at Hong 泓 (637 BCE), as recorded in the *Zuozhuan*. Though the order of the situations is reversed, a similar thread of thought concerning an advantage posed by disorganized troops is apparent. In this historical episode, at two separate instances Duke Xiang of Song (宋襄公, r. 650–637 BCE) forbids his own outnumbered armies from engaging the superior Chu forces as they ford a river, and then again as they slowly form up their battle arrays. Twice the officer of the horse (*sima* 司馬) advises him of the tactical advantage presented by the enemy's situation, especially considering Song's numerical disadvantage. However, Duke Xiang's refusal to seize these opportune moments results not only in the decimation of his armies, but in the duke himself being wounded.[20] Although this passage is frequently read within Confucian paradigms, emphasizing Duke Xiang's refusal to attack a disadvantaged enemy at his own expense as a model of *li* 禮, a military lesson can certainly be gleaned from the advice of his officer. That is, a military force should never squander a tactically advantageous occasion to attack a superior enemy force caught in a disadvantageous situation.

Such arguments are not reserved to historical texts. There are many references within early Chinese military texts advocating an immediate attack on an enemy that has not yet formed for battle or is traversing dangerous terrain. Perhaps the most vivid testament to this tactic can be found in the *Wuzi* chapter 'Assessing Enemies' 料敵. In this chapter Wu Qi 吳起 advises the Marquis of Wu on methods for determining whether or not an enemy should be engaged on the battlefield. He provides eight conditions under which one can 'without performing divinations engage the enemy in battle' 有不卜而與之戰.[21] The eighth condition is amazingly similar to the above statement from Cao Mie.

> 八曰：陳而未定，舍而未畢，行阪涉險，半陰半出，諸如此者擊之勿疑. The eighth states: When arraying for battle, yet not quite settled; when encamping, yet not quite finished; when traversing dangerous terrain and narrow defiles; being half concealed and half exposed; under all of these conditions strike them without doubt.[22]

Like Wu Qi, Cao Mie too advocates quick and decisive attacks on the enemy in such situations. Again, Cao Mie is concerned that the size of the army that the state of Lu is capable of mustering would not compare to that of Qi or its other large neighbours. However, in his reply to Duke Zhuang, Cao Mie assumes that troops have already been committed to battle, and, as such, a general must be ever aware of potential tactical opportunities that arise before a battle takes place. To let such opportunities slip past for any reason would potentially place Duke Zhuang in a situation quite similar to that of Duke Xiang of Song.

Opportune moments during battle (戰之機)

The previous two questions posed by Duke Zhuang refer, temporally, to situations and conditions that arise before two opposing armies formally meet on the battlefield. The first dealt with selection of generals that occurs even before the armies are mustered, while the second concerned opportune moments that present themselves as the enemy forces are moving towards the field of battle. Duke Zhuang's third question requests Cao Mie to expound on opportune moments found once the battlefield is set. In this scenario, both armies have successfully arrayed for battle and the tactical opportunities giving advantage to a smaller force mentioned in the previous questions have passed. Fielding an army of a small state, like Lu, against a much larger and better resourced army, like that of Qi, is extremely dangerous; yet, once committed, a general must continually scrutinize the unfolding battle and be ever conscious of any potential opportunity that could shift the advantage to his forces. Indeed, it is this tactical perspicacity for which the historical Cao Mie is renowned. As the following passage from the *Zuozhuan* demonstrates, his ability to assess and manipulate battlefield situations helped the smaller Lu army defeat the army of Qi at the battle of Changzhuo.

> 公與之乘，戰于長勺，公將鼓之，劌曰，未可齊人三鼓，劌曰，可矣。齊師敗績，公將馳之，劌曰，未可下視其轍，登軾而望之，曰，可矣遂逐齊師。既克，公問其故，對曰，夫戰，勇氣也，一鼓作氣，再而衰，三而竭，彼竭我盈，故克之，夫大國難測也，懼有伏焉，吾視其轍亂，望其旗靡，故逐之
>
> The duke rode with him [Cao Gui]. The battle was fought in Changzhuo. The duke was about to order the drum cadence, when [Cao] Gui said, 'Not yet.' The men of Qi had beat their drums three times, then he said, 'Now we can.' The army of Qi was severely defeated. The duke was preparing to give chase. Gui said, 'Not yet.' He alighted and examined the tracks left by their chariot-wheels. He remounted, got on the frontbar, and gazed at them [the enemy]. After this he said 'Pursue'; which the duke did. After the victory, the duke asked Gui the reasons for what he had done. He replied, 'In warfare, all depends on the courageous spirit. When the drums first beat, the spirit is excited. A second beat cause the spirit to wane; and with a third, it is exhausted. With our spirit at the highest pitch we fell on them with their spirit exhausted; and so we conquered them. Large states are difficult to predict; I feared a possible ambuscade. I observed that their chariot tracks were disorderly; I saw that their flags were drooping. Thus, I gave the order to pursue them.'[23]

In this historical episode, Cao Mie (a.k.a. Cao Gui) scrutinizes the battle spirit of the enemy troops looking for its diminution and examines the confused nature of the enemy chariot tracks in retreat. Cao Mie's ability to effectively manipulate Qi's forces and then read the signs related to Qi's retreat ensured that when the forces of Lu finally engaged in battle, they did so with the utmost advantage. Returning to the manuscript, one sees that Cao Mie's response to Duke Zhuang's question about opportune moments during battle echoes the need for such perspicacity. The commander on the battlefield must be able to accurately assess the nature of enemy troop movements as they advance and retreat from battle. Yet, what is deemed crucial is the ability to ascertain the level of *doubt* in enemy movements. This doubt is evinced by the enemy's vacillating commitment to attack. That is, they do not attack swiftly, and if they do advance on the opposing army, they typically do not draw near enough for combat. As Cao Mie concludes, 'doubt when arraying for battle will bring defeat; doubt in battle will bring death' (slip 44).

Cao Mie is certainly not alone in pointing out doubt as an inevitably fatal military flaw on the battlefield. Early military texts are replete with references that warn against engendering doubt within one's forces or selecting a general that lacks decisiveness.

> 善戰者，居之不撓，見勝則起，不勝則止 故曰：無恐懼，無猶豫 用兵之害，猶豫最大；三軍之滅，莫過孤疑
> One who excels in battle stays without flinching. When he sees victory, then he sets out. When he sees no victory, he desists. Thus it is said, one must not harbour fear or vacillation. Vacillation is the greatest threat to the use of forces. Of the potential disasters for the three armies, none surpasses doubt.[24]
>
> 進退多疑，其眾無依，可震而走
> If their advance and retreat are filled with doubt and there is nothing upon with they can rely, then they can be shocked into fleeing.[25]
>
> 若計不先定，慮不蚤決，則進退不定，疑生必敗
> If plans are not first determined and deliberations are not resolved, then advances and retreats will be unstable, doubts will occur, and defeat will be inevitable.[26]
>
> 廿一曰，多疑，眾疑，可敗也
> (On fatal flaws of a commander) number 21 states that if (the general) has much doubt, the masses will also be doubtful and can be defeated.[27]

For many military theoreticians in early China, doubt was often linked with defeat. In a worst case scenario, the general's doubt would be palpable to his troops and infect them like a plague. With doubts clouding their judgment, the troops' willingness to follow orders and their commitment to the engagement would be seriously undermined. Contra this, however, if the general of a small army was capable of ascertaining a high level of doubt in the larger opposing force, or in some instances even engendering such doubt, the odds in favour of victory would be greatly shifted to his forces. This seems to be the point that Cao Mie is trying to make to Duke Zhuang. By determining which sections of the enemy's forces appear weaker or have troops evincing a level of doubt over the battle, the commander of the smaller forces of Lu could focus his army's strength on these sections and cause disorder and retreat.

Opportune moments after battle (既戰之機)

The final question posed by Duke Zhuang concerns the feasibility of swiftly launching a subsequent attack after having already done battle with an enemy. This is a very dangerous proposition for many reasons. First, whether or not one can launch such an attack depends upon the outcome of the first battle. If one was defeated, then it would be difficult, though not impossible,[28] to regroup and reengage the victorious forces. Second, having already battled, one must be cautious of the well-being of one's own troops, regardless of whether or not the initial battle was won. Returning to the fray with beleaguered troops would not be a wise course of action for any army. In the event, however, a military force finds favourable indications for reengagement, its general must still be cautious of the post-battle status of the opposing army, especially if the general's army is smaller. This is precisely the meaning embedded into Cao Mie's response to Duke Zhuang's query. The opportune moment for reengagement arises if, and only if, the morale of the enemy soldiers is at its lowest. A lack of morale creates doubt amongst the ranks, weakens troop commitment to fight and engenders difficulty in maintaining order. As such, morale is the key to success in subsequent engagements.

Regardless of an army's capability to reengage, Cao Mie argues that whether or not it should do so depends upon the physical and mental condition of the enemy troops. As in other sections of the manuscript, responsibility for the maintenance of troop morale is assigned to the general (or ruler). Yet, how does one determine a lack of morale and order? In this particular section, Cao Mie sees a *lack* of morale stemming from two different sources, both of which have correlates in other military treatises from early China.

Cao Mie first elucidates the inverse relationship between rewards and punishments, deemed unfair, harsh or arbitrary, and the morale of troops. If having battled, the troops of the enemy forces are unsure as to whether or not they will be rewarded or punished, then their morale will certainly be weakened, as will their willingness to resume combat. As the following lines from elsewhere in the *Cao Mie* manuscript demonstrate, Cao Mie repeatedly advocates constancy in rewards and punishments as a necessary requirement not only for a general, but also for the ruler of a state.

> 則祿爵有常幾(機)莫之當
> Thus when emoluments are regular, there are no other opportune moments match it.[29]
>
> 莊公又問曰復敓戰有道乎答曰有收而聚之束而厚之重賞薄刑使忘其死而見其生
> Duke Zhuang further inquired saying, 'Is there a way to reengage after an X battle?' [Cao Mie] replied, 'There is. Regroup and assemble them [i.e.

troops]. Gather and reward them. Heavy rewards and light punishments cause [them] to forget of their deaths and focus on living.'[30]

> 莊公或聞為和如何答曰毋嬖於便嬖毋長於父兄賞均聽中則民和之
> Duke Zhuang further asked, 'How does one act harmoniously?' [Cao Mie] replied, '[The ruler] ought not pander to favoured ministers; ought not give precedence to father and elder brother. If his rewards are equitable, and adjudications are centred, then the people will harmonize with him.'[31]

The first two quotations deal specifically with the scenario that Duke Zhuang's question alluded to, namely opportune moments for reengagement. The initial statement emphasizes that after suffering a defeat one should ensure that rewards are doled out according to dictated principles. To do otherwise would be interpreted by other soldiers as pandering to favourites and could easily engender resentment. The second quotation emphasizes the inherent psychological value of heavy rewards and light punishments. Utilizing this method, the troops will be concerned less with death, and more with living so as to enjoy the rewards bestowed upon them. And although the last quote refers to attributes of a ruler necessary for ensuring harmony within his state, it pertains as well to a general. If a general arbitrarily dispenses rewards and punishments, then he will certainly have difficulty in maintaining harmony among the troops, and this lack of harmony will affect troop morale and the ability of a general to command his troops if battle recommences.

Other early Chinese military texts contain numerous references to the use of rewards and punishments. Interestingly, however, different texts evince differing opinions over their intrinsic value for maintaining order and morale. Further, none are as adamant as Cao Mie over a direct connection between their improper usage and the demise of troop morale.

> 田忌曰：賞罰者，兵之急者耶。孫子曰：非。夫賞者，所以喜眾，令士忘死也。罰者，所以正亂，令民畏上也。可以益勝，非其急者也。
> Tian Ji said, 'Are rewards and punishments that which is most pressing for the military?' Sunzi (aka Sun Bin) replied: 'No. Rewards are means of elating the masses and causing soldiers to forget their deaths. Punishments are means of rectifying disorder and causing the populace to be in awe of their superiors. [These] can be used to aid in victory, but they are not that which is most pressing.[32]

> 有不占而避之者六: … 三曰賞信刑察，發必得時 … 凡此不如敵人，避之勿疑.
> There are six situations in which without divining, you should avoid them [the enemy] … The third states: [Their] rewards are trusted and punishments are scrutinized, each is invariably executed at appropriate times … In these situation, you are no match for the enemy and should without a doubt avoid them.[33]

> 民非樂死而惡生也，號令明，法制審，故能使之前。明賞於前，決罰於后，是以發能中利，動則有功。
> The populace does not revel in death, nor detest life. When orders are clear and regulations are promulgated, then one is able to make them advance. When rewards are clear beforehand and punishments decisive afterwards, then when deployed they will be able to centre in on advantages and when in action will have meritorious deeds.[34]

Yet, balanced rewards and punishments are but one element required for the maintenance of troop morale, and Cao Mie also offers a second element linking a general's actions to a concept similar to *esprit de corps*. A general who fails to ensure the reclamation of dead bodies and fails to inquire about his wounded soldiers will impart both a sense of callousness and a sense of ambivalence towards the lives of his soldiers. Thus, for those soldiers who survived the initial battle, such a display from their commanding officer will likely elicit thoughts of their own fate and cause doubts over their own commitment to a subsequent military endeavour. Such a loss of morale would make an army quite susceptible to defeat if a second assault were launched. This emphasis on the causal link between troop morale and their relationship to their general is seen elsewhere in the manuscript.

> 率車以車率徒以徒所以同死
> Commanding chariots by chariot, and commanding foot soldiers by foot is the means to share death.
>
> (slip 58)

For Cao Mie, any army that has already seen battle must be especially cautious when contemplating reengaging the enemy. Even if one's own army is capable of mounting a second offensive, the true indicator of success or failure lies with the proper assessment of the morale level of the enemy troops. It is the responsibility of the general to maintain high levels of morale, and it is through his own actions that such morale can be quickly diminished. If the enemy general has failed in his duty to maintain troop morale, then even a small army, such as that of Lu, has a clear advantage when reengaging the enemy.

Conclusion

As China's oldest extant military treatise, *Cao Mie's Battle Array* provides the scholarly world with its first contemporary manuscript copy of a text on the most prevalent socio-political issue of the aptly named Warring States period, namely warfare. In particular, its systematic treatment of 'opportune moments', when contextualized with the writings of other military texts, allows for a deeper understanding of what constitutes an 'opportune moment' in early Chinese military thought. Before the discovery of this manuscript, what was known of opportune moments was primarily restricted to unsystematic references scattered throughout early Chinese military texts, as well as historical anecdotes. From the latter, we know that Cao Mie contributed to early Chinese strategic theories related to opportune moments as evidenced by his skill at manipulating the forces of Qi and ascertaining the appropriate moment for the troops of Lu to strike at the battle of Changzhuo. In the manuscript *Cao Mie's Battle Array*, similar concerns are clearly evident. Duke Zhuang of Lu is considering going to war with Qi, his largest and most formidable neighbour. Knowing that his forces will be at a disadvantage, the duke seeks advice from Cao Mie on ways to increase his odds of success. In Cao Mie's responses, he alludes to visible indications of potential problems within the enemy's armies that, if capitalized on, can become a tactical advantage to a smaller army. These problems include the enemy's selection of poor generals, their traversing of dangerous terrain, their hesitancy to do battle and their post-battle lack of morale. These concerns match well those found in other Chinese military texts. Yet, unlike these latter texts, the *Cao Mie* manuscript provides a rather concise, but systematic analysis of visible indicators for opportune moments to be found at different stages of battle. Each stage of battle offers new potentials for success, or failure. In keeping with the text's central theme of proper action, Cao Mie emphasizes that only perspicacious rulers and commanding officers can exploit such opportune moments. Failure to fully capitalize such moments will quickly transform one's advantage into tactical disadvantage.

Notes

1 *Han shu* 30.1762. All citations of traditional Chinese works will be based on *Wen Yuan ge si ku quan shu* 文淵閣四庫全書 (Taipei: Taiwan shangwu yinshu guan, 1983–1986) and citations to dynastic histories will follow the Zhonghua shuju 中華書局 editions (Beijing: Zhonghua shuju, 1959–).

2 The title of the manuscript is written in black ink on the back of slip 2. All citations to this manuscript and all others in the Shanghai corpus will be by slip numbers according to the arrangements found in *Shanghai bowuguan cang Zhanguo Chu zhushu* 上海博物館藏戰國楚竹書, 7 vols (Shanghai: Shanghai guji chubanshe, 2001–).

3 One notable exception to this would be the military text, *Wuzi* 吳子. The Liao di chapter of this text lists 20 situations in which an army can engage its adversary immediately without recourse to divination. See *Wuzi*, Liao di, 6a–b.

4 This chapter is primarily concerned with textual content and not issues of codicology; therefore, I will offer a transcription of the text in modern Chinese graphs and will attempt to keep notes related to issues of palaeography or phonology to a minimum.

5 This refers back to the previous section of the manuscript in which Cao Mie lists three social conditions which the ruler of a state brings about through proper indoctrination of the populace.

6 Determining the meaning of the graph *zhi* 廌 [OCM *dreʔ] in Warring States manuscripts is rather complicated. It is often understood to function as an auto-antonym representing either *cun* 存 [OCM *dzə̂n], 'to survive', or 'exist', or as *fa* 灋/廢 [OCM *pap/pats], 'to cast away', or 'disregard'. Prior to the discovery of Warring States manuscripts, we do not have any other excavated source materials in which the graph [廌] is used independently, but it frequently appears as a component of *fa/fei* 灋. In this case, however, scholars have opted to read the graph as representative of the word *jian* 薦 [OCM * tsə̂ns], 'to recommend'. Though phonologically plausible, such a reading

seems rather forced and can perhaps be better understood if the graph were read as *cun* 存 'exist'.

7 Here I understand *ji* {疾} (OCM *dzit) to mean 'fierce' following a phrase 'Grasp one's sword and looking fierce state, "Who dares to match me!"' 夫撫劍疾視曰彼惡敢當我哉 from *Mengzi* 1B.3.

8 Ma Chengyuan (ed.), *Shanghai bowuguan cang Zhanguo Chu zhushu* (*si*), p. 269.

9 Chen Jian, 'Shanghai zhushu <Cao Mo zhi Chen> xinbian shiwen (gao)', in Ji Xusheng (ed.), *Shanghai bowuguan cang Zhanguo Chu zhushu (si)* (Shanghai: Shanghai guji chubanshe, 2004), pp. 190–192.

10 Ji Xusheng (ed.), *'Shanghai bowuguan cang Zhanguo Chu zhushu (si)' duben* (Taipei: Wan quan lou tushu, 2007), p. 191.

11 *Wei Liaozi*, Shi'er ling 十二令: 2.2b.

12 *Wuzi*, Tu guo 圖國: 1a.

13 *Wuzi*, Lun jiang 論將: 11a.

14 *Wuzi*, Liao di 料敵: 6a.

15 For example, in slip 14 of this manuscript the graph *zhi* [廌] is read as *cun* {存}, 'to exist' or 'survive', in the line 三代之陳皆存 'The battle arrays of the three dynasties all survive.' It also occurs in slip 41: 周志是存 'This survives in the writings of Zhou.'

16 Examples emphasizing the necessity of a complete commitment to battle, even to the point of death, can be found in the *Wuzi* chapters: Lunjiang: 10a; Liao di: 5b; and Zhi bing 治兵: 8a.

17 Zhang Zhenze 張震澤, *Sun Bin bingfa jiaoli* 孫臏兵法校理 (Beijing: Zhonghua shuju, 1984), Xuan zu 選卒, slip 4, p. 54. My arrangement and numbering of the Yinqueshan *Sun Bin bingfa* manuscript slips follows that of Zhang Zhenze, *Sun Bin bingfa jiaoli*.

18 Zhang Zhenze, *Sun Bin bingfa jiaoli*, Jiang de, slip 3, p. 176.

19 *Liu Tao*, Long tao 龍韜, Li jiang 立將: 3.5b–6a.

20 *Zuozhuan zhushu* 左傳注疏, Xi gong 22: 14.4a–5a.

21 *Wuzi*, Liao di: 6a.

22 *Wuzi*, Liao di: 6a–b.

23 *Zuozhuan zhengyi*, Zhuang gong 10: 7.33b–36a.

24 *Liu Tao*, Long tao, Jun shi 軍勢: 3.9b–10a.

25 *Wuzi*, Lun jiang: 11a.

26 *Wei Liaozi*, Lei zu ling 勒卒令: 4.3a.

27 Zhang Zhenze, *Sun Bin bingfa jiaoli*, Jiang shi 將失, slip 8, p. 180.

28 Indeed, the next section of the *Cao Mie* manuscript deals precisely with methods for regrouping and reengaging the enemy after particular types of battles.

29 *Cao Mie zhi chen*, slip 50. This line comes from the section of text directly after the section dealing with opportune moments. This particular line is referring to the ability of an army to reengage battle after suffering a defeat.

30 *Cao Mie zhi chen*, slips 53b and 54.

31 *Cao Mie zhi chen*, slips 35 and 36.

32 Zhang Zhenze, *Sun Bin bingfa jiaoli*, Wei wang wen 威王問, slips 16–17, p. 27.

33 *Wuzi*, Liao di: 6b.

34 *Wei liao zi*, Zhi tan 制談, 1.3b–4a.

2 *Yuddha* and *Vijaya*

Concepts of war and conquest in ancient and early medieval India (up to CE 1300)

Krishnendu Ray

Introduction

From time immemorial, human beings in South Asia have been performing different kinds of war-like activities in order to protect their existence. Such activities have in different terrains of the subcontinent. These are the hilly area, the cultivated plain lands, the maritime space, the riverine zones, the coastal area and the forest area, etc.[1] Human beings interacted with their physical surroundings in order to settle in the space they selected and branched out. Their interactions with their physical environment developed in them impressions about it, by which they became conditioned at the same time. This conditioning in turn led them to develop their attitudes towards their environment; these attitudes shaped their mentality. Accordingly, they set their aims/objectives/plans in order to live by elimination of opposition at the physical level. In this connection, the prehistoric and proto-historic cultures of the Indian subcontinent may be discussed. These cultures point out that human beings have used stones as weapons[2] in order to defend themselves against the opposition put forward either by animals or by other human groups. As years passed, they gathered experiences, developed and polished their weapons with a view to face the challenges of the new situations. They developed metal technology. This resulted in segmentation and attendant complexities in their lives. The problem then was who was to control the new technology? For those who had the power were certainly not the same as those who produced the necessaries of life.[3] In other words, the early people witnessed a kind of difference among themselves. This stratification brought further complexities in their lives.

People initially formed groups under chiefs of their choice. Thus, they have established chiefdoms in order to ensure their protection under the chief who was later called the *raja* (king). The word *raja* means 'to shine' or 'to please' or 'to lead'. The *raja* was essentially a military leader who successfully led and defended the people under him. Gradually, the chiefdoms became territorial units with the settlement of clans, such as the Madras, the Kurus, etc., around 600 BCE in north India. With the passage of time, the issue of providing protection to the people of the kingdom became complex. In this connection, the concepts of *yuddha* (war) and *vijaya* (conquest) gradually became significant.

Yuddha essentially means not only the application of force for the possession of the material riches/wealth of a territory but also a part of the strategy as to how to live without opposition or enmity of any kind and thus the concept of *vijaya* is implied in that of *yuddha*. It appears that the concept of *yuddha* had been applied in order to ensure the possession of a territory or the riches thereof.

In this connection a few existing works by P.C. Chakravarti,[4] V.R. Ramachandra Dikshitar,[5] U.P. Thapliyal,[6] etc., in the context of early Indian warfare may be mentioned. These works point out the organizational aspect of war, tactics that were applied/followed for the conquest (*vijaya*) of the enemy. They have mentioned textual statements about conducting a war. Thus, it appears that the experiences people have gathered in the course of conducting a war (*yuddha*) from the days as old as pre-Harappa (prior to *c*.2500 BCE) might have led them to develop gradually their ideas about conducting a war against the enemy. The language of the Harappan civilization is yet to be deciphered. However, we will start our discussion regarding the historical evolution of the concepts of *yuddha* and *vijaya* especially taking into account the theoreticians' standpoints, beginning with the *Rig Veda* (*c*.1500–1000 BCE).

Vedic era

We start our discussion with the period from *c*.1500 to *c*.600 BCE, when the Indo-Aryans were engaged in clannish warfare for the possession of the cattle wealth. In this connection, Ranabir Chakravarti has drawn our attention to the relevant passages from the *Rig Veda*.[7] According to a hymn of the text, addressed to Soma (*Pavamana Soma devata*), a prayer was made in order to win riches by his might.[8] Soma is known to have been considered important enough to win material wealth.[9] It appears that winning material riches was the principal objectives of conflict and thus the aim of *vijaya* for the Rig Vedic people. This assumption is supported by another hymn of the *Rig Veda*. It points out that Soma was considered as the first strong leader during war, who won the cattle.[10] Thus, Soma was addressed to become successful in warfare for winning material/animal wealth. Chakravarti also focuses on a very interesting word, *gavishti*.[11] Now, the *Rig Vedic* hymn in which the word *gavishti* occurs brings out that the Rig Vedic people thought of Soma with his *ayudha* (weapons) to win/conquer the cow.[12] In other words, they desired to possess the cow by way of warfare. This brings one close to the word *gavishti* meaning 'to desire for fighting, ardour of battle'.[13] Thus, by way of successful fighting the leader/king became *gopati*. The term means one who possesses cows.[14]

In connection with warfare during the period of the *Rig Veda*, the most important was the Battle of Ten kings recorded as the *Dasarajna*.[15] The battle was fought between the two hostile groups of kings; Trtsu-Bharata Sudas represented one group and a confederacy of ten kings stood as the opponent. The Battle of Ten Kings may be indicative of the territorial expansion of the royal power. This resulted in subjection of the enemy to the king as well as their villages along with the cattle wealth.[16] Thus, it appears that the Vedic warfare (*yuddha*) signified the attempts towards possessing/conquering (*vijaya*) initially the cattle wealth and later securing territorial annexation. In the *Rgveda*, the Bharata chief is praised to have removed all the enemies from the west, north and east; in the south he found no enemy as there was a forest called Khandava. The Khandava forest was probably situated to the south of Kuruksetra (present day Haryana). In other words, we may say that the Bharatas extended their power territorially probably up to Kuruksetra by achieving victory in the Battle of Ten Kings.[17]

However, the time of fighting, the calculations of the conquering king and his opponent, the situations of the *vijigisu raja* (conquering king) as well as those of the enemy, etc., are not clear. As the rulers gathered experiences with the passage of time, the concepts of *yuddha* and *vijaya* started to evolve and expand. Thus, one is drawn to another phase in early Indian history which was represented by the *mahajanapadas* (large territorial kingdoms), which emerged mostly in the Gangetic valley around *c*.600 BCE. This phase witnessed royal powers aggressively pursuing the policy of territorial expansion. *Mahajanapadas* came into existence by the absorption of rival *janapadas* (populated territory) by the victorious *janapadas*. Such territorial absorption therefore brought about war and conquest. The concepts relating to such war and conquest are available from major literary textual sources such as the *Arthasastra* of Kautilya (composed between the third century BCE and the first century CE), *Manava-Dharmasastra* (composed between 200 BCE and CE 200), *Yajnavalkyasmrti* (CE 100–200), etc. Here we will confine our discussion to how the concepts of war and conquest changed between 600 BCE and CE 1300.

In the phase of the *mahajanapadas*, as mentioned in the *Anguttara Nikaya*, there were 16 *mahajanapada*s in north and north-west India. These were as follows: Anga (near Bhagalpur in east Bihar), Magadha (Patna and Gaya districts in Bihar), Kosala (Lucknow, Gonda, Faizabad and Baharaich districts of Uttar Pradesh), Kasi (Banaras, Uttar Pradesh), Vrji (modern Vaishali, or Basrah, Muzaffarpur district, Bihar), Malla (Pawapuri, Bihar), Chedi (area around Jabalpur, Madhya Pradesh), Vatsa (area around Allahabad, Uttar Pradesh), Matsya (eastern parts of Rajasthan), Surasena (Ganga–Yamuna *doab* area), Kuru (Delhi–Haryana region), Panchala/Pancala (present Rohilkhand and a part of central *doab*), Asmaka (Godavari Valley), Avanti (western part of Madhya Pradesh), Gandhara (north-western frontier part of the Indian subcontinent) and Kamboja (Hazara district of Pakistan). Finally, Magadha emerged as the most dominant power by annexing all the other *mahajanapadas* by around the late fourth century BCE.[18] In other words, the king who wanted to establish his power and authority over a place termed as an empire fought wars for territorial annexation.

In this connection, Bimbisara (*c*.545/44–493 BCE) of the Haryanka Dynasty followed a shrewd diplomatic policy. He established marriage alliances with the ruling dynasties of Madra, Kosala and Vaishali. Thus, he made those powers his friends (*mitra*). This certainly strengthened his political position and paved the way for the territorial advancement of his power and authority. Thus, the diplomatic move he adopted empowered Magadha with control over Kasi, which was economically rich.[19] Bimbisara is also known to have territorially annexed his enemy (*ari*) kingdom of Anga to his expanding realm.[20] This advancement/augmentation of power by Bimbisara may be compared to Kautilya's *yanam* (marching), which comprises one of the six foreign policies (*sadgunyam*).[21] Bimbisara's son and successor Ajatasatru (493–466 BCE) continued the expansionist policy. He followed the policy of aggressive diplomacy (*upalapana*) and disunion (*mithubhed*) and thus defeated the enemy kingdom of Vaishali and established the hegemony of Magadha.[22] Magadha had another rival enemy to reckon with and that was the kingdom of Avanti. The expansionist policy of Magadha continued. This enemy kingdom was destroyed by the Magadhan king Sisunaga.[23] Magadha gradually emerged victorious by depriving the glory (*yasah*) of other powers like the Sisunagas, Pancalas, Kurus, Kalingas etc. In this connection Mahapadmananda, the first ruler of the Nanda dynasty of Magadha (*c*. fourth century BCE), is said to have destroyed all the Kshatriyas (*sarva ksatrantaka*) and thus become the ruler with sole authority (*ekarat*).[24] The power and authority of the Magadhan king continued to expand territorially. Thus, Magadha became a formidable power in the Gangetic valley.

However, the Macedonian conqueror Alexander during his march through the north-western part of the Indian subcontinent encountered many powers like Assakenoi (region in Swat), Nysa (area between Kabul and Indus valley), Taksasila (Taxila) etc.; but he faced stiff resistance from the kingdom of Elder Porus, i.e. Puru. Probably, in response to deployment of Puru's army of 50,000 infantry, 3,000 cavalry, 1,000 chariots and 130 war elephants, the Macedonian king resorted to sophisticated tactics in the Battle of Hydaspes (*c*.326 BCE). Alexander's companions charged Puru's cavalry. And then his mounted archers, then the guards led by himself; together with these the two cavalry regiments under Koinos swept the front of Puru, galloped round his right wing, and then 'threatened the rear of the Indian cavalry and chariots' and, thus, ultimately, the Indian soldiers suffered a crushing defeat at the hands of the Macedonian soldiers. Puru was captured. The Greek hero emerged victorious in his territorial advancement in the north-west of the Indian subcontinent. Yet, the victor, respecting the vanquished's proud request to be 'treated as a king', showed political generosity by confirming the latter in his own territory. Strategically, thus, the victorious Alexander secured 'a grateful and faithful friend'.[25] However, the Macedonian conqueror is not known to have encountered the formidably growing Magadhan power in the Indian subcontinent. But the fact that Alexander's campaign weakened many powers in the north-western corner of the Indian subcontinent made the situation there advantageous for Chandragupta Maurya to advance his power territorially in that area.[26]

Asoka's *dhamma*

Chandragupta Maurya (324–300 BCE) and then Asoka (273–236 BCE) are historically known to have caused the territorial advancement of the Magadhan empire. The territorial conquests of the Magadhan ruler extended his power and authority to Afghanistan in the north-west, Saurashtra in western India, Karnataka in the south and Kalinga (Orisssa) in the east. Thus, historically it appears that both diplomatic and violent wars made Magadha nearly a pan-Indian power.[27] The warfare of the conquering king for wealth and land may be compared to what Kautilya referred to the conquering king as the greedy and the demonical (*lobhavijayi asuravijayi*).[28] In the rise of Magadha, Asoka appears to have had launched the most violent military campaign against Kalinga and annexed it to his own kingdom. In this war, many people of Kalinga were killed (*vadha/maranam*), and numerous people were carried away (*apavaha*) from Kalinga. Thus, Asoka through the application of high intensity organized violence acquired Kalinga (*Kalingah vijitah*).[29] In order to retain his hold over the vast Mauryan empire with subjects of different interests, Asoka preached *dhamma* which was, according to B.N. Mukherjee, 'a political philosophy based on social and religious (primarily Buddhist) codes and ancient custom'.[30] *Dhamma* also included 'the subjects' firm devotion to the king's (i.e. Asoka's) interest' (*didhabhatita* or *dridabhaktita*).[31] It seems to have been a novel technique and Asoka adopted it in order to hold his 'absolute political authority' over the empire.

It appears that the conquest that we have talked about so far is that by force of arms. In this connection, we should take note of Asoka's attitude towards war and conquest. According to him, the true conquest is the conquest through *dharma*. This is *dhammavijaya* (conquest by righteousness). It is true because it brings happiness in this world and in the next world (*hidalokika paralokika* Sanskritized as *aihalaukikah paralaukikah*). He advised his descendants not to think of conquest by the application of force of arms; even if they had to conquer by arms, they should achieve it with forbearance and light punishment (*ksantih ca laghudandata*).[32] Asoka's *dhammavijaya* did not mean territorial conquest. It recommended certain forms of behaviour that constituted his *dhamma* and he envisaged unification through *dhamma*. Thus, his *dhammavijaya* was different from *dharmavijaya* of Brahmanical texts, in which the conquered territory could exist autonomously in relation to the conquering king.[33] In raising Magadha successfully to its superior political position, it appears that the king established friendship with neighbouring kingdoms as well as also remaining careful about the activities or strengths of his enemies. Thus, he destroyed his enemies by conducting a war. At the same time, he had also to follow techniques for the security of his kingdom, as might have been the case with Asoka's *dhamma* for his empire. He changed the sound of war (*bherighosa*) to the reverberation of *dhamma* (*dhammaghoso*).[34] This was the most significant pacification technique for maintenance of the empire. For the king's primary objective was to keep his kingdom free from both internal and external dangers. It therefore appears from the above that the king who wanted to advance territorially the sphere of his power and authority at the cost of other kings pursued varied techniques like diplomacy and direct war, friendship/alliances, etc., in accordance with the changing political and military contexts. Therefore, we may say that the king had bravery, intellect, strength, far-sightedness, which enabled him to become successful in his military operations/territorial advancement. The king (*svami*), as one of the limbs (*angas*) of the state, could therefore wield *danda* or force. He is known to have protected the social order by putting an end to the situation of conflict.[35] Therefore, *danda* is the most visible aspect of the coercive power/authority of the king and his army. With this, he continued military operations for the advancement of his power and authority putting an end to opposition of different kinds.

Kautiliya *Arthasastra*

Thus, one comes to the conquering king of the Kautiliya (also known as Canakya or Visnugupta) *Arthasastra* (Science of Material Gain). According to Kautilya, the king who would be conqueror should be endowed with personal excellences like intelligence, spirit, memory, thorough understanding (*atmasampat*) as well as his material excellences like the minister (*amatyasampada*), the country with resources (*janapadasampat*), the fort (*durgasampat*), the treasury (*kosasampat*), the army (*dandasampat*) and the ally (*mitrasampat*) (*svamyamatyajanapadadurgakosadandamitrani*).[36] Therefore, Kautilya locates the position of such a potential king (*vijigisu*) in relation to other powers or kingdoms within the *mandala* (circle of kings). According to the author, the *vijigisu raja* is to be at the centre. In this connection, it should be noted that if a power with territory lies immediately next in front of such a powerful *vijigisu* king, that is the enemy (*bhumyantara ariprakriti*); if a neighbouring territorial power possesses the excellences of the enemy, that is also the enemy of the king (*arisampadyuktah samantah satruh*). However, if the territorial power is separated by another territory, it is called the ally/friend (*bhumyekantara mitraprakritih*). Thus, other neighbours are *arimitra* (friend of the enemy), *mitramitra* (friend of the friend), *arimitramitra* (friend's friend of the enemy), *parsnigraha* (rear enemy), *akranda* (rear friend), *parsnigrahasara* (friend of the rear enemy), *akrandasara* (friend of the rear friend), *madhyama* (middle) and *udasina* (indifferent or neutral power). Of these political powers, the *udasina* is the most powerful political entity. It remains outside the sphere of the enemy, the conqueror and the middle king. It is stronger than their constituents. It is capable of extending help to the enemy, the conqueror and the middle king when they are united or disunited. Even the neutral power can suppress them when they are disunited. Thus, the extremely powerful neutral king enjoys his elevated position (*urddhvam asina*) and because of his such high position, he can prefer to remain indifferent to the affairs of other political powers in the circle of kings (*mandalayoni*).[37] It is suggested that the *Arthasastra* might have been composed during the period from 300 BCE to CE 100.[38]

This period witnessed the attempts of the Mauryas and their successors to ensure the territorial augmentation of their power with control over the economic resources of different areas of the Indian subcontinent. Thus, the king who rose to the occasion and played the instrumental role in spreading his power and authority territorially has therefore been described as the *vijigisuraja* (the king desirous of conquest). He is advised in the Kautiliya *Arthasastra* to maintain varied types of relationships with his neighbouring powers, as mentioned earlier, in terms of six policies described as *sadgunyam* in the *Arthasastra*. These are: *sandhi* (alliance), *vigraha* (war), *asana* (status quo), *yana* (mobilization to deter the potential enemy), *samsraya* (seeking political asylum with a superior power) and *dvaidhibhava* (following both the policies of war and peace simultaneously).[39]

Except *vigraha* (the policy of war), the other policies are primarily aimed at achieving economic gains and territory without using outright organized violence. The king is advised that whatever policy he follows, in the first place he should think that he shall be able to promote his own interests concerning his forts, trade routes, forests, water-works and injure the same of his enemy. Thus, the *vijigisu raja* is advised to strengthen the territorial base of his power and authority by possessing resources and at the same time by destroying or ravaging the territorial and economic resources of the enemy. The king is directed to pursue *vijaya* in order to ensure economic gains and territorial expansion in return.

A weak king is advised to make peace instead of going to war. Peace is discussed by Kautilya at length in the *Samsrayavrtti* (a weak king making peace with a strong king) section of the *Arthasastra*. Kautilya suggests mainly three types of *sandhi* such as *Dandopanata sandhi* meaning making *sandhi* with a strong king by surrendering the army, *Kosopanata sandhi* in which peace is sought by surrendering the treasury and *Desopanata sandhi* by which the land was surrendered. We may support the surrender of land by *sandhi* (treaty) by referring to a historical case. We know that a treaty was signed between Chandragupta Maurya and Seleucus '*Nicator*' (the Victor), putting an end to hostilities between them. According to that treaty, Seleucus ceded Arachosia (Kandahar), Gedrosia (Baluchistan) and Paropomsadai (south-east of Hindu Kush) to Chandragupta Maurya.[40]

In addition to these three types of *sandhi*, Kautilya has also suggested subtypes of *sandhi* and these are: *atmaraksana* (self-defence), *parikraya* (releasing the royal officers by surrendering treasury to the enemy), *atyayasandhi* (making certain the payment of wealth in a certain land at a certain time), *kapalasandhi* (the *sandhi* that ensures the immediate surrender of wealth), and the gains *pratyadeya* (the wealth that is difficult to obtain), *prasadaka* (the wealth that a righteous king obtains from an unrighteous king), *hrsvakala* (the wealth that is obtained within a short time), *alpavyaya* (the wealth that is obtained by giving food), etc.[41] Overall, launching a royal campaign is approved only when financial gains are ensured. An economically motivated campaign had occurred during early fifth century BCE, when Ajatasatru of Magadha annexed economically productive Kosala with its capital Sravasti (in modern Uttar Pradesh).[42] We may assume that the ideas/concepts that are so articulately recorded in the *Arthasastra* in regard to war (*yuddha*) seem to have been rooted into the attempts towards the spread of power and authority territorially at the cost of other chiefs or *rajas* (kings). The same scenario is also noticeable in the period following the exit of the Mauryas from the Indian historical landscape after 200 BCE.

Manu's *Manava-Dharmasastra*

Ideas about war for the post-Maurya period are available from the *Manava-Dharmasastra*. The text points out that following the fall of the Mauryas certain rules had been framed regarding the king's attack on his enemy; the timing for launching a war against the enemy, treatment of the non-combatants, etc.[43] In connection with the policy of war, four measures such as *sama* (conciliation), *danda* (coercion), *dana* (gift) and *bheda* (dissension) are emphasized both in the Kautiliya *Arthasastra*[44] and in the *Manava-Dharmasastra*. According to Manu, the policies of conciliation and coercion always are the best to ensure the security and economic development of the kingdom[45] and the policies of gift and dissension might have been considered secondary in importance. The *Yajnavalkyasmrti* points out that the king acquires the same merit/virtue (*nrpaterdharmah*) by conquering/bringing another's (enemy) kingdom under his subjection (*pararastram vasam nayan*) as (*tameva*) by protecting his own kingdom (*svarastrapratipalane*). Thus, it may lead one to say that launching a pre-emptive war by the king for the protection of his kingdom, has been supported in the text.[46] In the *Yajnavalkyasmrti* launching a war has been regarded as a meritorious/spiritual act and therefore supported.[47] Simultaneously, the text also pushes a moderate view. It notes that if the king gives to the Brahmins the wealth acquired by war against the enemy's kingdom, then no other duty is higher than this act of giving wealth acquired by war.[48] Giving wealth to the Brahmins was a technique of legitimizing external aggression by a ruler among the audience of his domestic society (home front). Side by side, the text also suggests that the acquisition of a friend is more valuable than having gold and land.[49] As an example, it could be cited that Bimbisara's friendship with Madra, Kosala and Vaishali enabled him to conquer Anga.[50]

U.N. Ghoshal refers to a verse in the *Manava-Dharmasastra*.[51] It reads thus: *sarvopayaistatha kuryannitijnah prthivipatih/yathasyabhyadhika na syurmitrodasinasatravah*.[52] The verse points out that the king well versed in policy-making *nitijna* (policy-making) should particularly act in such a way that neither a friendly power/king nor neutral power, or even the enemy king, can become dominant in inter-state relations. On this point Manu differs from Kautilya. While Kautilya supports the expansionist policy of *vijigishu*, Manu is for the *dharmik* (just) ruler following a moderate (if not pacific) policy.

Keeping these textual ideas concerning a war in mind, attention may now be turned to the historical scenario of the period from about 200 BCE to about CE 300 in early India. The period witnessed 'confrontations' between different political powers like the Greeks, Saka-Pahlavas, Kusanas/Kushanas in the large parts of north India, the Satavahanas in the Deccan and the Colas, the Pandyas and the Ceras in the far south. They are known to have launched wars understandably in order to ensure their economic gains. The rulers concerned attempted to keep under their control trade routes, ports etc.[53] It seems that they endeavoured to ensure their own prosperity and progress. For, the king was required to ensure his own progress (*svavrddhi*) according to Kautilya.[54] We can refer to the fact that the Satavahana king Gautamiputra Satakarni (*c*. early CE second century) conquered eastern Deccan, Malwa, western and southern Gujarat and possibly the riches of his enemy the Saka Kshatrata king Nahapana (perhaps CE late first century to early second century).[55] Thus, we may assume that he became successful in increasing the strength of his wealth for progress. Understandably, the ruler had to show his might, authority and superiority in ensuring his own prosperity. We may compare it historically to the attempts of the Kushanas in early India. Their imperial titles like *Maharaja* (the Great King), *Rajatiraja* (King of Kings), their 'majestic representations on their coins' may indicate, according to B.N. Mukherjee, their 'might and authority'.[56] Their epithet *Devaputra*[57] may indicate the message of their 'might and authority' and superiority both to their subjects and opponents. Thus, their high-sounding titles might have been a kind of diplomatic move/message on their part towards spreading their power and authority territorially.

Coming to the *Mahabharata* (composed between *c*.400 BCE to *c*. CE 400), we find that an integral part of the king's duty was to have lessons on fighting for killing robbers or ensuring victory (*vijaya*) over his enemies. Launching a conventional war (for external security) as well as low-intensity war (for internal security) was approved for the good governance and expansion of the kingdom. However, it is emphasized that the policy of *sama* (conciliation) should be tried first[58] and if it is not successful then only the policy of war may be followed. Thus, it may imply that the security of the subjects was given a priority over expansionist urge of the *vijigisu* king. Ghoshal draws our attention to the Buddhist literature *Mahavastu* (the Great Event, dated to the early Christian era) which refers to the five categories of king's strength. These are as follows: *sahja-bala* (his personal power), *putra-bala* (strength of his son), *jnatimitra-bala* (strength of friends and relatives), *caturanga-bala* (strength of his army) and *prajna-bala* (strength of his wisdom).[59] The text points out that the strong point of the king is his wisdom because it brings happiness to himself, his realm as well as his friends and relatives. In this connection, we may refer to the policy of the Kushanas. The Kushana kings had the vision to strengthen financially the territorial base of the kingdom with commercial benefits. According to B.N. Mukherjee, the Kushanas' policy was 'probably that of *laisseze faire*'; they were in favour of 'the collection of custom-duties which would grow with unhindered development of trade'.[60]

Understandably, the king was expected to have such strength in order to ensure the conquest of obstacles or hurdles which might have come his way. More so, because the period that was addressed by the author of the text might have been the early Christian era when extra-regional political powers, like the Kushanas etc., were engaged in their mutual rivalries and struggle for existence. This is probably why the *Mahavastu* asks the king to be objective to 'the acquisition of what has not been gained and the preservation of what has been acquired'.[61] We may compare the statement with a similar one in the *Manava-Dharmasastra*. In the text, we find that the king should maintain the policies of acquisition, preservation, augmentation and distribution.[62] Probably for this, the king has been advised in the *Saundarananda* to follow the *Smrti* line of policy as that of conciliation, gift, dissension and coercion along with restraint (*niyama*).[63] We may suggest that the idea behind this might have been to overcome opposition or obstacles posed by the invading foreign powers.

Politics, warfare and theories of warfare in the Gupta age

We find that ideas in connection with launching a war changed a little from *c*. CE 300 onwards. In the case of treating the enemy, according to *Pancatantra*, (possibly written between CE 300 and 400)[64] one (or the king) should not trust him who was once enemy but turned into a friend.[65] Visnusarma, the author of the collection of tales in the *Pancatantra*, approves a campaign when the acquisition of a territory, a friend and gold is ensured[66] but if one of these three gains is not ensured, then war should not be waged.[67] In this connection reference may be made to the probable occupation by the Kushana king of the territory of Akara (eastern Malwa including Sanchi), rich in diamond mines.[68] Thus, we come to Visnusarma's suggestion that where there is no prospect for making substantial material gain, it is better to avoid aggressive warfare.[69]

Thus, Visakhadatta's drama *Mudraraksa* brings to light the debate between two political thinkers, Canakya (Kautilya) and Raksasa, on the application of intrigues against a foreign enemy. And in that debate Canakya emerged victorious with the idea that the battles of intrigues and treachery (*mantrayuddha* and *kutayuddha*) were appropriate against a foreign enemy.[70] In other words, the Kautilyan principle of *kutayuddha* (unjust war) over a set-piece battle as regards launching and conducting a war is supported.

The practice of showing 'might and authority' and superior position on the part of the ruler continued understandably diplomatically. The Gupta king Chandra Gupta I (early CE fourth century) adopted the epithet *Maharajadhiraja*[71] and thereby he might have shown his superior status to his contemporary powers. He also strengthened his superior position diplomatically by establishing marriage alliance with the Licchavis of Vaishali. The diplomatic attempts together with violent warfare are known to have become more acute during the reign of the next Gupta emperor Samudra Gupta (CE fourth century). The Allahabad Stone Pillar Inscription of Samudra Gupta,[72] popularly called Allahabad *Prasasti* (praise), by his court poet Harisena signifies that Samudra Gupta simultaneously conducted both violent and diplomatic warfare. He uprooted in a very short time (*ksanad unmulya*)[73] many kings such as Acyuta, Nagasena, Rudradeva, to mention a few of Aryavarta (the region bounded by the Himalayas in the north and the Vindhyas in the south) (*anek-aryyavatta-raja*).[74] It is to be noted that the territorial limits of the kingdoms of these uprooted kings are absent in the Allahabad *Prasasti*. In other words, Samudra Gupta annexed their kingdoms to his own kingdom. Thus, he established conquest (*vijaya*) over a large territory covering Punjab, western India, Uttar Pradesh, Madhya Pradesh and West Bengal.[75] He successfully conquered many kings along with their kingdoms by the arms of his own prowess (*sva-bhuja-bala-vijit-aneka-narapati*).[76] The Gupta emperor's (or 'digvijayi' as described) successful military conquest (*vijaya*) in the north was, according to H.C. Raychaudhuri, 'of Early Magadhan type'. This kind of conquest has been compared to *asura vijaya* of the *Arthasastra*. The Haryanka ruler Ajatasatru might have realized this kind of demonic conquest when he subjugated the Licchavis.[77] However, strategically, the Gupta emperor took the most significant diplomatic measure in his military advance in south India. The Allahabad *Prasasti* shows that Samudra Gupta conquered all south Indian kings (*sarvva-daksinapatha-raja*).[78] It is interesting to note that the author of the *Prasasti* has mentioned the territories of the kings defeated by Samudra Gupta in south India, but not the kingdoms of the Aryavarta kings conquered by the Gupta emperor. Therefore, it is said that the emperor seized (*grahana*) the south Indian kingdoms, but released (*moksa*) by favouring (*anugraha*) the captives taken.[79]

In the light of literary and archaeological evidence, Ranabir Chakravarti shows that the littoral area adjoining Orissa, Andhra Pradesh and northern Tamil Nadu was dotted with a number of ports which were involved in maritime trade and commerce along the eastern coast and also with Ceylon and South-East Asia. In this context it is to be noted that all the people of Sri Lanka and South-East Asian islands pleased the Gupta emperor by surrendering themselves to his service (*atmanivedana*), by presenting maidens (*kanyopayanadana*) and by seeking Garuda seal-bearing charters for ruling and enjoying their own territories (*garutmadanka-savishayabhukti-sasana*).[80] May be it was an exaggeration by Harisena about his lord. The emperor with his formidable rule (*pracandasasana*)[81] appears to have been a successful one with political far-sightedness. The commercial importance of these areas might have led the Gupta emperor to launch a victorious military campaign with such a unique diplomatic policy along the eastern seaboard of the Indian subcontinent. Therefore, it appears that on the one hand, he followed the policy of direct war in north India (*prakasayuddha* of Kautilya) and on the other, pursued the diplomatic policy for peace with other powers. On the basis of the numismatic evidence, Chandragupta II Vikramaditya is said to have conquered western Malwa and Gujarat-Kathiawar territories from the western Ksatrapas. Side by side, he also followed the policy of marriage alliances with the Nagas of central India and the *Doab*, the Vakatakas of the Deccan and a part of central India and the Kadambas of Kuntala (in Karnataka). The Gupta emperor diplomatically strengthened his own political position. As a result, he became successful in annexing to the empire the prosperous Gujarat-Kathiawar region with the hinterland of Malwa.[82] Thus, the reflections of the Kautilyan open war (*prakasayuddha*) and diplomatic war (*kutayuddha*) may be noticeable in the conquests of the Gupta emperors for economic gains for the sake of their power and authority.

Kamandaka's *Nitisara*

Almost the same principle is also supported by the Kamandakiya (or Kamandaka) *Nitisara*, a treatise on the science of polity (composed between *c.* CE 700 to 750) authored by Kamandaka. Kamandaka also conceptualizes war in terms of gains. Thus, he states that territory, allies and wealth are the three likely gains in a war. Therefore, a ruler should embark upon a war when there is a clear prospect of these gains.[83] According to Kamandaka, the wise/intelligent king should launch those campaigns which would bring both immediate as well as future gains.[84] As regards material wealth, Kamandaka considers that the acquisition of land is the most important; because wealth and friends are obtainable if a territory is under possession.[85] Thus, the conquest (*vijaya*) of a territory was stressed upon. In order to ensure it, the wise king, according to Kamandaka, should damage the enemy's transport (*yanapahara*), educational resources/institutions (*jnanapahara*) and the enemy's wealth (*dhanapahara*).[86] Thus, the *vijigisuraja* (king desirous of victory) has been advised to march to conquer that part of his enemy's territory/land which is rich in grains (*sampannasasyam*). In that case the conquering ruler will add those resources to his own army (*sasainya = upacayasca*) leading to loss in the enemy's logistical resources and material capacity for sustaining a long drawn-out war (*parasya vrtticchedah*).[87] Then, the army will be able to ensure victory for the ruler by being able even to march forward through the plains (*same pradese*), rough lands (*visame ca bhume*), low lands (*nimne sthale va*)[88] of the enemy.

Here again we may remember that Yajnavalkya had long back supported the matter of acquiring wealth by a war as meritorious, as mentioned above. In this connection, reference may be made to the rivalry between the Pusyabhuti king Harsavardhana (CE 606 to 647) and the Western Chalukya king Pulakesin II (CE 610 to 642) for the possession of some areas in western India. According to Ranabir Chakravarti, the economic advantages of Lata, Gujarat and Malava (Malwa) might have prompted both the contestant kings to occupy the areas.[89] Thus, the conquest (*vijaya*) of a territory with economic potentials was emphasized upon by Kamandaka as mentioned above.[90] In point of taking possession of the enemy's wealth, we may refer to the Kendur plates of the Chalukya king Kirtivarman II (*c.* CE 750).[91] The epigraphic record points out that the Chalukya king led an open war (*prakasa-yuddham*) against his political enemy Nandivarman Pallavamalla (*c.* CE 730 to 800) of Kanci/Kanchi (modern Canjeeveram near Chennai) and took possession of the huge wealth of the latter. The booties thus won included elephants, rubies and gold (*prabhuta-matta-matangaja-suvarna-manikya-kotir-adaya*).[92] The present record shows that the victorious king delivered (*samarppitavan*) the wealth to his father (*pitre*) and gradually (*kramena*) achieved a sovereign position (*prapta-sarvvabhauma-padah*).[93] Thus, it appears that the *vijigisuraja* conquered that part of his enemy's territory/land that was resourceful (*sampannasasyam*) as mentioned above. In that case, the conquering king caused the loss of the enemy's wealth and the latter's capacity for sustaining a long drawn war (*parasya vrtticchedah*).[94] It is likely that the victorious king Kirtivarman II added to the strength of his kingdom with the resources he had won. At the same time, he caused the loss of the same to his enemy. In this connection, we may remember the above mentioned advice of Kamandaka in regard to the conquering army's acquiring ability to move on through the land of the enemy.[95]

Kamandaka has also pointed out the techniques and timing of conducting a war. The author states that if the conquering king finds that he is endowed with the requisite powers, and time and place are favourable to him and the elements of his enemy are disaffected and having no coordination, then he will conduct an open war (*prakasayuddha*); otherwise he will recourse to deceitful war (*kutayuddha*).[96] If the king adopts the guileful or tactical warfare for the destruction of the enemy (*satrunasah*), then he does not transgress righteousness (*dharma*).[97] In other words, we may say that there was nothing unfair in warfare or in the act of destroying the enemy for the safety and security of the kingdom. It is necessary for the preservation of the welfare and development of the state and the people thereof. The king wielding *danda* can integrate and ensure the pursuit of the three branches of knowledge such as philosophy (*anvikiksiki*), the three *vedas* (*Rgveda*, *Samaveda* and *Yajurveda*) and economics (*vartta*) for the welfare of the state. Thus, the policy (*niti*) the king pursues/follows by means of his *danda* (*dandaniti*) maintains/keeps the worldly life of the people in order.[98] Kamandaka accepts this piece of Kautilyan advice.[99] Thus, that is *niti* or moral action, we may say, for the king, which enables or approves him to keep his kingdom and the subjects free from dangers both internal and external as well as to guide them towards further development and progress. However, this required the king to be equipped with three such powers (*sakti*): competence (*prabhusakti*), power of intelligence (*mantrasakti*) and dynamism, energy and creativity (*utsahasakti*).[100]

These powers, according to Kautilya, could ensure three kinds of successes (*Siddhi*); that by intelligence/counsel (*Mantrasiddhih*), that by might (*Prabhusiddhih*) and that by enterprise/energy (*Utsahasiddhiriti*).[101] Understandably, the king required these powers for the development of the security of the kingdom. We should note here that the *Mahavastu*, already noted above, recognized the king's power of intelligence as the best (*uttamam*) for the kingdom. Thus, in point of time, we may note that Kamandaka, with acknowledgement to learned Visnugupta (Canakya or Kautilya) (*namastasmai Visnuguptaya*), produced/composed an abridged treatise on the science of polity (*rajavidya-sankhiptagranthasamarthavat*) from the ocean of the *Arthasastra* (*Arthasastra mahodadheh*). He followed the opinions of the scholars on polity (*rajavidyavidam matam*). Therefore, the Kamandakiya *Ntisara* appears to have been a later text on polity.[102]

Kautilya's legacies: spies and internal security

In this connection, one may remember the *Nitivakyamrta* (the *Nectar of the Sayings of Polity*) of Somadeva (*c.* CE 900).[103] The author virtually echoes the tradition of the *Arthasastra* as regards the strategy for conducting war. The author advises the king that *sastrayuddha* (use of the armed force in accordance with tradition, i.e. just war) should be followed only when the enemy cannot be won by applying intellect.[104] The enemy can be destroyed without recourse to open warfare by giving him poison and overtures. This technique has been called as *tusnidandah*.[105] *Tusnidandah* is actually a modification of Kautilyan *kutayuddha*. This required the king to have also the power of wisdom.

This brings us to the *Kathasaritsagara* (Ocean of Stories) of Somadeva Bhatta (*c.* CE 1100). According to the *Kathasaritsagara*, the ruler is advised to have also the powers of *utsaha* (enterprise), *mantra* (counsel) and *prabhutva* (competence).[106] We may refer to the *Manasollasa* (the Delight of the Mind) of Somesvara (*c.* CE 1200). According to the author, conciliation (*sama*) brings success without injury and should be applied to those who look for advantage for themselves; bribing (*dana*) is to be used against those who are dissatisfied with their master, dissension (*bheda*) is to be followed up against those who cannot be won over by means of conciliation and the policy of force (*danda*) is to be applied when there is no other option for destroying or weakening the hostile power.[107]

Before and after launching a campaign, the king is warned to closely watch his dynamic enemy in order to acquire victory. This requires him to constantly maintain spies of different types, who, like the ears and eyes of the king, enable the rulers to face both internal as well as external dangers. In this connection, the *Arthasastra* is a classic text in the game of power for power's sake in South Asia. However, the rulers had developed the sense of spying for the sake of *vijaya* long before the days of Kautilya in the Indian subcontinent; an allusion towards this development may be noticeable in the *Rgveda*. The allusion is that *Agni* (Fire God) is invoked to send his rays in order to save the people either from their nearby or distant enemies.[108]

The idea of watching the enemy seems to have gained ground gradually as the territorial units emerged and their rulers gathered experiences regarding the consolidation of their power and authority through aggressive warfare. However, in connection with watching another kingdom the most brilliant ideas are available from the *Arthasastra* of Kautilya. The author advises the king to appoint persons as spies (*gudapurusah*)[109] for watching the movements/activities of the enemy. They were employed in order to strengthen the security of the kingdom. The author has also referred to the various types of spies. They were nine types such as *Kapatika* (sharp pupil), *Udasthita* (apostate monk), *Grhapatika* (householder), *Vaidehaka* (trader), *Tapasa* (ascetic), *Satri* (knowledgeable secret agent), *Tiksna* (brave person), *Rasada* (poison-giver) and *Bhiksuki* (begging nun).[110] The author has also stated the procedures of how the secret persons were asked to collect information for the king. They were to be moving (*sancarah*) and unknown to one another (*na ca anyonyam samsthaste va vidyuh*).[111] Thus, we find that the ideas about the activities of spying came to be articulated, expanded and systematized in the *Arthasastra* obviously for the security of the kingdom. The idea behind such an elaborate system of espionage in the *Arthasastra* might have been to pave the way towards the conquest of enemy territory or keeping constant watch over the enemy to prevent rebellions in the conquered territories. In this context, it is interesting to note that Kamandaka has also recorded almost the same types of spies (*cara*) and they were such as *Tiksna* (i.e. the Kautilyan type), *Prabrajita* (mendicant, i.e. Kautilyan *Tapasa*), *Visada* (poisoner, cf. Kautilyan *Rasada*), *Satri* (sacrificer or magician type, cf. Kautilyan type). They were also asked to remain moving (*samcarah*) for collecting information about the activities or movements of officials of the kingdom including enemies and their affairs within and without the kingdom.[112] Probably for this reason, Kamandaka also advised them to remain unknown to one another (*na anyonyavedinah*).[113] Thus, Kamandaka seems to have been indebted to Kautilya for his ideas on spies and their movement and activities.

In passing, we can refer to another type of spy called *ubhayavetana* (spies in the pay of both) in the Kautiliya *Arthasastra*. According to the author, the king should take help of such spies in order to get over the mixed dangers (*paramisra*) caused by the treasonable and the quasi-treasonable elements within and outside the kingdom. In such case, we learn that Kautilya advises the king to win success by dividing the ally from the enemy. Probably in case of the enemy we find that Kautilya advises the king to favour, with money and honour (*arthamana*), those who committed great crimes (*mahaparadha*) and to employ them with weapons, poison and fire (*sastrarasagni*) against the enemy to destroy him. If they failed to discharge their work as directed by the king, they were to be seized by the spies.[114] Such efforts were undertaken to ensure *vijaya* by tiding over various threats posed by the enemy.

However, it is to be noted that the *Manava-Dharmasastra* also asked the king to remain informed through spies (*caraih*) about the activities of the whole of his own kingdom (*samagrarastresu*) as well as also of the neighbouring kingdoms (*carena utsahayogena svasaktim parasaktim ca*).[115] Thus, according to the *Yajnavalkyasmriti*, the king should receive secret reports from his confidential spies in the evening.[116] Historically, we may refer to the Irda copper-plate of the Kamboja king Nayapaladeva (*c.* CE 1035 to 1055). The plate points to the presence of the officers/persons of the secret service (*gudhapurusan*) in connection with the grant of a village by the king to one Brahmana *Pandit* Asvatthasarmman. According to the epigraphic record, the officers of the secret service were probably under the office of some higher authority (*Adhyaksa-varggam karanaih sametan-sa-gudhapurusan*).[117] It is likely that the persons of the secret service had submitted their information to the concerned office before the grant was released.

The inhabitants of ancient India may have developed a kind of cultural identity of their own. This had helped them think about the world around them. Therefore, they have acted in order to ensure their living as free from opposition or dangers as possible. Thus, the early Indian theoreticians developed ideas relating to political security in real life situations. In this connection it may be said, according to the Kautiliya *Arthsastra*, that ideas relating to strategy have been cultivated and developed for the king. This may be applicable in the case of kings who are desirous of victory in ensuring their power and authority and the expansion of their kingdoms. The Kautiliyan ideas relating to strategy appear to have influenced also other authorities in the development/cultivation of their strategic ideas for *vijaya* for the kingdom in the subsequent years. These are virtually in common with Kautilya on determining strategic policy for the kingdom. Thus, it may be assumed that a kind of cultural trait relating to determining strategy for the kingdom might have grown in early India. This culture trait might have also suggested when, where and how coercive authority (*danda*) was to be used. Hence, a set of values relating to the use of force emerged. They have understood that *yuddha* causes the loss of life and property and therefore is an evil; but still they have to be resorted to occasionally as a necessary evil. The *acharyas* have incorporated the concept of *yuddha* as part and parcel of their life-saving strategies. They have used it in the manner of either *prakasayuddha*, or *kutayuddha*, or *mantrayuddha*, or *tusnimyuddha* in order to win victory (*vijaya)* in the game of power for the sake of power in real life situations. Understandably, the idea behind such a move was probably to ensure the stability, peace, progress as well as protection of the kingdom and its people. This in other words might have led the early Indian ruler to resort to the doctrine of the application of force (*danda*) for the sake of security of the kingdom. In this connection, the *Arthasastra* reference to *danda* may be mentioned and it is *yatharhadanda pujyah*. It signifies that one (i.e. the king) who applies *danda* properly is worshipped/honoured by all.[118]

According to the *Manava-Dharmasastra*, *danda* or the coercive authority of the king protects the subjects of the kingdom, is the law that disciplines the subjects.[119] Understandably, the king is asked to be careful about the security of his kingdom. This may be compared to another verse in the same text according to which the king is advised to keep his army force always ready (*nityam udyatadandah*) to protect his secrets as well as to constantly find out the weaknesses of his enemy (*nityam chidranusaryareh*).[120] According to the *Arthasastra*, *danda* is necessary for ensuring the security, existence and prosperity of the subjects of the kingdom.[121] Since the proper use of *danda* also generates wealth (*dandasampad*) with which the king organizes the army so as to ensure the safety and security of the subjects of the kingdom, thus, it was the duty of the king to apply *danda* to maintain his kingdom. In this connection, the *Arthasastra* served as a model of strategic culture for the later authorities of ancient and early medieval India like Kamandaka and Somadeva Suri.[122]

Kautilya's legacies: armies and external security

However, conquest (*vijaya*) in war (*yuddha*) requires the marching and logistical arrangement of the army. As a part of it, the construction of a fort is also important. Information about fortifications is available from as early as the prehistoric days.[123] However, coming to the age of Kautilya we find information about the construction of forts for the purpose of a war. In this connection, Kautilya suggests four types of forts such as water-fort, hill-fort, desert-fort and forest-fort. These forts should be caused to be built for conducting a war.[124] According to the *Manava-Dharmasastra*, the king should reside in a fort well guarded by forests, desert (*marudurgam*) or water (*abadurgam*) or on the hill (*giridurgam*); probably such forts were suitable for the king to conduct a war.[125] It is noticeable that in line with earlier authorities Kamandaka also noted the same types of forts (*audakam parvatam varksmam-airinam dhanvan*). According to the author, each of the forts was fitted with high gateways (*uccaprakaragopuram*).[126]

The conduct of war successfully requires the employment of different kinds of soldiers and proper tactical deployments. In this connection, Kautilya draws our attention to the different units of the army of the king. These fighting forces are such as the hereditary troops (*moulabala*) for the defence of the base; the hired troops (*bhrtakabala)* for military operations with diplomacy; the mercenary bands (*srenibala*); the ally's soldiers (*mitrabala*), for conquest/success in military operations is also dependent (perhaps to some extent) on the ally; foreign troops (*amitrabala*); and the forest troops (*atavibala* or the soldiers with the *Atavi* chiefs).[127]

Regarding the arrangement of the soldiers, we find information from the *Arthasastra*. The soldiers can be arranged in four such manners as the *Danda vyuha* (the staff array), *Bhoga vyuha* (the snake array), *Mandala vyuha* (the circle array) and *Asamhata vyuha* (the diffuse array).[128] Thapliyal has discussed these arrays at length.[129] In connection with the arrangement of soldiers for conducting a war against the enemy, Dikshitar draws our attention to the *Agnipurana*. According to the text, the commander of the army (*nayaka*) should march with the brave soldiers in the first line (*nayakah puratah yayat prabira purusavrtah*). The king, the treasury, the women should march in the middle (*madhye kalatram svami ca kosah ca yadbalam*). Both the flanks of the army should be protected by the horsemen (*parsvayorubhayorasva bajinam*). The horsemen should be protected by the charioteers (*parsvayo rathah*) laterally. The charioteers should be protected by the snakeforces (a type of order of battle) laterally (*rathanam parsvayornaga*). The snakeforces should be protected by the forest tribes (*naganam ca atavikabalam*). With all these forces in the front, the commander of the army was asked to remain in the rear (*pascat senapatih sarvvam puraskrtya*).[130] The *Agnipurana* prescribes that the battle should be fought in forests, along the river banks and near the forts (*nadyadrivanadurgesu*).[131]

However, the arrangement of soldiers alone cannot bring victory to the king in a battlefield. The army requires weapons with which they win the battle. Dikshitar has dished out a good number of literary, sculptural and numismatic references to weapons such as bows, arrows etc.[132] However, in the light of numismatic, sculptural and literary references, B.N. Mukherjee has also drawn our attention to the use of sword, dagger, mace, battle-axe, spear and bow and arrow by the army. In this connection he has referred to the Junagadh inscription of Rudradaman, a Kshatrapa ruler of western India (*c*. CE 150/152).[133] The epigraphic record points out that the Kshatrapa ruler was well versed in the use of sword and shield and knew how to decrease pugilistic fight (*asi-carma niyuddhadya bala-laghava sausthava-kriyena*).[134] It may be an indication towards the provision of training in the use of arms for fight.

Conclusion

It appears from our discussion that the concepts of *yuddha* and *vijaya* have been confined to the king and his realm. We have seen that the king has tried to defeat his political rival or enemy king. In doing this, he has resorted to diplomacy in more than one manner or sometime engaged himself in an open war against his enemy. Thus, the king has been active in different ways in order to become victorious in defeating or keeping away his enemy from the kingdom. For, the theoreticians have asked the king to ensure the development and progress of his kingdom. That which guides the kingdom/state may be a *Niti*. We have seen how the Magadhan ruler or the Gupta emperor became militarily victorious. Thus, we have tried to understand the concepts of *Yuddha* and *Vijaya* through certain historical cases from ancient and early medieval India. We have seen the kings following 'the practice of theory'[135] in attaining to victory in the affairs of his polity.

Acknowledgement

I gratefully acknowledge the scope that Professor Kaushik Roy, Jadavpur University, and Professor Ranabir Chakravarti, Jawaharlal Nehru University, my teachers, have generously given me to write this essay. I remain particularly grateful for Professor Chakravarti's valuable comments on the article.

Notes

1 B.D. Chattopadhyaya, *Studying Early India: Archaeology, Texts and Historical Issues* (New Delhi: Permanent Black, 2003), p. 50.
2 D.K. Chakrabarti, *India: An Archaeological History Palaeoloithic Beginnings to Early Historic Formations* (New Delhi: Oxford University Press, 1999), p. 41.
3 Romila Thapar, *The Penguin History of Early India from the Origins to AD 1300* (New Delhi: Penguin, 2003), p. 79.
4 P.C. Chakravarti, *The Art of War in Ancient India* (1941, reprint, Delhi: Low Price Publications, 2010).
5 V.R. Ramachandra Dikshitar, *War in Ancient India* (1944, reprint, Delhi: Motilal Banarsidass, 1987).
6 U.P. Thapliyal, *Warfare in Ancient India: Organizational and Operational Dimensions* (New Delhi: Manohar, 2010).
7 Ranabir Chakravarti, *Warfare for Wealth: Early Indian Perspective* (Calcutta: Firma KLM Pvt. Ltd, 1986), pp. 11–13.
8 R.C. Datta (tr.), *Rigveda-Samhita* (Calcutta: Haraf Prakashani, 1976), Bengali, vol. 2, p. 387, *Rik* IX.65.10 (*vrsa pavasva dharaya marutvate ca matsarah/visva dadhana ojasa//*).
9 R.C. Datta (tr.), *Rigveda-Samhita*, vol. 2, p. 406, *Rik* IX.78.4 (*gojinnah somo rathajiddhiranyajit–pavate sahasrajit/svarjidadvijat–*); Ranabir Chakravarti, *Warfare for Wealth*, p. 11.
10 R.C. Datta (tr.), *Rigveda-Samhita*, vol. 2, p. 421, *Rik* IX.89.3 (*simham nasantu–/suro yutsu prathamah prcchute ga asya//*); Ranabir Chakravarti, *Warfare for Wealth*, p. 11.
11 R.S. Sharma, *Perspectives in Social and Economic History of Early India* (New Delhi: Munshiram Manoharlal Publishers Pvt. Ltd, 1983), p. 111; Ranabir Chakravarti, *Warfare for Wealth*, p. 13.
12 R.C. Datta (tr.), *Rigveda-Samhita*, vol. 2, p. 404, *Rik* IX.76.2 (*suro na dhatta audha gabhastyoh sa sisasn rathiro govistisu*).
13 M.M. Williams, *A Sanskrit-English Dictionary* (1899, reprint, Delhi: Motilal Banarsidas, 1990), p. 351, cols I and II. Ranabir Chakravarti, *Warfare for Wealth*, p. 13.

14 R.S. Sharma, *Political Ideas and Institutions in Ancient India* (Delhi: Motilal Banarsidas, 2005), p. 195.
15 R.C. Datta (tr.), *Rigveda-Samhita*, vol. 2, p. 138, *Rik* VII.33.3.
16 Ranabir Chakravarti, *Warfare for Wealth*, p. 18.
17 Ranabir Chakravarti, *Exploring Early India: Upto c. AD 1300* (Delhi: Macmillan Publishers India Ltd, 2010), p. 57.
18 Ranabir Chakravarti, *Warfare for Wealth*, pp. 26–76; Ranabir Chakravarti, *Exploring Early India*, p. 89.
19 H.C. Raychaudhuri, *Political History of Ancient India: Commentary by B.N. Mukherjee* (Delhi: Oxford University Press, 1996), p. 182, also see fn. 3; F. Max Muller (ed.), *The Sacred Books of the East*, vol. 17 (1882, reprint, Delhi: Motilal Banarsidas, 1990), p. 294, X.2.3; Ranabir Chakravarti, *Warfare for Wealth*, pp. 30, 33.
20 Ranabir Chakravarti, *Warfare for Wealth*, pp. 34, 64.
21 R.P. Kangle, *The Kautiliya Arthasastra* (1972, reprint, Delhi: Motilal Banarsidass, 2003), Part II, p. 321, 7.1.9.
22 H.C. Raychaudhuri, *Political History of Ancient India*, p. 189.
23 H.C. Raychaudhuri, *Political History of Ancient India*, pp. 192–194.
24 H.C. Raychaudhuri, *Political History of Ancient India*, p. 206.
25 V.A. Smith, *The Early History of India* (Oxford: Oxford University Press, 1957), pp. 72–74; Ranabir Chakravarti, *Exploring Early India*, pp. 90–91.
26 Ranabir Chakravarti, *Exploring Early India*, p. 92.
27 Ranabir Chakravarti, *Exploring Early India*, pp. 121–132; Ranabir Chakravarti, *Warfare for Wealth*, pp. 54–55.
28 R.P. Kangle, *The Kautiliya Arthasastra*, Part II, p. 460 (12.1.10); R.G. Basak, *Kautiliya Arthsastra*, 2 vols (Calcutta: General Printers and Publishers Pvt. Ltd, 1989), vol. 2, p. 103.
29 D.C. Sircar, *Select Inscriptions Bearing on Indian History and Civilization* (Calcutta: University of Calcutta, 1965), vol. I, p. 37; Ranabir Chakravarti, *Exploring Early India*, p. 127.
30 B.N. Mukherjee, *The Character of the Maurya Empire* (Calcutta: Progressive Publishers, 2000), p. 127.
31 B.N. Mukherjee, *The Character of the Maurya Empire*, p. 120.
32 D.C. Sircar, *Select Inscriptions Bearing on Indian History and Civilization*, vol. I, pp. 37–38; U.N. Ghoshal, *A History of Indian Political Ideas* (Glasgow: Oxford University Press, 1959), p. 294.
33 B.D. Chattopadhyaya, *Studying Early India Archaeology, Texts and Historical Issues*, pp. 142, 150, fn. 26.
34 H.C. Raychaudhuri, *Political History of Ancient India*, p. 290; D.C. Sircar, *Select Inscriptions Bearing on Indian History and Civilization*, vol. 1, p. 21; Ranabir Chakravarti, *Exploring Early India*, p. 127.
35 B.D. Chattopadhyaya, *Studying Early India Archaeology, Texts and Historical Issues*, pp. 146–147.
36 R.G. Basak, *Kautiliya Arthsastra*, 2 vols (Calcutta: General Printers and Publishers Pvt. Ltd, 1989), Bengali, vol. 2, pp. 32, 77–80; R.P. Kangle, *The Kautiliya Arthasastra*, Part II, pp. 314–316, VI.1.
37 R.G. Basak, *Kautiliya Arthsastra*, vol. 2, pp. 33–35, 82–83; R.P. Kangle, *The Kautiliya Arthasastra*, Part II, pp. 318–319.
38 Ranabir Chakravarti, *Warfare for Wealth*, p. 188.
39 R.G. Basak, *Kautiliya Arthsastra*, vol. 2, pp. 86–171, *Adhikarana* VII, *Adhyaya* 1–18.
40 H.C. Raychaudhuri, *Political History of Ancient India*, p. 594; Ranabir Chakravarti, *Exploring Early India*, p. 124. Ranabir Chakravarti, *Warfare for Wealth*, p. 190.
41 R.G. Basak, *Kautiliya Arthsastra*, vol. 2, pp. 91–98, VI. 2–3.

42 Ranabir Chakravarti, *Warfare for Wealth*, pp. 35–36.
43 Patrick Olivelle (ed.), *Manu's Code of Law: A Critical Edition and Translation of the Manava-Dharmasastra* (New Delhi: Oxford University Press, 2006), pp. 645–649, vv. 181–200; U.N. Ghoshal, *A History of Indian Political Ideas*, pp. 181–182.
44 R.G. Basak, *Kautilya Arthasastra*, vol. 2, p. 148 (7.14); R.P. Kangle, *The Kautilya Arthasastra*, Part II, p. 367 (7.14.11).
45 Patrick Olivelle (ed.), *Manu's Code of Law*, p. 632, VII.109b (*samadandau prasamsanti nityam rastrabhivrddhaye*).
46 Gangasagar Rai (ed. and tr.), *Yajnavalkyasmriti* (Delhi: Chaukhamba Sanskrit Pratisthan, 2002), p. 149, *Acaradhyaya*, verse 342 (*ya eva nrpaterdharmah svarastra paripalane/tameva krtsnamapnoti pararastram vasam nayan//*); cf. M.N. Datta (tr.) and R.K. Panda (ed.), *Yajnavalkyasmriti* (Delhi: Bharatiya Kala Prakashan, 2011), p. 161, *Acaradhyaya*, verse 342; U.N. Ghoshal, *A History of Indian Political Ideas*, p. 183.
47 Gangasagar Rai (ed. and tr.), *Yajnavalkyasmriti*, p. 149, *Acaradhyaya*, verse 342 (*ya eva nrpaterdharmah svarastra paripalane/tameva krtsnamapnoti pararastram vasam nayan//*).
48 Gangasagar Rai (ed. and tr.), *Yajnavalkyasmriti*, p. 143, verse 323 (*natah parataro dharmohnrpanam yadranarjitam/viprebhyo diyate dravyam–//*); U.N. Ghoshal, *A History of Indian Political Ideas*, p. 183.
49 Gangasagar Rai (ed. and tr.), *Yajnavalkyasmriti*, p. 153, verse 352a (*hiranybhulabhebhyo mitralabdhirvara yatah*).
50 H.C. Raychaudhuri, *Political History of Ancient India*, pp. 182–183.
51 U.N. Ghoshal, *A History of Indian Political Ideas*, p. 184.
52 Patrick Olivelle (ed.), *Manu's Code of Law*, p. 644, VII.177.
53 Ranabir Chakravarti, *Exploring Early India*, pp. 168–188; Ranabir Chakravarti, *Warfare for Wealth*, p. 112.
54 R.G. Basak, *Kautiliya Arthsastra*, vol. 2, pp. 35–37; R.P. Kangle, *The Kautiliya Arthasastra*, Part II, pp. 321–324.
55 Ranabir Chakravarti, *Warfare for Wealth*, pp. 101–102; Ranabir Chakravarti, *Exploring Early India*, pp. 184–185.
56 B.N. Mukherjee, *The Rise and Fall of the Kushana Empire* (Calcutta: Firma KLM Pvt. Ltd, 1988), pp. 325–26.
57 B.N. Mukherjee, *The Rise and Fall of the Kushana Empire*, p. 325; Ranabir Chakravarti, *Exploring Early India*, p. 179.
58 U.N. Ghoshal, *A History of Indian Political Ideas*, pp. 234–235.
59 U.N. Ghoshal, *A History of Indian Political Ideas*, pp. 266, 269 fn. 6.
60 B.N. Mukherjee, *The Economic Factors in Kushana History* (1970, 2nd edn, rev'd, Kolkata: Progressive Publishers, 2002), p. 49.
61 U.N. Ghoshal, *A History of Indian Political Ideas*, p. 265.
62 Patrick Olivelle (ed.), *Manu's Code of Law*, pp. 159, 630 (*alabdham caiva labdham rakset prayatnatah/raksitam vardhayeccaiva vrddham patresu niksipet//*).
63 U.N. Ghoshal, *A History of Indian Political Ideas*, p. 267.
64 Maurice Winternitz, *History of Indian Literature*, vol. 3, Parts one and two combined (1967, reprint, Delhi: Motilal Banarsidass, 1985), p. 317.
65 M.R. Kale, *Pancatantra of Visnusarma* (Delhi: Motilal Banarsidass, 1986), p. 134, verse 2a (*ya upekseta satrum svam prasarantam yadrcchaya*).
66 M.R. Kale, *Pancatantra of Visnusarma*, p. 136, verse 16a (*bhumirmitram ca vigrahasya phalatrayam/*).
67 M.R. Kale, *Pancatantra of Visnusarma*, p. 136, verse 16b (*nastyekamapi yadyepam vigraham na samacaret*).
68 B.N. Mukherjee, *The Economic Factors in Kushana History*, pp. 29–30. Chakravarti Ranabir, *Warfare for Wealth*, pp. 94–95.

69 M.R. Kale, *Pancatantra of Visnusarma*, p. 136, verse 18 (*tasmanna syatphalam yatra pustam yuddham tu kevalam/na hi tatsvayamutpadayam kartavyam na kathamcana//*).
70 U.N. Ghoshal, *A History of Indian Political Ideas*, pp. 359–360.
71 H.C. Raychaudhuri, *Political History of Ancient India*, p. 468.
72 J.F. Fleet, *Corpus Inscriptionum Indicarum Inscriptions of the Early Gupta Kings and their Successors*, vol. 3 (1970, reprint, Varanasi: Indological Book House), pp. 1–17.
73 J.F. Fleet, *Corpus Inscriptionum Indicarum Inscriptions of the Early Gupta Kings and their Successors*, p. 6, line 13.
74 J.F. Fleet, *Corpus Inscriptionum Indicarum Inscriptions of the Early Gupta Kings and their Successors*, p. 7, line 21b; Ranabir Chakravarti, *Exploring Early India*, p. 233.
75 Ranabir Chakravarti, *Exploring Early India*, p. 233; Ranabir Chakravarti, *Warfare for Wealth*, p. 130.
76 J.F. Fleet, *Corpus Inscriptionum Indicarum Inscriptions of the Early Gupta Kings and their Successors*, p. 8, line 26c.
77 H.C. Raychaudhuri, *Political History of Ancient India*, p. 474, fn. 1.
78 J.F. Fleet, *Corpus Inscriptionum Indicarum Inscriptions of the Early Gupta Kings and their Successors*, p. 7, line 20b–c.
79 J.F. Fleet, *Corpus Inscriptionum Indicarum Inscriptions of the Early Gupta Kings and their Successors*, p. 7, line 20c (*grahana-moksh-anugraha-janita*); Ranabir Chakravarti, *Exploring Early India*, p. 234.
80 J.F. Fleet, *Corpus Inscriptionum Indicarum Inscriptions of the Early Gupta Kings and their Successors*, p. 8, line 24a; Ranabir Chakravarti, *Exploring Early India*, pp. 235–237; Ranabir Chakravarti, *Warfare for Wealth*, pp. 131–133.
81 J.F. Fleet, *Corpus Inscriptionum Indicarum Inscriptions of the Early Gupta Kings and their Successors*, p. 8, line 23a.
82 H.C. Raychaudhuri, *Political History of Ancient India*, pp. 489–490; Ranabir Chakravarti, *Exploring Early India*, pp. 237–238; Ranabir Chakravarti, *Warfare for Wealth*, pp. 133–135.
83 S.K. Mitra (rev'd and tr.), *The Nitisara of Kamandaka* (Kolkata: Asiatic Society, 2008), p. 215, X.31 (*bhumirmitram hiranyamca vigrahasya phalam trayam/ Yadaitanniyatam bhavi tada vigrahamacaret//*).
84 S.K. Mitra (rev'd and tr.), *The Nitisara of Kamandaka*, pp. 213–214, vv. 24–25.
85 S.K. Mitra (rev'd and tr.), *The Nitisara of Kamandaka*, p. 215, v. 32.
86 S.K. Mitra (rev'd and tr.), *The Nitisara of Kamandaka*, p. 208, v. 8.
87 S.K. Mitra (rev'd and tr.), *The Nitisara of Kamandaka*, p. 325, v. 4.
88 S.K. Mitra (rev'd and tr.), *The Nitisara of Kamandaka*, pp. 326–327, v. 6.
89 Thomas Watters (tr.), *On Yuan Chwang's Travels in India*, vol. 2 (Delhi: Munshi Ram Manohar Lal, 1961), pp. 241, 243–244, 248–249; Ranabir Chakravarti, *Warfare for Wealth*, pp. 143–146.
90 S.K. Mitra (rev'd and tr.), *Nitisara of Kamandaka*, p. 325, v. 4
91 K.B. Pathak, 'Kendur Plates of Kirtivarman II, Saka Samvat 672', *Epigraphia Indica*, vol. 9, no. 28 (1907–1908, reprint, 1981), pp. 200–206; Ranabir Chakravarti, *Warfare for Wealth*, pp. 156–157.
92 K.B. Pathak, 'Kendur Plates of Kirtivarman II, Saka Samvat 672', p. 203, lines 50–51.
93 K.B. Pathak, 'Kendur Plates of Kirtivarman II, Saka Samvat 672', pp. 203–204, line 51.
94 S.K. Mitra (rev'd and tr.), *The Nitisara of Kamandaka*, p. 325, v. 4.
95 S.K. Mitra (rev'd and tr.), *The Nitisara of Kamandaka*, pp. 326–327, v. 6.
96 S.K. Mitra (rev'd and tr.), *The Nitisara of Kamandaka*, p. 405, v. 54 (*visisto desakalabhyam bhinnariprakritirvali/kuryat prakasayuddhamca kutayuddham viparyeye//*).

97 S.K. Mitra (rev'd and tr.), *The Nitisara of Kamandaka*, p. 411, v. 71 (*niyatamiti nihanyat kutayuddhesu satrun na hi nirayati dharmacchadmana satrunasah*).
98 R.G. Basak, *Kautiliya Arthsastra*, vol. 1, pp. 4–5 (*anvikiksikitrayivartanam yogksemasadhano dandah*); R.P. Kangle, *The Kautiliya Arthasastra*, Part II, p. 9 (1.4.4).
99 S.K. Mitra (rev'd and tr.), *The Nitisara of Kamandaka*, p. 35 (*anvikiksiki trayi varta dandanitisca sasvati/Vidyascatasra evaivata yogaksemaya dehinam//*), (2.2).
100 S.K. Mitra (rev'd and tr.), *The Nitisara of Kamandaka*, p. 219, v. 43 (*iti sma dosanvitamasu–trisaktiyukto vijigisaya vrajet/*).
101 R.G. Basak, *Kautiliya Arthsastra*, vol. 2, p. 34 (6.2); R.P. Kangle, *The Kautiliya Arthasastra*, Part II, p. 319 (6.2.34).
102 Upinder Singh, 'Politics, Violence and War in Kamandaka's *Nitisara*', *Indian Economic and Social History Review*, vol. 47, no. 1 (2010), pp. 31–32; S.K. Mitra (rev'd and tr.), *The Nitisara of Kamandaka*, pp. 3–5, vv. 6–8.
103 Suri Somadeva, *Nitivakyamrtam* (ed. Pandit Pannalal Soni), Manikchandra Digambar Jain Granthamala Series (Bombay, 1922), p. 345, v. 5
104 Suri Somadeva, *Nitivakyamrtam*, p. 345, v. 4 (*buddhiyuddhena param jetumasaktah sastrayuddhamupakramet*).
105 U.N. Ghoshal, *A History of Indian Political Ideas*, p. 491.
106 U.N. Ghoshal, *A History of Indian Political Ideas*, p. 452.
107 U.N. Ghoshal, *A History of Indian Political Ideas*, p. 453.
108 R.C. Datta (tr.), *Rigveda-Samhita*, vol. 1, pp. 472–473, IV.4.3.
109 R.G. Basak, *Kautiliya Arthsastra*, vol. 1, p. 11; R.P. Kangle, *The Kautiliya Arthasastra*, Part II, p. 21 (1.11.1).
110 R.G. Basak, *Kautiliya Arthsastra*, vol. 1, p. 11; R.P. Kangle, *The Kautiliya Arthasastra*, Part II, p. 21 (1.11.1).
111 R.G. Basak, *Kautiliya Arthsastra*, vol. 1, p. 12 (1.12).
112 S.K. Mitra (rev'd and tr.), *The Nitisara of Kamandaka*, pp. 266–275.
113 S.K. Mitra (rev'd and tr.), *The Nitisara of Kamandaka*, p. 269, v. 34 (*Tiksnah pravrajitascaiva satri visada eva ca/ete jneyastu samcarah sarve nanyonyavedinah//*).
114 R.G. Basak, *Kautiliya Arthsastra*, vol. II, pp. 86–88, 225, particularly, p. 87; R.P. Kangle, *The Kautiliya Arthasastra*, Part II, p. 424.
115 Patrick Olivelle (ed.), *Manu's Code of Law*, p. 633 (VII.122), p. 802 (IX.298); S.L. Roy, *Diplomacy in Ancient India from the Early Vedic Period to the End of the Sixth Century A.D.* (Calcutta: Charu Publishing Co., no date), p. 193.
116 M.N. Datta (tr.) and R.K. Panda (ed.), *Yajnavalkyasmriti*, pp. 157–158, v. I.330 (*Sandhyamupasya srinuccaranam gudabhasitam/*).
117 N.G. Majumdar, 'Irda Copper Plate of the Kamboja king Nayapaladeva', *Epigraphia Indica*, vol. 22, no. 25 (1933–1934, reprint, 1984), p. 156, lines 34–35; D.C. Sircar, *Indian Epigraphical Glossary* (Delhi: Motilal Banarsidass, 1966), p. 123.
118 R.G. Basak, *Kautiliya Arthsastra*, vol. 1, p. 5.
119 Patrick Olivelle (ed.), *Manu's Code of Law*, pp. 155, 615, VII.18 (*danda sasti prajah sarva danda evabhiraksati/dandah suptesu jagarti dandam dharmam vidurvudhah//*).
120 Patrick Olivelle (ed.), *Manu's Code of Law*, pp. 159, 630, VII.102.
121 R.G. Basak, *Kautiliya Arthsastra*, vol. 1, pp. 5–6.
122 S.K. Mitra (rev'd and tr.), *The Nitisara of Kamandaka*, pp. 89–90; Suri Somadeva, *Nitivakyamrtam*, p. 103 (*prajapalanaya rajna dandah praniyate*).
123 P.C. Chakravarti, *The Art of War in Ancient India*, pp. 126–130.
124 R.P. Kangle, *The Kautiliya Arthasastra*, Part II, p. 61 (2.1).
125 Patrick Olivelle (ed.), *Manu's Code of Law*, p. 624, vv. 70–71.
126 S.K. Mitra (rev'd and tr.), *The Nitisara of Kamandaka*, pp. 86–87, vv. 57, 59.
127 R.G. Basak, *Kautiliya Arthsastra*, vol. 2, pp. 203–208, 79–81; R.P. Kangle, *The Kautiliya Arthasastra*, Part II, pp. 409–413.
128 R.P. Kangle, *The Kautiliya Arthasastra*, Part II, p. 450ff. (10.6).

129 U.P. Thapliyal, *Warfare in Ancient India*, p. 237.
130 V.R. Ramachandra Dikshitar, *War in Ancient India*, p. 242; Panchanan Tarkaratna (ed.), *Agnipuranam* (Calcutta: Nababharat Publishers, 1999), p. 468, vv. 1–6.
131 Panchanan Tarkaratna (ed.), *Agnipuranam*, p. 467, v. 3.
132 V.R. Ramachandra Dikshitar, *War in Ancient India*, pp. 93–152.
133 B.N. Mukherjee, *The Rise and Fall of the Kushana Empire*, p. 340, pp. 470, fn. 140, 596–597; F. Kielhorn, 'Junagadh Rock Inscription of Rudradaman: The Year 72', *Epigraphia Indica*, vol. 8, no. 6 (1905–1906, reprint, 1981), pp. 36–49.
134 F. Kielhorn, 'Junagadh Rock Inscription of Rudradaman: The Year 72', p. 44, lines 13c–d.
135 Sheldon, Pollock, 'The Theory of Practice and the Practice of Theory in Indian Intellectual History', *Journal of American Oriental Society*, Vol. 105, no. 3 (1985), p. 1.

Part II

Technology, geography and warfare

3 Chinese border garrisons in an international context

Liaodong under the early Ming dynasty

David M. Robinson

Introduction

This chapter examines the military garrisons of the Ming dynasty (1368–1644) in the international context of North-East Asia. It focuses on Liaodong (or southern Manchuria as it is sometimes known), a broad region located at the intersection of Chinese, Korean, Jurchen and Mongolian polities, with particular focus on the fourteenth and fifteenth centuries. The first section shows that a century of intensified regional integration under the Mongols convinced Chinese, Korean and Mongolian leaders that control of Liaodong was essential to their strategic interests. As a result, when Mongol rule deteriorated in the second half of the fourteenth century, all parties scrambled to lay claim to the region and its diverse populations. The second half of the chapter explores how the winner in that scramble, the Ming dynasty, used the Liaodong garrisons to secure horses, military labour and intelligence on the international market of North-East Asia. Despite oft-repeated characterizations as xenophobic, inward looking and committed to the primacy of the civil over the military, the Ming dynasty systematically exploited the porosity of Liaodong to achieve its strategic objectives in eastern Eurasia through the use of military garrisons.

Mongol transformation and legacy

Ming interest in Liaodong grew directly from the geopolitical conditions created by the Chinggisid polity, known in east Eurasia as the Great Yuan *ulus* or the Yuan dynasty. Since the 1230s, the Chinggisid polity had controlled Liaodong, establishing interlocking systems of governance that supervised an ethnically complex population that included people of Jurchen, Mongol, Parhae, Khitan, Chinese and Korean descent. Such policies produced unprecedented levels of economic, administrative, military and ethnic integration among Korea, Liaodong and northern China.[1] When in the mid-fourteenth century, ambitious men throughout east Eurasia challenged the weakened Chinggisid Mongols, warfare spread to Liaodong. The king of Koryŏ, Kongmin (1330–1374), whose family had been close allies of the Chinggisid house for nearly a century, saw an opportunity to strengthen his position in North-East Asia. In 1356, he purged rivals at his court who were tied through office and marriage to the Chinggisids, announced an end to his loyalty to the Mongols and launched military strikes against Liaodong's eastern border, justifying the attack as the recovery of Koryŏ lands. The vagaries of intensifying war quickly forced the Koryŏ and Chinggisid ruling houses to resume an uneasy alliance, but the fate of Liaodong remained an unsettled question. As the Chinggisid central government's effective control of Liaodong collapsed in the 1360s, Mongol commanders such as Naghachu, Gao-jianu and others who had held posts in the Liaodong administration grew in influence. They commanded tens of thousands of troops, enjoyed close ties with the Koryŏ throne and had organized themselves into a loose alliance to defend their interests.[2] In 1368, the shift in the geopolitical balance of east Eurasia accelerated; in January, the Hongwu emperor (1328–1398) announced from his capital in Nanjing the establishment of the new Ming dynasty and in September the Yuan emperor Toghan-Temür (1320–1370) fled his capital of Daidu (present day Beijing) for the steppe.[3] Within a few years, the Koryŏ and Ming courts challenged first the Mongol commanders and then each other for dominance in Liaodong.

Making Liaodong their own: Koryŏ and Ming

Early in January 1370, King Kongmin strengthened the military defences of his northern border, establishing additional garrisons and appointing several veteran officers to take command of the north-east and north-west armies.[4] The king was preparing to attack Dongning Prefecture, located in southern Liaodong and still under Chinggisid control. By the end of the month, the Koryŏ court had mobilized a force of 5,000 cavalry troops and 10,000 infantry soldiers from its northern provinces, begun the march westward and crossed the Yalu river into Liaodong.[5] After brief fighting, the Dongning commander, a Korean in the service of the Mongols, surrendered.[6] Over the next two months, as many as 'ten thousand households' similarly switched their allegiance from the Chinggisids to the Koryŏ dynasty.[7] Six months later, in September 1370, one of the commanders involved in the attack on Dongning, Yi Sŏng-gye, received orders for another assault; late in November, Koryŏ forces again crossed the Yalu river, seizing control of Liaoyang, the most important city in Liaodong.[8] Nearly a year later in October and November 1371, Koryŏ struck against Wulaoshan Fortress, seizing a Chinggisid official.[9] The following spring, perhaps in retaliation, several Mongol commanders based in Liaodong raided Korean border cities.[10]

The *Official History of the Koryŏ Dynasty* preserves two major justifications of Koryŏ's repeated attacks on Liaodong, especially such important cities as Liaoyang. The first emphasizes that prior to the Mongol irruption of the thirteenth century, Liaodong was Koryŏ territory. Thus, the Koryŏ attacks are framed in terms of a reassertion of sovereignty and royal authority. In the wake of their attacks in 1370 and 1371, Koryŏ officials distributed placards urging 'all the subjects of this country east of the Liaodong river, leaders great and small' to transfer their allegiance to the Koryŏ throne from which they could expect titles and emoluments. During the thirteenth and fourteenth centuries, the Chinggisid had directly and indirectly contributed to the development of a string of sizeable Korean communities through southern Liaodong, most especially in Liaoyang, Shenyang and the lands eastward to the Yalu river, that in theory were subject to joint Chinggisid and Koryŏ rule.[11] As Mongol governance in North-East Asia collapsed, King Kongmin and his court saw an opportunity to win back Korean lands and regain control of Liaodong's Korean population.

A second justification, elaborated at great length in official correspondence with the Ming court, involved the alleged villainy of the Ki family, especially one Ki Sayin-Temür. During the 1340s and 1350s, the Ki family had achieved considerable political influence in North-East Asia, primarily as a result of Empress Ki, Korean wife of the Yuan emperor Toghan-Temür, who used her position to secure official posts for her brothers in both the Yuan and Koryŏ governments. In 1356 King Kongmin had killed and exiled a number of influential men at his court with ties to the Yuan throne, including several of Empress Ki's brothers. She never forgave King Kongmin, seeking revenge even after the Chinggisid court retreated to the steppe.[12] Ki Sayin-Temür was her nephew, son of one of her fallen brothers. The Koryŏ throne insisted that not only was Ki using Dongning as a base from which he plotted against King Kongmin, seeking out conspirators in Liaodong, but that he had betrayed Toghan-Temür, his ostensible liege, abandoning him to the wilds of the steppe.[13] During a time of rapidly shifting geopolitical conditions, such a posture was politically deft if factually inaccurate.[14] King Kongmin and his advisers committed themselves to little, neither staking a claim to territory nor pledging clear alliance to either the Chinggisid or Ming courts.

Although some scholars have dismissed the threat of Ki Sayin-Temür as nothing more than vacuous post facto justification for Koryŏ's military adventures in Liaodong, King Kongmin's desire for some measure of control in Liaodong was firmly grounded in reality.[15] From 1359 to 1361, powerful rebel military forces passing through Liaodong had repeatedly raided Koryŏ lands, twice occupying the western capital (present day P'yŏngyang) and eventually driving King Kongmin and his court from the main capital of Kaegyŏng. As part of its attempt to replace King Kongmin with a more biddable king, in 1364, the Yuan court massed an army reputed to number 10,000 strong in Liaoyang before crossing the Yalu to make sure their man took the throne.[16] These dangerous challenges to his rule were launched from Liaodong when it was under control of an ostensible ally. The king no doubt worried what might happen if it fell into openly hostile hands.

Political and military consolidation of China proper was Hongwu's top priority, but by 1371 he had turned his formidable energies to Liaodong. Although Hongwu criticized the last Yuan emperor's misrule and decried the Mongols' deleterious influence on Chinese society and culture, he generally portrayed himself as the rightful successor to the Yuan, including lands not previously subject to Chinese rule (such as Yunnan in the south-west) and those long outside Chinese control, such as Liaodong, which had not been under sustained control of a Chinese dynasty since the fall of the Han in the third century. Given the formidable military capacity of North-East Asia, including powerful commanders like Naghachu and his army of several hundred thousand men in Liaodong, and former and potential future allies among Jurchen leaders and the Koryŏ throne to the north and east, Hongwu felt that control of Liaodong was vital in the wider war with the Yuan court.[17] Late in autumn 1370, Hongwu began sending envoys to meet with power-holders in Liaodong, Yuan military commanders who had formed a loosely organized league designed to protect their interests, stressing his military triumphs, his rightful succession to the Mandate of Heaven, the Yuan's declining fortunes on the steppe and his promise to reward all those who surrendered. He also pointed out that other rulers in Eurasia formerly tied to the Chinggisid polity had already transferred their allegiance to the Ming. 'Surely your intelligence matches theirs', Hongwu observed encouragingly.[18] The emperor's overture quickly bore fruit. Liu Yi, a senior Yuan official in the Liaoyang provincial government, who had established a power base along the south-western coast of the Liaodong peninsula indicated his desire for a deal. In exchange for turning over his lands, military forces, and population and tax registers to the Ming, Liu was to be appointed a Vice Commander in the planned Military Command of Liaoyang, which Hongwu envisioned as an institutional beachhead into Liaodong.[19]

Observers across North-East Asia were keenly interested in Ming intentions and Liu Yi's decision. Noting Liu Yi's worry that the Ming court might insist on relocating him to another region, *The Official History of the Koryŏ Dynasty* relates that 'since Liaodong was originally our territory', Liu approached the Koryŏ court to intervene on his behalf with Hongwu.[20] The Koryŏ court knew what lands Liu would turn over to the Ming and what rank he would receive in the Ming military.[21] It commanded such information, because although Liu Yi had transferred his allegiance to the Ming, he, like all Yuan commanders in Liaodong, continued to send envoys to the Koryŏ court, bearing gifts and felicitations on occasions like King Kongmin's birthday.[22] Other observers were not so cordial. In fact, before Liu Yi actually transferred his lands and subjects, two Yuan officials in the Liaoyang provincial administration assassinated him. In short order, several military men interested in a deal with the Ming killed one of the murderers; the second fled, seeking refuge with Naghachu in northern Liaodong.[23] By early in 1371, Hongwu had lined up several willing partners in Liaodong,[24] and by late summer, he had dispatched two senior, trusted military commanders from Nanjing to oversee operation of the forerunner of the Liaodong Military Commission, which over the next two decades would expand to include 25 garrisons and serve as the primary institution of Ming rule in North-East Asia. In these early days when Ming logistical capacity in the region was limited, both these senior commanders strongly favoured a sustained campaign of persuasion, displaying 'benevolence and awe' and 'broadly showing favor' to win over local power-holders rather than rely exclusively on military confrontation.[25]

In summary, when the Yuan court withdrew to the steppe in autumn 1368, several groups contested control of Liaodong. Based on their local bases of power and the authority of their office, military commanders from the Yuan's Liaodong provincial government initially appeared to be in the strongest position; chief among them was Naghachu. The Koryŏ court, however, quickly staked its claims, launching military strikes, opening a campaign of persuasion directed at the local Korean population, and welcoming Mongols and Jurchens who had once served the Yuan with offices and material incentives. The fledgling Ming dynasty was a latecomer: its base was far to the south, its ability to project power was initially limited and it faced a myriad of urgent challenges elsewhere. However, the Ming took advantage of the precarious situation in North-East Asia. Yuan military commanders were as suspicious of their ostensible allies as they were of the Ming. Further, although in name still loyal servants of the Mongol court, in reality influential commanders like Gaojianu, Naghachu and others were independent operators acutely attuned to shifting opportunities in the region. The Koryŏ court's ability to pursue its interests in Liaodong was handicapped by political infighting in Kaegyŏng, a distracted king and a deep reluctance to poison relations with the Ming.[26] Hongwu exploited Liaodong's fluid political landscape. On the one hand, he alternately cajoled and browbeat, warred and rewarded the major actors of Liaodong; on the other hand, he systematically built a formidable administrative, military, transportation and economic infrastructure. By the end of his reign in 1398, Hongwu had made Liaodong Ming territory, destroying or co-opting all the important military commanders. In fact, for the next two centuries Liaodong was the linchpin of Ming power in North-East Asia.

Paradoxes of a border region

Like most border regions, Liaodong was rich in paradox. As a bastion of Ming imperial power in North-East Asia, it was the focus of massive state investment. It was a garrison state, the only major region of the Ming empire governed entirely through a military administration, which oversaw 25 military garrisons, each manned by 5,000 to 6,000 men.[27] During the mid-sixteenth century, the Liaodong garrisons could muster approximately 100,000 combat troops and 45,000 soldiers devoted to farming.[28] It boasted an elaborate system of border fortifications complete with watchtowers, long-distance signalling and stores of grain, weapons and fuel.[29] The Ming state invested in a vast overland and maritime transportation network to facilitate the movement of troops and the delivery of grain, silver and government paperwork.[30] It established Confucian academies to educate and inculcate proper values in the local population, including the officer corps.[31] It opened new agricultural lands to feed imperial troops, improve people's livelihood (and of course cut costs to the dynasty) and support an expanding Chinese population.[32] It integrated local religious cults through economic support, official endorsement and sacrifices by the emperor (usually conducted by palace eunuchs travelled from the capital on the emperor's behalf) to the deities of Mt Iwulü.[33] The Ming state also sent convicts from elsewhere in China to Liaodong, where they were registered in military garrisons and worked on military agricultural fields (politically connected elite men usually passed their time more agreeably in urban centres socializing with men of similar status).[34] In other words, Liaodong neatly reflected the formidable organizational capacity of the Ming as an early modern empire to mobilize military, economic, political, ideological and personnel resources to the very edges of its territory.

Despite massive investment from the centre and thorough integration into the empire, its status as a border region created deep anxieties. Hongwu and his successors fretted about insufficient oversight over military commanders in Liaodong. Hongwu, especially, rebuked his officers for incompetence, poor judgment, corruption and unregulated contact with Korean envoys.[35] Hongwu worried about collusion between local Ming authorities and Mongols, Jurchens and Koreans just as he feared collusion elsewhere in empire. Fears of barbarization also surfaced periodically. The large number of Jurchens, Mongols and Korean who lived in Liaodong, the region's long history under foreign rule, the area's frequent warfare and supposed martial ethos all threatened proper Chinese values, observation of correct ritual practices and ultimately allegiance to the Ming throne, warned the compilers of the 1537 *Liaodong Gazetteer*.[36] The great state resources committed to Liaodong worried some. Military men of ambition and uncertain loyalty might (and in the event did) exploit the manpower, wealth and authority granted by the state to pursue their own private interests, or more dire still, challenge the dynasty.

Fears of barbarization, collusion with foreigners and betrayal by the military all seem to fit the enduring image of the Ming dynasty as xenophobic, inward-looking and anti-military, obsessed with rigid borders, limited contact with the outside world and the dominance of the civil over the martial. In fact, the discourse that highlights tensions between the civil and martial masks their deeply intertwined nature, whether considered in intellectual interests, career tracks or administrative structures. Occasional worries over barbarization should not obscure the extensive interaction, both official and unofficial, among Chinese, Jurchens, Mongols, Koreans and others. The Ming state's ability to secure horses, military labour and intelligence in North-East Asia rested squarely on Liaodong's garrisons and the porosity of the border.

The horse markets of Liaodong were one element of an enduring, pan-Eurasian pattern, whereby, as Jos Gommans has shown, sedentary polities used their superior economic and administrative resources to acquire good quality horses from the steppe. Given that maintaining sufficient numbers of mounted archers, the most dominant form of military technology for most of Eurasia during these centuries, was essential to the survival of such sedentary polities, the acquisition of horses formed a central pillar in administrations from Korea and China in the east to Rus and Hungary in the west.[37] The Ming state depended on a steady supply of horses from Tibetan, Mongolian, Jurchen, Korean, Timurid (or more broadly Central Asian), Japanese and Okinawan sources, again giving lie to stereotypes about Ming isolation. The Liaodong garrisons supervised the acquisition of horses from Mongolian, Jurchen and Korean sources.

In order to protect its territory and to project military power beyond the border, the Ming like other Eurasian states, needed large numbers of horses. This was especially true when the enemy commanded a steppe army, such as the Mongols, or one inspired by steppe warfare, such as many of the Manchurian powers. During times of military crisis, the Ming state needed to quickly augment its supply of horses. When for instance Hongwu intensified his campaign against Naghachu in Liaodong during the 1380s, he repeatedly pressured the kings of Koryŏ and Ryūkyū into selling him thousands of horses.[38] In the case of Koryŏ, officers from the Liaodong Military Commission supervised the purchase of as many as 10,000 horses, receiving the Korean officials who transported the steeds, inspecting the horses and handling payment. Hongwu used recently surrendered Mongol commanders who had been settled in Liaodong like Gaojianu who had formed ties to the Koryŏ court to facilitate the acquisition of horses from Koryŏ. During the warfare surrounding the succession crisis of 1400–1402, both contenders for the throne purchased horses from Koryŏ's successor, the Chosŏn dynasty; Jianwen secured 7,000 of a requested 10,000 horses in exchange for silks, and Yongle acquired more than 2,000.[39] During the next 20 years of his reign, Yongle acquired an additional 33,000 horses from Chosŏn.[40] In the wake of a disastrous defeat at the hands of the Oirat Mongols in 1449 and braced against further attacks, the Ming throne purchased slightly fewer than 2,000 (of a requested 30,000) horses from Chosŏn.[41] Personnel from the Liaodong Military Commission oversaw all these transactions.

The Ming dynasty acquired most of the 600,000 to 700,000 horses it needed through state-run border markets along its northern and western borders.[42] Alarmed by the severe dearth of government horses, shortly after taking power early in the fifteenth century, the Yongle emperor established the first of what in time would grow to be 14 horse markets in Liaodong, usually along important transportation trunk roads, to secure mounts from Jurchen and Mongol populations in exchange for textiles, grain and, increasingly over time, payments in silver.[43] The scale and longevity of the markets varied, but those at Kaiyuan (one north and one south of the city) and Fushun were the largest and most enduring – the so-called Three Great Horse Markets of Liaodong. The Ming invested in a substantial institutional infrastructure to secure a regular supply of reliable horses, constructing warehouses to store goods to be traded for horses and establishing bureaus to inspect and pasture horses, conduct financial transactions, entertain foreign envoys and traders, and verify their credentials. Security provisions were strict. Regulations forbade Jurchen and Mongol parties from carrying weapon into the markets, which were often located within imposing wooden stockades. Military units patrolled the surrounding area, in part to deter raiding parties and in part to discourage riotous behaviour at the market.[44] Personnel from the Liaodong Military Commission supervised the markets, which were open daily year round. They compiled written records on all transactions, levied taxes in kind and silver, and submitted quarterly reports to the Regional Inspector of Shandong, a civil official, for review before they were turned over to the central government in Beijing.[45] Aggregate numbers are not available, but scattered evidence suggests that a market like Kaiyuan regularly purchased 500 horses a year, often many, many more. The best quality mounts were transported to Beijing for use in the capital garrisons; ordinary horses were kept as war horses for the Liaodong soldiers, and those that did not pass minimum standards were retained for use by military personnel responsible for farming.

The Ming state had a strong interest in making sure that Jurchen and Mongol traders did business at the horse markets. To that end, it permitted private trade and offered bonuses. Taxes on transactions at the market were used to cover the expense of entertaining Jurchen and Mongol traders and to provide them with gifts of rice, salt, cooking woks and other goods. By the late sixteenth century, such costs far outstripped the revenue generated through the market taxes; officials diverted other funds to cover the discrepancy. The central government also contributed to such expenses. Sometimes local officials ended up using silver from the court designated for the purchase of horses to pay for bonuses and entertainment. The enduring appeal of the horse markets was due primarily to its thriving private trade, which in time overshadowed the state trade.[46] The Jurchens and Mongols from the Three Garrisons brought goats, cattle, sable furs, leathers, honey, ginseng and leathers to exchange for Chinese processed goods ranging from silks and linen textiles to cooking pots, porcelains, rice and salt. Such large-scale transactions (1,000 pounds of ginseng and 1,000 sable furs each year during the mid-sixteenth century) stimulated socioeconomic growth, agricultural development and the circulation of silver among border populations. Such developments sometimes subverted the interests of the Ming state. By the late sixteenth century, powerful military families abused their position and local influence to dominate the markets, manipulating prices, while ambitious Jurchen leaders like Nurhaci monopolized key trade items and access to the Ming economy to consolidate political power on both sides of the border.[47] Thus, an institution developed at great cost to bolster the dynasty's military defences ultimately facilitated abuses by leading military families and contributed to the rise of those who would eventually usher in its demise, including Nurhaci and his successors.[48] Although historians of many stripes have long delighted in this ironic twist, such a denouement was not at all clear to Ming officials, who like their counterparts throughout Eurasia, felt that the benefits of the horse markets outweighed its costs.

Just as it secured horses on the international market, so the Ming state recruited military labour beyond its borders. Throughout Eurasia, the military served as a great engine of demographic transformation. The state or regional rulers deployed soldiers where they needed them. Such demographic shifts might be seasonal, long term or permanent; they might involve a few dozen men or tens of thousands; they might occur within a single polity or extend beyond one or more political borders. In east Eurasia, the state frequently used military units as a way to bundle men and their families, registering them into garrisons, which were not only fighting forces but administrative units that monitored everything from battlefield performance and military rank to socioeconomic status, children, infirmity, welfare benefits and death. During the Ming dynasty, hereditary households with the obligation to provide one active service soldier and one or more helpers who contributed to his upkeep formed the core of the military system, but aboriginal tributary armies, militiamen, mercenaries and other forms of military labour grew increasingly important as socioeconomic conditions changed.[49] From its earliest days, the Ming state recruited non-Chinese men to serve in its military forces. In the 1360s and 1370s, it integrated tens of thousands of Mongol, Turkic, Jurchen, Khitan, Central Asian and Korean soldiers from surrendering Yuan units into the Ming military. Although many had been stationed in areas that the Ming now claimed as dynastic territory, many were recruited from beyond Ming borders.

During the fourteenth century, the Ming court competed directly with the Yuan and Koryŏ courts for the loyalty and services of Mongol and Jurchen military men. To this end, Hongwu recruited leading Mongol and Jurchen commanders, offering them official posts, generous salaries, housing, annual bonuses in textiles and rice, the opportunity to lead men in battle and chances to win further promotions.[50] He granted aristocratic titles and senior positions in his armies to the most influential and capable men, such as Naghachu and Gaojianu. In the wake of such elites were hundreds, sometime thousands, of lesser men who followed their leaders into Ming service, garrisoned in border, capital and hinterland units. Throughout much of the fifteenth century, the Ming state continued to recruit such men, as did the Chosŏn court. The Liaodong garrisons played an important role in both recruiting and employing Jurchen, Mongolian and Korean military labour from an international market.

During the last decades of the fourteenth century and the early decades of the fifteenth century, thousands of Jurchen and Mongol travelled to the Liaodong garrisons, where military officers supplied them with food and lodging. Most were settled locally. In June 1378, Shijianu 史家奴, a former Assistant Military Affairs Commissioner 樞密副使 under the Yuan, and 40 other men surrendered at Liaodong.[51] A month later, 106 recently surrendered Mongols arrived with 170 head of livestock.[52] In 1382, an officer from Dongning Garrison 東寧 accepted the surrender of troops of a former Yuan general and nearly 3,000 followers, who were sent to Liaoyang. The emperor ordered that they be supplied with clothing and grain.[53] The *Veritable Records of Taizu*, one of the richest sources for tracking the numbers of Mongols who migrated during the early Ming, usually indicates that a given number of families ‘offered submission’ in Liaodong and received varying amounts of silk, clothing and cash but often passes lightly where they were settled.[54] Sometimes the Liaodong authorities supervised the relocation of Mongol and Jurchen aristocrats to the capital. The Liaodong Regional Military Commission provided a protective escort for the Mongol noble, the Liao Prince, and nearly 50 of his administrators and followers as they traveled to Nanjing to present horses to the throne in 1388.[55]

Several Liaodong garrisons figured especially prominently in the integration of Mongol, Jurchen and Korean populations. In the 1380s, Hongwu created

Dongning Garrison 東寧衛 (in Liaoyang) and the Three Tümens Garrison 三萬衛 (in Kaiyuan) to accommodate the large numbers of men who had formerly been under the control of the Yuan general Naghachu.[56] The former was home to significant Korean and Jurchen populations, while in the latter Jurchens predominated. During the Hongwu reign, most immigrants were registered as hereditary military households, subject to the same obligations and restrictions as other Ming military households.[57] In an effort to make border garrisons more appealing than relocation to the capital, in 1408 the throne ordered that local authorities provide the group with 'cash, clothing, saddles and horses' as well as 'lodgings, sundry utensils, firewood for cooking, rice, oxen, and sheep'. This order served as a precedent for Jurchens 'who wished to settle in border garrisons'.[58] These émigrés were registered in military garrisons, where they would receive posts designed to reflect their relative importance in Mongolian or Jurchen society. Military appointment registers last updated in the 1590s indicate that nearly half of the active duty officers of the Garrison of the Three 10,000 were of Jurchen descent.[59] In time, a steadily growing Korean population registered in the Dongning Garrison spilled over into areas outside Liaoyang.

Mongols, Jurchens and Koreans registered in Liaodong garrisons played many roles. They defended the northern frontier, fought in lesser and greater campaigns on the steppe, rotated to the capital for training and occasionally suppressed rebellion in the hinterlands of the Ming empire. Because of their language skills and familiarity with Mongol, Jurchen and Korean lifeways, during the fourteenth and fifteenth centuries, they worked in the administration of the Liaodong horse markets, acted as liaison personnel for more recent migrants, received Chosŏn missions in Liaodong, worked in diplomatic bureaus in Beijing and served on diplomatic and trade missions throughout North-East Asia. Such skills explain much of the reason that during the fifteenth and sixteenth centuries, the Korean court also recruited Mongols and more especially Jurchens. Like the Ming court, the Chosŏn dynasty offered titles and sinecures in the Korean military, lands (or the revenue derived from such lands) and the opportunity to trade at border markets.[60] Some Mongol and Jurchen military men held substantive posts, commanding men in the field and acting as Korean representatives on diplomatic missions in North-East Asia. Although beyond the scope of this chapter, the Ming and the Koryŏ/Chosŏn courts understood such arrangements as part of broader strategies to eliminate predatory raids, project influence beyond their borders and prevent consolidation of power under a single Jurchen or Mongol leader by satisfying economic needs and fostering varying levels of economic and political dependence through patronage. As was true elsewhere in Eurasia, economics, politics and the military were inextricably linked in the Ming and Koryŏ/Chosŏn polities.

Finally, the state secured intelligence from an international market through the Liaodong garrisons. Because Mongols, Jurchens, Koreans and Chinese had relatively plentiful opportunities for direct and indirect contact in Liaodong, all groups cultivated sources of information about neighbouring groups. As Kawachi Yoshihiro and Hasumi Moriyoshi have shown and even a cursory glance at the *Veritable Records of Chosŏn Dynasty* confirms, the Korean court used Liaodong as a listening post.[61] Liaodong authorities also monitored wider developments in North-East Asia.

Consider a 1459 example preserved in the *Veritable Records of King Sejo*. A Korean envoy reported to the king a secret conversation he had with a centurion of Guangning Garrison in Liaodong. The centurion had heard that the Regional Commander of Liaodong had talked with a Jurchen leader about the visit of several Jurchen leaders to the Chosŏn court. The Jurchen told the Commander that King Sejo had lavishly feted the Jurchens every day and showered them with valuable gifts such as horse saddles. The Jurchen informant told the Liaodong commander that it seemed likely that the Chosŏn was trying to win their allegiance. The Liaodong Commander expressed some scepticism and asked his informant point blank 'How did you learn of this?' 'I saw it with my own eyes', responded the Jurchen leader. The Commander rewarded the Jurchen for the information. Having told one of the eunuch officials in Liaodong that the news must be kept confidential, the Commander then sent out an officer to secretly observe the Jurchen leaders reputed to be considering allegiance to the Chosŏn court.[62]

Here we see the flow of people and intelligence across dynastic borders, ethnic groups, within the ranks of the Ming military and back again. Whereas the *Veritable Records of Sejo*, a source compiled at the central court, has preserved the flow of information, the *Veritable Records of Yingzong*, however, contains no corresponding record. One might infer that the flow of intelligence and knowledge of its structure did not extend to the central court in Beijing. It is more likely that Ming imperial records omitted the messy, contested interactions along the frontier because they contradicted the lofty rhetoric of harmonious tributary relations.[63] The example also suggests that mid-level Ming officers and palace eunuchs were susceptible to Korean efforts to secure military and political intelligence.

In other cases, the Chosŏn throne shared intelligence with the Ming court. In 1460, the Korean king informed the Ming throne that a number of Jurchen nobles whose families had lived on his northern border for several generations had recently relocated to Jianzhou. The Korean king warned that they might be planning to join the Jurchen noble, Li Manzhu 李滿住 (1407–1467), in an attack on the Ming. The emperor ordered the Grand Defender of Liaodong to dispatch men to dissuade Li Manzhu from such a course of action.[64] The intelligence sharing may have been intended to diminish the Ming court's pique about the Chosŏn state's unilateral decision to kill a 'commissioner-in-chief' of Maolian Garrison and nine members of his family who had 'plotted to foment revolt'.[65]

The strategy seems to have worked. By late in December 1460, the Ming throne issued an edict to King Sejo commiserating with the king's difficulties with the Jurchens in his territory. The Tianshun emperor also informed King Sejo that he had dispatched military officers from Liaodong to proclaim imperial instructions to the Jurchens of Jianzhou to set aside their plans for retribution against Chosŏn and to turn over Chosŏn captives seized in raiding to Ming authorities in Liaodong.[66] In this case, the Ming and Chosŏn thrones exchanged information on a regular basis. Both sides took for granted that personnel from the Liaodong garrisons would perform much of the heavy lifting of communication, negotiation and relocating key figures.

Liaodong military personnel were not passive or neutral conduits of information from the field to the Ming throne. They revealed information to Korean representatives (whether for personal profit, friendship or indiscretion). They also fabricated intelligence assessments to protect their interests. For instance, in 1450, Yu Qian, the Minister of War, accused Liaodong commanders of providing the central government with grossly inflated numbers of Jurchens on the border. Yu complained to the newly enthroned Jingtai emperor that the commanders transformed 'one or two Jurchens on the move' into 'several thousand Jurchen attacking the border'. The commanders' goal, explained Yu, was to avoid the deployment of their troops. Yu noted that commanders from Gansu, Ningxia, Datong and Xuanfu – in other words in all the most critical garrison regions – resorted to similar strategies. Equally common was inflating the number of enemy troops killed in combat as a way to secure greater rewards and promotions. 'This sort of behavior is the same everywhere', he observed with asperity.[67]

Concluding remarks

This chapter has attempted to consider the Ming military in a more explicitly international context than is common. In fact, general accounts of China and in global histories that try to integrate the experiences of the East Asia too often characterize the Ming as inward looking and isolated. My emphasis on international connections is also intended as a reminder that the Ming court knew that it was one among many in Eurasia and that its authority and influence were subject to nearly constant challenge by rivals great and small. Many facets of Ming history are most profitably considered from a wider Eurasian perspective. For instance, as the first section of this chapter showed, the contours and perspectives of the early Ming's world had been profoundly shaped by the Mongol century; they explain why Hongwu and his Korean and Mongolian rivals were so deeply invested in control of Liaodong. The clear parallels between the Ming dynasty's military and those of other Eurasian powers in regard to securing steppe horses, military labour and intelligence on an international market similarly suggest fruitful lines of comparative and integrative military history.

The chapter's second objective is an effort to think about 'the military' as a massive and central feature of the Ming dynasty through the case study of Liaodong. The Ming state invested heavily in the Liaodong garrisons and expected them to do far more than fight on the battlefield. Personnel from the garrisons oversaw critically important horse markets, the recruitment and settlement of migrants from the north, and the gathering and evaluation of military, economic and diplomatic information from Mongol, Jurchen and Korean sources. The state used the garrisons as an administrative vehicle to organize, monitor and exploit the people and resources of Liaodong. As a border region, Liaodong illustrates the centrality of military institutions to the Ming state with unusual clarity, but the importance of the military to trade, demographic and diplomacy – not to mention agricultural development, ethnicity, land tenure, penal systems, religion and foreign relations among others – holds true for the rest of Ming territory.

Notes

1 David Robinson, *Empire's Twilight: Northeast Asia under the Mongols* (Cambridge, MA: Harvard University Asia Center, 2009). Zhang Xianhai 張現海 explains the importance of Liaodong to the Yuan, Ming and Koryŏ in terms of an eastward shift of China's geopolitical interests, which he contrasts with a focus on the north-west under early dynasties down to the Song, 'Hongwu chunian Ming, Bei Yuan, Gaoli de diyuan zhengzhi geju' 洪武初年明，北元，高麗的地緣政治格局, *Gudai wenming* 古代文明, vol. 4, no. 1 (2010), pp. 90–97, esp. p. 91. This is misleading; the Chinggisid polity was not equivalent to China. The Ming had no choice but to deal with a world deeply shaped by Mongol – not Chinese – perspectives and policies.

2 Yun Ŭnsuk 尹銀淑, 'Nagach'u ŭi hwaldong kwa sipsa segimal Tong Asia chŏngse' 나가추의 활동과 14세기말 돈아시아 政勢, *Myŏng Ch'ŏngsa yŏn'gu* 明清史研究, vol. 28 (2007), pp. 1–27; 'Pukwŏn kwa Myŏng ŭi taerip: Yodong munje rŭl chungsim ŭro' 北元과 明의 대립: 遼東 문제를 중심으로, *Tongyang sahak yŏn'gu* 東洋史學研究, vol. 105 (2008), pp. 81–112; 'Koryŏ ŭi Pukwŏn ch'ingho sayongkwa TongAsia insik' 고려의 北元칭호 사용과 동아시아 인식, *Chung'ang Asia yŭn'gu* 中央아시아 研究, vol. 15 (2010), pp. 189–216, esp. p. 202. See also her chart of visits between King Kongmin and leading Yuan military commanders such as Köke-Temür, Gaojianu, Naghachu, and Hong Baobao from 1361–1368 (p. 198).

3 Hongwu was the reign title of Zhu Yuanzhang (1328–1398). Secondary scholarship uses both his reign title and his personal name, often interchangeably. Toghan-Temür is better known as the Shundi emperor, the temple name selected by the Ming court in 1370.

4 *Koryŏ sa* 高麗史 (hereafter abbreviated as *KS*) (ed. Chŏng In-ji 鄭麟趾) (1454, reprint, Tokyo: Kokusho kankōkai, 1908), *kwŏn* 41, pp. 630–631, 恭愍十八年十二月辛未.

5 For brief overview, see Wada Sei 和田清, 'Minsho no Manshū keiryaku' 明初の満州經略, 2 pts, *Man Sen chiri rekishi kenkyū hōkoku* 満鮮地理歴史報告 14 (1934), pp. 177–298; 15 (1937), pp. 71–319, esp. pp. 204–208.

6 *KS*, *kwŏn* 42, pp. 631–632, 恭愍十九年正月甲午，甲辰.

7 *KS*, *kwŏn* 42, pp. 631–632, 恭愍十九年二月戊寅，壬午.

8 *KS*, *kwŏn* 42, p. 640, 恭愍十九年八月己巳，十一月丁亥. In 1392, Yi Sŏng-gye founded the Chosŏn dynasty.

9 *KS*, *kwŏn* 43, p. 644, 恭愍二十年九月辛亥，十月丙戌; Nam Su-min 南秀文 (ed.), *Koryŏsa chŏlyo* 高麗史節要 (1452, reprint, Seoul: Asea munhwasa, 1974), 29.15b, p. 738, 恭愍二十年九月，十月.

10 *KS*, *kwŏn* 43, p. 646, 恭愍二十一年正月甲戌，二月辛丑. For painstaking consideration of these events, see Ikeuchi Hiroshi 池内宏, 'Kōrai Kyōmin ōcho no Tōneifu seibatsu ni tsuite' 高麗恭愍王朝の東寧府征伐について, in Ikeuchi Hiroshi 池内宏, *Man Senshi kenkyū chūsei* 満鮮史研究 (Tokyo: Yoshikawa kōbunkan), vol. 3, pp. 197–234.

11 Robinson, *Empire's Twilight*, pp. 28–34.

12 Robinson, *Empire's Twilight*, pp. 98–129.

13 *KS*, *kwŏn* 42, p. 641, 恭愍十九年十二月丁巳; *kwŏn* 43, pp. 646–647 恭愍二十一年三月庚戌; Nam Su-min, *Koryŏsa chŏlyo*, 29.7b, p. 734, 恭愍十九年八月.

14 A chronicle compiled by a member of Toghan-Temür's entourage makes clear that Ki Sayin-Temür did visit the Mongol ruler after his retreat to the steppe.
15 Ikeuchi Hiroshi, 'Kōrai Kyōmin ōcho no Tōneifu seibatsu ni tsuite', pp. 236–238.
16 Robinson, *Empire's Twilight*, pp. 147–198, 220–251.
17 For Toghan-Temür's reliance on Liaodong commanders after his flight from Daidu (and relevant secondary Chinese scholarship), see Zhang Xianhai, 'Hongwu chunian Ming, Bei Yuan, Gaoli de diyuan zhengzhi geju', pp. 92–94.
18 *Ming Taizu shilu* 明太祖實録 (facsimile reproduction of *Guoli Beiping tushuguan cang hongge chaoben*, 133 vols (Taibei: Zhongyang yanjiuyuan lishi yuyan yanjiusuo, 1961–1966), 56.7b 洪武三年九月是月.
19 *Ming Taizu shilu*, 61.5a 洪武四年二月壬午.
20 *KS*, *kwŏn* 43, p. 642, 恭愍二十年閏月己未.
21 *KS*, *kwŏn* 43, p. 642, 恭愍二十年四月戊戌.
22 *KS*, *kwŏn* 43, p. 642, 恭愍二十年五月癸丑.
23 *Ming Taizu shilu*, 66.4a–b 洪武四年六月壬寅; 66.8a–9a 洪武四年六月是月.
24 *Ming Taizu shilu*, 66.4b 洪武四年六月壬寅; Gu Yingtai 谷應泰, *Ming shi ji shi ben mo* 明史紀事本末 (1658, reprint, Taibei: Sanlian shuju, 1985), ch. 19, p. 93.
25 Bi Gong 畢恭 *et al.* (eds), *Liaodong zhi* 遼東志 (compiled 1443, rev'd 1488, rev'd 1529, reprinted in Jin Yufu 金毓黻 (ed.), *Liaohai congshu* 遼海叢書 (Dalian: Liaohai shushe, 1932), 5.39a–b).
26 Ye Yuanhong 葉泉宏, *Mingdai qianqi Zhong-Han guojiao zhi yanjiu* 明代前期中韓國交之研究 (Taibei: Taiwan shangwu yinshuguan, 1991), pp. 79–118. For broader discussion of Ming relations with Koryŏ and more especially Chosŏn era Korea, see Donald Clark, 'Sino-Korean Tributary Relations under the Ming', in Denis Twitchett and Frederick Mote (eds), *Cambridge History of China, volume 8, The Ming Dynasty 1368–1644, Part 2* (Cambridge: Cambridge University Press, 1998), pp. 272–300.
27 Li Xinfeng 李新峰, 'Mingchu Liaodong zhanzheng jincheng yu weisuo shezhi shiyi' 明初遼東戰爭進程與衛所設置拾遺, *Mingshi yanjiu luncong* 明史研究論叢, vol. 9 (2011), pp. 50–62; Zhu Chengru 朱誠如, 'Ming Liaodong dusi ershiwu wei jianzhi kaobian' 明遼東都司二十五衛建置考辨, in Zhu Chengru 朱誠如, *Guan kui ji: Ming Qing shi sanlun* 管窺集: 明清史散論 (Beijing: Zijincheng chubanshe, 2002), pp. 279–288. See also Chen Wenshi 陳文石, 'Mingdai qianqi Liaodong de bianfang' 明代前期遼東的邊防, in Chen Wenshi 陳文石, *Ming Qing zhengzhi shehui shilun* 明清政治社會史論 (Taibei: Xuesheng shuju, 1991), pp. 177–282.
28 Wang Tingyuan 王廷元, 'Luelun Mingdai Liaodong junhu' 略論明代遼東軍戶, *Anhui shida xuebao* 安徽師大學報 (zhexue shehuixue ban), vol. 4 (1981), pp. 74–82, esp. pp. 75–76.
29 Liu Qian 劉謙, *Ming Liaodongzhen changcheng ji fangyu kao* 明遼東鎮長城及防禦考 (Beijing: Wenwu chubanshe, 1989).
30 Cong Peiyuan 丛佩遠, *Mingdai Dongbeibian* 明代東北編, in Tong Dong 佟冬 (ed.), *Zhongguo Dongbei shi* 中國東北史 (Changchun, Jilin wenshi chubanshe, 1998), vol. 3, pp. 630–654.
31 Yang Yang 楊暘, *Mingdai Dongbei shigang* 明代東北史綱 (Taibei: Xuesheng shuju, 1993), pp. 414–422.
32 Cong Peiyuan, *Mingdai Dongbeibian xu* 明代東北編續, in Tong Dong 佟冬 (ed.), *Zhongguo Dongbei shi* 中國東北史 (Changchun, Jilin wenshi chubanshe, 1998), vol. 4, pp. 940–972.
33 For stele inscriptions commemorating such trips by the emperor's envoys to offer or announce the enthronement of new rulers, see Sonoda Kazuki 園田一亀, *Manshū kinseki shikō* 満洲金石志稿 (Dairen: Minami Manshū tetsudō kabushiki kaisha, 1939), vol. 2, pp. 7–9, 35, 39, 101–103, 106–107, 126, 136, 143, 168–169, 181–183.
34 Yang Yang, *Mingdai Dongbei shigang*, pp. 43–86.
35 *Ming Taizu shilu*, 124.2a 洪武十二年四月庚申, 125.2b–3a 洪武十二年閏五月甲戌, 132.6a 洪武十三年六月庚子, 132.5a 洪武十三年六月甲午.

36 Bi Gong 畢恭 *et al.* (eds), *Liaodong zhi*, 1a of introduction, 1.21a, 4.2a.

37 J. Gommans, 'Warhorses and post-nomadic empire in Asia, *c.*1000–1800', *Journal of Global History*, vol. 2, no. 1 (2007), pp. 1–21.

38 Kageki Motohiro 蔭木原洋, 'Kōbutei shoki no taiRyūkyū seisaku – uma, Kōrai, Nahashutsu o tōshite' 洪武帝初期の対琉球政策 – 馬 高麗 納哈出を通して –, *Tōyōshihō* 東洋史訪, vol. 14 (2008), pp. 1–14.

39 Pak Wŏn-ho 朴元熇, '"Jingnnan zhi yi" yu Chaoxian' 靖難之役與朝鮮, *Mingshi yanjiu* 明史研究, vol. 1 (1991), pp. 227–247, esp. pp. 231, 236–238, 243; Pak Wŏn-ho 朴元熇, *Myŏng-ch'o Chosŏn kwan'gyesa yŏn'gu* 明初朝鮮關係史研究 (Seoul: Iljogak, 2002), pp. 141–145.

40 Kitajima Manji 北島万次, 'Eirakuteiki ni okeru Chōson ōkoku no sakuhō to kōeki' 永楽帝期における朝鮮王国の冊封と交易, in Tanaka Takeo (ed.), *Zenkindai no Nihon to Higashi Ajia* 前近代の日本と東アジア (Tokyo: Yoshikawa kōbun, 1995), pp. 197–215, esp. pp. 200–202.

41 Hasumi Moriyoshi 荷見守義, 'Minchō no sakuhō taisei to sono yōtai' 明朝の冊封体制とその様態, *Shigaku zasshi* 史学雑誌, vol. 104 (1995), pp. 37–72, esp. pp. 42–44.

42 Morris Rossabi, 'The Tea and Horse Trade with Inner Asia during the Ming', *Journal of Asian History*, vol. 4, no. 2 (1970), pp. 136–168; Tani Mitsutaka, 'A Study on Horse Administration in the Ming Period', *Acta Asiatica*, vol. 21 (1971), pp. 73–98, esp. pp. 73–79.

43 Ejima Hisao 江島寿雄, 'Ryōtō bashi kigen' 遼東馬市起源, *Tōyō shigaku* 東洋史学, vol. 9 (1954), reprinted in Ejima Hisao, *Mindai Shinsho no Jochokushi kenkyū* 明代清初の女直史研究 (Fukuoka: Chūgoku shoten, 1999), p. 217. Ejima notes that late-sixteenth-century government records indicate a total of approximately 420,000 imperial horses empire-wide and that the mid-sixteenth century *Liaodong Gazetteer* gives a total of just more than 55,000 horses maintained by the Liaodong Regional Commission (p. 217). At its height under Yongle, the Ming state maintained an estimated 1.7 million horses. In 1389, the Ming state established the Three Garrisons for Mongol chiefs on the eastern slopes of Khinggan mountain range; they were composed of the Uriyanghan clan, the Mongolized Tungusic Üjiyed and the Ongni'ud (so named because they were under the rule of two princes of Chinggisid descent appointed by the Ming). See Christopher Atwood, *Encyclopedia of Mongolia and the Mongol Empire* (New York: Facts on File, 2004), p. 535.

44 Cong Peiyuan, *Mingdai Dongbeibian xu*, vol. 4, pp. 1063–1090, esp. p. 1072.

45 Hasumi Moriyoshi 荷見守義, 'Mindai Ryōtō bashi tōankō' 明代遼東馬市档案考, *Jinbunken kiyō* 人文研紀要, vol. 44 (2002), pp. 27–59; Hasumi Moriyoshi 荷見守義, 'Mindai Ryōtō bashi chūginkō' 明代遼東馬市抽銀考, *Jinbun shakai ronsō jinbun kagakuhen* 人文社会論叢人文科学篇, vol. 8 (2002), pp. 1–18.

46 Lin Yanqing 林延清, 'Lun Mingdai Liaodong mashi cong guanshi dao minshi de zhuanbian' 論明代遼東馬市從官市到民市的轉變, *Minzu yanjiu* 民族研究, vol. 4 (1983), pp. 50–57.

47 Tian Jing 田靜, 'Mingdai Liaodong de mashi maoyi' 明代遼東的馬市貿易, *Shixue yuekan* 史學月刊, vol. 6 (1960), pp. 11–17, esp. pp. 15–16; Cong Peiyuan, *Mingdai Dongbeibian xu*, pp. 1082–1085.

48 Iwai Shigeki, 'China's Frontier Society in the Sixteenth and Seventeenth Centuries', *Acta Asiatica*, vol. 88 (2005), pp. 1–20.

49 David Robinson, 'Military Labor Markets in China, *c.*1500', in Erik-Jan Zürcher (ed.), *Fighting for a Living: A Comparative History of Military Labour in Europe and Asia, 1500–2000* (Amsterdam: Amsterdam University Press, 2014), pp. 41–77.

50 Focused on the early fifteenth century but useful for the Hongwu reign is Henry Serruys, *Sino-Jürced Relations during the Yung-lo Period (1403–1424)* (Wiesbaden: Harrassowitz, 1955). Henry Serruys's vast body of work on Mongols in the Ming is still essential a half century and more after its publication. A sampling of his work includes: 'The Mongols in China during the Hung-wu Period', *Mélanges chinois et*

bouddhiques, vol. 11 (1959); 'Land Grants to the Mongols in China, 1400–1460', *Monumenta Serica*, vol. 25 (1966), pp. 394–405; 'Were the Ming Against the Mongols' Settling in North China?' *Oriens Extremus*, vol. 6, no. 2 (1959), pp. 131–159; 'A Mongol Settlement in North China at the end of the 16th Century', *Central Asiatic Journal*, vol. 4 (1959), pp. 237–278; 'The Mongols of Kansu during the Ming', *Mélanges chinois et bouddhiques*, vol. 10 (1955), pp. 215–346; 'Mongols Ennobled during the Early Ming', *Harvard Journal of Asiatic Studies*, vol. 22 (1959), pp. 209–260.

51 *Ming Taizu shilu*, 118.4b 洪武十一年四月戊寅.

52 *Ming Taizu shilu*, 119.1b 洪武十一年六月辛丑.

53 *Ming Taizu shilu*, 144.4b 洪武十五年四月辛丑.

54 For further examples, see *Ming Taizu shilu*, 132.3b 洪武十三年六月癸酉, 137.3b 洪武十四年四月辛巳, 138.4b 洪武十四年七月癸卯, 143.3b 洪武十五年閏二月乙卯.

55 *Ming Taizu shilu*, 193.6a 洪武二十一年九月乙酉.

56 Kawachi Yoshihiro 河内良弘, 'Mindai Ryōyō no Tōneiei ni tsuite' 明代遼陽の東寧衛について, *Tōyōshi kenkyū* 東洋史研究, vol. 44, no. 4 (1986), pp. 693–697; Sun Chunri 孫春日, 'Mingdai Chaoxianren zai Dongningwei juju zhuangkuang de lishi kaocha' 明代朝鮮人在東寧衛聚居狀況的歷史考查, in Li Xun (ed.), *Mingshi lunji* 明史論集 (Changchun: Jilin wenshi chubanshe, 1993), pp. 462–475; Dong WanX 董萬X, 'Mingdai Sanwanwei chushedi yanjiu' 明代三萬衛初設地研究, *Beifeng yanjiu* 北方文物, vol. 39, no. 3 (1994), pp. 52–57.

57 Chosŏn observers often remarked that Ming (or at least Dongning) taxes and service obligations were lower than at home, which probably explains Liaodong's appeal for fifteenth-century Korean immigrants (Sun Chunri 孫春日, 'Mingdai Chaoxianren', p. 468).

58 *Ming Taizong shilu*, 78.1b 永樂六年四月戊子.

59 *Sanwan wei* 三萬衛, *Wuxuanbu* 武選簿 14, Ming Archives held at the First Historical Archives, Beijing. Using different material, Kawagoe Yasuhiro 川越泰博 has argued that Jurchens constituted more than 70 per cent of the upper ranks of the officer corps in the Three Tümens Garrison (Kawagoe, 'Mindai Jochoku gunkankō josetsu' 明代女直軍官考序説, *Shien* 史淵, vol. 38, no. 1–2 (1977), pp. 1–24, 4). As Qi Wenying 奇文瑛 has demonstrated, such aggregate numbers mask important differences in status among the Jurchens of the Three Tümens Garrison ('Lun *San wan wei xuan bu* zhong de junji nuzhen' 論《三萬衛選簿》中的軍籍女真, *Xuexi yu tansuo* 學習與探索, vol. 5 (2007), pp. 205–210).

60 Kenneth Robinson, 'From Raiders to Traders: Border Security and Border Control in Early Chosŏn, 1392–1450', *Korean Studies*, vol. 16 (1992), pp. 94–115.

61 Kawachi Yoshihiro, 'Mindai Ryōyō no Tōneiei ni tsuite'; 'Minchō no sakuhō taisei to sono yōtai', pp. 48–52.

62 *Sejo taewang sillok* 世祖大王實錄, 世祖五年二月己巳, 世祖五年四月己未 in Wu Han 吳晗, *Chaoxian Lichao shilu zhong de Zhongguo shiliao* 朝鮮李朝實錄中的中國史料 (Beijing: Xinhua shudian, 1980), ch. 8, vol. 2, pp. 507, 508).

63 Extant Ming records such as the Veritable Records understate tensions between the Ming and Chosŏn courts, especially from the mid-fifteenth century onward (David Robinson, 'Korean Lobbying at the Ming Court: King Chungjong's Usurpation of 1506', *Ming Studies*, vol. 41 (1999), pp. 37–53). For a similar articulation based on early eighteenth-century Qing–Chosŏn relations, see Andre Schmid, 'Tributary Relations and the Qing-Chosŏn Frontier on Mount Paektu', in Diana Lary (ed.), *The Chinese State at the Borders* (Vancouver and Toronto: University of British Columbia Press, 2007), pp. 126–50.

64 *Ming Yingzong shilu*, 320.3b 天順四年十月乙卯.

65 The Ming court asked that the Chosŏn government turn over the wife of one of those killed to Liaodong authorities so that she might be reunited with her son. The Ming court argued that this might defuse an explosive situation and demands for retribution (*Ming Yingzong shilu*, 314.6a–b 天順四年四月甲戌).

66 *Sejo taewang sillok*, 世祖六年閏十一月戊午 in Wu, *Chaoxian Lichao shilu zhong de Zhongguo shiliao*, ch. 8, vol. 2, pp. 532–533.

67 Yu Qian, 'Bing bu wei jian yan shi' 兵部為建言事, in Yu Qian, *Yu Qian ji* 于謙集 (Beijing: Zhongguo wenshi chubanshe, 2000), p. 245.

4 Elephants in pre-modern India

Debraj Chakraborty

Introduction

The association of man and elephant has been described by ecologist Raman Sukumar in his book entitled *The Asian Elephant: Ecology and Management*:

> An object of worship, a target of hunters, a beast of burden, a burden to the people, gentle in captivity, dangerous in the wild, the pride of kings, the companion of mahouts, a machine of war, an envoy of peace, loved, feared and hated.[1]

It can be assumed when man was only hunter–gatherer they used meat of elephant as food and subsequently their bones and ivory had been utilized as weapons. It is well evident that elephant hunting was a common practice from the Neolithic/Chalcolithic period.[2] It is estimated by zoologists that there were probably 17 species of elephants in ancient times, but now the number is limited to two species of elephant, one is the Asian elephant (*Elephans maximus*) and another is the African elephant (*Loxodonta africana*).[3] During the Neolithic period agricultural activities and animal husbandry were both considered as the most significant ways of livelihood, though hunting–gathering was still considered common practice. The earliest reference relating to man–elephant interaction comes to us from the cave paintings of Bhimbetka in present Madhya Pradesh. Most of the paintings depict scenes of elephants with riders, sometimes they are carrying weapons.[4] As this chapter will show, the elephant was an integral part of the imperial armies throughout pre-modern Indian history.

Elephants in pre-Mughal India

It has been considered by scholars that the earliest evidence of domestication of the elephant appears to us from the Harappan cultural zone. The elephant motif is depicted in the discovered Harappan seals. Possibly, the elephant was tamed and engaged in various activities during the mature Harappan phase.[5]

The introduction of Aryan settlements has been treated as a landmark in Indian history. The Aryans were familiar with a wide range of animals and the elephant appeared in the Vedic texts in different names such as *mriga*, *hastin*, *ibha* and *varana*.[6] In the *Rigveda* the term used for elephant was *mriga-hastin*, which means a beast with hand. On the other hand the term *mirga* is also found in the later works on elephants such as *Gajasastra*, which means 'a particular breed of elephant'. A hymn to Soma (the moon god) has described him as 'the king of the elephant'. It clearly indicates that he possessed a good number of elephants with him. Even hunting of wild elephant was a common practice in those days. Evidence relating to the use of the elephant appears to us from the hymns of the *Atharvaveda* when it clearly writes about 'taming of a wild elephant'. Most likely the elephant was used for riding purposes during that period as a verse of the *Atharvaveda* mentions that the elephant has become chief of all the pleasant beasts to ride. During the later Vedic phase, maintenance of the elephants emerged as a regular profession. The *Yajurveda* mentions a professional class called *hastipa* or 'elephant-keeper'. In the *Upanishads*, the elephant has been described as the wealth of state and also as a symbol of power, prestige and glory. The term *chaturanga* used in the *Rigveda* means 'human body' but in a later period this term was commonly used to denote a fourfold division of the army in ancient India, namely, foot soldiers, cavalry, chariots and elephants.[7] By the later Vedic period elephants were engaged in warfare regularly. During the period of epics, the elephant was an essential object to the kings' establishments and it was an indispensable part of a royal army.[8]

The sixth century BCE is an important phase in the political history of India because the process of state formation was initiated.[9] The rulers became passionately attached to the elephants and usually elephants were captured from the different forests under their control. The elephant forests were especially prized, for ivory was valuable and victory in the battle depended mainly on the elephant wing of the army. The rulers had great numbers of war-elephants in their army. It is reported that the Kurus of the *Mahabharata* had around 30,000 elephants in stable and these were used not only for war, but also for hunting and ceremonial rides.[10]

The elephant became the most important arm of the Indian armies about the time of Alexander's invasion in 326 BCE. In the course of battle between Alexander and Porus (Paurava), elephants were used in a decisive manner. And in the battlefield, Porus deployed the elephants in the front line. The fighters were placed on their *howdas*. Jadunath Sarkar writes that Porus had 200 war-chariots with him and these were posted in two equal divisions of 100 each on his right and left.[11] Contemporary literary evidence testifies the fact that the Nandas, the predecessor of the Mauryas, had around 3,000 elephants at their own establishment for the purpose of war.[12] Undoubtedly, the elephant became an integral part of royal life and statecraft.

Kautilya's *Arthasastra* confirms perhaps the first geographical locations of the distribution of wild elephants in the forests of the subcontinent. Supplies of elephants may have come largely from hill forested areas. Some of the forests were strictly protected for the elephants and commenting on that subject *Arthasastra* writes, 'on the border of the kingdom he shall establish a forest for elephants, protected by other forests and under the supervision of the Chief Elephant Forester'.[13] The *Arthasastra* has referred to as many as eight *hasti-vanas* (forests for elephants) and their accurate locations. The eastern Gangetic plain was the natural habitat of wild elephants, and rulers in these areas have enjoyed monopoly rights over it. The *Arthasastra* equally gives a detail insight about the royal stables. Trained staffs with specific duties were appointed at the imperial stables to look after the elephants. Some elephants were given interesting pet names. Proper maintenance and health care of the royal and war elephants was taken. Doctors were appointed at their service.[14]

Even in the post-Mauryan period the elephant remained a symbol of power, glory and strength of the army. Under the imperial Guptas (CE 300 to 550), war elephants continued to be used as an important part of the army. In his *Raghuvansham*, Kalidasa has mentioned about the mighty elephantry of the rulers of Kalinga (Orissa) and Anga (Bihar).[15] Another literary personality, Vakpati, refers to the king of Vanga (Bengal), who had a large number of war elephants.[16] *Brithat Samhita* of Varamihira, an encyclopedic work composed during the CE sixth century contributes a short chapter classification of the elephant as well as treatment of them.[17]

The elephant was not only used as a war machine in the ancient period, but it was also related to the culture and sentiment of the people. The worship of the elephant as a form of Lord Ganesha (elephant god) originated in the CE third or fourth century and it attained special importance particularly in the south of India during the CE seventh century at the time of the Pallava rulers. The scholars have often called this culture that developed in south India as the 'elephant culture'. The Tamil people developed expertise in capturing elephants and subsequently in training them for use in battle. Under the different dynastic rule in south India through the ages, elephants were used not only as an effective instrument in war, but elephants were also a part of socio-cultural life. Numerous references in this context are available in the Sangama literary works as it provides glimpses into its distribution, ecology and the behaviour of this animal. The Western Chalukya ruler Somadeva III (r. 1126–1138) composed an encyclopedic work entitled *Manasollasa* or *Abhilashitarthachintamani*, which consists of detailed observational information on the elephant. It provides interesting information about their characteristics, natural habitat, method of trapping, preservation and training of these beasts. The elephant was used in two ways; it was considered as a war machine, and, also, as a beast of burden.[18]

The establishment of the Delhi Sultanate in the early thirteenth century was undoubtly considered as a landmark in the military history of India. There is a general idea among historians for long time that medieval society was nothing but a static society. It was considered a period which lacked rational thought and scientific bent of mind. But the works of Irfan Habib and some other notable experts in this field have used the contemporary sources to denote technological progress achieved during this phase.[19] The reason behind the victory of the Turkish-Islamic armies over the Hindu-Rajput rulers was due to their superior military skill and technology.

The Sultans of Delhi paid noteworthy attention for maintaining and preserving of animals especially those required for the purpose of war. So they took interest in the elephant. But, it is often argued by recent scholars that cavalry received more importance in comparison to elephantry in this phase.[20] However, contemporary chronicles show that large numbers of war elephants (*pil-i jangi*) were deployed in the battlefields under the Sultanate rule. The Delhi Sultans were fond of the elephant and naturally they had a large number of elephants at their imperial stables (*pil-khana*). Simon Digby writes, 'The princes of India, including the Delhi Sultans, put their faith in elephants in warfare; besides which the actual possession of elephants was an important symbol of royalty and independent power.'[21] The rulers were interested in increasing the number of elephants in their stable as it was considered both as a prestige symbol and constituted the backbone of the army.

Sultan Giyasuddin Balban had a fascination for elephants. He considered that 'five hundred horsemen were equal to one elephant'. Commenting on the use and necessity of the elephant in medieval war Simon Digby writes that elephants were principally esteemed and treasured in medieval India for their performance in the battlefields. The elephants were also used as living battering rams to break open the gates of fortresses. Further, elephants also served as pack animals carrying large loads.[22] Though, Barani has pointed out both the advantages and the disadvantages of the employment of the elephant in war.[23]

Elephants were also used in conducting warfare outside India. The Ghaznavid Sultanate had engaged a considerable number of elephants for the purpose of war. Sultan Mahmud had 1,300 elephants in CE 1023–1024 while this number again increased in CE 1031 when his son and successor Masud was in power. The elephants under him numbered 1,670.[24]

It is said that among the rulers of the Delhi Sultanate, Sultan Alauddin Khilji had great interest in the elephant. Zia-ud-Din Barani has mentioned in his account that there were around 1,500 elephants in his *pil-khana*.[25] The number was almost double during the reign of Sultan Muhammad Tughluq in around the 1340s, as Al-Umari wrote, 'The Sultan of Delhi has 3,000 elephants of different kinds and sizes, and for the expenditure on their food the revenue of the large kingdom would hardly suffice.'[26] But, the figure reported by Al-Umari is to some extent exaggeration of the accurate number. Sultan Muhammad Tughluq had 750 war elephants in his establishment. Sultan Firuz Tughluq organized two major campaigns against Bengal and Sindh and most likely he employed 470 to 480 war elephants. Digby writes that a maximum number of from 750 to 1,000 war elephants in the possession of the Delhi Sultans in the period of their greatest power in the early fourteenth century appears not unlikely. The number of elephants in the imperial *pil-khana* (elephant stable) began to decline after the 1350s and at the time of Firuz Shah it was only 500. The figure reduced to 120 by the end of the fourteenth century.[27]

One needs to enquire about the source or supply line of these elephants. Sind, Orissa and Bengal were the places where elephants were found in great numbers. The rulers during the Sultanate period attacked these places in order to get elephants. The elephants used to come to the imperial service in two ways, first, as a tribute or booty and, second, wild elephant trapping from different parts of the country. It is said that during his expedition against the ruler of Jajnagar (in Orissa) in 1281–1282, Sultan Balban killed four elephants in war and captured 20 elephants and these were brought to Delhi with him. When Sultan Firuz Shah made his campaign against the ruler of Bengal, he captured 47 war elephants and out of this 44 were dispatched to Delhi as war booty.[28] The author of *Tarikh-i Mubarakshahi*, Abdullah Sirhindi, informs us that in 1359 Sultan Firuz Shah received five elephants from the ruler of Bengal as tribute. Similarly Sultan Firuz Shah received 25 elephants as a tribute from the ruler of Jajnagar.[29]

During the thirteen and fourteenth centuries, different parts of northern India were full of wild elephants. The author of the *Tarik-i Firuzshahi*, Afif, has mentioned an elephant trapping that was conducted by Sultan Firuz Shah during his departure from Jaunpur to Jajnagar. Afif narrates this incident and technique of trapping elephants in the following words:

> Having come to know of the elephants, the Sultan with great majesty, power and manliness, marched into the forest. It was known that the elephants were located about fifteen *kos* inside the forest. As ordered by the Sultan, all the workers, and the army men, along with *Maliks* and *Khans* and courtiers surrounded the forest from all the four sides. A wooden circle was raised all along the forest camp. Except two exit routes, all the ground was sealed with mud. The wooden circle so erected was inspected by the Sultan every day. A few blood-thirsty elephants along with their drivers (*Mahawat*) were brought in. They were directed to enter the forest from one side and from the other side, band-masters and drum-beaters were pushed in to make excessive noise and alarm though their drum-beating and playing of the bands. The elephants came out of the hiding and ran towards forest. It is reported that the elephants were so mighty that during their fight, several trees of the jungle were uprooted or fell down. The wild elephants arrived at the end of the forest near the wooden circle when the troopers beat the drums with pitched tones. The elephants ran from one end to the other until they were badly exhausted within a few days. No fodder was made available during the pursuit. The brave elephant drivers jumped over their backs and tamed these elephants. Thus these elephants were hunted down and captured.[30]

Elephants in Mughal India

It is said that the introduction of gunpowder in war reduced the effectiveness and necessity of the elephants in war. Elephants were used in direct battle lines along with explosive weapons (incendiary weapons like rockets, etc.) during the fourteenth century. But, the sixteenth century is the landmark in Indian military history as artillery and hand guns came into use in a large way in war for the first time. It was Babur (r. 1526–1530, in Hindustan), who fought the first battle of

Panipat in CE 1526 against Sultan Ibrahim Lodi and used field artillery and hand-held firearms. In this connection it is important to note that the elephant was still an important arm in war. In his autobiographical writings Babur records that Lodi fielded a much larger army than his, including 'nearly a thousand elephants'.[31]

The Mughal rulers were fond of the elephant. Contemporary chronicles, memoirs of the rulers and some visuals clearly represent the fondness of the Mughal emperors for these beasts. Like the sultans of Delhi, the Mughals continued the tradition of maintaining large numbers of elephants in the imperial establishment. Elephants occupied an important position both in civil and military life in Mughal India. Elephants were significant in all affairs such as battles, pompous court life and ceremonies as well as hunting and sport. They were widely used for transport and communication and had become an important part of court life. The efficiency and strength of the army to a great extent depended on elephantry throughout the early modern period. The Mughals learnt and applied the conventional Indian art of elephant trapping.[32]

Babur was a keen lover of nature. His graphic description on the wildlife of India in his memoirs helps us to form an idea of the natural history of the pre-colonial period. Babur was impressed by the huge elephants, their sharp intelligence, faithful nature and indispensability in the battlefield. He found them useful to carry heavy loads across large rivers, thick jungles and mountainous terrain. Big elephants were found in Eastern Kalpi (eastern part of Uttar Pradesh) that was covered with dense forest. Babur writes in his autobiography:

> The elephant which Hindustanis call *hat(h)i*, is one of the wild animals peculiar to Hindustan.... The elephant is an immense animal and very sagacious. If people speak to it, it understands; if they command anything from it, it does it.[33]

He further writes,

> Its value according to its size; it is sold by measures (*qarilab*); the larger it is, the higher its price. People rumour that it is heard of in some islands as 10 *qari* high, but in this tract it is not seen above 4 or 5.... It eats and drinks entirely with its trunk; if it lose the trunk, it cannot live ... one elephant eats corn (*bughuz*) of two strings (*qitar*) of camels.... It has two great teeth (tusks) in its upper jaw, one on each side of its trunk; by setting these against walls and trees ... people call these ivory.... The elephant has no hair ... its stomach is large.[34]

A painting entitled *Elephants by the River* from the *Baburnama* depicts a beautiful elephant herd. Presently this painting is preserved in the Aga Khan Museum, Toronto in Canada. It was drawn by Kanha and painted by Ikhlas. They were the master painters at the court of Akbar and it was dated *c*.1589. The painting reveals naturalistic representation of the elephant's figure and it equally shows similarity with the description that was offered by Babur in his memoirs. In this painting the female elephant is represented somewhat in the background keeping watchful eyes, with her ears turned and trunk curled up. On the other hand the tusker nudges with its front left foot the young one as he moves towards the water hole as if instructing him firmly but with affection and care.[35]

Zain Khan's *Tabaqat-i Baburi* also gives us a graphic description of India's wildlife. About elephants he writes,

> Some put forward a rumour before His Majesty that an elephant had been seen in a certain jungle or islands which measured so high ten yards from its pillar-like feet up to the roofs of its elevated chamber of its backs.... But what came up into the view of His Majesty, the *Nawab*, after the conquests of these tracts, and what the people said in the stables of the ruler and nobles of His Majesty, they were not higher than five or six yards.[36]

Humayun (r. 1530–1540/1555–1556), the son and successor of Babur, was interested in nature and wildlife. He was fascinated in the elephant as it is told that he had captured eight elephants from Ibrahim Lodi's advance force, prior to the First Battle of Panipat in 1526. The elephants were deployed as important war machines under the Sher Shah Suri, an Afghan chief and contender of Humayun. Jouher, the author of *Tezkereh-al Vakiat* (also known as *Memoirs of Humayun*), has mentioned that Khaus Khan, one of the army chiefs under Sher Shah assured him (Sher Shah) victory against the Mughals 'if he would give him a detachment of good soldiers and a number of war elephants, he would attempt it, and exert himself to the utmost'.[37] Abbas Khan Sarwani, author of the *Tarikh-i-Sher Shahi* has mentioned that there were around 5,000 elephants at the stable of Sher Shah.[38] Hemu, the leader of the Turkish-Afghan army under Adil Shah fought against the Mughal governor of Delhi just before the Second Battle of Panipat in 1556, was reported to be leading 1,000 elephants, 50,000 horses, 51 cannon and 500 falconets.[39]

Akbar's (r. 1556–1605) interest in the elephant is well known as his official historian Abul Fazl in his *Ain-i Akbari* has offered detailed descriptions of the elephants. Abul Fazl writes, 'This wonderful animal is in bulk and strength like a mountain; and in courage and ferocity like a lion.'[40] This animal was considered as an indispensable part of the army and provided strength and glory to the army. Abul Fazl continues,

> It adds materially to the pomp of a king. Experienced men of Hindustan put the value of a good elephant equal to five hundred horses.... When guided by a few bold men armed with matchlocks, such an elephant alone is worth double that number.[41]

Abul Fazl mentions various breeds and as well as price of the elephants. He writes, 'The price of an elephant varies from a lac to one hundred rupees; elephants worth five thousand, and ten thousand rupees, are pretty common.'[42]

Abul Fazl mentions four categories of elephants like *Bhaddar*, *Mand*, *Mirg* and *Mir*. This classification was probably based on physical appearance and character of the elephants. *Bhaddar* was long in height, well proportioned and brave with 'an erect head, a broad chest, large ears, a long tail, and [it] is bold, and [it] can bear fatigue'.[43] *Mand*, was wild, black and unmanageable; it had yellow eyes. *Mirg* had the whitest skin with black spots and its eyes are a 'mixture of red, yellow, black and white'. The last one called *Mir* had a small head and obeys readily. In India, the skin of the elephants is generally black and grey. Abul Fazl has also recorded their habits, diet, mating and gestation period.[44] Abul Fazl notes different stages of an elephant's life as well as changes in appearance. A five year old calf is called a *bal* (or child) when ten it is known as *put*, at the age of 20 it is called *bikka* and at 30 *kalba*.[45]

Each category of elephants had been further subdivided into three broad classes, i.e. large, middle and small. Abul Fazl also mentions the food-habit, size, physical structure, upkeep and kinds of staff appointed to train the elephants.[46] Commenting on the quantity of food consumed by an elephant, G.P. Sanderson, a renowned hunter and chronicler of British Indian sport, found that a full grown elephant consumes between 600 and 700 lbs of fodder per day.[47]

Akbar had 101 elephants for his personal use. Each elephant had been given a name, an exalted position, trained staff for its proper upkeep and even some special diet. They were used mostly for ceremonial rides, hunting expeditions, elephant fights and for warfare. Some of the royal elephants were given the names Girdbaz, Dhokar, Madhukar, Jangia, Hawai, Bhirun, Gajpati, Ran Bagha, Sabdaliya and Kadirs.[48] Writing about Jangia Abul Fazl says,

> The elephant Jangia displayed great deeds. One of them was that a Rajput ran and struck his trunk with his sword and cut it off. Though his trunk was severed, which makes life difficult.... It had killed thirty distinguished men before it was wounded, and it slew fifteen afterwards.[49]

Adventure-loving and fearless Akbar could ride the most fierce and deadly elephant even when the *mahout* declined to ride. Abul Fazl writes, 'he rode more than a hundred times on *mast* elephants which had killed their drivers and were men-slayers, and were capable of smiting a city or perturbing an army and engaged them in fighting'.[50] About Hawai, Abul Fazl writes,

> The elephant Hawai was a mighty animal and reckoned among the special elephants. In colour, passion, fierceness and wickedness he was a match for the world. Strong and experienced drivers who had spent a long life in riding similar elephants mounted him with difficulty.... One day without hesitation Akbar mounted this elephant, in the very height of its ferocity.... After that he pitted him against the elephant *Ran Bagha* which nearly approached him in his qualities.... The lion-hearted Shahinshah calmly went on with his terrifying pursuit until the elephant *Hawai* by the strength

> of a hidden arm, and the divine fortune, got the victory over his opponent. *Ran Bagha* let fall the strong cable of steadfastness and turned to flee. *Hawai* looked neither behind nor before and disregarded highlights and hollows and went like the wind in pursuit of the fugitive.... Owing to the great weight of those two mountain forms the pontoons were sometimes submerged and sometimes lifted up.[51]

It is interesting to know that the elephant on which Hemu rode during the Battle of Second Panipat (1556), was none other than the *Hawai* of Akbar.[52]

Akbar's personal elephants were called *khasa*, i.e. special elephants. A number of servants were appointed for the proper maintenance of these *khasa* elephants. A special superintendent called *Dahaidar* was appointed to look after the stable of the *khasa* elephants.[53] The *Ain-i Akbari* also informs us about the different types of servants and keepers appointed at the royal stable for upkeep and training of the elephants.[54]

Akbar was extremely fond of elephant hunting. He took vigorous part in an elephant catching venture at Sipri, a place between Mandu and Agra. He undertook several expeditions in the dense jungles of Malwa and adjoining regions to capture wild elephants. During his elephant trapping foray at Sipri he was impressed by a wild *mast* (intoxicated, wild) young elephant from a herd of 70 elephants. Akbar called his trained elephant, *Bhirun* or *Ran Bhirun* to fight with it in order to exhaust this wild elephant, when all attempts at capturing it had failed. Ultimately, the wild elephant was captured and brought into the fort of Sipri. Interestingly this male elephant broke the strong stone wall of the fort and ran away. It was, however, later recaptured.[55]

Elephant fighting was a favourite pastime for the Mughals. Akbar enjoyed elephant fighting when he went to Nagaur. Niccolao Manucci says,

> When the king makes them (the elephants) fight, the wives of the drivers remove their ornaments, smash their bracelets, and put on mourning, just as ... they were widows. If their husbands come back alive they give a great feast, just as if newly married; for in these encounters and combats the drivers put their lives in great jeopardy.[56]

Training was an important exercise for the war elephant. Writing about the purpose of elephant training Abul Fazl mentions, 'Elephants were trained to be bold, and to stand firm at the sight of fire and at the noise of artillery; and he [*mahawat*] was responsible for their behavior in this respect.'[57] Different ornaments and trappings were used to control harness the royal elephants.[58]

Jahangir maintained a large *pil khana* (elephant stable) containing many elephants collected from far and near. He himself took part in an elephant-trapping venture near Dohad during his travels to Gujarat and central India and captured many elephants. The number of captive elephants in the subcontinent perhaps reached its highest level during the time of Jahangir. He also noticed the behaviour of the elephants. He says,

> From the day on which the elephant Nur-bakht was put into the special elephant stables, there was an order that he should be tied up in the public place (court). Among animals elephants have the greatest liking for water; they delight to go into the water, not-withstanding the winter and the coldness of the air, and if there should be no water into which they can go, they will take it from a water-bag (*mashk*) with their trunks and pour it over their bodies. It occurred to me that however much an elephant delights in water, and it is suited to their temperament, yet in the winter the cold water must affect them.[59]

Jahangir had an inquisitive mind and strong observational power. In his memoirs Jahangir has offered a description on elephants, which were known for their size. He writes,

> In the elephant-stables of H.M. Akbar the largest elephant I saw was Durjan Sal. It was long the premier elephant. Its height was 4 yards (*dara*), and 3½ quarters of the *Ilahi gaz*, which is 8 yards and 3 fingers of the ordinary *gaz*.[60]

He further provides information about another special elephant, which was in his stable. Jahangir writes,

> At present, among the elephants of my establishment the largest athlete is 'Alam Gajraj', which H.M. Akbar himself had caught. It is the chief of my special elephants. Its height is 4⅛ yards, or 7 yards and 7 fingers of the ordinary yard.

Therefore, from this statement it can be calculated that the height of Alam-Gajraj was about 11 ft and Durjan Sal was about 12¼ ft.[61] The Mughals were also aware about the variety of African elephants. Jahangir had two African elephants in his royal establishment. Two images of his African elephants have come to our notice.[62]

The elephant was thus important for court ceremony and in the battlefield under the Mughals. In fact it was an integral part of the Mughal royal life. So the rulers and their nobles paid noteworthy attention for the proper upkeep and preservation of this huge animal. The Sultans of Delhi kept a good number of elephants in their royal establishments. Around 750 to 1,000 elephants were maintained by the Delhi Sultans during the early fourteenth century for military campaigns. This number declined to 500 during the reign of Firuz Shah Tughlaq. The size of the *pil khana* increased sharply under the Mughals. Akbar maintained 5,000 elephants. In addition, his nobles had 7,709 elephants. Abul Fazl informs that there were nearly 2,000 elephants in the establishments of the zamindars and the highest number (possibly 1,170) belonged to the zamindars of Bengal. Abu'l Fazl has estimated that there were around 14,709 elephants at the Mughal establishments and 25,000 in the entire empire during the last part of the sixteenth century. It is notable that Abul Fazl's statistics did not cover Assam and Peninsular India. These two areas were full of wild elephants. The author of *Alamgirnama*, Muhammad Kazim reports that the ruler of Assam alone possessed 1,000 elephants. It has been estimated that around 500 to 600 elephants were caught only in Assam every year. Elephants were supplied to Delhi mainly from Bengal, Assam and Malwa where the animal was found in huge numbers.[63] On the basis of statistical data provided by Abul Fazl, scholars like Shireen Moosvi have estimated that the number of elephants in the Mughal stables increased from 14,709 to 25,000 at the early seventeenth century. This number again increased during Jahangir, Shahjahan and Aurangzeb's time and finally reached 40,000.[64] Irfan Habib has focused on the 'Great Central Indian Forest zone' which was the most prominent natural habitat of wild elephants.[65] When the forest area shrank due to cutting of jungles for human settlement the supply of elephants was also reduced in great numbers. *Mirat-i Ahmadi* notes that sometime before 1761 wild elephants no longer reached Rajpipla from other areas because the forests between Rajpipla and Malwa had been cut off due to expansion of human settlement in those areas.[66]

The average price of an elephant during the time of Akbar was Rs2,500. Hakim Abu'l Fath Gilani, a noble under Akbar could expect to buy a *tusker* at price of Rs2,000 or 3,000. During Shahjahan's time the ruler of Golkunda Abdullah Qutub Shah wrote to his agent at the imperial court that superior Indian elephants were valued at rupees one lakh and they was no longer available at half or third of that price. The elephants that were brought from South-East Asia were costlier.[67]

Several methods were adopted to catch wild elephants. Abul Fazl records four methods of elephant hunting. They were: *Kheda*, *Chor Kheda*, *Gad* and *Bar*. *Kheda* was the very old and simple practice to trap an elephant. Abul Fazl records this technique in detail. According to this process drums were beaten inside the jungle and which made the elephants quite frightened. Generally, they started to rush about and it made them tired. They usually took shelter under a tree and at this time some experienced hunters threw a rope and soon after that they tied up the elephant with the tree. Abul Fazl has also recorded information about the wages of the hunters. He writes, 'One-fourth of the value of an elephant thus caught is given to the hunters as wages.'[68] *Chor Kheda* was another technique of elephant hunting in Mughal India. In this process a hunter usually sat on a trained female elephant and stretched himself secretly towards the wild elephant. When the female elephant commenced fighting with the wild elephant the hunter immediately threw a rope around the foot of the wild elephant and thus managed to capture it. *Gad* was the most common and popular practice of elephant trapping and even today it is followed by the people in India. In accordance with this technique a large pit was dug and covered with grass. When the elephant came close to it, the hunters used to commence great noise. This loud noise made the elephant quite puzzled and ultimately it fell into the pit. The fourth technique was called as *Bar*. In this process generally the hunters dug a ditch around the resting place of elephants, and left some level ground for allowing the wild elephant to come out. A gate was suspended with a rope was put at the level open space. Then, the hunters placed some favourite food at the door. As soon as the elephant reached near the door to take food the rope was cut and the door was closed, and finally the wild elephant was held captive. These were some conventional techniques.[69] Abul Fazl informs us that Akbar employed a new technique to trap the elephants. He writes,

> His Majesty has invented a new manner, which admits of remarkable *finesse*. In fact, all excellent modes of hunting are inventions of His Majesty. A wild herd of elephants is surrounded on three sides by drivers, one side alone being left open. At this place several female elephants are stationed. From all sides, male elephants will approach to cover the females. The latter then go gradually into an enclosure, whither the males follow. They are now caught.[70]

Jahangir has, in fact, left a detailed description about hunting in his autobiography, *Tuzuk-i Jahangiri*. Jahangir records his elephant hunting expedition which took place near Dohad in Gujarat where he had encamped in 1618 in the course of a military expedition. The hunt began in a *qamargah* and with the help of some trained elephants two wonderful male elephants were captured. He writes, 'Although many elephants were not caught, at least two excellent ones were captured, very handsome in shape, of good breed, and perfect marks.' These elephants were named as *Ravan Sar* and *Pavan Sar*.[71]

Aurangzeb was equally interested in elephant hunting. He had a large number of elephants at the imperial stable. Some detail about these elephants is available in the travelogue of Niccolao Manucci, a Venetian traveller to India during the reign of Emperors Shahjahan (r. 1627–1658) and Aurangzeb (r. 1658–1707). Manucci has recorded the names of chief elephants maintained in the royal stable under Aurangzeb who gave some interesting names like Dil-Pasand, Madan-Mast, Latif, Fateh-Mabarak, Fateh-Jung, Nur and Lal to his elephants.[72] Manucci also mentions that the 'king gives these names according to his fancy or some aptitude he detects in these animals'.[73] Some experienced persons were appointed to look after these elephants at the royal stable. It is reported by Manucci that 'Usually the king has one hundred very tall elephants which he uses himself.'[74] The imperial elephants were well trained and Manucci wrote about this aspect as, 'All the elephants he rides are trained to stand fire of both artillery and musketry, of rockets, and other fireworks.'[75] Besides that elephants were specifically trained for hunting also. There is a delightful description by Manucci about the training of elephants for hunting lion and tigers: he writes,

> Others are taught not to be frightened of tiger or of lions, so that they may be used in hunting. To teach them, they take a tiger-skin or lion-skin, and stuff it with straw. Then, just as if it were alive, they move it here and there by a rope. The driver encourages the elephant, and urges him towards the dummy, which with feet and trunk he tears to pieces. The elephants are well looked after; they are given spirits to drink to increase their courage in a fight.[76]

Manucci further informs us that apart from these elephants, another 300 to 400 elephants were employed at the royal service to carry the queen, princess and prince and even for royal household works.[77]

The elephants were associated with military exercises in India until nineteenth century. Sir Sydney Cotton has mentioned about the usefulness of the elephant in the following words:

> Of all the animals of the creation there are none so intelligent and so useful in military operations as the elephants, and sufficient value and importance is not attached to them in the British service. It might be an immense saving to European troops if elephants were employed in transporting them, when long distances are required. Each elephant can carry six men, with arms, ammunition, and bedding, besides rations. The author resorted to this expedient with great advantage to the service, although he never heard of its being done before, or since ... they raise heavy guns by simultaneous movements, and the power can thereby be increased to any extent, and this is certainly one of the most extraordinary parts of the elephant's value. In throwing down trees of great size, their services are most valuable. The pioneers having cut the roots, the elephants undertake the remainder of the work. Their heads are armed with bullock hides, and to work they go. By simultaneous movements, they push the tree from them, very gradually at first, and so on more and more, (to and fro,) and having got the tree on the swing, at one desperate push, down it goes.[78]

Conclusion

Growth in human population was undoubtly a major cause for clearing the forest. As a result of which the elephants were removed in large numbers from their natural habitats. A late-eighteenth-century regional chronicle of Gujarat mentions that elephants were no longer to be found at Dohad as their routes of migration were now under human habitation. Even in recent periods, scholars have indentified certain factors, such as loss of natural habitat, capture in large numbers for domestication and poaching of male elephants for ivory, etc., that are largely responsible for the gradual decline in the elephant population.[79] Concentrating on that subject eminent elephantologist Ramam Sukumar writes,

> In a part of the Asian continent that has witnessed among the most rapid rates of economic growth globally post-World War II, attrition of the natural habitat of the elephant has been accompanied by not just a reduction in elephant populations but also sharp increases in elephant-human conflicts.[80]

Another element was sport hunting, introduced by British officials by the eighteenth century onwards, and was responsible for the gradual decline in the elephant population. It is surprising to note by the end of the nineteenth century the British Government (1873 onwards) announced rewards for the killing of elephants in order to control damage to crops. With the introduction of commercial crops, mining, irrigational works, railways, telegraph and process of urban development, the process of decline of the natural habitat of elephants has accelerated. One cannot deny that the Mughals also indulged in the practice of hunting on a large scale. They were fond of trapping elephants and domesticated them. But, it is also true that they equally made an attempt to protect this huge wild beast. It was Akbar who took great initiatives for breeding of elephants. Akbar was personally fond of the elephant. The tradition was continued even at the time of his son and successor the 'naturalist' Emperor Jahangir. Their practices never hampered the natural balance and natural habit of the animals. Living with the natural world was the nature of living in entire Mughal India. But, from the nineteenth century onwards the elephant population has been in deep crisis due to the gradual decline of the elephants' natural habitat, which creates an imbalance in the eco-system.

Notes

1 Raman Sukumar, *The Asian Elephant: Ecology and Management* (Cambridge: Cambridge University Press, 1989), p. 1.

2 Irfan Habib, *Prehistory: People's History of India*, 1 (2001, reprint, New Delhi: Tulika Books, 2012), pp. 14–17; B.P. Sahu, *From Hunters to Breeders: Faunal Background of Early India* (New Delhi: Anamika Prakashan, 1988).

3 Raman Sukumar, *The Asian Elephant*, p. 2.

4 Yashod Mathpal, *Prehistoric Rock Paintings of Bhimbetka* (New Delhi: Abhinav Publications, 1992), pp. 46, 52, figs 14, 40, 53.

5 Raman Sukumar, *The Story of Asia's Elephant* (Mumbai: Marg, 2011), p. 31; R. Carrington, *Elephants: Their Natural History, Evolution and Influence on Mankind* (New York: Basic Books, 1959), p. 162; Dhriti Kanta Lahiri-Choudhury, 'History of Elephants in Captivity in India and their Use: An Overview', *Gajah*, vol. 14 (June 1995), pp. 28–31.

6 R.T.H. Griffiths (ed. and tr.), *The Hymns of the Rigveda* (Benaras: E.J. Lazarus & Co., 1890–1896), Book 1, Hymn, 64,7; Book 4, Hymn, 16,14; Book 8, Hymn, 33,8; Sarva Daman Singh, *Ancient Indian Warfare* (1965, reprint, New Delhi Motilal Banarasidass, 1997), pp. 72–84.

7 Some scholars admit the use of elephants in war was introduced in the Rig Vedic period. See, P.C. Chakravarti, *The Art of War in Ancient India* (1941, reprint, Delhi: Low Price Publication, 2010), p. 47; B.K. Majumdar, *The Military System in Ancient India* (Calcutta: Firma KLM, 1960), p. 16; Sarva Daman Singh, *Ancient Indian Warfare*, pp. 72–84; V.R. Ramachandran Dikshitar, *War in Ancient India* (1948, reprint, New Delhi: Motilal Banarasidass, 1987), p. 167; Sarva Daman Singh, 'The Elephant and the Aryans', *Journal of the Royal Asiatic Society of Great Britain and Ireland*, nos 1/2 (April 1963), pp. 1–6.

8 G.N. Pant, *Horse and Elephant Armour* (Delhi: Agam Kala Prakashan, 1997), pp. 89–91.

9 Romila Thapar, *From Lineage to State: Social Formations in the Mid-First Millennium B.C. in the Ganga Valley* (1984, reprint, New Delhi: Oxford University Press, 2011); Romila Thapar, *The Penguin History of Early India from the Origins to AD 1300* (New Delhi: Penguin Books, 2002), pp. 150–173.

10 Romila Thapar, 'Perceiving the Forest: Early India', *Studies in History*, vol. 17, no. 1, New Series (2001), pp. 1–16, esp. p. 5.

11 Jadunath Sarkar, *Military History of India* (1960, reprint, New Delhi: Orient Longman Ltd, 1970), p. 18.
12 Romila Thapar, *History of Early India*, p. 156.
13 Kautilya, *The Arthashastra* (ed., rearranged, tr. and introduced by L.N. Rangarajan) (New Delhi: Penguin Books, 1992), p. 156.
14 G.N. Pant, *Horse and Elephant Armour*, pp. 183–185.
15 Kalidasa, *Raghuvamsha*, (ed. Gajanana Mahadeva Pandit), Bombay Sanskrit Series (Bombay: Bombay Government Central Book Depot, 1905), vol. 4, 40.43; vol. 6, 27.54.
16 Vakpati, *Gaudavaho: A Historical Poem in Prakit* (ed. Shankar Pandurang Pandit), Bombay Sanskrit Series No. 34 (Bombay: Bombay Government Central Book Depot, 1887), p. XXVI.
17 Varahamihira, *Brihat Samhita* (ed. and tr. M.R. Bhat), 2 Parts (New Delhi: Motilal Banarasidass, 1987), Part 2, pp. 595–597.
18 E.S. Varadarajaiyer, *The Elephant in the Tamilland* (Annamalainagar: Annamalai University, 1945), pp. 49–50, 59–69; Raman Sukumar, *The Story of Asia's Elephant*, p. 139; Nalini Sadhle and Y.L. Nene, 'On Elephant in Manasollasa: Characteristics, Habitat, Methods of Capturing and Training', *Asian Agro-History*, vol. 8, no. 1 (2004), pp. 5–25.
19 Irfan Habib, 'Technological Changes and Society: 13th and 14th Centuries', Presidential Address, Medieval India Section, 31st Session, Indian History Congress, Varanasi, 1969, pp. 1–25; Irfan Habib, 'Akbar and Technology', *Social Scientist*, vol. 29, nos 9–10, (September/October 1992), pp. 3–15; Irfan Habib, 'Pursuing the History of Indian Technology: Pre-Modern Modes of Transmission of Power', *Social Scientist*, vol. 20, nos 3–4 (March/April 1992), pp. 1–22.
20 Jadunath Sarkar, *Military History of India*, pp. 163–168; D.K. Lahiri-Chaudhuri, 'History of Elephants', pp. 28–31; Raman Sukumar, *The Story of Asia's Elephant*, pp. 179–209.
21 Simon Digby, *War-Horse and Elephant in the Delhi Sultanate: A Study of Military Supplies* (Karachi: Oxford University Press, 1971), p. 58.
22 Simon Digby, *War-Horse and Elephant*, pp. 51–52.
23 Ziya ud-din Barani, *Tarikh-i Firuzshahi* (ed. S.A. Khan), B.I. Series (Calcutta: Asiatic Society of Bengal, 1862), pp. 260–261.
24 C.E. Bosworth, *The Ghaznavids: Their Empire in Afghanistan and Eastern Iran* (Edinburgh: Edinburgh University Press, 1963), pp. 116–117; Simon Digby, *War-Horse and Elephant*, p. 55.
25 Ziya ud-din Barani, *Tarikh-i Firuzshahi*, Ms. No. 546, Khuda Bakhsh Oriental Public Library, Patna, f. 97A.
26 Ibn Fazl Allah al-Umari, *Masalik al-Absar* (ed. K.A. Fariq) (Delhi: Kitab Bhawan, 1961), pp. 51–52.
27 Simon Digby, *War-Horse and Elephant*, pp. 58–59; Shams-i Siraj Afif, *Tarikh-i Firuzshahi* (ed. M. Vilayat Husain), B.I. Series (Calcutta: Asiatic Society of Bengal, 1891), pp. 144, 197.
28 Ziya ud-din Barani, *Tarikh-i Firuzshahi*, pp. 80–81, 593–594; Amir Khusrau, *Ijaz-i-Khusravi*, 5 vols (Lucknow: Nawalkishore Press, 1876), vol. 5, pp. 9–11; Simon Digby, *War-Horse and Elephant*, p. 60, fn. 190; Shams-i Siraj Afif, *Tarikh-i Firuzshahi*, p. 118.
29 Yahya Bin Sirhindi, *Tarikh-i Mubarakshahi* (ed. M. Hidayat Hosain), B.I. Series (Calcutta: Asiatic Society of Bengal, 1891), p. 128.
30 Shams-i Siraj Afif, *Tarikh-i Firuzshahi*, pp. 168–169.
31 Iqtidar Alam Khan, 'Coming of Gunpowder to the Islamic world and North India: Spotlight on the role of the Mongols', *Journal of the Asian History*, vol. 30, no. 1 (1996), pp. 26–45; Iqtidar Alam Khan, 'Origin and Development of Gunpowder Technology in India: AD 1250–1500', *Indian Historical Review*, vol. 4, no. 1 (1977), pp. 26–27.

32 Jadunath Sarkar, *Military History of India*, pp. 163–168. Elephants were placed in the front rank of the army in medieval India. Isami, *Futuh al-Salatin* (ed. A.S. Usha), Islamic Series No. 9 (Madras: University of Madras Publication, 1948), p. 260; Amir Khusrau, *Tughluq Nama* (ed. S.H. Faridabadi) (Aurangabad: Urdu Publishing House, 1933), pp. 92–93.
33 Zahir-ud-din Muhammad Babur, *Babur-Nama or Memoirs of Babur* (ed. and tr. A.S. Beveridge), 2 vols (bound in one) (1921, reprint, Delhi: Low Price Publication, 2006), p. 488.
34 Zahir-ud-din Muhammad Babur, *Babur-Nama*, pp. 488–489.
35 Asok Kumar Das, 'The Elephant in Mughal Painting', in S.P. Verma (ed.), *Flora and Fauna in Mughal Art* (Mumbai: Marg, 1999), pp. 44–45, fig. 9; Zaheda Khatun, *Birds and Animals in Mughal Miniature Paintings* (New Delhi: D.K. Print World, 2009), p. 27.
36 Zain Khan, *Tabaqat-i-Baburi* (ed. and tr. S.H. Askari, annotation B.P. Ambastha) (Delhi: Idarahi-i-Adabiat-i-Delhi, 1982), p. 116.
37 Jouher, *The Tezkereh al Vakiat or Private Memoirs of the Moghal Emperor Humayun* (ed. and tr. from original Persian into English by Major Charles Stewart) (London: Oriental Translation Fund of the Great Britain and Ireland, 1882), p. 17.
38 Abbas Khan Sarwani, *Tarikh-i-Sher Shahi* (tr. into English from original Persian by B.P. Ambashthya) (Patna: K.P. Jayaswal Research Institute, 1974), p. 772.
39 Jadunath Sarkar, *Military History of India*, p. 67.
40 Abul Fazl Allami, *The Ain-i-Akbari* (tr. into English by H. Blochmann, ed. Col. D.C. Phillot), 3 vols, B.I. Series (1927–1949, reprint, Delhi: Low Price Publication, 2006), vol. 1, p. 123.
41 Abul Fazl Allami, *Ain-i-Akbari*, vol. 1, pp. 123–124.
42 Abul Fazl Allami, *Ain-i-Akbari*, vol. 1, p. 124.
43 Abul Fazl Allami, *Ain-i-Akbari*, vol. 1, pp. 124–125.
44 Abul Fazl Allami, *Ain-i-Akbari*, vol. 1, pp. 124–125.
45 Abul Fazl Allami, *Ain-i-Akbari*, vol. 1, pp. 125–126.
46 Abul Fazl Allami, *Ain-i-Akbari*, vol. 1, p. 126.
47 G.P. Sanderson, *Thirteen Years Among the Wild Beasts of India* (Edinburgh: John Grants, 1907), pp. 19, 22, 25, 66, 201, 243, 365.
48 Abul Fazl Allami, *Ain-Aikbari*, vol. 1, p. 137.
49 Abul Fazl, *Akbarnama* (ed. and tr. by H. Beveridge), 3 vols (1902–1939, reprint, Delhi: Low Price Publication, 2002), vol. 3, pp. 473–474.
50 Abul Fazl, *Akbarnama*, vol. 2, pp. 115–116.
51 Abul Fazl Allami, *Ain-i-Akbari*, vol. 1, p. 138.
52 Abul Fazl Allami, *Ain-i-Akbari*, vol. 1, pp. 132–133.
53 Abul Fazl, *Akbarnama*, vol. 2, pp. 355–356; Asok Kumar Das, 'The Elephant in Mughal Painting', p. 42.
54 Abul Fazl Allami, *Ain-i-Akbari*, vol. 1, p. 128.
55 Abul Fazl, *Akbarnama*, 3 vols (ed. and tr. H. Blochmann), B.I. Series (Calcutta: Asiatic Society of Bengal, 1907–1939), vol. 3, p. 870.
56 Niccolao Manucci, *Mogul India or Storia Do Mogor (1653–1708)* (ed. and tr. William Irvine), 4 vols (bound in two) (1907–1908, reprint, Delhi: Low Price Publication, 2005), vol. 2, p. 340.
57 Abul Fazl Allami, *Ain-i-Akbari*, vol. 1, p. 133.
58 Abul Fazl Allami, *Ain-i-Akbari*, vol. 1, pp. 134–137.
59 Jahangir, *Tuzuk-i-Jahangiri or Memoirs of Jahangir* (tr. Alexander Rodgers and ed. by H. Beveridge), 2 vols (1909–1914, reprint, Delhi: Low Price Publication, 1999), vol. 1, p. 410.
60 Jahangir, *Tuzuk-i-Jahangiri*, vol. 2, p. 18.
61 Jahangir, *Tuzuk-i-Jahangiri*, vol. 2, p. 18; Salim A. Ali, 'The Mughal Emperors of India as Naturalists and Sportsmen', in J.C. Daniel (ed.), *A Century of Natural History* (Bombay: Bombay Natural History Society, 1983), pp. 1–16.

62 D.J. Ehrenfeld, *Indian Miniatures: The Ehrenfeld Collection* (New York: Hudson Hills Press, 1985), no. 22; Asok Kumar Das, 'The Elephant in Mughal Painting', p. 51, fig. 16.
63 Father Anthony Monserrate, *The Commentary of Father Monserrate* (tr. J.S. Hoyland, annotated by S.N. Banerjee) (Cuttack: Oxford University Press, 1922), p. 89; R. Sukumar, *The Living Elephants: Evolutionary Ecology, Behaviour and Conservation* (Delhi: Oxford University Press, 2003), pp. 76–77; Shireen Moosvi, *The Economy of the Mughal Empire* c.*1595: A Statistical Study* (Delhi: Oxford University Press, 1987), p. 289.
64 Shireen Moosvi, 'Ecology, Population Distribution and Settlement Pattern in Mughal India', in Shireen Moosvi, *People, Taxation, and Trade in Mughal India* (Delhi: Oxford University Press, 2008), pp. 89–102; Shireen Moosvi, 'Man and Nature in Mughal Era', Indian History Congress, Mysore 1993 (Delhi: Proceedings of the Indian History Congress, 1993), p. 21.
65 Irfan Habib, *An Atlas of the Mughal Empire: Political and Economic Maps with Detailed Notes, Bibliography and Index* (1982, reprint, New Delhi: Oxford University Press, 1986), Sheet nos 4B, 8B–11B, 13B–16B and notes.
66 Ali Muhammad Khan, *Mi'rat-i Ahmadi* (ed. by Nawab Ali), 3 vols (Boroda: Gaikwar Oriental Studies, 1927), vol. 1, p. 214.
67 Abdul Fath Gilani, *Ruqaat-i Abul Fath Gilani* (ed. Muhammad Bashir Hussain) (Lahore: Punjab University, 1968), pp. 46–47; Shireen Moosvi, 'Man and Nature in Mughal Era', p. 33, fn. 124.
68 Abul Fazl Allami, *Ain-i-Akbari*, vol. 1, p. 295.
69 Abul Fazl Allami, *Ain-i-Akbari*, vol. 1, p. 295.
70 Abul Fazl Allami, *Ain-i-Akbari*, vol. 1, pp. 295–296.
71 Jahangir, *Tuzuk-i-Jahangiri*, vol. 2, pp. 4–5.
72 Niccolao Manucci, *Mogul India or Storia Do Mogor*, vol. 2, pp. 337–339.
73 Niccolao Manucci, *Mogul India or Storia Do Mogor*, vol. 2, p. 339.
74 Niccolao Manucci, *Mogul India or Storia Do Mogor*, vol. 2, p. 339.
75 Niccolao Manucci, *Mogul India or Storia Do Mogor*, vol. 2, p. 339.
76 Niccolao Manucci, *Mogul India or Storia Do Mogor*, vol. 2, p. 339.
77 Niccolao Manucci, *Mogul India or Storia Do Mogor*, vol. 2, p. 240.
78 Sydney Cotton, *Nine Years of the North-West Frontier of India from 1854 to 1863* (London: Richard Bentley, 1868), pp. 90–92.
79 T.N.C Vidya, P. Fernando, D.J. Melnick and R. Sukumar, 'Population Genetic Structure and Conservation of Asian Elephants (*Elephas maximus*) across India', *Animal Conservation*, vol. 8 (2005), pp. 377–388, esp. p. 377.
80 Raman Sukumar, *The Story of Asia's Elephant*, p. 299.

5 British-India and Afghanistan

1707–1842

Kaushik Roy

Introduction

Afghanistan has an area of about 250,000 square miles. The north and the east consist of a mountain system which emerges from the tablelands of Pamirs.[1] North and southern parts of Afghanistan were known in the medieval era as Kabulistan and Zabulistan.[2] The eastern fringe of the country is mountainous especially in the north-east where the protruding Wakhan Corridor joins the Pamirs. Elevations frequently exceed 10,000 feet and forests appear at medium elevations. The central mountainous region, the Hindu Kush, forms an obstruction as regards travel across the centre of the country. The Turkoman Plain characterized by sandy desert and scattered scrub grasses dominates the northern edge of Afghanistan. The Herat-Farah Lowlands in the west are part of the Iranian plateau and some regions of it are suitable for cultivation. South-western Afghanistan is mostly sandy desert.[3] Persia's (Iran) frontier lies west of Herat, from Zulfiqar in the north to Koh-i-Malik Siah in the south, a distance of about 400 miles.[4] The Kabul river forms part of the Indus river system, whereas the dry western and south-western parts of Afghanistan geographically are a continuation of the Iranian Plateau. The area north of Hindu Kush known as Afghan Turkestan is part of the Central Asian plains.[5] In the absence of navigable rivers, the principal means of transport in Afghanistan until *c.*1900 were mules, camels and the pack horses.[6]

The origin of the Afghans is still debated. The Iranians, Greeks and Scythians (Aryan stock) settled in Afghanistan between the third and first century BCE. The nineteenth-century British scholar-officials of the Raj believed in the Armenian and Jewish origin of the Afghans.[7] H.W. Bellew, the Sanitary Commissioner of Punjab in an article dated 1881 asserted that the Afghans were partly Israelites. The Afghans, according to him, belonged to the 'lost tribes of Israel'. The Assyrian ruler Tiglath Pilesar captured many Israelites and they were settled in Ghor in around 700 BCE. The Israelites intermixed with the Arians (Bellew meant Aryans) who were original inhabitants of Afghanistan. Bellew divided the Arians into eastern (Indian) Arian and western (Persian) Arian. The Persian Arians were Tajiks and inhabited western Afghanistan. And the eastern Arians who inhabited eastern Afghanistan were later known as Pathans. They were very similar to the Kshatriyas (Rajputs) of India. In addition, in the mountainous fastness, lived the Turanian (Mongol) people. Further intermixture occurred with the coming of the Scythians (Sakas, Parthians, Huns, etc.) from north of the river Oxus (Amu Darya) into Afghanistan after the Common Era. The Scythians especially settled in Seistan and in the Kandahar region. Overall, in Bellew's format, the Mongols were in north Afghanistan, Arians in southern Afghanistan and the Semitic people were sandwiched between them in central Afghanistan.[8]

Olaf Caroe writing in the 1950s neglects the 'Bani Israel' concept.[9] Caroe divides the Afghans into Eastern and Western Afghans. Among the Western Afghans, he includes the Saduzois (one branch which ruled Afghanistan from 1747 to 1818) and the Muhammadzais (a branch which ruled Afghanistan from 1826 until the 1950s). And among the Eastern Afghans he includes the Shinwaris (around Khaibar), Mohmands, Daudzais and Khalils (inhabiting the region around Peshawar) and the Yusufzais (settled in Dir, Swat and Buner region).[10]

One Muslim chronicler who wrote in around CE tenth century, noted that Kabul was inhabited by the Buddhists, Jews and the Muslims. The Arabs entered Afghanistan after CE seventh century. Nirodbhusan Roy asserts that the Afghan 'race' sprang from the intermixture of Scythian and Jewish colonists.[11] The tribes who inhabited eastern Afghanistan especially the region west of the Indus river were called Pathans by nineteenth-century British officials. Now, they are called Pashtuns. According to one tradition, the word Pathan is a corrupted version of Pahtan which was derived from the word Batan. The word means delivered, set free. The Pathans would prove it by repeatedly rising against all sort of foreign regimes throughout history.[12]

Islam accompanied the Arab force which under Qutaiba entered Afghanistan by the end of the CE seventh century. Many of the Arabs were from the Prophet's own family. The many groups of Pathans who claim to be Sayyids or Quraish, probably have sprung from them.[13] Conversion of the Afghans into Muslims was not complete even during the CE twelfth century.[14] Most of the inhabitants from the medieval era have been Muslims and the majority belongs to the Sunni faith. Among the various tribes, the Durranis from Zamindwar and Ghor were dominant. Many of them settled in Derajat and Multan in India.[15] Until the nineteenth century, the Afghans were addicted to a local wine, referred to as 'grape juice'. It was purified in clay vats and then boiled and poured into goatskins. It was said to improve after two years of keeping and one British officer declared during the early nineteenth century that it tasted like Madeira.[16] Most of the Afghans were intensely Islamic and generally maintained long beards.[17] Jos Gommans writes that the pre-modern Afghans had the triple identities of being shepherds, traders and cavalrymen.[18] The Afghans of Gilzai and Lodi tribes participated in long distance trade with Central Asia and India.[19]

There have been several studies of British involvement in Afghanistan[20] and some scholars have also analysed British policy towards the Indus tribes.[21] Most of them have analysed British policy towards the Indus tribes as the product of 'Forward' and 'Masterly Inactivity' policies. Christian Tripodi notes that there were four stages as regards British north-west frontier policy. According to

Tripodi, between 1843 and 1874, the Raj followed a Closed Border Policy (i.e. Masterly Inactivity). In the mid 1870s, the Forward Policy took over. In 1901, Lord Curzon (Governor-General/Viceroy of India) initiated modified Closed Border Policy. From 1922 until 1947, Modified Forward Policy operated.[22] The reality was probably messy and could not be categorized so nicely. What the above mentioned scholars have missed is the interconnections between British policy towards Afghanistan and the British policy vis-à-vis the tribes along India's north-west frontier. This chapter attempts to fill this historiographical gap. Further, while Tripodi focuses mostly on the role of the 'politicals' in the making of frontier policy, this chapter focuses on tactics, technology and logistics of conventional and sub-conventional warfare conducted by the British and Indian troops in Afghanistan and the north-west frontier. Now let us have a flashback to the rise of Afghan monarchy on the ruins of the Mughal Empire which ultimately would lead to British intervention.

Mughal decline and Afghan revival: 1707–1830

By the beginning of the eighteenth century, the three Islamic empires of Central and South Asia (the Safavids, Uzbeks and the Mughals) were declining. Decline of Central Asian trade and reduction in agricultural productivity were key factors in the weakening of the landed agrarian bureaucratic Islamic empires of Asia. The Mughal Empire was racked by agrarian crisis which in turn intensified faction fight among the *mansabdar*s resulting in the decline of the power of the Mughal *durbar* (court) and rise of semi-autonomous *subadar*s (governors) in the outlying regions of the empire.[23] Especially after the death of the last great Mughal Emperor Aurangzeb in 1707, the Mughal central government lost control over the *suba*s (provinces). During the late seventeenth century, the Uzbek Khanate was economically impoverished because large scale movement of the caravans along the Great Silk Road had stopped.[24] Decline of Persia's silk export to the West weakened the financial foundations of the Safavid monarchy. In addition, unchecked debauchery by Shah Hossein and centrifugal tendencies of the Qizilbash chieftains further weakened the central government at Isfahan. In 1719, Asadullah Khan's 15,000 Afghans of Herat defeated a 30,000 strong Persian Army under Saif Quli Khan.[25] In 1722, the Afghans invaded Persia when the latter country was under the rule of weak Shah Hossein. On 8 March 1722, the Persian royal army was defeated by 11,000 Afghans at the Battle of Golnabad. Victory went to the Afghans because of the tactical retreat by Amanullah's Gilzais and then the pursuing Persian cavalry was destroyed by fire of the *zamburak*s. The *zamburak*s fired balls weighing one to two pounds. As the *zamburak* volley dislocated Persian cavalry, the Afghan horsemen led a counter-attack, and victory was complete.[26] Seven years of Afghan domination of Persia resulted in the death of one million inhabitants of the country.[27] Afghan control over Persia was finally ended by Nadir Shah.

Nadir Shah was born on 6 August 1698 in north Khorasan. He belonged to the Qereqlu Afshar semi-nomadic Turkoman tribe.[28] At a young age, he led a band of robbers. The collapse of the Safavid monarchy enabled him to pillage the Persian frontier province of Khorasan. He seized the Kalat Fort by murdering the governor who happened to be his own uncle. Then, he defeated the Afghans in Khorasan and captured Nishapur District. After defeating the Afghans, he became the national hero. And in 1727, he entered the service of Shah Tahmasp.[29] On 26 August 1732, Shah Tahmasp was deposed by Nadir. The latter became the regent of eight month old son of the Shah. When four years later the infant died, Nadir ascended the throne on 26 February 1736 as *Shahenshah* (king of kings).[30]

Nadir gained Armenia and Georgia from the Ottomans and the provinces bordering the Caspian from the Russians. The Island of Bahrain was taken from the Arabs. Next, the predatory Bakhtiari Tribe of the Shuster Hills was pacified and some were enrolled in his army.[31] In 1737, Nadir with 80,000 men marched towards Kandahar. The Afghans at Kandahar posed a threat to Khorasan in particular and to Persia in general. Moreover, recovery of Kandahar was necessary in order to reclaim the full heritage of the Safavids. At that time, Kandahar was under the Afghan chief Hussein. On 30 March 1737, the Persian Army invested Kandahar. The fort fell on 12 March 1738.[32]

Nadir then advanced towards Mughal-held Afghanistan and India for several reasons. First, the faction-ridden Mughal government despite Nadir's repeated requests failed to cooperate with the Persians against the rebellious Afghans in Kandahar. Actually, the faction-ridden Mughal *durbar* suffering from financial troubles and Maratha threat in central India was in no position to launch a military expedition against the Afghans along the Indus. Second, the Mughal *durbar* refused to recognize Nadir as the *Shahenshah* of Persia. The Mughals were not sure whether Nadir would last long in Persia's tortuous political scene. Third, the weak Mughal Empire motivated Nadir to invade India to plunder the country of its riches, a traditional motive of many Central Asian warlords including *Amir* Timur. Nadir commanded about 55,000 cavalry.[33] In June 1738, Nadir occupied Mughal Kabul.[34] At the Battle of Karnal (24 February 1739), the Mughals were defeated by the Persians. Rustam Ali who composed his chronicle in 1741–1742 noted: 'The Indian warriors … fought with their cruel swords.… The Iranis … firing their guns from a distance and from different quarters, made heaps of corpses of Indians.'[35] Another Mughal chronicler, named Muhammad Muhsin Sadiki also noted the hailstorm of bullets unleashed by the Persians and the uselessness of the heavy Mughal cannons which could not be used effectively in accordance with the changing tactical scenarios.[36] Superior Persian cavalry, long range muskets and *zamburak*s (camel swivel guns) destroyed the indisciplined and ill-led Mughal force which depended on elephants and slow firing immobile large cannons. On 20 March 1739, Nadir entered Delhi.[37] After extracting a heavy indemnity in cash from the hapless Mughal Emperor Muhammad Shah, Nadir annexed the Mughal provinces west of Indus to his own empire. On 16 May 1739, Nadir left Delhi for Persia.[38]

After Nadir's murder by his own soldiery in June 1747, one of his famous Afghan generals named Ahmad Shah Abdali (ruler of Afghanistan from 1747 to

1773) of the Durrani tribe established an independent realm centred round Afghanistan. Ahmad Shah could be regarded as the founder of the first 'national' Afghan monarchy. In 1747, near Kandahar, the Abdali tribes held a *jirga* (tribal assembly) and selected Ahmad Khan then aged 25 as their ruler. A holy man declared him as *Dur-i-Durrani* (pearl of pearls).[39] Ahmad Khan later known as Ahmad Shah Durrani/Abdali was born in 1722 at Multan.[40] He was from the Saduzoi sub-tribe of the Abdalis.[41] In his sixteenth year, Ahmad Khan was appointed in Nadir's personal staff as a *yasawal* (orderly officer). Later, he rose to the rank of *bank-bashi* (treasury officer). As commander of the Afghan contingent, Ahmad accompanied Nadir in his Indian and Turkish campaigns.[42] After Nadir's death, Ahmad Shah built up a well disciplined Afghan Army. Besides Afghans, Ahmad Shah also employed Qizilbash cavalry. These troopers were Persians who had settled in western Afghanistan. He organized his heavy cavalry in 24 corps and each of them numbered up to 1,200 horsemen. In addition, Ahmad Shah also maintained *yatim*s (light cavalry) used for reconnaissance and harassing the enemy army. Camels were used for transportation of military baggage and for the *zamburak*s. Each *zamburak* was worked by two men and each such gun was carried on the back of a camel.[43]

Ahmad Shah led several raids into India. In December 1747, he left Peshawar with 18,000 men. Besides the Afghans, the frontier Pathans encouraged by the prospect of loot also joined Ahmad Shah. With 30,000 Afghan cavalry, Ahmad Shah entered Lahore on 12 January 1748 and after staying there for a month moved towards Delhi.[44] The Mughal *Wazir* (Prime Minister) Qumaruddin came with an army of 60,000 to confront Abdali. On 11 March 1748, the Afghans and the Mughals met each other at Manupur, about ten miles from Sirhind. The Mughals enjoyed artillery superiority over the Afghans.[45]

In the ensuing battle, Qumaruddin was killed but his son Mir Mannu who took over command, forced the Afghans to retreat. The retreating Afghans were harried all the way back to the Indus by the Sikhs. Mir Mannu was appointed *Subadar* of Punjab in April 1748. However, the Afghans still held Multan and the *Raja* of Jammu had declared his independence from the Mughals and created an independent Dogra kingdom. Worse, Mir Mannu received no support from the Mughal *Wazir* Safdar Jang (leader of the Irani/Persian *mansabdar*s) who was trying to undermine the power of the Turani party whose leader was Mir Mannu. Nevertheless, Mir Mannu ordered Adina Beg to curb the power of the Sikhs. And Kaura Mal (a feudatory of Mir Mannu) with a force of Sikhs and Turani mercenaries was ordered to eject the Afghans from Multan. In December 1748, Abdali left Peshawar again for India. Mir Mannu send frantic appeals to Delhi for aid but Safdar Jang did not send him any aid. Mannu with his own force stopped the Afghans at Chenab, at a point four miles east of Wazirabad. Abdali realized that the coming summer would cause hardship for his soldiers in India in case he had to go for a long drawn-out campaign. While with the main army Ahmad Shah faced Mir Mannu, the former sent a strong column under Jahan Khan to Lahore. Mannu's deputy at Lahore moved out from the city to confront Jahan Khan. This enabled Kapur Singh and the Sikhs under him to occupy

Lahore temporarily. Jahan Khan was unsuccessful in Lahore and the Afghans wanted to return with whatever gain was possible. Since Mir Mannu had not received any help from Delhi, he decided to open negotiations. Abdali was given all the territory west of Indus which was previously ceded to Nadir Shah. In addition, the revenue of the four districts of Punjab – Sialkot, Aurangabad, Gujarat and Pasrur, which amounted to 14 lakh (one lakh = 100,000 rupees) – was also assigned to Abdali. Theoretically, Mir Mannu from a Mughal official became a feudatory of the Afghan ruler. Mir Mannu then returned to Lahore and engaged in the task of chastising the Sikhs. When Mir Mannu failed to pay the revenue of the four districts of Punjab which had been ceded to the Afghans, Ahmad Shah prepared for the third invasion. In the middle of November 1751, advance units of the Afghan Army under Jahan Khan crossed the Indus and Ahmad Shah with the main body followed behind. Mir Mannu paid Rs900,000 to buy off the invaders. The Afghans took the money but moved further into Punjab. No help was forthcoming from the Delhi *durbar*. Mir Mannu concentrated his force by calling Kaura Mal from Multan and Adina Beg Khan from Jalandhar. Mir Mannu's army comprised mainly 20,000 Sikhs. In December 1751, Mir Mannu crossed Ravi to check the Afghans under Jahan Khan. Instead of joining Jahan Khan, Abdali made a detour and closed in towards Lahore from the north-east. Mannu returned to the city and entrenched himself outside the city walls. This tactic proved to be a mistake as the Afghans controlled the countryside from where they could draw provisions for their army. In contrast, Mannu had to depend on the supplies stocked in the city in peacetime. After six weeks, all the grain and fodder in his camp was exhausted. Then, Mir Mannu called a council of war. Kaura Mal counselled patience as it was already March and Abdali was bound to go back to Afghanistan with the onset of the summer season in April. But, Adina Beg was for immediate action. Mir Mannu accepted Adina Beg's advice. On 5 March 1752, Mannu attacked the Afghans. Kaura Mal fell in the battlefield and Adina Beg with his contingent retreated. Mir Mannu fought on as long as possible and then surrendered. The Afghans extracted Rs30 lakhs in cash from Mannu and by the terms of the treaty which was ratified by the Mughal Emperor on 13 April 1752, Lahore and Multan were ceded to Ahmad Shah. Thus, Punjab broke away from the declining Mughal system to the rising Abdali Empire. Before returning to Afghanistan, Ahmad Shah also conquered Kashmir. Mir Mannu was appointed as Governor of Punjab by Ahmad Shah and ordered to check the rising power of the Sikhs. Mir Mannu died in May 1754. At that time, the scenario in Punjab was becoming chaotic. Mir Mannu's widow Mughlani Begum controlled Lahore. Multan and the four districts – Gujarat, Sialkot, Pasrur and Aurangabad – continued to pay the revenue to Abdali. Jalandhar and Sirhind were under Adina Beg.[46]

By this time, the Mughal Empire had disintegrated into a series of semi-independent *nawabi*s (like Awadh, Hyderabad, Bengal, etc.) and the Marathas had emerged as the premier power in the subcontinent. The *subadar*s had declared independence and styled themselves *nawab*s. They, however, accepted the *de jure* authority of the Mughal Emperor in Delhi. In 1760, Ahmad Shah with 60,000 cavalry, 20,000 infantry armed with *jezail*s and 2,000 *zamburak*s and 200 pieces of heavy cannon moved from Afghanistan with the objective of destroying the Maratha power in north India. The Rohillas (45,000 cavalry and infantry with 90 guns) of Rohilkhand and the *Nawab* of Awadh Shuja-ud-Daulah with 30,000 infantry and cavalry and 50 artillery pieces joined Ahmad Shah's side against the Marathas. On 12 January 1761, Ahmad Shah engaged the Marathas in a conventional set piece battle known in history as the Third Battle of Panipat.[47]

The Jat Raja Suraj Mal advised the Marathas to conduct guerrilla warfare against Ahmad Shah for several reasons. First, the 51,000 strong Maratha light cavalry force was suited for harassing attacks against the Afghans' lines of communication with Kabul. But, against the heavy disciplined Afghan cavalry, the Maratha ponies would prove useless. Again, Abdali was bound to go back in Afghanistan in summer. So, it was better for the Marathas to bide time by conducting attritional guerrilla warfare from their base at Jhansi. The Maratha *Sirdar* Malhar Rao Holkar agreed with Suraj Mal's assessment of the scenario. However, the headstrong Maratha commander Sadashiv Rao Bhau and the *Peshwa* Baji Rao II's (Prime Minister of the Maratha Confederacy) son Viswas Rao did not pay any attention to Suraj Mal and decided to engage Abdali in a conventional battle. Bhau and Viswas Rao put their faith on the 9,000 Western trained troops led by Ibrahim Khan in the Maratha force.[48] When the Maratha force encamped at Panipat in late 1760, Abdali used his superior cavalry to cut all communications between the Maratha camp and the surrounding regions. Soon, the Maratha soldiers and horses started suffering due to lack of provisions.[49] Abdali's generalship, infantry–cavalry cooperation, *zamburak*s and superior logistics resulted in an Afghan victory over the Marathas at Third Panipat.

Ahmad Shah conquered Sind, Baluchistan and Kashmir. The total population of his empire came to about five million and of them half were Pathans and the rest were a mixture of Hazaras, Qizilbash, etc. When Ahmad Shah died in 1772/1773, he was succeeded by his son Timur Shah (ruler of Afghanistan from 1773 to 1793).[50]

In order to reduce the power of the Pashtuns/Pathans, Timur changed the capital from Kandahar to Kabul and made Peshawar, on the eastern i.e. Indian side of the Khyber Pass, his winter capital.[51] Under Timur's weak rule, the Durrani Empire lost Sind, Balkh and Afghan Turkestan. There were revolts in Kashmir and Turkestan. When he died after a reign of 20 years, he left behind 23 sons. Timur Shah belonged to the Saduzoi clan of the Durrani tribe. The next in importance were the Barakzai clan of the Durrani tribe. The Barakzai clan inhabited the region south of Kandahar and the banks of Helmand. Those who inhabited the fertile banks of Helmand practised agriculture but the rest of the tribe were shepherds. Under Timur Shah, the Barakzai clan numbered some 30,000 families.[52] Haji Jamal Khan of the Barakzai clan was a noble of Ahmad Shah. When Haji died, he left behind four sons. Timur made the eldest one, named Rahimdad Khan, leader of the Barakzai clan in place of Haji. However, due to pressure from the other tribal leaders, Timur was forced to remove

Rahimdad, who was appointed as Keeper of the Government Records. In place of Rahimdad, Powindah Khan, the second brother was appointed.[53]

Powindah Khan, the leader of the Barakzais served Timur Shah with efficiency. When Azad Khan, the Governor of Kashmir rebelled, *Sirdar* Madad Khan and Powindah Khan led an expedition and chastised the rebellious governor. A grateful Timur Shah appointed Powindah as leader of a section of Glizai tribe in addition to his other posts.[54] When Prince Abbas, son of Timur Shah rebelled, the rebellious prince was joined by Arsalan Khan Mohmand. Powindah Khan was ordered to crush the rebellion. At Lalpurah, the rebel force was destroyed and Timur Shah gave his loyal servant the title of Sarfaraz (Lofty) Khan.[55] Meanwhile, the Uzbek tribes threatened north Afghanistan. Timur Shah lost his nerve and decided to leave Kabul for Herat. Powindah Khan dissuaded his master from taking such a cowardly step. Powindah with an army marched towards Balkh and patched up a fragile peace with the Uzbek chief and then returned to Kabul. For the time being, the Uzbek threat was contained.[56]

When Timur Shah died in 1793, his fourth son named Zaman Shah (ruler of Afghanistan from 1793 until 1800) by his first wife ascended the throne with Powindah Khan's support. The Durranis actually wanted to place Prince Abbas on the throne. Zaman Shah rewarded Powindah by giving him command of a part of the Qizilbash force and an annual salary of Rs80,000.[57] But, Powindah Khan who hoped to become the *wazir* was sidelined and in disappointment he engaged in treason against his master. Powindah got in touch with Shah Shuja. Wufadar Khan, who wanted to become the *wazir* instigated Zaman Shah to take drastic action against Powindah Khan. Powindah was caught and executed. The result was that the powerful Barakzai clan now became hostile to the royal Saduzois.[58] Powindah's son Fateh Khan along with Zaman Shah's half-brother Shah Muhammad took refuge with the ruler of Bokhara Shah Murad Beg.[59]

Though Zaman Shah ascended the throne at Kabul, his half-brothers ruled various parts of Afghanistan in a semi-independent manner. The situation is somewhat similar to that of the Mughal Empire after the death of Babur and rise of Humayun on the Mughal throne. Humayun (not to be confused with Babur's son and Mughal Emperor Humayun) seized Kandahar, Abbas ruled Peshawar, Haji Firuz-ud-Din and Mahmud took Herat. Zaman Shah was able to defeat Humayun and occupied Kandahar.[60] Abbas was forced to give up Peshawar and thrown in prison.[61]

Long distance trade between Central Asia and India during the eighteenth century was conducted through Afghanistan.[62] Grain trade in Afghanistan was in the hands of the Hindu Punjabi Khatris/Arora merchants. They gave loans at very high rates (50 per cent interest) to the Afghan rulers for launching military expeditions into India.[63] In order to solve the perennial economic problem of Afghanistan, Zaman Shah decided to invade Punjab. Moreover, the disgruntled princes of India encouraged Zaman Shah to invade the subcontinent and overthrow the British. Two of Tipu Sultan's (ruler of Mysore until 1799) agents named Mir Raza Ali and Mir Habibullah visited Kandahar in May 1797 and encouraged Zaman Shah to invade India to overthrow the British and the

Marathas. Tipu agreed to provide a contingent of 20,000 soldiers to cooperate with Zaman Shah and Rs3 crores (100 lakhs=one crore) as expense for his Indian invasion. One *Shaikh* Rahim Ali, an intelligence agent for the British reported that the rulers of Jammu, Kangra, Awadh and Rohilkhand also encouraged Zaman Shah to attack India.[64] In 1798, the Shah requested the *Nawab* of Awadh's assistance for the impending military operations against the Sikhs and the Marathas.[65] Zaman Shah's objective was to annex Punjab and to retain the Mughal Emperor Shah Alam II as a puppet ruler in charge of the region between Jamuna river (including Delhi) and Chambal.[66] In November 1798, the East India Company (henceforth EIC) collected grains and troops at Allahabad to protect Awadh in case Zaman was able to break through the Sikh resistance in Punjab.[67] Zaman Shah left Kabul in September 1798 and reached Peshawar. On 13 October, he left Peshawar after receiving tributes. The Indus was crossed at Attock. On 3 November, the Afghan Army reached Rawalpindi and on 16 November, the Shah was at Rohtas.[68]

Due to lack of available cash, Zaman Shah was able to muster only 12,000 miserably equipped men. His corps of 500 *zamburak*s lacked *sarwan*s (camel drivers). Zaman's attempt to crush the Sikh resistance by engaging them in a decisive set piece battle failed due to Ranjit Singh's strategy. The latter encouraged the Sikh *misl*s to conduct guerrilla attacks on the invading Afghan Army. This time a conventional Afghan Army was at the receiving end of guerrilla resistance by the Sikh light cavalry. Due to dispersed but intense Sikh resistance, Zaman Shah failed to raise adequate revenue to sustain his army. When Zaman Shah was engaged in Punjab against the Sikhs, the Barakzais instigated rebellions in Afghanistan. Hence, Zaman was forced to return to Afghanistan. Back in Afghanistan, Zaman was arrested and blinded and Fateh Khan took over in Kabul.[69]

Shah Mahmud, Timur Shah's son by his second wife was placed on the throne and he ruled from 1800 until 1803. In 1801, Shah Shuja, brother of Zaman Shah was in charge of Peshawar, now assembled a military force and marched towards Kabul. Shah Shuja commanded 10,000 Afridis and Yusufzais. He was opposed by Shah Mahmud's 3,000 strong cavalry and infantry force. The battle was fought at Surkh-Rud, west of Jalalabad. Indiscipline resulted in Shah Shuja's defeat. The eagerness of Shuja's troop to loot the royal treasure enabled Fateh Khan to launch a charge by the Barakzais which discomfited Shuja's force. Shah Shuja was defeated and took refuge first among the *sirdar*s of Khyber and then among the Afridis.[70] In 1802, Shuja with 12,000 Afridis attacked Peshawar which was under Mahmud's control. The confrontation which occurred in mid-summer ended with large slaughter of the Shuja's soldiers. Shuja escaped to Chora and then took refuge into the Maidan of Tirah.[71] Shuja went from Zhob to Quetta and gained money from some caravan traders. Then, he again advanced to Kabul with his ragtag force and found Mahmud besieged in Bala Hissar. Fateh Khan with his 10,000 strong army retreated from the vicinity of Kabul leaving the city to Shuja.[72] In July 1803, Shah Shuja became the ruler. Shuja ruled from 1803 until 1809 when he was driven out by Shah

Mahmud. In 1809, Shah Mahmud returned from Herat and got support from his powerful *Wazir* Fateh Khan of the Barakzai tribe. Shah Mahmud defeated Shah Shuja at the Battle of Nimla fought between Kabul and Peshawar. Though Shah Mahmud became the ruler of Afghanistan, Shuja (whose mother was of Yusufzai tribe) retained support among the tribes which bordered between India and Afghanistan.[73] Shuja would later become a pawn in the British attempt of controlling Afghanistan. Shah Mahmud continued to rule until 1818.[74] In 1809, Mountstuart Elphinstone arrived in Peshawar. At that time, Ranjit Singh's *Khalsa* Kingdom's frontier ran between Hasan Abdal and Rawalpindi on the Margalla Pass.[75]

Prelude to British intervention in Afghanistan

The successful conclusion of the Third Anglo-Maratha War (1817–1818) bought the whole of the subcontinent, except Punjab, under British control. As a paramount power, the British debated whether an independent Punjab should be retained as a strong buffer against the 'unruly' tribes who inhabited the region between Indus and Afghanistan or whether Punjab should be annexed to the British-Indian Empire and a secure frontier along Indus should be established. After British victory in the Second Anglo-Sikh War (1848–1849), British-India's frontier reached Indus and British officials in India realized that they had to deal with the Afghan 'problem' in a much more systematic manner. By this time, as far as the Great Game for the control of Afghanistan was concerned, the Uzbek Khanate and Iran were down and out. The big players were Tsarist Russia (which then was in the process of absorbing the Uzbek realm) and the British Empire (which had succeeded the Mughals and was in the process of reducing the autonomy of Iran). By 1830, Tsarist Russia had replaced France as the principal power threatening the Raj through Afghanistan. Throughout the nineteenth century, the British made an error in thinking that the Russians could move large number of troops through the deserts and mountains of Central Asia to Afghanistan to attack India.[76]

Unlike the indigenous powers, the British freed themselves from supply of war horses through Afghanistan by importing horses from Britain and Australia.[77] In March 1809, Elphinstone was sent to Shah Shuja's winter capital at Peshawar in order to seal off possible Russian and French influence in Afghanistan. At that time, Shah Shuja was engaged in reconquering the rebellious province of Kashmir.[78] The rebels defeated Shuja's two commanders Akram Khan and Madud Khan.[79] Shah Shuja lost power due to shifting Afghan politics and in November 1815 finally took refuge in the British cantonment at Ludhiana.[80] As early as the 1820s, several Afghan *sirdar*s realized that the British were a rising power in South Asia and attempted to deal with the '*feranghi*s'. Dost Muhammad, the ruler of Ghazni, captured Kabul in 1826. In 1827, Dost Muhammad became the ruler of Afghanistan.[81] In 1829, Sultan Muhammad Khan of Peshawar wanted to negotiate with the British as an independent chief. When in 1834, Shah Shuja advanced on Kandahar, and the *Khalsa* Kingdom increased pressure on the Indus frontier, Dost Muhammad appealed to Great Britain for aid in order to recover Peshawar. His nephew, son of Jabbar Khan, was sent to Ludhiana to establish diplomatic relationship with the EIC state.[82]

In 1834, Ranjit Singh was able to capture Peshawar from Afghanistan. The Sikhs remembered the atrocities committed by Abdali's troops in Punjab. Hence, Sikh rule in the Afghan city was indeed harsh. In the 1830s, one Colonel Leslie/Rattray was the *Khalsa* Kingdom's Governor of the Khyber Pass. He converted to Islam, accepted the title of Fida Muhammad Khan. He established his headquarters at the Ali Masjid Fort and levied tolls on the caravans which moved through the Khyber Pass.[83] In 1837, the *Khalsa* Kingdom's Governor of Peshawar General Paulo Avitabile (an Italian mercenary) set up open gallows in the city to deter miscreants.[84] On the road between Rohtas to Peshawar (200 miles east of Kabul), Ranjit Singh built a mud fort at Khairabad. Its garrison was ordered to maintain control over the surrounding heights.[85] Another mud fort for the same purpose was built at Naushera.[86] Overall, Ranjit Singh took care to make the road to Peshawar safe for the travellers. Through fire and sword against the marauders and the villages around the city of Peshawar, which harboured them, the *Khalsa* Kingdom was able to maintain a semblance of order along the highway.[87] In 1837, when Dost Muhammad Khan made an unsuccessful attempt to wrest Peshawar from the Sikhs, Alexander Burnes was at Kabul on a commercial mission. Burnes was ordered by the governor-general to establish British influence in Afghanistan. However, Burnes perceived that Russian and Persian influence were dominant in Kabul.[88]

First Anglo-Afghan War: 1839–1842

Henry Durand came to India in May 1830, as Lieutenant of the Corps of Engineers. In 1836, he was in charge of bridging the Indus near the Fort of Bhakkar.[89] This was done in anticipation of a campaign in Afghanistan. In December 1837, a Russian envoy named Vitkevich arrived at Kabul.[90] Herat was under the Saduzoi chieftain Kamran Shah, who was Shah Shuja's nephew. The real power was in the hand of his *Wazir* Yar Muhammad Khan.[91] In early 1837, Kamran Shah and Yar Muhammad were campaigning in Seistan in eastern Persia. The Shah of Persia decided to besiege Herat. On 17 September, Kamran and Yar Muhammad returned to Herat. On 23 November 1837, a large Persian Army (10,000 infantry, 2,000 cavalry and 30 guns) backed by Russian advisers laid siege to Herat.[92] And this increased the anxiety of the British about probable Russian influence in Afghanistan, despite the fact that Herat is 2,000 miles from Calcutta. The Raj pressurized Ranjit Singh to cooperate in the British invasion of Afghanistan. Ranjit demanded some more territories west of Indus in addition to Peshawar as the price for cooperation. This demand was refused by the Raj.[93] As a result, *Khalsa* cooperation with the British during the First Afghan War was half-hearted.

On 6 April 1838, the British envoy John McNeill arrived at Herat to initiate a ceasefire.[94] On 4 June 1838, a squadron of five warships and two steamers with detachments from three regiments and a marine battalion set sail for the Island of Karrack (Kharg), a few miles off the Persian Coast. The Persian governor surrendered to the British force.[95] On 9 September 1838, the Persian Shah raised the siege of Herat.[96]

Two regiments of regular infantry, two troops of horse artillery and two regiments of irregular cavalry were raised under the command of the British officers for serving with Shah Shuja. In November 1838, Shuja left Ludhiana for Shikarpur. The Army in India assembled at Ferozepur was ordered to join Shuja at Shikarpur.[97] By this time, the Persians had raised the siege of Herat due to the occupation of Kharg Island in the Persian Gulf by the troops of the Bombay Army. Nevertheless, the Raj under Governor-General Lord Auckland proceeded with the plan to install Shah Shuja at Kabul.[98] Patrick Macrory asserts that instead of trying to install the discredited Shah Shuja, Auckland should have tried to strengthen Dost Muhammad's regime. Dost Muhammad was suspicious of Russian expansion in Central Asia and would have welcomed British aid in this respect. However, Dost was an avowed enemy of Ranjit Singh of Punjab since the latter had occupied Peshawar. Auckland could not support Dost due to the good relations which existed between the *Khalsa* Kingdom and the Raj.[99]

In February 1839, the Bombay column was at Larkana, and the two brigades of the Bengal Army under Major-General Willoughby Cotton moved from Shikarpur towards Kandahar. On 4 March 1839, John Keane assumed command of both the columns. These two columns united at Quetta in Shaul Province during the beginning of April 1839.[100]

In 1839, British-India mobilized about 39,000 soldiers against *Amir* Dost Muhammad.[101] Dost Muhammad's kingdom yielded an annual revenue of about Rs2,262,943 (226,294 sterling).[102] Theoretically, Dost Muhammad commanded 38,000 feudal cavalry (of them 8,000 were Durranis). Most of the regular units plus the tribal levies were put under command of Dost's son Akbar Khan.[103] Under his direct command, Dost Muhammad had 12,000 cavalry organized in two divisions. One division known as *Khud Aspah* comprised personnel who rode their own horses. And the other division known as the *Amlah Sarkari* comprised personnel who rode on the horses provided by the government. These two divisions were subdivided into parties comprising 200 horses. Several such parties were put under the charge of various sons of the *amir*. For instance, Muhammad Akbar Khan was given command of 2,000 such cavalry, Muhammad Afzal Khan commanded 600 cavaliers and Ghulam Haidar Khan commanded 1,000 cavaliers.[104]

The Army of Indus comprised 9,500 men from the Bengal Army, 5,600 from the Bombay Army and an auxiliary force of 6,000 raised for Shah Shuja.[105] The infantry was equipped with the Brown Bess musket. It was a muzzle loader effective to 150 yards and one could fire two rounds per minute. In contrast, the Afghan *jezail* was effective up to 800 yards when fired from a rest, and could kill a horse when fired from a distance of 600 yards.[106] Each *jezail* was about seven feet long and a fork was used while firing.[107] The Afghans had an advantage in hand-held firearms which would vanish only during the late nineteenth century.

The Army of Indus entered Afghanistan through the Bolan Pass and a small British-Indian and *Khalsa* force moved through the Khyber Pass under Claude Wade. Due to objections of Ranjit Singh, the principal invasion force moved into Afghanistan through Sind. From the very first, heat and lack of adequate supply of water troubled the British-Indian military column.[108] Lieutenant William Barr of the Bengal Artillery who was with the column writes about the 'notorious Khyberries':

> Their dress generally consisted of a long chupkun of a light brown colour, reaching to their knees, loose trousers, and grass sandals, or shoes with hob-nails.... Their usual weapons are a long *jezail*, or rifle, with a wooden fork attached to its extremity, on which the piece is rested to secure a better aim; a sword; and a large knife stuck into the sash: some had a pistol in addition.[109]

In March 1839, the British-Indian detachment which from Peshawar marched towards the Khyber Pass numbered to about 3,950 soldiers (including the Sikh allies).[110]

The Afghans assumed that the principal invasion force like that of the Mughals would invade Afghanistan through the Khyber Pass. On 17 May, Dost Muhammad's son Akbar Khan with a large train of artillery was at the vicinity of the Khyber Pass.[111] On 5 July 1839, a COIN operation was undertaken by the British-Indian force. Colonel Wade and Ferris with a detachment went to the region near Kabul river, where Sadat Khan was inciting the population on the left bank of Kabul river against the British. On 6 July, Lieutenant William Barr with a howitzer of the *Khalsa* Army, and *Shah Zada* Timur's bodyguard joined Wade and Ferris' force. Barr found out that from the shelter of the rocks, the Afghans fired their *jezail*s whose range outranged the matchlocks in the hands of Barr's contingent. Only a few rifles in the hands of the British soldiers were able to hit the boldest of those Afghans who refused the cover of the *sangar*s.[112] By 20 July 1839, Wade had at his disposal 10,000 men (including the 6,000 strong *Khalsa* contingent) for moving through the Khyber Pass. The British were anxious about the loyalty of the *Khalsa* contingent as all of its personnel were Muslims.[113] Barr was also worried about the Afghan levies raised by the British. He feared that most of them had joined the British bandwagon for economic gain and would desert at the earliest opportunity. Further, many of them in secret espoused the cause of Dost Muhammad rather than the British stooge Shah Shuja.[114] Barr's anxiety would become a reality in the next year.

For marching through Baluchistan, Lieutenant Mackeson constructed a military road through the Bhawalpur state.[115] In order to procure a supply of grain and fodder and to ensure the safety of the lines of communications stretching back to India, the British entered into political negotiations with the Baluchi chiefs. Conducting politics and diplomacy simultaneously while waging military operations was a characteristic of Small War. Mir Mehrab Khan of Kalat was acknowledged by the British as the ruler of his principality and in return he agreed to accept the authority of Shah Shuja. Further, Mehrab Khan was to keep the Bolan Pass open for the British and provide supply for the British-Indian troops. Mohal Lal recounts that though he paid Mehrab Khan Rs2,000 in cash (his annual subsidy as determined by the British), still Mehrab failed to provide forage and grain to the soldiers and the beasts of burden.[116] The Baluchis and the neighbouring Kakar tribes intensified their predatory raids on the British *dak*s (posts), camels and supply columns. Either Mehrab Khan was unable or was unwilling to check his followers' plundering activities.[117] Further, Mir Mehrab Khan's governor in Shal instead of protecting the camels of the British-Indian columns stole them when they went for grazing and these animals were then sold in Seistan rather than in Kalat to prevent any suspicion on the part of the British authority.[118]

Soon, the British got wind of Mehrab Khan's disloyalty and decided to teach him a lesson. Captain Bean, the Political Agent, decided that force needed to be used against Mehrab Khan. In response, on 31 October 1839, Thomas Wilshire with a detachment was sent to Quetta. Simultaneous conventional military operations against the hostile regular force and low-intensity policing operations against the recalcitrant tribal chiefs were a characteristic of Small War. On 12 November, Wilshire reached the Jiryani village, about eight miles from Kalat. Wilshire then made his disposition to storm the town and the citadel. The storming party was led by Brigadier Bomgardt. Mehrab Khan died fighting heroically with sword in hand.[119]

As the British-Indian force marched towards Kandahar, *Sirdar* Kohandil Khan with some 4,000 cavalry encamped at Dandi Gulai with the objective of checking the advance of the *kafir*s.[120] However, the lesser chiefs of the Kandahar region were willing to treat with the British. On 20 April 1839, a messenger from a Kakar chief named Haji Khan came to Mohan Lal with a letter addressed to Alexander Burnes. Thanks to Mohan Lal's political management, Haji Khan deserted to the British. The coalition of the Barakzai chiefs in Kandahar started to fall apart. One *Mulla* Nasu advised that the best course for them was to retreat to Persia and ask the aid of the Persian monarch Muhammad Shah or to get in touch with the shadowy Russian Count Simonich.[121] Haji Khan's accommodation by the British encouraged certain other chiefs also to submit. For instance, Abdulmajid Khan, son of Shah Pasan Khan, the Governor of Lash and Ghulam Akhund Zadah, a priest who previously preached hatred against the British, now submitted.[122]

On 25 April, Shah Shuja with the British envoy and minister reached Kandahar. Lord Keane with the 7,500 strong detachment followed the Shah next day and encamped in the neighbourhood of the city.[123] On 2 June, one of Shah Shuja's sons was appointed as the Governor of Kandahar and Major Leech was nominated as the Political Agent. That day proved to be an eventful one as Ranjit Singh breathed his last.[124]

Dost Muhammad's eldest son Afzal Khan with 5,000 cavalry conducted 'hit and run' attacks against the British columns.[125] During the Second Anglo-Afghan War, Afzal Khan's role would be replayed by Ayub Khan. In July 1839, the British captured the Ghazni Fort. Of its garrison, 600 were killed and 1,600 men became prisoners.[126] The storming of Ghazni cost the Army of Indus 17 killed and 175 wounded.[127] Henry Durand was in charge of blasting the gate of Ghazni Fort in 1839 which led to the capitulation of the fort's garrison.[128] Durand was then appointed as engineer of the Shah's troops but he resigned his commission and returned to Punjab. Then, Durand was ordered by the chief engineer to utilize his cartographical skill in drawing maps with reports on the Afghan campaign.[129]

On 1 November 1838, Colonel A. Roberts (father of the future Field-Marshal Frederick 'Bob' Roberts) of the Bengal Army was appointed to the command of the 4th Brigade of the Army of Indus (which comprised the 1st European Regiment and the 35th and 37th Indian Infantry Regiments). In a report submitted on 1844, he explained the reasons for the disaster in Afghanistan in the following words:

> I entreated that the number of Afghan levies may be limited until we had officers better qualified by a more perfect knowledge of their language, customs and feelings, to command them, and until from experience and observation to be well prepared to meet any emergency deep inside a hostile country. I was regarded as an alarmist, hence removed from the command of Shah's troops. Our Afghan troops became our bitter enemies and murdered the officers.[130]

A. Roberts also pointed out the logistical failure in his report. He penned: 'My plan to store grain for six months along with munitions and to strengthen the fortifications of Bala Hissar with more artillery was turned down.'[131]

Vincent Eyre was the Commissary of Ordnance to the Kabul Field Force in 1840.[132] The principal logistical vehicles for supplying the troops in Afghanistan both in Mughal and the British era were the mules, bullocks, elephants and camels. For the details regarding the carrying capacity, cost and forage required by the beasts of burden, refer to Table 5.1. In 1869, the cost of a mule driver came to about Rs7 per month. For three mules, a grass cutter was appointed at the wage of Rs7 per month.[133] The elephant could drag heavy siege guns, but the disadvantage with the elephant was that it consumed more than either the bullock or the mule. Moreover, elephants required lot of green forage and water and these two commodities were not easily available in the dry terrain of Afghanistan. During 1581–1582, the Mughal Army which accompanied Akbar during the invasion of Afghanistan had 500 elephants.[134]

However, camel was considered best for traversing the rocky and mountainous region of Afghanistan. Unlike elephants, camels thrive on dry scrubs and bushes available in Afghanistan. The British commanders considered camels as best equipped for the rocky terrain between Peshawar and Kohat.[135] A camel was able to drag a 18-pdr gun over the sandy terrain.[136] The British route of entry for moving into southern Afghanistan lay through Baluchistan. The mountain tract along the Bolan Pass which runs north along the Suleiman Mountains and south towards the coast was inhabited by the Baluch clans.[137] James Atkinson (of the

Table 5.1 Carriage animals of the Army in India for operation in Afghanistan

Animal	*Carrying capacity*	*Food required per day*	*Marching speed*	*Approximate price of a single animal in Rs*	*Remarks*
Donkey	1.5 *maunds*	–	1.5 miles per hour	4	Price in 1839 in Bhawalpur
Bullock	2 *maunds*	2 *seers* of gram and 14 pounds of dry forage	1.5 miles per hour	70	Price in 1878
Mule	2.5 *maunds*	3 *seers* of gram daily	3 miles per hour	200	Cost in 1869
Pony	2.5 *maunds*	–	5 miles per hour	–	–
Camel	5 *maunds*	–	2 miles per hour	–	–
Elephant	12 *maunds*	–	–	580	Price in 1871

Source: From Brigadier-General G. Bourchier, to the QMG, Army HQ, Simla, 22 June 1869, Records of Chief Commands, 1865–1876, Notes and Minutes by Napier, from Lieutenant J. Cookesley, RE, Eurasian Battery of Artillery, to the commissioner of Assam, Shillong, 4 June 1869, Enclosure B, Records of Chief Commands, Notes and Minutes by Napier, MSS.EUR.F.114, 5(2), IOR, BL; The commander-in-chief to the viceroy, Letter to John Lawrence, 27 June 1865, John Lawrence Collection, MSS.EUR.F.90/59, IOR, BL, London; From secy. to govt. Punjab, MD, to commissioner, Rawal Pindi, 13 September 1879, Copy of a telegram dated 13 September 1879, from the secy. to the Punjab Govt., MD, to all the commissioners in Punjab, Supply and Transport, December 1879, MD, NAI; James Atkinson, *Afghan Expedition: Notes and Sketches from the First British Afghan War of 1839–1840* (1842, reprint, ISBN 1-59048-280-8: Long Riders' Guild Press, 2007), p. 44; Brigadier R.C. Butalia, *The Evolution of the Artillery in India: From the Battle of Plassey (1757) to the Revolt of 1857* (New Delhi: Allied Publishers, 1998), p. 231; T.A. Heathcote, *The Indian Army: The Garrison of British Imperial India, 1822–1922* (London: David and Charles, 1974), p. 63.

Note
As a point of comparison, infantry marched on an average at three miles per hour and cavalry's trot was seven miles per hour.

Bengal Army) Superintending Surgeon of the Army of Indus describes the Bolan Pass in the following words:

> a rough and pebbly road, between sand hills, studded with flint and lime stones, small and large, the space in breadth between the hills varying from 300 or 400 yards to about 30. As the torrent runs deviously from one side of the gorge to the other, we had to cross it six or seven times. The bottom consisted of large round stones, which gave an uncertain footing for a horse, though the camels with their large spread of foot got through without much difficulty. The depth was not more, generally, than eighteen inches.[138]

For this terrain, camels were best suited. The Army of Indus had 30,000 camels and 38,000 camp followers. The expedition ran into logistical troubles as soon as the British and Indian units entered Baluchistan. For instance, on 7 March 1839, the 2nd Brigade was halted at Shikarpur due to shortage of camels and increasing mortality of the available camels.[139]

The British officers travelled in style. A senior British officer had six servants, crockery, silver plate, wine chest, portable bathtub, etc. In fact, for carrying the officers' baggage, some of the medical stores for the troops had to be left behind.[140] One brigadier required 60 camels to carry his personal kit. And the junior subalterns per regiment took up to 40 servants. The 16th Lancers took even their pack of foxhounds.[141]

On 7 August 1839, the Army in India reached Kabul and Shah Shuja was restored to the throne.[142] There was considerable opposition to Shah Shuja, which escalated into a full-blown insurgency by 1841.[143] As early as April 1840, the British garrison at Kahan, the principal town of the Marri tribe of Baluchistan, was attacked. By the end of September 1840, the British garrison commander Captain Lewis Brown retreated from Kahan and reached Pulaji on 3 October 1840.[144] In November 1840, Dost Muhammad surrendered to William H. MacNaghten (previously Chief Secretary to the Bengal Government), the British envoy and minister in the court of Shah Shuja.[145] But, this act did not stop the Afghan insurrection from getting stronger and wider. The Afghans were now reacting strongly against the presence of *kafir*s in their country. And Dost Muhammad's son Akbar Khan remained north of Hindu Kush and bided his time for making a bid to the throne. In April 1841, Major-General Elphinstone assumed command of the troops in Afghanistan.[146]

The Gilzai tribes who occupied a large portion of the region between Ghazni and Kandahar were not subdued. The Gilzai *sirdar*s were further enraged due to reduction of the subsidies, a measure which was forced upon Macnaghten by Lord Auckland.[147] Especially damaging was the cessation of the 3,000 sterling pound subsidy annually paid to the Eastern Gilzais for keeping the Khyber Pass open. The immediate result was that they rebelled and severed British communication between Kabul and Peshawar. Further, they occupied the passes on the road to Jalalabad and conducted systematic pillage and plundering on the convoys moving between Kabul and Peshawar.[148] Robert Sale's brigade which was on the point of marching back to India was ordered to bring the Eastern Gilzais back into line. Monteith was ordered to move with the 35th Indian Infantry Regiment, a cavalry squadron and some guns. Sale (nicknamed 'Fighting Bob') had fought in the First Anglo-Burma War (1823) and was the colonel commandant of the 13th Battalion. After the capture of Ghazni, he was promoted to the rank of major-general. Sale followed with the 13th Light Infantry Battalion. Suppressing the Eastern Gilzais proved to be a tough task. Sale himself was wounded. However, the Khurd-Kabul Pass was cleared and Sale halted at Gandamak.[149] On 12 October 1841, the 13th and the 35th with two guns under Dawes moved forward. Sale's objective was to go through the Khurd-Kabul Pass and deploy the 35th Indian Infantry Regiment in an advanced position at Khurd-Kabul after which the 13th was to retire to Bhudkak. The Khurd-Kabul is a narrow defile enclosed by high and rugged rocks. The Afghans in this pass did not exceed 60 men but they knew the terrain. Concealed behind the rocks, stones and *sangar*s, they targeted especially the British officers. In fact, Sale was wounded in the left leg.[150] On 18 October 1841, a 400 strong *lashkar* launched a night attack on the Khurd-Kabul Pass.[151] In the west of Kandahar, *Sirdar* Aktar Khan collected about 7,000 cavalry and infantry, but was defeated at Girhisk in July 1841 by a detachment of Shah Shuja's regular troops commanded by Captain Woodburn. Woodburn's detachment comprised one infantry regiment, two horse artillery guns and two regiments of Afghan cavalry.[152]

By the end of October 1841, Macnaghten had received enough warnings about a possible uprising in Kabul. However, suffering from the 'victory disease', he refused to take such warnings seriously. On 1 November, the shopkeepers in Kabul *bazaar* (market) put their shutters down and refused to sell any ware to the British troops.[153] Major-General Sale's wife Lady Florentina Sale noted in her journal on 26 October 1841: 'The general impression is that the Envoy is trying to deceive himself into an assurance that the country is in a quiescent state.'[154] Mohan Lal, the Indian interpreter in British service had warned Burnes that Abdullah Khan was instigating the tribal chiefs who had assembled at Kabul to rebel against the British. Burnes influenced by Macnaghten refused to believe the warnings by the 'croakers'.[155] Actually Macnaghten had been selected for this post partly for the fact that he was a 'moderate'. Moreover, Macnaghten had been appointed as Governor of Bombay and he was eager to leave Kabul for his new posting. By late October 1841, both Macnaghten and Burnes agreed that Afghanistan was settling down under the British supported ruler Shah Shuja and a significant number of British troops could be withdrawn from this country.[156]

On 2 November 1841, a popular insurrection broke out in Kabul.[157] British and Indian troops were stationed in the indefensible cantonment which was constructed in 1840. It was situated in the low swampy ground about two miles from the citadel. The cantonment was defended only by a waist height mud rampart and a narrow ditch. Further, the commissariat supplies were stored in a small fort without a proper wall.[158] The cantonment without any natural and manmade protection did not offer much scope for defence to the British and Indian troops against attacks by the armed populace of Kabul. Worse, the Afghans easily captured the commissariat supply fort. The British engineers urged that the troops must be stationed in the Bala Hissar citadel. However, Shah Shuja's *harem* was in the citadel.[159] Ultimately, respect for the cultural sensibilities of their political client, Shah Shuja, overrode military requirements of the British-Indian force. British response to the Kabul uprising between 2 and 25 November was fragmented due to inadequate coordination between Elphinstone, Macnaghten and Brigadier John Shelton (second in command of the Army in India's units in Kabul after the departure of Sale).[160] Lack of unified command of the COIN force had come to haunt the British in Kabul. Sale was ordered to bring up his brigade from the Khurd-Kabul Pass and Eldred Pottinger was asked to join the Kabul garrison with his detachment from Charikar, which was 60 miles north of Kabul. However, Sale had already moved towards Jalalabad in the opposite direction and of Pottinger's detachment only he along with one other person reached Kabul. To cap it all, Akbar Khan with a force moved out from Bamian for Kabul.[161]

The Afghan *sirdar*s promised that they would allow the British and Indian garrison in Kabul to retreat safely to India, however the British-Indian column on its way out was attacked. The British-Indian column that left Kabul on 6 January 1842 comprised of about 5,000 combatants and roughly 12,000 camp followers.[162] On 13 January 1842, the last remnant of the retreating British-Indian force was wiped out at Gandamak. Of the 16,500 British and Indians who left Kabul, roughly about 100 men survived.[163] The single British survivor was Surgeon William Brydon who rode alone to Jalalabad.[164]

Meanwhile, the Conservatives came to power in Britain and they appointed Lord Ellenborough as governor-general in place of Lord Auckland. When, on 25 November, Auckland received letters from Mr George Clerk and Captain Mackeson about the news of the disaster in Afghanistan, Auckland wrote to Jasper Nicolls, the commander-in-chief depreciating any plan of reconquest of Afghanistan. Nicolls was also against the invasion of Afghanistan. However, Mr Robertson, the lieutenant-governor of the North-West Provinces and Mr Clerk, the agent on the North-West Frontier pushed forward troops to Peshawar. On 16 November 1841, Clerk wrote to Colonel Wild, commanding at Ferozepur and to Colonel Rich at Ludhiana, urging them to send the 30th, 53rd, 60th and the 64th Indian Infantry Regiments to Peshawar.[165] While Nicolls was for withdrawing the troops east of Indus, Clerk was for pushing British-Indian troops west of Indus deep into Afghanistan.[166] Clerk wrote to General Boyd at Sirhind for the dispatch of another brigade. The Sikhs were also informed to collect the boats in Sutlej for transportation of the EIC's reinforcements. And 5,000 Sikh troops under Kunwar Partab Singh were ordered to march from Chach Hazara. Mackeson also requested the Sikh authorities to send 6,000 troops to Jalalabad but General Avitabile argued that he needed all the Sikh troops for protection of the *Khalsa* Kingdom. Auckland believed that instead of two brigades, one brigade would be able to maintain British presence in Afghanistan. Wild declined to push into Afghanistan without adequate number of guns. On 3 January 1842, four guns were provided to him. The camel drivers started deserting and the Afridi *malik*s were yet to be bribed into submission by Mackeson. Further, the loyalty of the Sikhs was fragile. Moreover, the Sikhs tampered with the loyalty of the sepoys and played on their fears.[167] The 9th Foot was ordered to be ready, and the 26th Indian Infantry Regiment, some irregular cavalry, two 9-pdrs and a howitzer were ordered to follow. Later, the 10th Cavalry was ordered to join this brigade of troops. On 4 January, a brigade comprising 3,000 combat troops crossed the Sutlej. General George Pollock commanding at Agra, who had entered the Indian Army as a lieutenant of artillery in 1803, was appointed commander of this avenging force.[168] Meanwhile, Wild at Peshawar faced lot of difficulties. His four Indian infantry regiments comprised a large number of young soldiers, who were demoralized due to Sikh propaganda. Wild had only one troop of irregular horse and four indifferent pieces of artillery. Ammunition was scarce and carriage was beginning to fail. Worse, Sale and Macgregor were requesting immediate advance of the brigade.[169]

The fortress at Ali Masjid lies five miles within the Khaibar Pass and 25 miles from Peshawar. It consisted of two small forts connected by a small wall and standing upon a rock and commanded on the south and west by two hills. It is called the 'Key to Khaibar'. This fort was held by Yusufzais and resisted the attack of the Afridis.[170] On 15 January, Colonel Moseley with the 53rd and 64th Indian Infantry Regiments accompanied by Mackeson started at night and reached Ali Masjid next morning. However, instead of 350 bullocks, only 60 bullocks came with the rearguard. This meant that the two Indian infantry regiments were without adequate provisions. On 19 January, the 30th and 60th Indian Infantry Regiments with some *Khalsa* guns started their march towards Ali Masjid. Under *jezail* firing by the Afghans, one gun was abandoned and the Indian infantry regiments retreated to Jamrud.[171] Meanwhile, the troops at Ali Masjid were suffering from lack of provisions, water, bedding and tents. On 24 Janaury, Ali Masjid was abandoned to the Afridis and Colonel Mosley retreated to Jamrud.[172] On 31 March, Pollock pitched the camp at Jamrud. Brigadier Wild was ordered to command the advance guard and M'Caskill the rearguard.[173] The flanking columns were ordered to advance in detachments of two companies at 500 yards interval.[174]

After receiving reinforcements, Pollock with some 8,000 soldiers marched west from Jalalabad. At Tezeen Valley, Akbar Khan's force of 16,000 men was defeated. Akbar Khan escaped to Goriband Valley. On 15 September 1842, Pollock reached Kabul and on 17 September, he was joined by Nott's detachment from Kandahar.[175] On 9 and 10 October, the Army of Retribution burnt the *bazaar* of Kabul as a mark of revenge. It was in this very *bazaar* the headless and limbless body of Macnaghten had been suspended by the unruly Afghan mob. On 11 November, the British-Indian force started retreating.[176] In the end, Akbar Khan seized the throne for his father and Dost returned from India to reclaim his hereditary possession.[177] The first British adventure in Afghanistan proved to be futile.

Small War along the Indus frontier: 1843–1860

The region between British Punjab and eastern Afghanistan was a no man's land. This region roughly along the banks of the Indus was inhabited by Pathan tribes who pillaged and plundered western Punjab. When the British and Indian troops pursued them, they escaped into eastern Afghanistan through the porous border. At times, these tribes also received official and unofficial encouragement from the Afghan government and the Pathan tribes in Afghanistan to pillage and plunder British territory. As a result, the GOI (henceforth Government of India) until the end of the Raj in 1947 had to conduct frequent 'Butcher and Bolt' expeditions (also known as Small War) against these frontier tribes. The fluid nature of tribal dynamics shaped the tribes' behaviour towards the British-Indian Empire. Each clan had their own perspective which was distinctive, unwritten and unpredictable, built on a series of precedents. Further, the perspective of the clans was also shaped by the requirements of the moment.[178]

In general, the Mughal chroniclers described all the tribes between Indus and Herat as Afghans. But, in contrast, the British attempted to identify and categorize the important tribes along the Indus frontier. In general, Tripodi writes, the British regarded the Pathans as 'noble savages'.[179] The British officers concluded that the Pathan tribes were divided into many clans and each clan was further subdivided into several sub-clans. Each sub-clan was in turn divided into families. And all the Pathans spoke Pushtu. The British officers concluded that many Pathans were Afghans and the rest were Rajputs who had converted into Islam. In British format, life in the rugged mountains, difficulties of livelihood and the Islamic ethos, made the Pathans and the Afghans good soldiers. The British officers tried to map the territorial localities of the different types of the frontier tribes. The Yusufzais inhabited the Peshawar Valley, Buner, Swat, Malakand Hills east of Indus in the Hazara District (Black Mountain). The Afridis inhabited the Khyber Hills and the British belief was that they were originally Rajputs who had accepted Islam. The Orakzais inhabited the valley north of the Samana Range. The Khattaks inhabited the Khattak Hills. Since their territory lay within British-India's border, they were more manageable. The Bangash were in Miranzai. And the Turis were in Kurram Valley. Most of the tribes were Sunnis but a few like the Turis, Bangash, Hazaras, etc., were Shias.[180] The Waziris were divided into three groups: Ootamanzis, Ahmudzais and Mashuds. The Ahmudzais took partly to agriculture. Some of them also engaged in salt trade and grazing.[181] The Ahmudzais passed the summer in the region round Wana and moved during the winter to the grazing land on the western border of the Bannu District. And the Ootamanzais inhabited the Tochi Valley.[182] The northern Waziris had very few villages. They generally inhabited the banks of the rivers and made temporary habitations with loose stones. In the mountains, they made an encampment known as *Kizdhee*. These encampments were constructed with woollen blankets supported by sticks with coarse matting at the sides. These blankets were impervious to rain and not easily inflammable. The cattle and sheep guarded by dogs were also kept within such tents.[183] Iron ore was available in the Waziri Hills. The principal site was Koh-i-Mahsood near Mulkin. The ore was dug out and then crushed. Every village had a smelting apparatus. The smelting furnace was constructed with a conical roof planted nearly vertically on the ground with long holes. Iron was extracted by smelting the ore with charcoal. This was used for making swords, daggers and barrels of *jezail*s. In the Mashud region, the artisans for making weapons were located at Kanirgorum Town.[184] And Khanki Valley was identified as the region inhabited by the Orakzais.[185] The three main branches of the Mashuds were Alizai, Shaman Khel and Bahlozai. Each was divided into numerous sections and sub-sections. They occupied the region around Razmak, Wana and Jandol.[186]

In 1844, mountain batteries were raised specifically for service along the north-west frontier and beyond it if required. Each battery had four 12-pdr howitzers and four 3-pdr guns along with two farriers, 168 syces and 168 mules.[187] The howitzer had a curved trajectory, hence useful for firing behind the enemy fortification like the *sangar*.[188] Quetta town in northern Baluchistan constituted the southern-most post of British-India's scientific frontier.[189] And the northern anchor of the British defensive line was Peshawar. Like the Mughals 250 years ago, in order to secure the southern end of the scientific frontier, the EIC annexed Sind in 1843. The Baluchis of Baluchistan were believed to be a Muslim hill community of Arab descent. They had entered Baluchistan from the Persian Gulf.[190] Their favourite weapon was their 28-inch-long knives and engaged frequently in nocturnal raiding.[191] At that time, the Baluchi chiefs were ruling Sind and they had Afghan retainers. Against the EIC troops' artillery and muskets, the matchlocks and knives of the Afghan retainers proved inadequate.[192]

After the First Anglo-Sikh War, the *Khalsa durbar* was dominated by the British officials of the EIC. Henry Lawrence was selected by Governor-General Lord Hardinge as the political agent of Punjab.[193] In 1847, Henry Lumsden raised the Corps of Guides from the younger sons and relations of the headmen of the villagers. The initial recruits were Yusufzais, Khattaks and Muhammadzai tribes. Later, the Guides also recruited Afridis, Waziris, Hazaras, Kafirs, Sikhs and Gurkhas. The Guides Corps included both cavalry and infantry. A *malik* (head of clan/*khail*) or a Sikh *sirdar* with his retainers constituted a *risala*.[194] Herbert Edwardes pacified Bannu between 1847 and 1848, by balancing the Sikh soldiers of the *durbar* with political agreements reached with the Muslim Khans (tribal chiefs). In order to conciliate the tribal *sirdar*s, the British resorted to symbolic compromise. Herbert Edwardes dressed like the Pathans and talked their language. The submission of the influential chiefs was stage-managed for the maximum symbolic effect.[195] However, permanent submission of the tribes remained a chimera. For instance, in 1850, the Kabul Khel section of the Ootamanzi Waziris attacked the salt mines at Bahadur Khel.[196]

Meanwhile, the GOI kept the restored Dost in good humour. Afghanistan was given a gift of 8,000 muskets and Rs500,000. During the Anglo-Persian War (1856–1857), a further Rs100,000 was given to Dost. This kept Dost neutral, when the 'Devil's Wind' blew across north India in the summer of 1857.[197]

In the summer of 1857, most of the Hindustani sowars and the sepoys of the Bengal Army rebelled. Along the north-west frontier, the rebellion of the Bengal Army units stationed at Peshawar, Rawalpindi, Sialkot, etc., put the British in a difficult situation.[198] In order to replace the manpower loss and also to augment their military strength in a time of crisis, the British recruited the Pathans in auxiliary units as well as in the regular units.[199]

However, some Pathan *sirdar*s tried to raise the religious war cry and attempted to raid the border of British-India thinking that as the British were moving most of their military assets into Hindustan to confront the rebellious *Pandie*s, the raiders would have easy pickings. The British detachment was attacked at Khyber Pass on 27 August 1857 by an insurgent force led by Shahzada Mahmud.[200] Luckily for the British, the Punjabis remained loyal to the Raj. After the Bengal Army's mutiny was crushed in 1859, the British transferred military units back to the Indus frontier. John Lawrence believed in following a Closed Border Policy. He wanted a 'no intervention' policy towards Afghanistan and minimum interaction with the tribes on the west side of the Indus.[201] In John Lawrence's view the Russian danger to India was merely a phantom menace.[202]

Nevertheless, sporadic attacks by the Pathan tribesmen continued. In retaliation, in 1860, Brigadier-General Chamberlain led an expedition comprising 5,196 men against the Mashud Waziris. As punishment, livestock from their villages were seized.[203] And, the north-west border tribes had to be managed. In the post-Mutiny period, the units designated for fighting along the north-west frontier started recruiting the 'unruly' Indus tribes in greater numbers in a systematic manner. For instance, in 1862, the Corps of Guides had Qizilbash, Pathans (Yusufzais, Khattak, Mohmand and Gilzai clans), Afridis (Zakka Khel, Basi Khel, Sepai and Koki Khel) and Kafirs in its ranks.[204] Obviously, the British were anxious about the loyalty of the tribes beyond Indus. The British were aware of the game of playing off the different 'races' against each other to prevent the emergence of any anti-British coalition among the soldiery. In 1860, Major-General S. Cotton commanding Peshawar Division spelled out the '*divide et impera*' policy in the following words:

> To produce and bring into operation that feeling of antagonism naturally existing between the castes or races of which the Indian Army is composed, by the formation of corps of distinct races, so that one would become an effective check on the other.... Thus regiments of Malwa Sikhs, Manjha Sikhs, Dogras, Mazbis, Afridis, Punjabi Muslims etc should be established.[205]

This policy continued with some success until the late 1870s when again the clouds of war gathered over the sky of Afghanistan.

Conclusion

From the nineteenth century onwards, extra-Asian powers (like Britain) also participated in the Great Game. Britain like the Mughals maintained a permanent power base (i.e. formal empire) in southern Asia for conducting warfare in Afghanistan. British-India was dependent on South Asian military manpower for projecting power in Afghanistan. The British relied on Indian and Indus tribes for manning the Indian Army which was the principal instrument for projecting power in Afghanistan. The British realized that management of the frontier tribes was inseparable from the related and greater Afghan 'problem'. Despite the passage of time and changes in military hardware, certain COIN techniques remained operational during the last five centuries. The British used subsidies to placate the Afghan tribes. When the subsidy amount was cut, rebellions broke out which in turn cost the imperial power much more. So, reduction of tribal subsidies proved to be a false economy. Second, British-India through limited military recruitment tried to co-opt the potentially rebellious manpower of Afghanistan. Nevertheless, with time, certain innovative COIN techniques also evolved. For instance, the British for the first time introduced 'politicals', frontier police and paramilitary units geared especially for acquiring real time intelligence and conducting specialized COIN duties.

The British experience during the First Afghan War was that it was easier to defeat the Afghan regular army in conventional battles and sieges but almost impossible to permanently subjugate the militias and the armed tribesmen who conducted hit and run attacks. In other words, it was easier to win the conventional campaigns in Afghanistan but extremely difficult to maintain a lasting peace. The British found out that it was logistically difficult and financially expensive to maintain a large force in Afghanistan for a long period. But, fighting the dispersed groupings of armed Pathans to secure peace required the maintenance of a large force for a long period. The Americans would certainly agree with such views in the beginning of the new millennium.

Notes

1 George MacMunn, *Afghanistan from Darius to Amanullah* (1929, reprint, Lahore: Sang-e-Meel Publications, 2002), p. 3.
2 George MacMunn, *Afghanistan*, p. 2.
3 Robert F. Baumann, *Russian-Soviet Unconventional Wars in the Caucasus, Central Asia, and Afghanistan*, Leavenworth Papers, no. 20 (Fort Leavenworth, KS: Combat Studies Institute Press, 1993), p. 134.
4 George MacMunn, *Afghanistan*, p. 6.
5 Jos J.L. Gommans, *The Rise of the Indo-Afghan Empire,* c.*1710–1780* (Leiden: E.J. Brill, 1995), p. 11.
6 George MacMunn, *Afghanistan*, p. 13.
7 Nirodbhusan Roy, *Niamatullah's History of the Afghans* (n.d., reprint, Lahore: Sang-e-Meel Publishers, 2002), p. 17. Mohan Lal considered that the theories of Israelite or Armenian origin of the Afghans as spurious. Lal, *Life of the Amir Dost Mohammaed Khan of Kabul with his Political Proceedings towards the English, Russian, and Persian Governments including the Victory and Disasters of the British Army in Afghanistan*, 2 vols (1846, reprint, New Delhi: Asian Educational Services, 2004), vol. 1, pp. 1–7.
8 H.W. Bellew, 'A New Afghan Question', *Journal of the United Service Institution of India*, vol. 47 (1881), pp. 49–97.

9 Olaf Caroe, *The Pathans: 550 BC–AD 1957* (London: Macmillan, 1958), p. 10.
10 Olaf Caroe, *The Pathans*, pp. 12–13, 15.
11 Nirodbhusan Roy, *Niamatullah's History of the Afghans*, pp. 20–21.
12 Nirodbhusan Roy, *Niamatullah's History of the Afghans*, p. 8.
13 Olaf Caroe, *The Pathans*, p. 10.
14 Olaf Caroe, *The Pathans*, p. 9.
15 Major G.F. MacMunn, *The Armies of India* (1911, reprint, New Delhi: Heritage Publishers, 1991), pp. 153–154.
16 Patrick Macrory, *Kabul Catastrophe: The Story of the Disastrous Retreat from Kabul, 1842* (Oxford: Oxford University Press, 1986), p. 19.
17 Patrick Macrory, *Kabul Catastrophe*, p. 19.
18 Jos J.L. Gommans, *The Rise of the Indo-Afghan Empire*, p. 16.
19 Jos J.L. Gommans, *The Rise of the Indo-Afghan Empire*, p. 21.
20 T.A. Heathcote, *The Afghan Wars: 1839–1919* (1980, reprint, Stroud: Spellmount, 2007); Brian Robson, *Crisis on the Frontier: The Third Afghan War and the Campaign in Waziristan, 1919–20* (Staplehurst: Spellmount, 2004).
21 T.R. Moreman, *The Army in India and the Development of Frontier Warfare: 1849–1947* (Basingstoke: Macmillan, 1998); Charles Allen, *Soldier Sahibs: The Men who made the North-West Frontier* (London: John Murray, 2000); Alan Warren, *Waziristan: The Faqir of Ipi and the Indian Army: The North West Frontier Revolt of 1936–7* (Karachi: Oxford University Press, 2000); Andrew M. Roe, *Waging War in Waziristan: The British Struggle in the Land of Bin Laden, 1849–1947* (Lawrence, KS: University Press of Kansas, 2010). The most recent monograph by Christian Tripodi titled *Edge of Empire: The British Political Officer and Tribal Administration on the North-West Frontier, 1877–1947* (Surrey: Ashgate, 2011) emphasizes the political aspect of border management.
22 Christian Tripodi, *Edge of Empire*, pp. 15–17.
23 Satish Chandra, *Parties and Politics at the Mughal Court: 1707–40* (1959, reprint, New Delhi: People's Publishing House, 1979).
24 M. Athar Ali, 'The Passing of the Empire: The Mughal Case', in M. Athar Ali, *Mughal India: Studies in Polity, Ideas, Society, and Culture* (New Delhi: Oxford University Press, 2006), p. 340.
25 Ganda Singh, *Ahmad Shah Durrani: Father of Modern Afghanistan* (Bombay: Asia Publishing House, 1959), p. 9.
26 Michael Axworthy, *The Sword of Persia: Nader Shah from Tribal Warrior to Conquering Tyrant* (London and New York: I.B. Tauris, 2006), pp. 45–49.
27 William Irvine, *Later Mughals*, 2 vols (reprint, New Delhi: Taj Publication, 1989), vol. 2, p. 318.
28 Axworthy, *The Sword of Persia*, pp. 17–18.
29 William Irvine, *Later Mughals*, vol. 2, pp. 317–318.
30 William Irvine, *Later Mughals*, vol. 2, pp. 318–319.
31 William Irvine, *Later Mughals*, vol. 2, p. 319.
32 William Irvine, *Later Mughals*, vol. 2, p. 319.
33 H.M. Elliott and J. Dowson (eds), *The History of India as told by its Own Historians: The Muslim Period, The Posthumous Papers of the Late H.M. Elliot, ed. and continued by John Dawson*, 8 vols (1876–1877, reprint, Delhi: Low Price Publications, 2001), vol. 8, *Tarikh-i-Hindi* of Rustam Ali, p. 61.
34 Paddy Docherty, *The Khyber Pass: A History of Empire and Invasion* (London: Faber & Faber, 2007), p. 180.
35 *The History of India as told by its Own Historians*, vol. 8, *Tarikh-i-Hindi* of Rustam Ali, pp. 61–62.
36 *The History of India as told by its Own Historians*, vol. 8, *Jauhar-i-Samsam* of Muhammad Muhsin Sadiki, p. 74.

37 M. Athar Ali, 'Recent Theories of Eighteenth Century India', in M. Athar Ali, *Mughal India*, p. 350.
38 Michael Axworthy, *The Sword of Persia*, p. 15.
39 Stephen Tanner, *Afghanistan: A Military History from Alexander the Great to the Fall of the Taliban* (New York: Da Capo Press, 2002), pp. 117–118.
40 Ganda Singh, *Ahmad Shah Durrani*, p. 15.
41 Stephen Tanner, *Afghanistan*, p. 118.
42 Ganda Singh, *Ahmad Shah Durrani*, p. 18.
43 *The History of India as told by its Own Historians*, vol. 8, *Nigar-i-Nama-i Hindi* by Sayid Ghulam Ali, pp. 398–399.
44 Ganda Singh, *Ahmad Shah Durrani*, p. 57.
45 Ganda Singh, *Ahmad Shah Durrani*, pp. 59–60.
46 Khushwant Singh, *A History of the Sikhs*, vol. 1, *1469–1839* (1963, reprint, New Delhi: Oxford University Press, 1989), pp. 131, 133–136, 137–141.
47 *The History of India as told by its Own Historians*, vol. 8, *Tarikh-i-Manazilul Futuh* by Muhammad Jafar Shamlu, pp. 148–149, 152–153.
48 *The History of India as told by its Own Historians*, vol. 8, *Tarikh-i-Ibrahim Khan, Tamadus Saadat* of Mir Ghulam Ali, pp. 274–275, 400.
49 *The History of India as told by its Own Historians*, vol. 8, *Tarikh-i-Ibrahim Khan*, p. 280.
50 Patrick Macrory, *Kabul Catastrophe*, pp. 17–20; Stephen Tanner, *Afghanistan*, p. 122.
51 Stephen Tanner, *Afghanistan*, p. 124.
52 Mohan Lal, *Dost Mohammad Khan*, vol. 1, pp. 10–11.
53 Mohan Lal, *Dost Mohammad Khan*, vol. 1, pp. 11–12.
54 Mohan Lal, *Dost Mohammad Khan*, vol. 1, pp. 12–13.
55 Mohan Lal, *Dost Mohammad Khan*, vol. 1, pp. 13–14.
56 Mohan Lal, *Dost Mohammad Khan*, vol. 1, pp. 14–15.
57 Mohan Lal, *Dost Mohammad Khan*, vol. 1, pp. 15–16.
58 Patrick Macrory, *Kabul Catastrophe*, p. 21; James Atkinson, *Afghan Expedition: Notes and Sketches from the First British Afghan War of 1839–1840* (1842, reprint, ISBN 1-59048-280-8: Long Riders' Guild Press, 2007), p. 7; Mohan Lal, *Dost Mohammad Khan*, vol. 1, pp. 16–18.
59 James Atkinson, *Afghan Expedition*, p. 7.
60 James Atkinson, *Afghan Expedition*, p. 6.
61 James Atkinson, *Afghan Expedition*, p. 7.
62 Jos J.L. Gommans, *The Rise of the Indo-Afghan Empire*, p. 5.
63 Stephen Frederick Dale, 'Indo-Russian Trade in the Eighteenth Century', in Sugata Bose (ed.), *South Asia and World Capitalism* (New Delhi: Oxford University Press, 1990), pp. 140–150.
64 Hari Ram Gupta, *History of the Sikhs*, vol. 4, *The Sikh Commonwealth or the Rise and Fall of Sikh Misls* (New Delhi: Mushiram Manoharlal, 1982), p. 486.
65 Hari Ram Gupta, *History of the Sikhs*, vol. 4, p. 487.
66 Hari Ram Gupta, *History of the Sikhs*, vol. 4, p. 493.
67 Hari Ram Gupta, *History of the Sikhs*, vol. 4, p. 491.
68 Hari Ram Gupta, *History of the Sikhs*, vol. 4, pp. 494–495.
69 Patrick Macrory, *Kabul Catastrophe*, pp. 23, 27.
70 James Atkinson, *Afghan Expedition*, p. 8; Olaf Caroe, *The Pathans*, p. 272.
71 Olaf Caroe, *The Pathans*, p. 272.
72 Olaf Caroe, *The Pathans*, p. 273.
73 Stephen Tanner, *Afghanistan*, p. 125.
74 Victoria Schofield, *Afghan Frontier: At the Crossroads of Conflict* (2003, reprint, London: I.B. Tauris, 2010), p. 33.
75 Olaf Caroe, *The Pathans*, p. 274.

76 Pierce G. Fredericks, *The Sepoy and the Cossack* (London: W.H. Allen, 1972), p. 28.
77 Breeding of Horses, Remount Horses, Minutes by Napier, Notes and Minutes by Lord Napier of Magdala, Commander-in-Chief of Bombay, 1866–1869, MSS.EUR. F114, 1C, India Office Records (IOR), British Library (BL), London.
78 Victoria Schofield, *Afghan Frontier*, p. 63.
79 James Atkinson, *Afghan Expedition*, p. 10.
80 James Atkinson, *Afghan Expedition*, p. 19.
81 George MacMunn, *Afghanistan*, p. 87; Jules Stewart, *Crimson Snow: Britain's First Disaster in Afghanistan* (2008, reprint, Stroud: History Press, 2010), p. 11.
82 George MacMunn, *Afghanistan*, p. 102.
83 Paddy Docherty, *The Khyber Pass: A History of Empire and Invasion* (2007, reprint, London: Faber & Faber, 2008), p. 199.
84 George MacMunn, *Afghanistan*, p. 109.
85 Lieutenant William Barr, *Journal of a March from Delhi to Peshawar and from thence to Kabul with the Mission of Lieutenant-Colonel C.M. Wade, including Travels in the Punjab, a Visit to the City of Lahore and a Narrative of Operations in the Khyber Pass undertaken in 1839* (1844, reprint, New Delhi: Munshiram Manoharlal, 2003), p. 114.
86 William Barr, *Journal of a March*, pp. 120–121.
87 William Barr, *Journal of a March*, pp. 97–08.
88 James Atkinson, *Afghan Expedition*, p. 2.
89 From Major Henry Durand to the President of the Board of Control, 15 February 1834, paras 1 and 7, Memorandum of Services of Henry Durand, Major, Engineer, Bengal Army, Letters from Miscellaneous Correspondence, Charles Wood Collection, MSS.EUR.F.78/1b-2, IOR, BL.
90 Paddy Docherty, *The Khyber Pass*, p. 200.
91 Stewart, *Crimson Snow*, p. 13.
92 Stewart, *Crimson Snow*, pp. 19–21.
93 James Atkinson, *Afghan Expedition*, p. 3.
94 Stewart, *Crimson Snow*, p. 22.
95 Stewart, *Crimson Snow*, pp. 24–25.
96 Stewart, *Crimson Snow*, p. 25.
97 James Atkinson, *Afghan Expedition*, p. 3.
98 James Atkinson, *Afghan Expedition*, p. 3.
99 Lady Florentina Sale, *A Journal of the First Afghan War* (ed. Patrick Macrory) (1969, reprint, Oxford: Oxford University Press, 2002), Introduction by Macrory, pp. xiv–xv.
100 James Atkinson, *Afghan Expedition*, p. 5.
101 T.A. Heathcote, *The Afghan Wars*, p. 25.
102 Mohan Lal, *Dost Mohammad Khan*, vol. 1, p. 236.
103 Rob Johnson, *The Afghan Way of War: Culture and Pragmatism, A Critical History* (London: Hurst & Co., 2011), p. 51.
104 Mohan Lal, *Dost Mohammad Khan*, vol. 1, p. 240.
105 Pierce G. Fredericks, *The Sepoy and the Cossack*, pp. 65–66.
106 Pierce G. Fredericks, *The Sepoy and the Cossack*, p. 66.
107 William Barr, *Journal of a March*, p. 144.
108 Mohan Lal, *Dost Mohammad Khan*, vol. 2, p. 185.
109 William Barr, *Journal of a March*, p. 139.
110 William Barr, *Journal of a March*, pp. 127–128.
111 William Barr, *Journal of a March*, p. 177.
112 William Barr, *Journal of a March*, pp. 177–179.
113 William Barr, *Journal of a March*, p. 183.
114 William Barr, *Journal of a March*, p. 184.

115 James Atkinson, *Afghan Expedition*, p. 40.
116 Mohan Lal, *Dost Mohammad Khan*, vol. 2, p. 193.
117 Mohan Lal, *Dost Mohammad Khan*, vol. 2, p. 197.
118 Mohan Lal, *Dost Mohammad Khan*, vol. 2, pp. 199–200.
119 Mohan Lal, *Dost Mohammad Khan*, vol. 2, pp. 200–201.
120 Mohan Lal, *Dost Mohammad Khan*, vol. 2, p. 205.
121 Mohan Lal, *Dost Mohammad Khan*, vol. 2, pp. 206–207.
122 Mohan Lal, *Dost Mohammad Khan*, vol. 2, p. 208.
123 Mohan Lal, *Dost Mohammad Khan*, vol. 2, p. 210.
124 Mohan Lal, *Dost Mohammad Khan*, vol. 2, p. 215.
125 T.A. Heathcote, *The Afghan Wars*, p. 33.
126 T.A. Heathcote, *The Afghan Wars*, p. 34.
127 Lady Florentina Sale, *A Journal of the First Afghan War*, Introduction, p. xv.
128 From Major Henry Durand to the President of the Board of Control, 15 February 1834, para 8, Letters from Miscellaneous Correspondence, Charles Wood Collection.
129 From Major Henry Durand to the President of the Board of Control, 15 February 1834, paras 9 and 10, Charles Wood Collection.
130 Statement of the Services of Colonel Roberts, Bengal Army, 27 July 1853, Charles Wood Collection.
131 Statement of the Services of Colonel Roberts, Bengal Army, 27 July 1853, Charles Wood Collection.
132 Lieutenant Vincent Eyre, *The Military Operations at Kabul: The Retreat and Destruction of the British Army, 1832* (1843, reprint, Stroud: Nonsuch, 2005), p. 8.
133 From Brigadier-General G. Bourchier, to the QMG, Army HQ, Simla, 22 June 1869, Records of Chief Commands, 1865–1876, Notes and Minutes by Napier, MSS. EUR.F.114, 5(2).
134 Vincent Smith, *Akbar: The Great Moghul, 1542–1605* (n.d., reprint, Delhi: S. Chand, 1962), p. 142.
135 John Lawrence to Blane, Miscellaneous Letters to Colonel L.S. Blane, Military Secy., 1865–1866, John Lawrence Collection, MSS.EUR.F.90/60, IOR, BL.
136 Brigadier R.C. Butalia, *The Evolution of the Artillery in India: From the Battle of Plassey (1757) to the Revolt of 1857* (New Delhi: Allied Publishers, 1998), p. 225.
137 George MacMunn, *Afghanistan*, p. 122.
138 James Atkinson, *Afghan Expedition*, p. 56.
139 James Atkinson, *Afghan Expedition*, pp. 52–53.
140 Pierce G. Fredericks, *The Sepoy and the Cossack*, p. 68.
141 Lady Florentina Sale, *A Journal of the First Afghan War*, Introduction, p. xv.
142 Lady Florentina Sale, *A Journal of the First Afghan War*, Introduction, p. xv.
143 Vincent Eyre, *The Military Operations at Kabul*, p. 7.
144 T.A. Heathcote, *The Afghan Wars*, p. 41.
145 T.A. Heathcote, *The Afghan Wars*, p. 43.
146 Vincent Eyre, *The Military Operations at Kabul*, p. 21; Lady Florentina Sale, *A Journal of the First Afghan War*, Introduction, p. xv.
147 Vincent Eyre, *The Military Operations at Kabul*, pp. 21, 29.
148 Mowbray Morris, *The First Afghan War* (London: Sampson Low, 1878), pp. 74–75.
149 Mowbray Morris, *The First Afghan War*, p. 75; Lady Florentina Sale, *A Journal of the First Afghan War*, Introduction, pp. xviii–xix.
150 Lady Florentina Sale, *A Journal of the First Afghan War*, pp. 6–7.
151 Lady Florentina Sale, *A Journal of the First Afghan War*, p. 7.
152 Vincent Eyre, *The Military Operations at Kabul*, p. 21.
153 Mowbray Morris, *The First Afghan War*, pp. 75–76.
154 Lady Florentina Sale, *A Journal of the First Afghan War*, p. 9.
155 Mowbray Morris, *The First Afghan War*, p. 76.
156 Lady Florentina Sale, *A Journal of the First Afghan War*, p. 9.

157 Vincent Eyre, *The Military Operations at Kabul*, p. 11.
158 Mowbray Morris, *The First Afghan War*, p. 72.
159 Mowbray Morris, *The First Afghan War*, pp. 72–73.
160 Mowbray Morris, *The First Afghan War*, p. 79; Lady Florentina Sale, *A Journal of the First Afghan War*, p. 11.
161 Mowbray Morris, *The First Afghan War*, pp. 80–81.
162 Vincent Eyre, *The Military Operations at Kabul*, p. 11.
163 Vincent Eyre, *The Military Operations at Kabul*, pp. 7, 11.
164 Lady Florentina Sale, *A Journal of the First Afghan War*, Introduction, p. xiii.
165 *Frontier and Overseas Expeditions from India*, vol. 3, *Baluchistan and the First Afghan War* (1910, reprint, Quetta: Nisa Traders, 1979), p. 385.
166 *Frontier and Overseas Expeditions from India*, vol. 3, p. 387.
167 *Frontier and Overseas Expeditions from India*, vol. 3, p. 386.
168 *Frontier and Overseas Expeditions from India*, vol. 3, p. 386.
169 *Frontier and Overseas Expeditions from India*, vol. 3, p. 387.
170 *Frontier and Overseas Expeditions from India*, vol. 3, p. 388.
171 *Frontier and Overseas Expeditions from India*, vol. 3, p. 388.
172 *Frontier and Overseas Expeditions from India*, vol. 3, p. 389.
173 *Frontier and Overseas Expeditions from India*, vol. 3, p. 397.
174 *Frontier and Overseas Expeditions from India*, vol. 3, p. 398.
175 Mowbray Morris, *The First Afghan War*, pp. 99–101.
176 Mowbray Morris, *The First Afghan War*, p. 104.
177 Mowbray Morris, *The First Afghan War*, p. 105.
178 Christian Tripodi, *Edge of Empire*, p. 12.
179 Christian Tripodi, *Edge of Empire*, p. 3.
180 G.F. MacMunn, *The Armies of India* (1911, reprint, New Delhi: Heritage Publishers, 1991), pp. 146–148, 150–153.
181 Waziris, Description of the Various Sections of the Three Great Divisions of Waziris, no. 4, pp. 1–3, MSS.EUR.F.163/3, IOR, BL.
182 *Official History of Operations on the N.W. Frontier of India 1936–7* (n.d., reprint, Uckfield, East Susex: Naval and Military Press, n.d.), p. 2.
183 Waziris, no. 4, p. 13.
184 Waziris, no. 4, Report by Lieutenant-Colonel C.W. Macgregor.
185 George MacMunn, *Vignettes from Indian War* (1901, reprint, New Delhi: Low Price Publications, 1993), p. 187.
186 *Official History of Operations on the N.W. Frontier of India 1936–7*, p. 2.
187 R.C. Butalia, *The Evolution of the Artillery in India*, pp. 225–226.
188 R.C. Butalia, *The Evolution of the Artillery in India*, p. 222.
189 Victoria Schofield, *Afghan Frontier*, p. 53.
190 G.F. MacMunn, *The Armies of India*, p. 155.
191 Pierce G. Fredericks, *The Sepoy and the Cossack*, p. 69.
192 G.F. MacMunn, *Vignettes from Indian War*, pp. 103–111.
193 Charles Allen, *Soldier Sahibs*, pp. 69–70.
194 Charles Allen, *Soldier Sahibs*, pp. 94–96, 98, 103.
195 Charles Allen, *Soldier Sahibs*, pp. 126–127, 139, 149.
196 Waziris, no. 4, p. 2.
197 Rob Johnson, *The Afghan Way of War*, p. 87.
198 From offg. secy. to chief commander, Disturbances among the Native Troops in Punjab, no. 2, 3 October 1857, List of Letters addressed to govt. subsequent to 29 June 1857, Foreign Secret Political (FSP), NAI, New Delhi.
199 From offg. secy. to chief command of Punjab, no. 20, FSP, 1857.
200 List of Letters, no. 45, 28 August 1857, FSP.
201 Charles Chenevix Trench, *The Frontier Scouts* (London: Jonathan Cape, 1985), Foreword by Philip Mason, p. xii.

202 K.M.L. Saxena, *The Military System of India: 1850–1900* (New Delhi: Sterling, 1974), pp. 127–128.
203 Waziris, no. 4, pp. 7–8.
204 Memo of the Composition of the Corps of Guides, Lieutenant-Colonel H.B. Lumsden, 17 January 1862, Records of Chief Commands, Notes and Minutes by Napier, MSS.EUR.F.114, 5(2).
205 From Major-General S. Cotton, Commanding Peshawar Division, to the ADG of the Army, Calcutta, 26 July 1860, no. 723, Records of Chief Commands, Notes and Minutes by Napier.

6 Technological advance in the war against the Mongols

Kai Filipiak

Chinese strategies to fight the Mongols

The conflict between the Chinese and the Mongols originated in the thirteenth century when the Mongols conquered China, ruling it for nearly 100 years. Although the Chinese, under the leadership of Zhu Yuanzhang, were able to defeat the Mongols and establish the Ming dynasty in 1368, the conflict continued until the second half of the sixteenth century. Due to the military power of the Mongols, the enormous costs of the conflict for the Chinese state and the persistence of the conflict, the northern border formed the key element of the Chinese defence strategy.

Even though the conflict remained a constant presence throughout the Ming dynasty, the historical background changed together with the structural causes of conflict, its intensity, Chinese strategies as well as the structure of the Mongolian confederations. With regard to Chinese military strategies, three stages can be distinguished.

First stage (1368–1398). After a series of successful offensive operations against the Mongols during the first five years of the Ming reign, the renowned Chinese generals Xu Da, Tang He and Li Wenzhong suffered serious defeats in 1372. As a result, the government initiated the change towards a more defensive strategy aimed at securing the border. In this context, the system of guards and posts along the border expanded and people were resettled to the northern periphery. A double defensive line consisting of outposts and strong garrisons behind them was designed to prevent the Mongols from raiding Chinese territories. This is not to say, however, that the Chinese completely gave up offensive operations. For instance, in 1388 Chinese forces launched a surprise attack on Tögüs Temür's camp near Buyuerhai, almost 1,000 kilometres north of Beijing.[1] Zhu Yuanzhang, the first Ming emperor, changed his strategy after 1372 and focused on establishing control over a limited territory without any claim to ruling the steppe and the people who lived there.

Second stage (1403–1424). The third Ming emperor, Zhu Di, modified the strategy by a combination of offensive and defensive measures, which included pre-emptive strikes and the expansion of border defence. From 1409 until 1424 he conducted six campaigns against the increasing power of the East Mongols under Arughtai and the West Mongols under Mahmud.[2] Due to the relocation of the capital from Nanjing to Beijing, the bulk of the Chinese military force was concentrated in the north.

However, Zhu Di's pro-active policy against the Mongols should not be considered a change of Zhu Yuanzhang's defence strategy. It is apparent, for example, that his campaigns were not aimed at conquering land from the nomads. The withdrawal from the outposts, which formed the first defensive line, did not help to improve the offensive capacities of the empire. Furthermore, the construction of signal towers near Juyongguan (1410),[3] in Shanxi[4] and Liaodong (1413),[5] the building of fortresses along the border to protect men and cattle, and the increasing number of soldiers positioned at border crossing points strengthened the system of border defence during the Yongle reign (1403–1424).

Third stage (since 1449). The defeat of the Chinese army near Tumupu,[6] led by the inexperienced emperor Zhu Qizhen in 1449, initiated the transition to the static defence of the northern border to one in which various parts were combined to form the largest defensive wall the world has ever seen. Based on a dissociative strategy that called for a clear-cut separation between the Chinese and nomadic world, the systematic expansion of the so-called Great Wall had been pushed ahead. Even worse, the tea and horse markets along the border that served to satisfy Mongolian demands for Chinese goods were often closed when nomadic tribes did not accept the Chinese world order.

As a result, the Mongols were cut off from the flow of goods and raided the Chinese provinces in search of products they wanted, such as grain, textiles, luxury goods, as well as armour and weapons. The conflict escalated in the sixteenth century when the harsh emperor Zhu Houcong (Jiajing 1522–1566) strongly condemned any compromise between China and the Mongols. In response, the Mongols under the leadership of Altan Khan invaded China almost every year in order to enforce trading opportunities. Under these circumstances, the Chinese government made a concerted effort to enhance the border defence. The following investigation demonstrates how the war against the Mongols improved Chinese technological progress in the field of firearms and fortification during the sixteenth century.

The impact of European cannons on the border war

Since the eleventh century firearms and gunpowder had been widely used in the Chinese military. In terms of modern warfare, however, gunpowder weapons became a regular tool of war in the battles against the Jurchen and Mongols during the twelfth and thirteenth centuries.[7] In other words, when Chinese armies fought the Mongols in the sixteenth century both sides had long been familiar with this kind of warfare.

In the sixteenth century, however, Chinese firearms advanced due to the integration of foreign technology. The progress was related to the war against the pirates during the sixteenth century, when the Chinese were faced both with well-armed Portuguese traders and Japanese who had imitated and improved on the Portuguese arquebuses.

The Chinese recognized the importance of firearms. The famous Ming general Qi Jiguang (1528–1588) states in his *New Book of Effective Discipline* (*Jixiao xinshu* 紀效新書), dated from approximately 1560, that firepower is the most important part of naval warfare.[8] According to Wang Zhaochun, 30 per cent of his army was armed with European cannons and arquebuses. Another 20 per cent used Chinese types. All in all, 50 per cent of Qi Jiguang's troops were armed with fire weapons.[9]

However, the continued war against the Mongols in the north created a higher demand for new military technology that caused the government to change the strict regulations for the production of firearms from the beginning of the Ming dynasty. The Ming government originally established the armoury (*junqi ju* 軍器局) as part of the Ministry of Works and the Office of Armament (*bingzhang ju* 兵丈局) to manage the production of weapons, including firearms.[10] Efforts to decentralize the structure to increase efficiency failed. When the two border generals Huang Zhen and Yang Hong opened an office in Dushi in 1441 that supervised the production of firearms in the Xuanfu section of the northern border, the emperor ordered the closure of the office because he feared the betrayal of state secrets.[11] Only after the serious defeat near Tumupu in 1449 did the government permit the production of firearms within the different defence areas of the northern border, starting with Sichuan.[12]

Another issue concerns the performance of guns in terms of range and firepower. The advancement of firearms was an important point in light of the decreasing number of soldiers in the frontier garrisons. High officials recognized the shortcomings of Chinese guns. Wang Hong (?–1536), Senior Censor-in-Chief, considered conventional measures to be insufficient to defend the border against the Mongols. The main reason for this development, he believed, was the limited range of Chinese guns.[13]

Weng Wanda (1498–1552), Supreme Commander of the defence areas Xuanfu, Datong and Shanxi, supported the further development of firearms. In his *Memorial on the Production of Firearms* (*Zhizao huoqi shu* 置造火器疏), presented in 1546, Weng Wanda lists several reasons why he modified the Chinese firearms. He argues that the knowledge of old firearms was not handed down to the present day. In addition, the old firearms could not be easily used in warfare due to their heavy weight. After firing one time, the guns had to be reloaded – at which point the enemy had the advantage. Moreover, when the aim was not exact, the projectile could not hit the target. For this reason, the soldiers needed extensive training. Otherwise, it was too difficult to use these weapons.[14]

According to Weng Wanda, another important reason for the development of firearms is the fact that the Mongols had adapted to Chinese warfare. He said:

> Today, the weapons and armament of the northern barbarians are excellent. China cannot defeat them. Several years ago, our soldiers fought against the barbarians. After the fighting had begun, our soldiers, using bone flower maces and wolf tooth clubs, defeated them. The barbarian blades were too short to hit us. At this time, China controlled the barbarians. However, last

> year, when the barbarians invaded Baoan in Shaanxi, they had adapted to bone flower maces and wolf tooth clubs. China could not defeat them. If we had firearms such as guns, European cannons, and poison-fire-cannons, then we would have something that the barbarians do not have and also what they fear.[15]

It seems that the Mongols were superior in the use of cold weapons. China, however, had the advantage of firearms. Weng Wanda, therefore, favoured the technical development of firearms, which could help to maintain China's hegemony. For this reason, the government forbade the trade of sulphur, an important component of gunpowder. Although the Mongols were able to produce saltpetre, they lacked sulphur resources.[16]

The adoption of European firearms enhanced the quality of Chinese weapons. Several historiographical sources outline the usage of European cannons, called *folang ji* (Frankish machines) in Chinese.[17] However, the time when the Chinese first learned of this weapon is the subject of controversy.

Well-armed Portuguese trade ships using 'modern' firearms with impressive range and firepower reached the Chinese coast at the beginning of the sixteenth century. The conventional view is that in 1523 the Chinese captured two Western ships. The Portuguese firearms they found were presented to the emperor and given the name *folang ji*. According to Needham, there is evidence that the Chinese used cannons of this type as early as 1510. For this reason, it is doubtful whether it reached China from the Portuguese, because Malacca did not fall until 1511, as Needham points out.[18]

Needham refers to an entry in the *Fujian tongzhi* (*Local Records of Fujian*) saying that rebels from Guangdong invaded Xianyou 仙游 county in 1506, but were defeated by the righteous man (*yimin* 義民) Wei Sheng 魏升, who lured the rebels into an ambush where he had hidden several hundred *folang ji*. In 1510 he used more than 100 *folang ji* to attack brigands from Tingzhang 汀漳, who invaded Xianyou.[19]

Although the *Fujian tongzhi* was compiled long afterwards, Needham has 'no reason for doubting [it]'.[20] In contrast to Needham, Wang Zhaochun questions the credibility of the source for several reasons. He argues, for example, that it seems highly unlikely that a group of local patriots had enough material and gunpowder and were also skilled enough to produce the cannons.[21]

However, there is no doubt that European cannons became popular in China at least starting in the first half of the sixteenth century. Wang Hong as well as Weng Wanda appreciated the quality of these weapons and advocated for their production. Compared to their Chinese counterparts, European cannons provided many benefits. Qi Jiguang described the main advantage in his *Jixiao xinshu* as follows:

> The advantage [of the *folang ji*] is the long barrel [*muchong*]. Its length and straightness enhance the maximum firing range. The belly [of the gun] houses the chamber [*zichong*], which consists of two parts that prevent the

> accumulation of heat. Behind the chamber there is a tenon clamped. Each time one fires the gun, the tenon flies several meters away from the chamber, injuring people. The best way is to use an iron bolt. Another advantage [of the *folang ji*] is the two front sights [*zhaoxing*] placed on the cannon one behind the other. Pressing the rear handle slightly down, [the gunner's] field of vision remains adequate and he targets his aim with his eyes.[22]

In his later treatise *Lianbing shiji* (*Practical Handbook on Military Training*) from 1571 Qi Jiguang states that the *folang ji* 'is the most effective weapon and without equal in terms of handling and the rate of fire'.[23] The rapid rate of fire, an important feature of the European breech-loader, was made possible through the use of several chambers prepared with gunpowder and balls. After firing the gun, the chamber was replaced by another one. To sum up, the larger firing range, the greater rate of fire, the proto sight for better targeting and the easy handling were the special features of the breech-loader, which made the cannon the most effective firearm at this time.

The production of the *folang ji* had already started in 1524 in Nanjing. In 1529 Wang Hong suggested the government produce cannons of the type *da jiangjun* (Generalissimo), a copy of the European cannon, which should be distributed to all sectors of the northern border.[24] By the end of the year 300 cannons of this type had been produced and distributed to the border areas.[25]

One year later Wang Hong asked for more European-styled cannons to defend observation towers (*duntai*) and fortresses (*chengbao*). Each observation tower should be defended by three soldiers using a small cannon that could fire cannon balls of 20 *jin* (12.0 kg) within a range of 600 *bu* (921 m). Each fortress should be defended by ten soldiers using three big cannons that could fire cannon balls of 70 *jin* (42.2 kg) within a range of five to six *li* (2.75–3.33 km).[26]

It is evident that Ming officials advocated the use European firearms. The question, however, is whether the government put these proposals into effect. In other words, can we find evidence for technological progress in the war against the Mongols?

There are three important indications of the development of European firearms in China. First, the increasing number of European cannons demonstrates that Chinese craftsmen acquired competence and the know-how to produce these weapons. Second, the increasing variety of European cannons made in China indicates the adaption of foreign firearms to Chinese needs. Third, new types of Chinese cannons prove the further development of Chinese firearms. Historical sources provide evidence supporting this view.

The *Da Ming huidian* (*Collected Statutes of the Great Ming*) lists, for example, quantities of different *folang ji* types. In 1528, 4,000 cannons of the small type were produced and distributed to garrisons and fortresses. One year later 300 additional cannons of this type were produced. In 1544, 1,000 guns of a portable *folang ji* were made to be used from horseback. In 1564, 100 additional hand arms of this type were made.[27] Considering the huge number of garrisons the amount of European cannons sounds not as impressive, but we have to take into account that this type was only one among many firearms.

Other sources provide information on the local distribution of European-styled firearms. For example, in 1536 the three border areas of Shaanxi received 2,500 *folang ji* cannons. In the following year Shaanxi received an additional 3,800 small *folang ji* cannons.[28]

Finally, military handbooks inform us about the quantitative distribution of firearms in Chinese barracks. In his *Liangbing shiji*, Qi Jiguang provides an overview of military equipment distributed to camps of charioteers, infantry and cavalry. According to Qi Jiguang, a charioteer camp of 3,190 soldiers and 128 chariots was armed with 256 *folang ji* cannons.[29] Cavalry and infantry camps did not have this type of cannon. How can we explain this?

Qi Jiguang states that *folang ji* cannons were not very mobile, because they were quite heavy. Due to their weight these firearms were not suitable for troops on the march. For that reason, *folang ji* were originally only used to defend the border walls.[30] Only after the establishment of chariot troops could the cannons be moved by chariots. The infantry and cavalry, however, did not have such chariots.

It seems that the Ming military made extensive use of heavy artillery to defend the so-called Great Wall against the Mongols. This corresponds to the static defence strategy adopted by the government after the Tumupu incident of 1449. High officials and the military recognized the importance of providing artillery with far-reaching projectiles and strong firepower. The great demand for *folang ji* at the northern border is illustrated in Wang Tao's (?–1587) memorial on firearms for towers and chariots from 1575. Wang Tao, Supreme Commander of the border areas Jizhou and Liaodong, calculated that 13,370 *folang ji* were needed for towers and 3,328 for chariots to defend Jizhou and Changping. On the basis of existing *folang ji*, he determined that 6,981 of these cannons were needed.[31]

Apart from the quantities, European cannons had a great impact on the qualitative improvement of Chinese firearms. For example, new types, such as *wudi dajiang jun* 無敵大將軍, *feishan shenpao* 飛山神炮, *baichu xianfeng pao* 百出先鋒炮, were developed imitating European cannons. In addition, the *folang ji* itself was produced in various types including small, middle and large cannons. Illustrations of the *folang ji* found in Qi Jiguang's *Jixiao xinshu* as well as Zheng Ruozeng's *Chouhai tubian* (*Illustrated Naval Strategy and Tactics*), published in 1562, prove another remarkable innovation. The cannon was flexibly mounted on a rack, which enabled the gunner to fire in many directions.[32] Furthermore, Weng Wanda's *Memorial on the Production of Firearms* caused the Ministry of Works to produce cannons that fired projectiles of different sizes. The range of large-sized projectiles was 700 *bu* (*c*.1,075 m). Small-sized projectiles had a range of 100 *bu* (*c*.153 m).[33] In short, the introduction of European cannons contributed to advances in range, rate of fire, explosive power and the serial production of highly developed firearms.

Did the development of more efficient firearms lead to military success against the Mongols? Some sources indicate that the increased firepower had an effect on the fighting. For example, when the Mongols under the leadership of Altan launched an attack on Yongning 永寧 in 1549, Ming troops led by the Supreme Commander Weng Wanda and the Regional Commander Zhou Xiangwen defeated them by using firearms. Several thousand Mongols died.[34] It seems that the Mongols were afraid of Chinese firearms. At the end of the year 1554, for example, when the Mongols attacked the area of Jizhou, the Supreme Commander Yang Bo chose fearless soldiers who went in the Mongolian camp and frightened them by using firearms.[35] Another source includes a statement attributed to the effect of Qi Jiguang's presence in Jizhou during the second half of the sixteenth century. It is said that when Qi Jiguang was in charge of Jizhou, the Mongols did not attack him, because they were afraid of his firearms.[36]

Progress in fortification

The second important technological advance in the war against the Mongols concerns fortification. During the sixteenth century, wall segments on the northern border were expanded and unified to become the Great Wall. Chinese and Western literature provide a wealth of information about the dimensions of this work.[37] However, little is known about the qualitative progress in fortification at this time.[38] General references to walls and beacon towers do not give any idea of innovation and progress, because they had both been elements of fortification for many centuries. Furthermore, walls, moats, pitfalls, fortresses, watch and beacon towers were already parts of border defence in the second half of the fifteenth century.

Progress in fortification during the sixteenth century can be attributed to the fact that the defence infrastructure was in bad condition. Many of the walls that had been constructed earlier had eroded, such as the north-west wall constructed by Yu Zijun in 1474.[39] In particular, the walls had proven unable to meet the military's needs. As Qi Jiguang states:

> In the past the border walls were low and thin and often collapsed. In between there were small platforms made of brick and stone that were not integrated [into the wall] and system of mutual assistance. Soldiers, standing there in the sun, rain, frost and snow, had no place for rest and protection. When military equipment and firearms were temporary needed they could not be transported forward [to the frontier]. In the case that [equipment and firearms] should be stored behind the wall, there was no place to store them.[40]

Chinese efforts to expand the fortification system provoked Mongolian raids, which increased during the Jiajing reign (1522–1566). The Mongols invaded the Chinese border areas to enforce trading opportunities nearly every year. Under the strain of increasing warfare, the impressive Great Wall was developed as a symbol of technological advance in fortification. The Great Wall exemplifies three important innovations. The first involves building materials and new construction techniques. Walls made of bricks and stones replaced the former walls of stamped earth.

Second, this kind of work required new forms of organization, more logistic efforts to provide materials, higher costs and a better quality of labourers to produce the bricks and carve the stones.

The third aspect of technological advance is related to new elements of fortification or, rather, the new design of the conventional defence architecture. Many of these elements we only find in Xuanfu, Datong and Jizhou, the key centtes of border defence in the north. The following investigation focuses on some of the important modifications in Chinese fortifications.

It is striking that fortifications during the sixteenth century were much larger than before. Supreme Commander Yang Yiqing (1454–1530), who reconstructed the old wall of Xu Tingzhang in 1506, was one of the first who designed huge walls. His walls were over seven metres high and thick and provided sleeping facilities for 900 soldiers. The moats behind the wall were also over seven metres wide and deep.[41] In 1555, high platforms for firearms were constructed on the walls.[42] The platforms became more and more robust indicating the increasingly use of heavy firearms. There were also changes outside the wall. Qi Jiguang (see below) provided the wall base with slopes preventing enemies from climbing the wall and hindering Mongolian horsemen from approaching the wall.[43]

Fortresses formed a significant part of border defence. They were located in the Chinese hinterland, serving as a base for soldiers and military equipment. Yin Geng's *Xiangyue* (*Community Contract*) from 1550 indicates that fortresses underwent important changes. Criticizing the old system of low parapets, he states that one could easily get over the walls and the parapet could not provide adequate protection against projectiles. In addition, towers were located on the same level as the walls, which made it difficult to defend the parapet. The small size of platforms hindered the firing of guns and arrows. Another problem related the spacing of towers. Often the large gap between two pinnacles allowed two Mongolian shooters to fire between the gap, with the result that Chinese soldiers did not dare to look out.[44]

According to Yin Geng, the system was modified during the sixteenth century. He describes the nature of the construction as follows:

> Today, the [corner] towers are three zhang [10.74 m] high and one zhang [1.53 m] higher than the walls. The tower platforms are two square zhang in size and protrude three chi [1.07 m] beyond the parapet from East to West. Wooden piles of two chi [0.71 m] in height bearing a crosspiece are erected on the parapet similar to a wall. Wooden plates fill the gaps [between the pinnacles]. There is a transverse batten on the outside and a hinge inside [the wooden plate] for opening and closing the plate as needed.[45]

Apart from architectural changes, the massive attacks of Mongolian horsemen required a more stable construction of the parapets. Although there is little information available about structural aspects, Yin Geng points out that in contrast to the old system, the parapet should be built with unbaked hand-formed mudbricks (*nipi* 泥坯) which must be horizontally placed. A mix of straw and clay keeps the mudbricks tight together.[46]

Yin Geng offers a new and cheap method to stabilize the fortification. First, the mix of straw and clay increased the strength of parapets. Second, the original vertically placed mudbricks were designed as a stretcher course enhancing its stability.

Another progress in fortification is related to the construction of fighting towers (*ditai* 敵臺). It seems that the fighting tower formed a particular element of Ming fortification, because the term is difficult to find in earlier sources. Furthermore, historical sources describe it as a part of the military urban architecture. The *Bamin tongzhi* (*Gazetteer for the Province of Fujian*), for example, states that in 1443 guards and posts, such as Yongning and Fuquan, protected their city gates with fighting towers.[47] Both cities were located on the coast. Although pirates did not threaten Fujian at that time, the development of fortification must be seen in the context of the war against pirates.

In 1450 fighting towers were also constructed in the north. Twenty towers were built in the county seat of Jiaocheng, south-west of Taiyuan.[48] Fighting towers as part of border fortifications became popular during the Jiajing reign (1522–1566). At the beginning of this reign Liang Duansu suggested the reconstruction of the approximately 70 kilometre long wall between the border areas Yansui and Ningxia. The proposal included the construction of fighting towers every 500 metres that were to be seven *chi* (2.5 m) higher than the wall with a ground plan of three square *zhang* (115 m^2) and two divided floors. The bottom floor should used for military equipment, rest and food. The upper floor consisted of a small chamber that served as a lookout.[49]

Several decades later, Qi Jiguang presented an advanced model that was described as a 'fighting tower with a hollow core' (*kongxin ditai* 空心敵臺). Qi Jiguang, who had been in charge of military training in Jizhou, Changping and Baoding since 1567, asked in his *Request for Fighting Towers* for the construction of 3,000 fighting towers in Jizhou and Changping.[50]

In his *Lianbing shiji* he introduces this new type of tower based on new tactical principles. According to the *Lianbing shiji*, the tower should have a height of three to four *zhang* (10.74 to 14.32 m) and a perimeter of 12 (42.9 m) to 17 or 18 *zhang* (60.87 to 64.45 m).[51] The towers, linked by a wall, protrude up to one *zhang* and four to five *chi* (4.98 to 5.33 m) beyond and up to five *chi* (1.79 m) behind the wall.[52] Wall and tower bases should be on the same horizontal axis. In his *Principles for the Construction of Fighting Towers*, Qi Jiguang paid special attention to the base:

> The base should be made of stone, but I am afraid it will be difficult to obtain squared stones. Quarry stones are certainly not stable enough. Should we find stones, use them. If not, use bricks. If there is good soil, we will use mortar for grouting [the base]. If the base is strong, the mortar should be thick. A strong base should have a wall thickness of two chi [0.7 m].[53]

Serruys states that most of the towers seem to have been surrounded by a wall, often with a ditch or moat.[54] Neither Qi Jiguang's description nor the illustration

Table 6.1 A quantitative survey on the existence and construction of fighting towers

Area	*Number of fighting towers*	*Additional information*
Jizhen	1,240	–
Changping	250	–
Liaodong	31	–
Baoding	291 old 359 new	–
Datong	89	Middle of the Jiajing era order to build 89 *ditai*
Shanxi	3,711	
216	1575 order to build 216 *ditai*	
Yansui	239	1575 order to build 216 *ditai*

of fighting towers prove his statement. The illustration shows clearly that the fighting tower was integrated in the wall. But it could be that watchtowers (*duntai* 墩臺), which were defended by a larger crew, had these characteristics.

According to Qi Jiguang, fighting towers should have an empty and open middle floor with crenels in all directions. The upper floor provides a platform surrounded by a battlement.[55]

In short, fighting towers provided many tactical and logistical benefits. Qi Jiguang emphasized their role for blocking and defending the border crossing points.[56] They should be built close together so that one might assist the other. In the event of an attack, the tower crew of 30 to 50 men[57] inside the tower is well protected against arrows and can fire guns at the attackers.[58]

In addition, the construction of the towers offers space to store large amounts of firearms and gunpowder. Qi Jiguang suggested the storage of ten different types of firearms, equipment and 200 kilograms of gunpowder.[59]

Did the government follow Qi Jiguang's suggestions? According to the *Da Ming huidian*, extensive work was dedicated to fortification in 1568 in Jizhou and Changping, where border walls were strengthened. The passage mentioned the construction of 1,500 watchtowers (*duntai* 墩臺), which were twice as high as the wall and offered room for 50 soldiers.[60] It is doubtful, however, that these towers were fighting towers. We have also taken into account that the sources sometimes lack a clear differentiation between 'fighting towers', 'fighting towers with hollow cores' and 'observation towers'.[61] But the fact is that numerous sources use the term 'fighting tower' and 'fighting tower with hollow core' indicating a new trend in fortification at the northern border. Shen Shixing's version of the *Da Ming huidian* provides a quantitative survey on the existence and construction of fighting towers (see Table 6.1).[62]

Conclusion

It has been shown that the technological progress in the war against the Mongols during the sixteenth century was a result of the clear political preference for a military solution to the conflict. This included a predominantly defensive strategy which focused on the expansion and strengthening of the fortification system and the increase of firepower.

Furthermore, the military performance of the Mongols initiated learning processes in which the supporters of technological progress criticized the old system of Chinese firearms and analysed the advantages of new technologies. It becomes evident that the civil-military elite, which promoted the development and adoption of non-Chinese technologies, was open to technological progress.

As demonstrated above, the knowledge that generated advances came from different sources. Technological progress in the north was the result of knowledge transfers from west to east (European cannons) and from south to north (weapons and military experiences) and possibly from urban to frontier space (fighting towers).

New weapons and elements of fortification, new materials and construction methods, mass production and the further development of European technology define the quantity and quality of technological change.

In European history, there was a close relationship between artillery and fortification, the one driving the other forward. But in contrast to Europe, where the increasing firepower required new types of protection architecture, we find another form of this relationship in China. Progress in fortification was not caused by advanced Mongolian firearms, but by heavier Chinese artillery that required a stable platform for firing.

As has been shown, technological advances brought about many changes not only in the military, but also other areas. New technology normally has a deep impact on organization, training, tactics and many other related military fields. For example, gun crews of various sizes were assigned to cannons of different types and had to be trained for varying purposes. They stored large amounts of gunpowder in new types of towers that were easy to defend through a system of mutual assistance.

In addition, the military change also effected bureaucratic changes (decentralization of production) and economic changes (ban of the export of gunpowder ingredients). Moreover, the technological change became part of the cultural memory by numerous texts which outline different aspects of the change.

Of course, many questions remain open. We need to know more about other actors of technological progress, such as the craftsmen and technicians who produced weapons or constructed fighting towers. We also know little about the social structures and training of the gun crew. In this context, it would be also interesting to know how effective military innovations actually were. More detailed investigations are necessary with regard to the political, economical and social consequences of military technological innovations. In this respect, the Ming dynasty seems to be a period of great potential for further research.

Notes

1 Gu Yingtai, *Mingshi jishi benmo*, in *Siku quanshu* (Shanghai: Shanghai guji chubanshe, 1987), vol. 364, pp. 216–217.

2 For details see Wolfgang Franke, 'Yunglo's Mongolei-Feldzüge', in *Selected Papers on Ming and Qing History and on the Overseas Chinese in Southeast Asia 1942–1988* (Singapore: South Seas Society, 1989), pp. 38–54.
3 *Ming shilu* (Taibei: Zhongyang yanjiuyuan lishi yuyan yanjiusuo, 1962), vol. 7, p. 1304.
4 *Ming shilu*, vol. 8, p. 1709.
5 *Ming shilu*, vol. 8, p. 1723.
6 The campaign is described in Frederick W. Mote, 'The T'u-mu Incident of 1449', in Frank A. Kierman and John K. Fairbank (eds), *Chinese Ways in Warfare* (Cambridge, MA: Harvard University Press, 1974), pp. 243–272.
7 Peter A. Lorge, *The Asian Military Revolution* (Cambridge: Cambridge University Press, 2008), p. 1.
8 Qi Jiguang, *Jixiao xinshu*, in *Zhongguo bingshu jicheng* (Beijing: Jiefangjun chubanshe, 1995), vol. 18, p. 1065.
9 Wang Zhaochun, *Zhongguo huoqi shi* (Beijing: Junshi kexue chubanshe, 1991), p. 181.
10 Zhang Tingyu *et al.*, *Mingshi* (Beijing: Zhonghua shuju, 1984), vol. 8, p. 2265.
11 Zhang Tingyu *et al.*, *Mingshi*, p. 2264.
12 Zhang Tingyu *et al.*, *Mingshi*, p. 2265.
13 Wang Zhaochun, *Zhongguo huoqi shi*, p. 125.
14 Chen Zilong *et al.*, *Ming jingshi wenbian* (Beijing: Zhonghua shuju, 1997), vol. 3, p. 2343.
15 Chen Zilong *et al.*, *Ming jingshi wenbian*, p. 2345.
16 Song Yingxing, *Erschließung der himmlischen Schätze* (tr. Herrmann Konrad) (Bremerhaven: Wirtschaftsverlag NW, 2004), p. 242.
17 *Folang ji* is a transliteration of the term Farangi that points to the Frankish Empire. *Folang* or Franken was a general designation of Europeans derived from Arabs talking of Frankistan. See Joseph Needham, *Military Technology, the Gunpowder Epic*, in *Science and Civilisation in China*, vol. 5, part 7, p. 372.
18 Joseph Needham, *Military Technology, the Gunpowder Epic*.
19 Chen Shouqi *et al.*, *Fujian tongzhi* (Taibei: Huawen shuju, 1968), vol. 10, p. 5061.
20 See Joseph Needham, *Military Technology, the Gunpowder Epic*, p. 372.
21 Wang Zhaochun, *Junshi gongcheng jishu shi* (Taiyuan: Shanxi jiaoyu chubanshe, 2007), pp. 276–277.
22 Qi Jiguang, *Jixiao xinshu*, p. 500.
23 Qi Jiguang, *Lianbing shiji*, in *Zhongguo bingshu jicheng* (Beijing: Jiefangjun chubanshe, 1994), vol. 19, p. 644.
24 Zhang Tingyu *et al.*, *Mingshi*, vol. 8, p. 2264.
25 *Xu wenxian tongkao* (Hangzhou: Zhejiang guji chubanshe, 2000), vol. 2, p. 3996.
26 *Xu wenxian tongkao*, p. 3996.
27 Li Dongyang, *Da Ming huidian* (Taibei: Xinwenfeng chubanshe, 1976), vol. 5, p. 2620.
28 *Xu wenxian tongkao*, p. 3997.
29 Qi Jiguang, *Lianbing shiji*, p. 705.
30 Qi Jiguang, *Lianbing shiji*, p. 644.
31 Liu Xiaozu, *Sizhen sanguan zhi* (e-edition Original Wanli edition, CrossAsia.org: Zhongguo jiben guji ku, 1576), p. 415.
32 See for example Zheng Ruozeng, *Chouhai tubian* (Beijing: Zhonghua shuju, 2007), p. 901.
33 Zhang Tingyu *et al.*, *Mingshi*, vol. 8, p. 2264.
34 Tan Qian, *Guoque*, 6 vols (Beijing: Zhonghua shuju, 2005), vol. 4, p. 3729.
35 *Ming shilu*, vol. 46, p. 7212.
36 Shen Defu, *Wanli yehuo bian* (Beijing: Zhonghua shuju, 1997), vol. 2, p. 433.
37 See for example Drafting Committee for Cultural Treasures, *Zhongguo Changcheng*

yiji diaocha baogao ji (Beijing: Wenwu chubanshe, 1981); Fan Zhongyi, Wang Zhaochun, Zhang Wencai and Feng Dongli, *Mingdai junshi shi*, in *Zhongguo junshi tongshi* (Beijing: Junshi kexue chubanshe, 1995), vol. 15 parts 1–2; Arthur Waldron, *The Great Wall of China* (Cambridge: Cambridge University Press 1990).

38 One noteworthy example is Henry Serruys' excellent study 'Towers in the Northern Frontier Defenses of the Ming', *Ming Studies* no. 14 (1982), pp. 9–76.

39 Arthur Waldron, 'The Problem of the Great Wall of China', *Harvard Journal of Asiatic Studies*, vol. 43, no. 2 (1983), p. 661.

40 Qi Jiguang, *Liangbing shiji*, p. 684.

41 Wei Huan, *Huang Ming jiubian kao*, in *Siku quanshu cunmu congshu* (Shandong: Qi-Lu shushe, 1996), vol. 226, part shi, p. 13.

42 *Xu wenxian tongkao*, vol. 2, p. 3953.

43 Fan Zhongyi *et al.*, *Mingdai junshi shi*, vol. 15/2, p. 747.

44 Yin Geng, *Xiangyue*, in *Zhongguo bingshu jicheng* (Beijing: Jiefangjun chubanshe, 1994), vol. 40, p. 57.

45 Yin Geng, *Xiangyue*, vol. 40, pp. 57–58.

46 Yin Geng, *Xiangyue*, vol. 40, p. 59.

47 Chen Dao, *Bamin tongzhi* (e-edition Original Hongzhi edition, CrossAsia.org: Zhongguo jiben guji ku, 1491), p. 231.

48 Hu Mi, *Shanxi tongzhi* (e-edition Original Chenghua edition, Cross Asia.org: Zhongguo jiben guji ku, 1474), p. 88.

49 Chen Zilong *et al.*, *Ming jingshi wenbian*, vol. 2, p. 936.

50 Chen Zilong *et al.*, *Ming jingshi wenbian*, vol. 5, p. 3749.

51 Qi Jiguang, *Liangbing shiji*, p. 684.

52 Qi Jiguang, *Liangbing shiji*, p. 685.

53 Chen Zilong *et al.*, *Ming jingshi wenbian*, vol. 5, p. 3759.

54 Henry Serruys, 'Towers in the Northern Frontier Defenses of the Ming', p. 23.

55 Qi Jiguang, *Liangbing shiji*, p. 685.

56 Qi Jiguang, *Liangbing shiji*, p. 684.

57 Qi Jiguang, *Liangbing shiji*, p. 685.

58 Qi Jiguang, *Liangbing shiji*, p. 685.

59 In detail see Qi Jiguang, *Liangbing shiji*, pp. 687–688.

60 Li Dongyang, *Da Ming huidian*, vol. 4, p. 1837.

61 See for example Henry Serruys, 'Towers in the Northern Frontier Defenses of the Ming', p. 23.

62 Data based on Shen Shixing, *Da Ming huidian* (e-edition, Cross Asia.org: Zhongguo jiben guji ku, 1587), pp. 1210–1222.

7 Bringing in the big guns

On the use of artillery in the Ming–Manchu war

Kenneth M. Swope

Not unlike the Ottoman Empire, the Chinese Empire has traditionally suffered from bad press when it comes to discussing the development and deployment of military technology, particularly in the early modern era. While the countries of Western Europe supposedly went through what is called a 'Military Revolution' from the sixteenth to the eighteenth centuries which enabled them to colonize the world through the application of superior technologies and tactics, the 'Oriental despotisms' of China and the Ottoman Empire allegedly suffered from technological and institutional ossification that left them open to foreign exploitation and domination.[1] This revolution was deemed even more remarkable by virtue of the fact that the great Muslim Empires of the Ottomans in Anatolia, Europe and Africa, the Safavids in Persia and the Mughals in India were acknowledged by their contemporaries as the world's first great 'Gunpowder Empires', an appellation a Chinese scholar also bestows upon Ming China, calling it the world's first gunpowder empire.[2] This in turn has caused generations of scholars to question exactly how and why the West was able to surpass the empires of the Middle East and Asia in this realm, many positing that it had something to do with the inherently superior nature of Western society and its emphasis upon capitalist competition and democracy, among other things.[3] More recently, scholars have adopted a more fruitful approach, attempting to discern exactly how and why certain cultures adopted firearms for their own military needs rather than assuming cultural superiority by one group or another.[4]

Moreover, recent scholarship from the Chinese and Ottoman perspectives has thankfully challenged such notions, demonstrating that in fact both these great empires were technologically innovative and tactically and institutionally dynamic throughout their later histories.[5] For example, during the Japanese invasion of Korea (1592–1598) the Ming Chinese and their Korean allies were able to defeat one of the world's most disciplined military forces by virtue of superior cannon technology at land and on the seas. And this was after the initial Japanese invasion force ran roughshod over their Korean foes by utilizing large numbers of Portuguese-derived arquebuses.[6] Likewise, the Qing (1644–1911) Empire was finally able to quell the nomadic challenge that plagued all previous Chinese dynasties by successfully melding Chinese bureaucratic and logistical capabilities with the adroit use of firearms in conjunction with mobile steppe warfare.[7] Thus, rather than reading back from nineteenth-century defeats at the hands of Western industrial powers who were enjoying the fruits of a sustained period of economic and political expansion as the result of a unique confluence of factors, it is instructive for us to evaluate the military technology of these empires on their own terms within their own specific contexts. This chapter will attempt to do that by examining the use of heavy artillery by the Ming Chinese and their Manchu (Jurchen) adversaries in the early to mid-seventeenth century. Although a comparative analysis is beyond the scope of the present piece, it should nonetheless provide grounds for further discussions along these lines.

Interestingly enough, while it has generally been widely acknowledged that Jesuit advisers aided both the Ming and the Manchus in casting cannon and in training gunners for their war, the role of firearms in the Qing conquest of the Ming has been downplayed.[8] This is doubly surprising in light of the fact that the foundation of the Ming itself was greatly facilitated by the use of firearms and the Ming state pioneered the use of training divisions specifically tasked with instructing soldiers in the use of firearms in the early fifteenth century.[9] There appear to be three major reasons for this. The first is that, as shall be discussed below, it was recognized that from the early sixteenth century, Western firearms were generally, though not always, superior to native manufactures. It seems that scholars have presumed that because of the vast size of Chinese armies, there simply were not that many firearms available, or at least not many really superior Western-style models. Therefore, such weapons are presumed to have had a negligible impact. Added to this is the fact that the numerically superior Ming were in fact defeated by the Manchus, which suggests that the Ming must not have made adroit use of the few weapons they did have. Along these lines Di Cosmo draws attention to the fact that scholars have tended to focus too much attention on the supposedly superior martiality of the Manchus as the primary reason for their success as opposed to examining more concrete factors such as the adoption of new tactics and technologies.[10] Such explanations also incidentally ignore the significance of the massive peasant uprisings which were concurrently raging throughout north-west and central China and which greatly impeded the Ming's ability to meet the Manchu threat with all its resources. In fact, as will be seen below, the ability to manufacture and deploy heavy artillery was critical to both belligerents in this war, and most of the major engagements were won by the side that made better use of these weapons.

It is well known that the Chinese developed gunpowder weapons though their subsequent widespread use and dissemination throughout Asia is less well documented, particularly by English language scholars.[11] This may be due to the fact that Western firearms were embraced by Asians upon their introduction to the scene in the early sixteenth century. It should be stressed, however, that Asians' willingness to purchase and copy European models was not due to the fact that they were ignorant of such technologies but rather the simple recognition that these models were in fact more efficacious. There was also the fact that the Ming themselves were reluctant to disseminate a technology they desired to monopolize so other Asian polities often had to seek alternative sources for their weapons. At any rate, as soon as Europeans, particularly the Portuguese, started coming to East Asian shores in large numbers, locals were impressed by their firearms and sought to integrate them into their own armies.

The first weapon to achieve widespread use was the *folangji* (feranghi), so named for the Chinese term for Franks and thereby denoting 'Frankish gun'. The date of its arrival in China is generally believed to have been the Zhengde reign (1506–1521). These early guns were approximately five to six feet in length and weighed anywhere from 50–100 catties (*jin*). This made them smaller and more portable than existing weapons and offered users a greater range of applications from mounting them on platforms and city walls to use on coastal patrol boats. They also had better rates of fire and were easier to load than native models. Thus they quickly found favour with capital officials.[12] They were among the weapons ordered manufactured by the Jiajing emperor (r. 1522–1566) as a result of his directive to the Ministry of War to establish a bureau for the manufacture of 'red barbarian cannon'.[13] According to the *Ming huidian* some 5,800 *folangji* were manufactured between 1523 and 1564 alone, in addition to over 30,000 specimens of other types of firearms.[14] Their widespread adoption is also attested by the prominent place their use occupies in the late Ming General Qi Jiaguang's (1528–1588) training manual, *Jixiao xinshu.*[15] Qi endeavoured to attain a ratio of 37.5 per cent of firearms use in his units, though such ratios were not always achieved. One source, for example, indicates that a division of 2,048 troops possessed 256 *folangji*, a ratio of only 12.5 per cent, though this source might not indicate other types of firearms.[16] Nonetheless, such records indicate that from an early date the Ming recognized the utility of firearms as force multipliers and encouraged their widespread use, in contrast to popular misconceptions about the Ming military's technological backwardness.[17]

In addition to these smaller calibre weapons, the Ming state eagerly adopted larger cannon for use in fixed defensive positions. The general term for the larger cannon was *hongyi pao*, or 'red barbarian cannon', but this blanket term could sometimes refer to several different weapons. The original *hongyi pao* were about two *zhang* (approximately 20 feet) in length and weighed around 3,000 catties. They fired blasts which could be heard as far away as ten *li*, or about 3.5 miles.[18] One important repercussion of their introduction to China was that by manufacturing these guns in greater quantities, the Chinese became more aware of the importance of casting guns that had the proper ratios of length to calibre and they made models that lasted longer and fired more consistently. It thus introduced more 'scientific' methods of cannon manufacture to China. The Chinese became so adept at casting cannon that the Portuguese at Goa later specifically requested Chinese cannon founders from Macao. The Chinese apparently specialized in casting iron cannons; the Portuguese themselves, bronze.[19] Chinese capabilities in cannon casting are evidenced by the fact that many of the home-cast Chinese weapons were used through the late Ming era (as opposed to those cast by Jesuits) and some remained in use up until the Opium War.[20] They also saw service in Korea and were copied by the Ming's Korean allies in battling the Japanese.

Turning now to the Ming–Manchu war, the first major military clash between the two sides was the Battle of Sarhu in early 1619. Nurhaci (1559–1626), the Jurchen founder of the Latter Jin state that later became known as the Qing, had earlier articulated his 'seven grievances' against the Ming and the Chinese had finally decided to send a punitive expedition to wipe out the threat burgeoning on their north-eastern frontier.[21] From the Ming perspective, this campaign was a debacle from the start as the Ming columns were poorly coordinated, hampered by late spring snowstorms and allegedly hurt by the greed of their respective commanders who contested with one another for the lion's share of the battle glory. As a result of the latter factor the Ming commander Du Song started off in search of the enemy three days ahead of schedule. Though urged to slow down by his subordinates lest his armies be cut off as they crossed the Hun river near where the Ming would engage the Jin, Du ignored their advice and left his weapons carts on the north bank of the river. His forces were hit as they were in the midst of crossing the river and were thus unable to bring their superior fire-power to bear.[22] A similar fate befell one of Du's co-commanders, the famous and highly decorated General Liu Ting (1554–1619).[23] A third commander, Ma Lin, was also ambushed and killed. Only Li Rubo, suspiciously a boyhood compatriot of Nurhaci, managed to get the orders to withdraw and saved himself and his troops, though he later committed suicide amidst suspicions of colluding with the enemy.[24] All told the Ming lost an estimated 310 officers, 45,870 men and some 28,000 livestock at the Battle of Sarhu.[25]

In the wake of the crushing Ming defeat the Ming Emperor Wanli (r. 1573–1620) cashiered several officials, including the supreme commander of the expedition, Yang Hao (d. 1629), who was replaced by Xiong Tingbi (d. 1625).[26] Xiong's perspicacity cost him his job, however, and he was soon replaced by one Yuan Yingtai.[27] In addition to calling for the mustering of some 130,000 additional troops to meet the Jurchen threat, the Ming court directed the Ministry of Works to construct 300 'caitiff exterminating cannon', 1,000 100-character cannon and approximately 7,000 three-eyed guns and bird guns (*niao chong*, which also sometimes denotes arquebuses) in addition to other weapons, armour, shields and military carts.[28] It was believed that these weapons were essential to restoring the fighting spirit of the Ming. But because the new touring pacification commissioner of the region, Wang Zaijin, feared the loss of all of Liaodong, he suggested that the Ming should emphasize defending fixed strong points with heavy use of firearms.[29] His suggestions would be echoed by other officials, including Censor Fang Zhenru in 1622, who held that firearms in stoutly defended cities could be the key to Ming success.[30]

The use of overwhelming firepower to overawe foes had been standard Ming practice, especially in the previous 50 years, when the Ming state had managed to reverse more than a century of military decline by not only quelling internal challenges but also by projecting Ming power outside the empire's formal boundaries. In any event, Yuan Yingtai proved unable to reverse the state of affairs during his short tenure as the primary Ming commander in the region and he died in the Latter Jin attack on the city of Liaoyang, despite achieving some initial success fighting with cannon.[31] In the fall of both Liaoyang and Shenyang the Jin relied upon superior mobility to negate Ming firepower advantages and managed to lure the defenders out to fight in terrain that better suited the attackers. But even though there were other Jin victories in these early engagements in the 1620s, it is worth noting that in several cases even isolated Ming fortresses held out if they were properly outfitted with cannon and had sufficient stores of gunpowder.[32] In fact, citing images from the *Manzhou shilu*, Nicola di Cosmo argues that the Ming might well have initially deployed their cannon *outside* city walls, rendering more open to attack and making it more difficult for the defenders to continually reload and keep up steady fire due to lack of protection. The Jurchen cavalry had time to close after initial volleys were fired. But once the Ming altered their tactics and brought their heavy guns inside their fortresses, they were much more successful. Furthermore, even though they were successful in most of these early clashes, the Latter Jin demonstrated their awareness of the importance of firearms by issuing directives to outfit Liaodong soldiers with guns as early as 1622.[33]

Meanwhile, for their part, the Ming were now turning to outside sources to aid their war effort. By the 1620s, as the result of the diligent efforts of such luminaries as Matteo Ricci (1552–1610), several prominent Ming officials had converted to Catholicism. Some of these men, most notably the famous Ming officials Xu Guangqi (1562–1633), Li Zhizao (1565–1630) and Sun Yuanhua (d. 1632) used their religious contacts to procure weapons and trainers in their use for the Ming state.[34] Xu Guangqi pioneered these efforts, initially arranging for four padres and four guns to be sent from the Portuguese colony of Macao to Beijing in 1621.[35] Xu was empowered to make this request because he had just been placed in charge of training troops for fighting the Jurchens. Xu stressed that the Ming needed to manufacture large numbers of guns and use them in defensive positions and obtaining proper guidance and training in their use was critical.[36] Actually the Portuguese were more valued as instructors in the art of gunnery than as casters of cannon. But because some at the Ming court distrusted the motives of the Portuguese, the priests were detained in Guangzhou and only the guns went on to Beijing.

Experts in their use arrived in the capital two years later and their expertise was valued even after one of them was killed by a recoil during a demonstration. As one Jesuit observed at the time,

> These guns were highly esteemed and carried to the frontiers against the Tartars; who were not knowing of this new invention, and coming on many together in a close body received such a slaughter from an iron piece that they were not only put to flight at that time, but went on ever after with more caution.[37]

The Ming received 26 more cannon from 1623–1625. Eleven of these went to Shanhaiguan. The rest, with the exception of the one that blew up (noted above) were deployed in Beijing.[38] Ming successes, detailed below, would prompt further requests for aid from Macao. In 1630 Emperor Chongzhen (r. 1628–1644) received approval from the Senate in Macao for the dispatch of 200 soldiers with assistants, ten field pieces and two captains to Beijing. But most only made it to Nanchang when they were sent back to Macao, supposedly because officials in Guangzhou had bribed officials in the Ministry of War to cancel the request for Portuguese troops because they feared a loss of the monopoly on trade with Macao that they then enjoyed.[39] A few though, went to Dengzhou where they helped Sun Yuanhua deal with the mutiny of Kong Youde, though most of them died in that affair.[40] Incidentally, the Ming court was not alone in requesting cannon from Macao. By the early seventeenth century Macao was the cannon-exporting centre of Asia. Some weapons were even shipped back to Portugal, where they were in use as late as the Peninsular War against Napoleon![41]

The Christian convert Sun Yuanhua, stationed in the critically important post of Denglai in Shandong adjacent to Manchuria, was one of the most enthusiastic Ming advocates of using firearms against the enemy. Sun repeatedly sent forth memorials stressing the need to manufacture more guns and train soldiers in their use. In a memorial from 1626 Sun argued (erroneously) that the Ming had to use Western firearms because Chinese models lacked the range of Jurchen bows. He finally obtained the backing of the influential adviser to the emperor, Sun Chengzong.[42] Sun believed both in using firearms and in conducting limited offensive operations against the Jurchens under the maverick commander Mao Wenlong (d. 1629) and others.[43] A directive was issued calling for the training of 8,000 men in firearms use over a period of three months.[44] Sun Chengzong emphasized the utility of firearms in defensive positions as force multipliers.[45] Unfortunately, Sun Chengzong soon retired due to differences of opinion with certain elements at Tianqi's court. But the subsequent Ming victory over Nurhaci's forces at Ningyuan in 1626 served to reinforce the court's faith in the efficacy of these foreign arms.

The Battle of Ningyuan commenced on 19 February 1626. Though prodded by Nurhaci to come out of the city and engage the Latter Jin on the plains outside the city, the town's defender, Yuan Chonghuan (1584–1630) elected to remain inside and trust to the strength of his Portuguese cannon.[46] The attackers hit the walls with scaling ladders and were repulsed by the defenders who set bedsheets on fire and poured boiling oil on the attackers.[47] These were combined with heavy artillery fire that inflicted massive casualties upon the assailants, as Yuan arrayed his 11 great Western cannon in such a way that they covered all the ground around the city.[48] Allegedly everywhere the cannon blasts struck smoke filled the sky for several *li* and the battering rams and carts of the Jin could not bring the stout walls of Ningyuan down. One shot could supposedly slay upwards of 100 men.[49] Jurchen corpses piled up outside the city walls and some 3,000 were allegedly killed in about six hours. Two more assaults the next day were also turned back with heavy losses.[50]

Nurhaci and his men retreated to the north and the Jin Khan would die several months later, possibly due to the lingering effects of wounds sustained in this battle. The red barbarian cannon of the Ming are credited with having a profound effect on the battle.[51] The following year one Zhao Shuajiao enjoyed similar success in using cannon to repulse a Jin assault on the city of Jinzhou.[52] In this battle thousands of Jin troops besieged Jinzhou and attacked the north and west sides of the city but Zhao pounded the attackers with catapults, cannon fire and arrows and their corpses riddled the field outside the town. They pulled back in the middle of the night. The next day their cavalry galloped around the city again and tried to attack but again were cut to ribbons by Ming cannon blasts, the enemy feeling 'as if heaven itself was falling on them'. They continued the siege for 25 days and fought three major battles, all won by the Ming.[53]

These defeats showed how superior firepower and defensive positional warfare could allow the Ming to negate the mobility advantage theretofore enjoyed by the Jin. But it also provoked the latter to retreat to Shenyang and reconfigure their own strategy. From this point on, under Nurhaci's son and successor, Huang Taiji (Abahai, 1592–1643), the Jurchens would explicitly train and prepare for siege warfare and more aggressively integrate firearms into their arsenal.[54] Certainly most authors seem to agree that it was his experiences at the battles of Ningyuan and Jinzhou that convinced Huang Taiji to definitively alter his tactics and by doing so this may well have facilitated the eventual Qing conquest of the Ming. For Huang Taiji realized that while the Ming could not match the Jurchens in the open field, his eight banner forces could not hope to prevail in extended siege warfare and that was required if the Jin were to eventually supplant the Ming.[55] So Huang Taiji endeavoured to level the technological playing field and for the next several years the Jurchens sought to acquire firearms and even impressed Koreans into service as trainers and gunsmiths after invading Korea in 1627. In 1630 the Jin captured a crew of Chinese artillery experts at the city of Yongping. These men were subsequently put under the command of a Chinese officer named Tong Yangxing who had served the Jin for over a decade. As of February of 1631, 40 European style cannon had already been cast.[56] Starting in 1631 Huang Taiji commenced building cannon in Shenyang.

As he prepared for his next major assault on the city of Dalinghe, Huang Taiji built a formidable network of siegeworks and refused to mount the kind of suicidal frontal assaults that had doomed the Jin at Ningyuan and Jinzhou. The Jin also used their own siege guns to take out outlying Ming defensive towers. Huang Taiji allegedly had some 40 'generalissimo' and red barbarian cannon with him. Tong Yangxing was given oversight of the red barbarian cannon while Huang Taiji himself took the generalissimo models to Jinzhou to guard against Ming flanking actions. Some were used defensively while others were apparently used as mobile field pieces. They were quite effective and the Jin enjoyed their greatest success to date in reducing city walls. Tong Yangxing also drove off multiple Ming relief columns with his cannon.[57] Tong then used six *hing yi pao* and 14 generalissimos against the fortified redoubt of Zizhangtui on 2 November 1631, taking the fort in four days and killing 57 defenders.[58] The next month, plagued by supply problems, the Ming commander there, Zu Dashou, eventually surrendered after 80 days of increasing desperation and starvation.[59] With their victory at Dalinghe in 1631, the Jin acquired some 3,500 guns of various sizes, accelerating their programme.[60] Moreover, the victory at Dalinghe reified Huang Taiji's new approach to empire building.

Soon thereafter the Latter Jin would create their first firearms division under the leadership of the aforementioned Tong Yangxing. In a memorial to Tong, he was told that he should manufacture as many cannon as possible as 'even one hundred cannon were not too many and even 100,000 catties of gunpowder was still too little'.[61] The Jin also realized they needed to build more carts to transport the cannon so urgent orders were issued to procure the necessary firewood to start building them. They were initially hampered, however, by the harsh climate of Manchuria so they eagerly attempted to attract skilled settlers to the region so as to aid their efforts to build up a support structure for such an ambitious operation. The Jin also hoped to attract more potential soldiers and were eager to keep their operations relatively secret so that they might enjoy the advantage of surprise when they unveiled their weapons against the Ming.

With an emphasis upon training soldiers, this allowed the Jin to create their first integrated divisions. Huang Taiji envisioned these units as flanking wings to be deployed outside the banners and aid the cavalry in concentrated assaults.[62] Originally the unit comprised solely Han Chinese soldiers outside the regular banners. As the Han banners evolved they became more closely integrated into the Jin military structure, initially in units of 1,580 men.[63] In 1637 they were organized into left and right wings under the Chinese defector commanders Kong Youde and Geng Zhongming (d. 1649).[64] Interestingly enough, in defecting to the Latter Jin after his mutiny failed against the Ming, Kong stressed that his knowledge of firearms could be very useful to his new masters.[65] These Chinese artillery wings were later expanded into four units and eventually became part of the eight Han banners.

Huang Taiji's efforts would bear fruit in what some consider to be the decisive engagement of the Ming–Qing conflict: the Battle of Song-Jin in 1642.[66] Over the previous several years, successive Qing (they had assumed this dynastic name in 1636) assaults on the cities, most notably Songshan, had been repulsed by the Ming with superior firepower.[67] Up until the start of the siege Ming officials remained confident that they could hold out with sufficient reserves of food, water and gunpowder. They were probably correct in this assessment. But Huang Taiji now had an impressive array of heavy artillery at his disposal. He used this to isolate and destroy Ming supply wagons and deter further relief columns from arriving. Still, the Ming defenders resisted valiantly, again relying upon their red barbarian cannon to drive off the attackers.[68] The defenders had a total of 37 of these guns and managed to repulse repeated attacks even when food supplies dwindled. The Ming cities of Songshan and Jinzhou held out for an amazing six months and actually only capitulated due to internal betrayal. Nonetheless, this proved a great windfall for the Qing. As a result of the epic battle of Song-Jin the Qing acquired 2,363 cannon at Songshan, 488 from Jinzhou, 452 from Tashan and 80 at Foshan. In addition to these 3,683 cannon, they acquired 1,515 smaller guns of various sizes.[69] So now over a period of a mere 13 years, Huang Taiji had turned weakness into strength and the Manchus had a fully operational military system capable of toppling the preeminent power in Asia.

Over the next year-and-a-half the Qing would increase their pressure on the Ming in the north-east, even as the Ming peasant rebellions raged in central and western China. In the end the Great Wall was not breached by cannon, but by treason. Nonetheless it is questionable as to whether the Manchus ever would have got to that point had they not been flexible in their willingness to adopt and deploy significant numbers of heavy artillery in their war against the Ming. In this they were not unlike their Mongol forebears who were willing to adopt all manner of new technologies to facilitate their conquest of a dizzying array of foes. And it's important that we also do not forget the adaptability of the Ming in this war. The Ming were not simply turning to the Jesuits out of desperation, as some scholars have asserted. In fact they were operating completely within their own flexible, even aggressive, military tradition when it came to exploring new technologies and deploying them on the battlefield. Far from being an insular passive, non-competitive power, the Ming state was looking to retain its hard-won preeminence in the Asian world order. Its defeat was not due to technological shortcomings or cultural predispositions against military innovations, but rather far more deep-rooted personal animosities and factionalism that transcended simple civil–military dichotomies. In keeping with the theme of this gathering I'd encourage scholars to reopen their eyes to military technological developments in Asia and other parts of the world and urge them to view these developments on their own terms. At the same time, given the Jesuit involvement in the events described herein, it is also useful to take note of these global connections and their foreseen and unforeseen ramifications.

Notes

1 The most influential recent proponent of the Military Revolution has been Geoffrey Parker, *The Military Revolution: Military Innovation and the Rise of the West, 1500–1800*, 2nd edn (Cambridge: Cambridge University Press, 1996). For a collection of essays on the subject, see Clifford J. Rogers (ed.), *The Military Revolution Debate* (Boulder, CO: Westview Press, 1995).

2 See Sun Laichen, 'Military Technology Transfer from Ming China and the Emergence of Northern Mainland Southeast Asia (*c.*1390–1527)', *Journal of Southeast Asian Studies*, vol. 34, no. 3 (Ocober. 2003), pp. 495–517.

3 See, for, example, Charles Tilly, *Coercion, Capital and European States, 990–1990* (London: Basil Blackwell, 1992).

4 Kenneth Chase, *Firearms: A Global History to 1700* (Cambridge: Cambridge University Press, 2003), serves as a good example.

5 See, for example, Gabor Agostan, *Guns for the Sultan: Military Power and the Weapons Industry in the Ottoman Empire* (Cambridge: Cambridge University Press, 2005), esp. pp. 1–95; and Peter A. Lorge, *The Asian Military Revolution: From Gunpowder to the Bomb* (Cambridge: Cambridge University Press, 2008), pp. 1–44.

6 For details on military technologies used in this war, see Kenneth M. Swope, 'Crouching Tigers, Secret Weapons: Military Technology Employed During the Sino-Japanese-Korean War, 1592–1598', *Journal of Military History*, vol. 69, no. 1 (January 2005), pp. 11–43. For a full treatment of this war, see Kenneth M. Swope, *A*

Dragon's Head and a Serpent's Tail: Ming China and the First Great East Asian war, 1592–1598 (Norman, OK: University of Oklahoma Press, 2009).

7 On Qing military logistics, see Yingcong Dai, '*Yingyun shengxi*: Military Entrepreneurship in the High Qing period, 1700–1800', *Late Imperial China*, vol. 26, no. 2 (December 2005), pp. 1–67. On the Qing conquest of Central Asia, see Peter C. Perdue, *China Marches West: The Qing Conquest of Central Eurasia* (Cambridge, MA: Balknap Press of Harvard University Press, 2005). In contrast somewhat to Perdue, the Chinese historian Xie Lihong has examined the rise and decline of Qing military power by examining the Manchus' use of firearms from their pre-dynastic phase through their defeat in the Opium War of 1839–1842. See Xie Lihong, 'Hongyi dapao yu Manzhou xingshuai', *Manxue yanjiu*, vol. 2 (1994), pp. 102–118.

8 For an overview of the issues of guns in the Qing formation, see Nicola di Cosmo, 'Did Guns Matter? Firearms in the Qing Formation', in Lynn A. Struve (ed.), *The Qing Formation in World Historical Time* (Cambridge, MA: Harvard University Asia Center, 2004), pp. 121–166. For an older piece on the Portuguese role in particular, see C.R. Boxer, 'Portuguese Military Expeditions in Aid of the Mings Against the Manchus', *T'ien-hsia Monthly* (August 1938), pp. 24–36. For a comprehensive recent history of the Ming–Manchu war, see Sun Wenliang and Li Zhiting, *Ming-Qing zhanzheng shilue* (Nanjing: Jiangsu jiaoyu chubanshe, 2005).

9 On the early Ming use of firearms, see Zhang Tingyu *et al.* (compilers), *Mingshi* 12 vols (Taibei: Dingwen shuju, 1994), pp. 2263–2267 (hereafter cited as *MS*). Also see Long Wenbin, *Ming huiyao*, 2 vols (Beijing: Zhonghua shuju, 1993), pp. 1188–1189. For a brief overview of the evolution of the Ming military with reference to the use of firearms, see Peter A. Lorge, *The Asian Military Revolution*, pp. 69–81.

10 Nicola Di Cosmo, 'Did Guns Matter?' p. 126.

11 The work of Sun Laichen is a notable exception. Di Cosmo notes that European adventurers are often given disproportionate credit for spreading firearms technology to places such as Korea. See Nicola Di Cosmo, 'Did Guns Matter?' p. 124. On the Dutch involvement in disseminating firearms to Korea, see Shin Dongkyu, 'Oranda jin hyōryūmin to Chōsen no seiyō shiki heiki no kaihatsu', *Shi'en*, vol. 61, no. 1 (November 2000), pp. 54–70.

12 Cheng Dong, 'Mingdai houqi youming huopao gaishu', *Wenwu*, vol. 4 (1993), p. 79.

13 *MS*, p. 2264. On the Ming propensity to adapt superior firearms whenever possible, see Fan Zhongyi, 'Mingdai junshi sixiang jianlun', *Ming-Qing shi*, vol. 24, no. 1 (1997), p. 38.

14 Cited in Cheng Dong, 'Mingdai houqi youming huopao gaishu', p. 81.

15 See Qi Jiguang, *Jixiao xinshu* (Taibei: Wuzhou chubanshe, 2000), pp. 207–218. For more information on Qi Jiguang, see Fan Zhongyi, *Qi Jiguang zhuan* (Beijing: Zhonghua shuju, 2003). For a short biography of Qi in English, see L. Carrington Goodrich and Chaoyiong Fang (eds), *Dictionary of Ming Biography*, 2 vols (New York: Columbia University Press, 1976), pp. 220–224 (hereafter cited as *DMB*). For his official biography, see *MS*, pp. 5610–5617.

16 Cheng Dong, 'Mingdai houqi youming huopao gaishu', p. 81.

17 Such notions are particularly common in general textbook treatments. See, for example, Albert Chan, *The Glory and Fall of the Ming Dynasty* (Norman, OK: University of Oklahoma Press, 1982); Ray Huang, *1587: A Year of No Significance* (New Haven, CT: Yale University Press, 1981), pp. 156–188.

18 For a table detailing specific characteristics and locations of surviving examples of these cannon, see Cheng Dong, 'Mingdai houqi youming huopao gaishu', pp. 82–83. For other types of cannon, see p. 85.

19 See C.R. Boxer, 'Portuguese Military Expeditions in Aid of the Mings Against the Manchus', pp. 26–27.

20 Cheng Dong, 'Mingdai houqi youming huopao gaishu', p. 83. Cheng notes that inscriptions on the guns normally indicate where and by whom they were cast.

Cannon bearing the inscription '*shenji*' for example, were cast under the direction of the capital firearms divisions (*shenji ying*).

21 The rise of Nurhaci is well chronicled. For a brief biography of him, see Arthur O. Hummel (ed.), *Eminent Chinese of the Ch'ing Period*, 2 vols (Washington, DC: Library of Congress, 1943), pp. 594–599 (hereafter cited as *ECCP*). Also see Pamela Kyle Crossley, *The Manchus* (Cambridge, MA: Blackwell, 1997), pp. 47–74. The definitive account of the Manchu conquest of China in English remains Frederic Wakeman Jr, *The Great Enterprise*, 2 vols (Berkeley, CA: University of California Press, 1985).

22 Li Jiuqi, *Mingji beilue*, 2 vols (Beijing: Zhonghua shuju, 1984), p. 11 (hereafter cited as *MJBL*).

23 For biographies of Liu Ting, see *DMB*, pp. 964–968; *MS*, pp. 6389–6396. Known by his nickname of 'Big Sword Liu', Liu Ting was also a veteran of the Ming war in Korea.

24 The story of the Li clan and their ties to Nurhaci is treated in Kenneth M. Swope, 'A Few Good Men: The Li Family and China's Northern Frontier in the Late Ming', *Ming Studies*, vol. 49 (2004), pp. 34–81.

25 Xie Lihong, 'Hongyi dapao yu Manzhou xingshuai', p. 102.

26 For a brief biography of Xiong, see *ECCP*, p. 308. For the official biographies of Yang Hao and Xiong Tingbi, see *MS*, pp. 6685–6688 and pp. 6691–6695 respectively.

27 On Yuan, see *MS*, pp. 6689–6691.

28 *MJBL*, p. 25.

29 Xie Lihong, 'Hongyi dapao yu Manzhou xingshuai', p. 103. On Wang Zaijin, see *ECCP*, p. 839.

30 See Shen Guoyuan, *Liang chao congxin lu*, in *Qing ru guan qian shiliao xuanji*, 2 vols (Beijing: Renmin daxue chubanshe, 1989), vol. 2, p. 237 (hereafter *QRGQSL*).

31 See *MS*, p. 6690. Incidentally Shenyang, located north-east of Liaoyang, had earlier fallen in part because the Ming defender there exhausted his gunpowder stores before the full battle was joined. See *MJBL*, p. 27.

32 An example would be Luo Yigui's defence of Xipingbao. See *MJBL*, p. 32.

33 See Nicola Di Cosmo, 'Did Guns Matter?' pp. 136–139.

34 On Xu and Li, see Chun-shu Chang and Shelley Hsueh-lun Chang, *Crisis and Transformation in Seventeenth-Century China: Society, Culture, and Modernity in Li Yu's World* (Ann Arbor, MI: University of Michigan Press, 1992), pp. 287–290. On Sun Yuanhua, see Huang Yi-long, 'Sun Yuanhua yu Mingmo chuanhua de xiyang huopao', *Lishi yuyan yanjiusuo jikan*, vol. 67, no. 4 (1996), pp. 911–966. For a brief biography of Sun, see *ECCP*, p. 686; for a biography of Xu Guangqi, see *ECCP*, pp. 316–319.

35 C.R. Boxer, 'Portuguese Military Expeditions in Aid of the Mings Against the Manchus', p. 24.

36 Huang Yi-long, 'Sun Yuanhua yu Mingmo chuanhua de xiyang huopao', pp. 913–916.

37 C.R. Boxer, 'Portuguese Military Expeditions in Aid of the Mings Against the Manchus', p. 26.

38 Xie Lihong, 'Hongyi dapao yu Manzhou xingshuai', pp. 103–104.

39 C.R. Boxer, 'Portuguese Military Expeditions in Aid of the Mings Against the Manchus', p. 32. Also see Li Hongbin, 'Huang Taiji yu huopao', *Lishi dang'an*, vol. 2 (1997), p. 88.

40 See C.R. Boxer, 'Portuguese Military Expeditions in Aid of the Mings Against the Manchus', pp. 31–32. Macao later supplied a gun and cannoniers to Guangzhou and Nanjing to defend those cities against the peasant rebel leader Li Zicheng. On Kong Youde, see *ECCP*, pp. 435–436. On his mutiny, see Huang Yi-long, 'Sun Yuanhua yu Mingmo chuanhua de xiyang huopao', pp. 939–946.

41 C.R. Boxer, 'Portuguese Military Expeditions in Aid of the Mings Against the Manchus', p. 29.
42 Excerpts from Sun Yuanhua's memorials can be found in Huang Yi-long, 'Sun Yuanhua yu Mingmo chuanhua de xiyang huopao', pp. 919–920. On Sun Chengzong, see *ECCP*, pp. 670–671; *MS*, pp. 6465–6477.
43 Fan Shuzhi, *Wan Ming shi*, 2 vols (Shanghai: Fudan daxue chubanshe, 2003), p. 771 (hereafter cited as *WMS*). On Mao Wenlong, see Kenneth M. Swope, 'Loyalist or Desperado? The International Context of the Execution of Mao Wenlong', unpublished paper presented at the 2008 Association for Asian Studies Annual Meeting.
44 Huang Yi-long, 'Sun Yuanhua yu Mingmo chuanhua de xiyang huopao', p. 920.
45 *WMS*, p. 773. On Sun Chengzong's plans for defending Liaodong incorporating firearms in defensive positions, see Tan Qian, *Guoque*, 10 vols (Taibei: Dingwen shuju, 1978), pp. 5235–5236.
46 For biographies of Yuan, see *ECCP*, pp. 954–955; *MS*, pp. 6707–6719.
47 *MJBL*, pp. 40–42; *Qing taizu shilu*, in *QRGQSL*, 2, pp. 386–387.
48 *MS*, pp. 6709–6710; Li Hongbin, 'Huang Taiji yu huopao', p. 92.
49 Xie Lihong, 'Hongyi dapao yu Manzhou xingshuai', p. 104.
50 *MJBL*, p. 42.
51 *WMS*, p. 773.
52 *MJBL*, p. 43.
53 *MJBL*, p. 43.
54 *MJBL*, p. 41; Xie Lihong, 'Hongyi dapao yu Manzhou xingshuai', p. 105. On Huang Taiji, see *ECCP*, pp. 1–3.
55 Institute of History and Philology (compilers), *Ming-Qing shiliao jia bian*, 15 vols (Taibei: Zhongyang yanjiuyuan Lishi yuyan yanjiusuo, 1972), vol. 1, pp. 48a–49a.
56 Frederic Wakeman Jr, *The Great Enterprise*, p. 169.
57 Li Hongbin, 'Huang Taiji yu huopao', pp. 88–89.
58 Li Hongbin, 'Huang Taiji yu huopao', p. 89.
59 For a biography of Zu, see *ECCP*, pp. 769–770.
60 Xie Lihong, 'Hongyi dapao yu Manzhou xingshuai', p. 105.
61 Li Hongbin, 'Huang Taiji yu huopao', p. 89.
62 Xie Lihong, 'Hongyi dapao yu Manzhou xingshuai', p. 106.
63 Xie Lihong, 'Hongyi dapao yu Manzhou xingshuai', pp. 106–107.
64 On Geng, see *ECCP*, pp. 416–417.
65 Huang Yi-long, 'Sun Yuanhua yu Mingmo chuanhua de xiyang huopao', p. 945.
66 For a full account of this battle, see Sun Wenliang and Li Zhiting, *Ming-Qing zhanzheng shilue*, pp. 337–392. Also see Chen-main Wang, *The Life and Career of Hung Ch'eng-chou (1593–1665): Public Service in a Time of Dynastic Change* (Ann Arbor, MI: Association for Asian Studies, 1999), pp. 94–110.
67 Fang Yujin, 'Ming jun shouwei Songshan deng chengbao de liujian zhanbao', *Lishi dang'an*, vol. 2 (1981), p. 5.
68 Fang Yujin, 'Ming jun shouwei Songshan deng chengbao de liujian zhanbao', p. 3.
69 Li Hongbin, 'Huang Taiji yu huopao', p. 90; and Chen-main Wang, *The Life and Career of Hung Ch'eng-chou (1593–1665)*, p. 110.

8 Battles, boats and bridges

Mughal amphibious warfare, 1571–1612

Pratyay Nath

A few years back, Jos Gommans, in an article in the *Journal of Global History*, identified the early modern empires of the Ottomans, Safavids, Mughals and Manchus as post-nomadic. He argues that each of these empires was based on an identifiable (semi-) nomadic heritage, linked, among other things, with the use of the warhorse. He argues that post-nomadic empires were formed when (semi-) nomadic powers conquered and settled down in sedentary lands. Their (semi-) nomadic heritage, however, lingered on in them. In the field of military affairs, the (semi-) nomadic way of war-making, with its central focus on the deployment of the warhorse, remained the dominant form of military technique in the field. In the case of the India, this pivotal role of the cavalry was gradually eclipsed only in the course of the eighteenth century due to the rise of the disciplined, firearm-bearing infantry and field artillery, based on the European model, among various post-Mughal polities of the subcontinent.[1] Together with his other publications on the nomadic frontier of the Indian subcontinent and its connections with the larger arid zone of Eurasia, Gommans puts forward a sophisticated framework for the study of the post-Timurid dynasty.[2] This chapter seeks to extend and nuance this framework for the analysis of Mughal warfare further.[3]

The intervention of Gommans marks a strong departure from the dominant mode of Mughal historiography that portrays the entire history of the Mughal state as that of a highly sedentarized agrarian bureaucratic power. Gommans establishes a connection between Mughal history and the historiography of other post-nomadic states, a genre enriched by the works of Andre Wink, Anatoly Khazanov, Thomas Barfield and others who analyse issues like the nature of political, geographical and cultural adaptation that polities with nomadic backgrounds undergo after settling down among a sedentary people.[4]

However, the main problem with Gommans' framework is that having labelled the Mughal Empire as a post-nomadic one, he tends to see the rest of its history as a somewhat unevolving phenomenon. For example, after he identifies unceasing royal mobility as the principal royal exercise of power and a major post-nomadic cultural hangover, he substantiates his argument about the centrality of royal mobility in the Mughal psyche throughout the sixteenth and seventeenth centuries by quoting from two texts – one from the late-seventeenth century and another from the eighteenth.[5] This betrays his underlying assumption that the particular post-nomadic trait in question and its understanding by the Mughal psyche had remained somewhat unchanged through a period of almost 200 years of the Mughals' residence in the subcontinent. In another place, Gommans holds that the main military legacy of post-nomadism – the mounted archer – retained unchallengeable centrality in Mughal warfare over a period of three centuries, right down to the eighteenth, when it was finally overshadowed by the handgun-bearing infantry and field artillery.[6] Such an understanding does not appreciate that Mughal warfare was moulded in these three centuries by diverse circumstances which they faced in the Indian subcontinent. In fact, contemporary Mughal chronicles hardly mention mounted archery in Mughal pitched battles after the 1570s. Thus, it may be pointed out that in this respect also, Gommans' model does not take into account the evolution of the Mughal post-nomadic experience.

A potent analytical category like post-nomadism needs to be seen as a dynamic one, constantly conditioned by various factors. My contention is that physical geography is an important factor among these. The present brief study of Mughal amphibious warfare in the late sixteenth and early seventeenth centuries points to the diverse, and often difficult, geographical conditions which the Mughals had to negotiate in the Indian subcontinent and the consequent adaptations that they had to make to their post-nomadic way of war-making. Once we shift our attention away from the vast plains of north India, we come to appreciate the fact that, contrary to the propositions by Gommans, Mughal warfare was not all about mounted archers, or even heavy cavalry for that matter. At the same time, they were also not only about gunpowder artillery, as some scholars have tended to suggest. They were as much about boats, war elephants and infantry, among various other things. Cavalry may have dominated pitched battles, but when we extend our gaze beyond the pitched battles towards other forms of military endeavours, we see that the cavalry was only a part of a larger body of military participants. This chapter tries to demonstrate how the physical geography of South Asia imposed certain conditions upon the Mughals, which, among other factors, was responsible in the mutation of Mughal cavalry-centric warfare in these regions and thereby contributed to the process of evolution of Mughal post-nomadism.

The Mughals were first pushed into amphibious engagements when, under Babur, they chased the Afghans across the Middle Ganga Basin. Here, we find two landmark engagements – one on the river Ganga (1528) and the other over Ghaghra (1529). The Mughals, after displaying fabulous cavalry tactics in the pitched battles of Panipat (1526) and Khanua (1527), were compelled to fight with boats during the latter two engagements. However, these were more like sporadic incidents and only signalled what the future had in store for them. In the following decade, the Mughal army under Humayun laid siege to and ultimately occupied the river fort of Chunar. Although the actual modalities in this operation are unclear, we can safely argue that the siege entailed an assault from a river-borne position on the Ganga.[7]

Apart from these three engagements, the Mughals under Babur and Humayun did not have to militarily negotiate with aquatic space much. This was precisely because they did not lead any expedition into the two regions of the Indian subcontinent dominated by rivers – Sind and Bengal. Having consolidated his empire in the Punjab Plains and the Gangetic Basin and then successfully subjugating the Rajasthan region and Malwa-Bundelkhand theatre, Akbar, for the first time, aimed at the military subjugation of these regions in the 1570s. Hence, from this time onwards, the Mughals were embroiled in amphibious warfare in these two regions. Campaigns started in the Indus Basin first, against the principality of Bhakkar in the Lower Indus Basin during 1571. This year is taken as the starting point for the present chapter. Military operations ended in this region with the subjugation of Thatta in 1592. However, in Bengal, amphibious operations continued right into the middle of the seventeenth century, first in the delta itself, then in the Brahmaputra Basin and coastal regions. This chapter stops in 1612, when the Mughals achieved decisive military supremacy over larger parts of the Ganga-Brahmaputra Delta for the first time and neutralized all of its major opponents in the region. This also marks the end of what we may call the first phase of Mughal riverine warfare. After 1612, the Mughals ventured into even more difficult riverine terrain as they entered the Brahmaputra Basin to fight the Ahoms.

As far as the sources for this present study are concerned, the Persian chronicles from this period – the *Akbarnama*, the *Tabaqat-i Akbari*, the *Muntakhab ut-Tawarikh*, the *Tuzuk-i Jahangiri* – are the main ones providing information regarding these campaigns. For Sind, we have Mir Masum's *Tarikh-i Sind*, which is a history of sixteenth-century Sind written from first-hand experience. It furnishes valuable details regarding Mughal campaigns in this region. For Bengal, a detailed source for this period is the *Baharistan-i Ghaybi*, written by Mirza Nathan, a Mughal *mansabdar* posted on the Bengal front. While early modern Europe saw the composition of a huge number of military manuals, military memoirs, normative texts about the conduct of war, war diaries etc., for the students of South Asian warfare, there is a dearth of such material for their region of study. Narrating the Mughal campaigns in Bengal and Assam in the first half of the seventeenth century, the *Baharistan* is one such rare Persian chronicle dedicated primarily to recording military campaigns. The memoir left by Father Monserrate, who visited Akbar's court in the 1580s, is also a valuable source, as it takes into account several details, particularly about military logistics, that the Persian court chroniclers have not considered worth mentioning.

Historiographically speaking, amphibious and riverine warfare has been explored at an appreciable length in the context of the early modern world. In the monsoon region of South-East Asia, riverine warfare was widely practised. Michael Charney has written extensively on this area.[8] However, in the context of the Mughal experience, the realm of amphibious warfare is a relatively underworked area. Apart from some oblique references in works dedicated to Mughal activities in Bengal, not much work has been done in this direction.[9] Historians of Mughal warfare have mostly focused their attention on pitched battles, military technology and military institutions. Consequently, several spheres of

Mughal military endeavour have gone relatively unattended. The present chapter intends to throw some light on one such area.

Geography and geopolitics of Mughal amphibious warfare

The two regions where amphibious engagements occurred intensely during the sixteenth and early seventeenth centuries were the Lower Indus Basin and the Ganga–Brahmaputra Delta. Geopolitically, these two regions were quite different. In the east, the Ganga River Basin constituted an open frontier, one which could be explored and expanded to include more and more agriculture based fertile land. The west–east orientation of the river basin also gave Mughal expansion a definite expansionist direction towards the east. Controlling the west–east stretch of this riverine highway by controlling the extremely fertile region on both sides of it was central to Mughal geopolitical considerations in this region.[10] On the contrary, the drive to control the Lower Indus Basin was more about gaining access to the flourishing sea ports at the mouth of the Indus and securing the passage between the Indus delta, where these ports were located. The Mughals also aimed to control the fort of Qandahar to secure their western front. As the Middle and Lower Indus Basin lay between the Mughal political heartland in north India and Kandahar, its pacification was imperative for ensuring lasting control over the fort.[11] In fact this factor is cited as the prime cause behind the Mughal advance on Thatta in 1591. The *Khan-i Khanan*, given the charge of occupying Qandahar, decided to subdue the principality of Thatta first, before going any further.[12]

Throughout the sixteenth century, the Mughal political heartland was located in the Punjab plains, the Indo-Gangetic Divide and the Upper and Middle Gangetic Basins. It was here that the cities of Lahore, Delhi, Fatehpur Sikri and Agra, imperial capitals at different points of time in the sixteenth and seventeenth centuries, were located. The Punjab plains were formed by the deposition of immense volumes of alluvium by the Indus and its five tributaries – from west to east the Jhelum, the Chenab, the Ravi, the Beas and the Sutlej. The whole area is a massive plain, some 565 kilometres north-west/south-east by 725 kilometres north-east/south-west with the altitude mostly under 183 metres. Near the Himalayan foothills in Kashmir, the land is broken by the innumerable torrents descending on the plains from the hills. Apart from some small elevations on either side of Chenab, the whole region presents a flat topography. The riverine belts housed dense population and were under intense agriculture. The region receives some rainfall from the monsoon winds. Temperatures are quite extreme. Under the Mughals, hunting was a regular affair in many parts of this region, giving the impression that there was considerable forest in certain parts.[13]

To its south-east, the Indo-Gangetic divide represents a transitory zone between the Indus system and the Ganga system. It is bounded by the sharply rising Himalayan foothills in the north, the Sutlej in the west, the Yamuna in the east, the low broken Aravalli outliers in the south-east and beyond Ghaggar, graduates into the Thar Desert. The region represents a rapid transition from a noticeable aridity of the Punjab plains in the west to the more humid Gangetic basin in the east. While in the north, the foothills receive fairly high rainfall, especially from the Western Disturbances, rainfall decreases sharply towards the south, making cultivation dependent on irrigation. The region presents a flat topography except for the few outlying ridges of the Aravalli in the south-east, which runs right into Delhi. The Ghaggar and the few Himalayan torrents that join it form the drainage system of the area.[14] This is an important strategic area, due to the convergence of the trade routes going towards Central Asia in the north-west, those commanding the whole of the rich and fertile Gangetic basin to its east as well as the trade routes towards the thriving ports on the west coast of the subcontinent towards the south-west.

The Bhakkar region would be approached by taking a southwards route from the Mughal bases in Lahore and Multan in the Punjab. Thatta, on the other hand, had two routes for an invasion force – along the Indus river from Bhakkar, or journey overland from Agra westwards through Jaisalmir. Thus, when Mirza Khan was besieging Jani Beg, the ruler of Sind in 1592, Akbar sent him cannons downstream via the Indus. Obviously it was easier to carry the guns down a river in boats than drag them overland, which would have entailed a bigger expense in terms of time, manpower and money. Money and musketeers, however, were dispatched by the land route via Jaisalmir.[15]

For campaigns in the Ganga Delta, the Mughals used the Patna–Hajipur zone of Bihar as the launch-pad. These had been occupied by the mid-1570s from the Afghans and although the cities faced a constant threat from the raiding Afghans, the Mughals managed to hold on to them successfully. South of the Patna-Hajipur region lay the densely forested tracts of the Chotanagpur Plateau. These were extensions of the forested highlands of the central parts of the Indian subcontinent. While the usual route to the Delta went from Patna-Hajipur eastwards through the Teliagarhi pass towards the Gaur–Tanda region, only thin traffic passed through the forests of Jharkhand, traversed usually by pilgrims travelling between the Vaishnava sites of Mathura and Vrindavan in the Middle Ganga Basin and Puri on the Orissa coast. Man Singh commanded Mughal forces through these dense forests into Orissa in the 1590s. But, in general, this route through the forests was not usually used for military traffic and the usual route to Orissa passed through Bengal.[16] This meant that the Patna–Teliagarhi–Tanda route was the only way the Mughals could supply their campaigns in the Bengal Delta. The command of Tanda hence became supremely important because this was the Mughal bridgehead in Bengal during the initial phase of their conquests.

Of these, in terms of resources and time consumed, as well as scale of operations, the wars in Bengal were of a greater dimension than those in Sind. In Sind, the struggle was over the control of one single central waterway – the Indus river – and the towns located near it. The land surrounding the Indus was far less productive than that in the Bengal Delta. In the Indus region, the concern was to control this waterway, linking the ports of the Indus Delta and the Mughal political and economic centre in the Punjab, which was of supreme importance. In contrast, the rich alluvial plain of the Lower Ganga Basin, especially the Ganga–

Brahmaputra Delta was proverbial. Up to Patna-Hajipur, the Ganga served as the all important waterway connecting the Delhi–Agra region with the eastern lands. Its tributaries, mainly on the left bank, connected the plains north of the Ganga with this waterway. However, further east, the Ganga lost this significance as the central highway owing to a complex meshwork of rivers, rivulets, canals piercing the low-lying land from all directions. In this region, the objective was to control the fertile land, which had potential for yielding very high revenues.

Outline of the Mughal campaigns in Indus and Ganga-Brahmaputra Deltas

In the Lower Indus Basin, the Mughals started campaigning against Sultan Mahmud Khan, the Governor of Bhakkar and Muhammad Jani Beg, the ruler of Thatta in 1571. These campaigns comprised a series of sieges and riverine engagements. Bhakkar capitulated in 1574 after the death of Sultan Mahmud Khan.[17] After the conquest of Bhakkar, the Mughals started engaging Jani Beg of Thatta over the control of the Indus. Armies going up and down the Indus would consist of *ghurabs* armed with artillery, accompanied by the land force that would march along the river. Thus on one occasion, while the Mughal army besieged the fort of Siwistan, Jani Beg advanced with a relief army to defend his fort. The Mughal general, Muhammad Sadiq Khan, raised the siege and moved to confront Jani Beg. When the two armies came face to face, the artillery on Jani Beg's *ghurabs* opened fire. His musketeers also started shooting from the boats. Faced by such aggression, the Mughals withdrew to Bhakkar. In 1592, with the surrender of Jani Beg and his incorporation into the Mughal imperial corps, Mughal military dominance over the entire Lower Indus Basin was established.[18]

As for the conquest of Bengal, we can indentify three distinct phases of military offensive spread over almost three decades. The first phase stretched from the time the Mughals first occupied Tanda (1574) to the appointment of Man Singh to the *subahdari* of Bengal (1594). Circled on all directions by their Afghan rivals and their Bengali allies, the Mughals were mostly holed up in their political base in Tanda–Rajmahal, although occasional campaigns towards the north and east did take place. Initially, these were no more than plundering campaigns and display of strength. Ralph Fitch, travelling amidst these Afghan and Bengali courts in Bakla, Sripur and Sonargaon in the east-central parts of the delta in 1586 observed: 'They be all hereabout rebels against their king Zelabdim Echeber [Jalaluddin Akbar]; for here are so many rivers and islands, that they flee from one to another, whereby his [Akbar's] horsemen cannot prevaile against them.'[19] This was a time, about a decade after the initial Mughal military forays into the delta, when Mughal presence was limited only to the western fringes of Bengal and militarily was still very much dependent on its cavalry. In the same year, the Magh ruler of coastal North Burma sent the Mughal governor of Bengal presents, including elephants, which were held valuable militarily and symbolically in both Bengal and Burma, as a diplomatic gesture. Abul Fazl says that this friendship with the Magh ruler was a matter of relief for the Mughals because '[N]either was there a [Mughal] fleet, which is the chief means of making war in that country, whereas the enemy had a large supply of war-boats'.[20] The comment betrays the relief at the settlement of peace with a strategically and logistically superior power. It also shows that even after a decade long presence in the Delta, a proper Mughal war fleet was yet to be built.

The second phase of Mughal campaigns started with Man Singh's appointment as the *subahdar* of Bengal (1594). Having defeated the Afghan leaders of Orissa two years back, the new *subadar* was all set to focus on the pacification of the eastern parts of the Delta. This, however, was an extremely difficult project. In front of rivals who had completely adapted their warfare to the local geographical factors, the Mughals were still heavily dependent on their cavalry. Although they seem to have built some sort of a riverine fleet, this does not seem to have matched the bigger fleets of their rivals. However, this was also the time when Mughal territories in the east started to be raided by the Arakanese and the Portuguese pirates and other freebooters. One such major raid happened in 1603, and much to the credit of the Mughals, they succeeded in faring pretty well in the riverine battle that took place at this time, and in spite of suffering reverses, held their ground. This indicates that a decade into Man Singh's *subadari* of Bengal, the Mughals had developed a war fleet of their own, which could take on, if not subvert, even the mightiest river fleet based powers of the region.

The third and final wave of conquest was begun by the appointment of Islam Khan Chishti to the *subahdari* of Bengal. His campaigns in the far eastern and southern parts of the Ganga–Brahmaputra Delta involved massive deployment of artillery, gunners as well as the building of a formidable and sizeable war fleet. Early during his tenure as the *subahdar* (1607–1613), Dhaka was made the Mughal capital in Bengal. From here, the *subahdar* sent out his armies in all directions, often personally accompanying them. Very soon all of the Bengali *bara bhuyans* or 12 war-lords, as well as Afghan leaders like Isa Khan, Musa Khan and Usman were defeated and larger parts of the Delta were brought under direct Mughal control. This phase of expansion involved riverine warfare on a very large scale. In the end, the already numerically inflated Mughal war fleet was boosted by the war boats of the chiefs whom the Mughals defeated and absorbed into their *mansabdari* system.

Military aspects of Mughal amphibious engagements

Interestingly, the theatres of riverine war – the Lower Indus Basin, the Lower Ganga Basin and the Ganga–Brahmaputra Delta – saw the use of gunpowder artillery at a far greater scale than what could be deployed elsewhere, the reasons being majorly logistical. Sixteenth-century heavy artillery was extremely difficult to be carried overland and entailed an enormous cost in terms of money, manpower as well as time. The waterways were the most convenient for transporting these heavy pieces. While arguing a case for the centrality of gunpowder artillery in Mughal state formation, it can be suggested that in central India and further south, Mughal centralization was incomplete and shaky precisely due to the absence of waterways to carry the artillery in there. Even if one does not accept the central role assigned to gunpowder artillery in Mughal state formation, it is true that the Mughals found it extremely difficult to carry their heavy artillery into central India and the Deccan from Agra during this time. Consequently, the sieges in this region saw a limited use of gunpowder artillery. In these regions, the Mughals conducted siege warfare with traditional means, like blockade, mining and escalades. Compared to this, amphibious wars saw much greater deployment of artillery because they could be easily mounted on boats and then deployed at will. The position of gunpowder artillery in Mughal military history, hence, is much more complex than the historians arguing in favour of a Gunpowder Empire would suggest. The imagery of cannons blasting away one fort after another, and thereby acting as the vanguard of empire, as the case had been in France or Spain, is alien to the Indian subcontinent, even in the riverine regions which saw heavy deployment of gunpowder artillery.[21] However, not only artillery, but all types of projectile throwing branches of the army – musketeers, rocketeers and archers – were much valued in this mode of warfare.

Command of fortified locations has always been the key to long term success in pre-modern warfare. While a victory in a pitched battle might vanquish the enemy on one day, control over land could not be ensured unless one secured the nearby forts and garrisoned them, so as to repulse any possible return of the enemy. Controlling strong fortresses was hence the key to the control of territory. Since controlling riverine channels was crucial to achieving domination over the enemy in amphibious warfare, fortified locations were extremely important in amphibious warfare as well. Thus, the control of the Indus Basin was entirely about controlling certain key towns, like Bhakkar, Sehwan and Thatta. In the case of the Mughals, the conquest of these cities proved to be coterminous with the dominance over the river basin. On the other hand, on a terrain infested much more with rivers and other water bodies, as in the Ganga–Brahmaputra Delta, rivers were the key to all types of communication, military as well as logistical. As such, even makeshift fortified locations could give a side enormous advantage in controlling riverine spaces. In the Ganga–Brahmaputra Delta, the river bed and its surroundings being abundant in soft mud, temporary forts could be quickly erected at strategic locations in order to control waterways. Although temporary and built within a very short period of time, and that too mostly by the cheap labour of local villagers and boatmen, the mud fort was famous for its strength. Built of riverside mud, its walls would not break when struck by a cannonball unlike stone ramparts; instead, the force would be absorbed by the mud and the shot rendered quite harmless. Owing to such strength, these mud forts could be defended well even by a small garrison and could be occupied by a besieger only after great exertions. In their initial phase of campaigns, the Mughals could bring more and more river channels under their control only by demolishing the mud forts of their enemy.[22]

They, in turn, were also quick to learn the strategic and military value of the mud fort and soon adopted the tactic themselves. That the Mughals understood the necessity of controlling waterways in the Ganga–Brahmaputra Delta by commanding river forts is clearly proven by several surviving Mughal riverine forts. These forts, in modern Bangladesh, are built of masonry and clearly represent an evolved form of the mud fort. These were probably built around the mid-seventeenth century by Mir Jumla to command river channels. They are provided with artillery platforms, which the Mughals used to mount artillery pieces and thereby command the transport on the river. River forts like these, equipped with artillery, gave the Mughals strategic and military strength.[23]

While marching along a river, the usual order of the march would be the war fleet sailing along the river with the land army accompanying it on both the banks. During battles, these two wings, the land force and the fleet, would assist and defend each other. The following example of a Mughal amphibious engagement with Udayaditya, the son of Pratapaditya of Jessore, may be cited here to illustrate a few typical features of amphibious warfare in Bengal at this time.

When the Mughal Army under Ghiyas Khan was marching down to Jessore against Pratapaditya, they were first opposed at Salka by his son, Udayaditya, who had built a strongly fortified mud fort for himself. In response to this, the Mughals decided to send their war fleet along the river, accompanied on either sides by a contingent marching along the banks. Mirza Nathan was placed at the head of one of the detachments moving on land, while Ghiyas Khan commanded the other. When Udayaditya's force did not march out as the Mughals had expected them to, the two Mughal detachments on land started fortifying their own positions with mud forts. When this building process was midway, Udayaditya came out of his fort with his war fleet. Khwaja Kamal led the fleet in the van while Udayaditya commanded the centre, with the major contingent. The *Baharistan* mentions a variety of boats in his fleet, some among them being *piara*, *kusa*, *balia*, *ghurabs* and *jaliya*, and probably quite a few among them carried artillery pieces. Jamal Khan remained inside the fort as a back up with the contingent of elephants.[24]

Udayaditya's boats first met the 20 boats of the Mughal naval vanguard. The *ghurabs* of Khwaja Kamal's vanguard and two *piaras* surrounded ten of the Mughal boats and pushed them back towards the bank where Ghiyas Khan was encamped. Some *mansabdar*s from here dismounted and came down near the bank of the river and showered the enemy boats with arrows, to allow their own boatmen and soldiers to take cover. They succeeded in seizing a *piara* and a *ghurabs* in the process. In the face of this Mughal counterattack, the crew of these boats of Udayaditya jumped into the water to save themselves.[25]

While Udayaditya's fleet was harassing the Mughal naval van, Mirza Nathan and his land force came to their rescue. Finding that the land route for their cavalry was blocked, they started showering arrows on the enemy boats from the shore. In this way, he kept Udayaditya's naval boats busy, and even pushed them back while his van moved forwards, and thus succeeded in dividing the enemy's naval forces. In the engagements that followed, Khwaja Kamal was killed by a

Mughal bullet. While both parties continued to exchange projectiles, Udayaditya's men supposedly lost heart at the death of Khwaja Kamal and their boats retreated. As Udayaditya's fleet headed in the opposite direction, six Mughal *kusas* chased him. Just when the Mughal *kusas* seemed to be closing in on Udayaditya, one *piara*, four *ghurabs* and one *machua* slowed down, anchored and engaged the *kusas*, giving the main fleet of Udayaditya some time to flee. Among these, the *machua* carried some 'Firingis', probably meaning Portuguese renegades in service of Udayaditya. A combined attack by the Mughal *kusas* along with a shower of arrows from Nathan's land force defeated these boats and four of the Mughal boats engaged in looting them. The remaining two *kusas* renewed their pursuit, shooting arrows. Soon they neared Udayaditya's fleet as the latter came to a narrow part of the river which compelled him to slow down to facilitate passage of the boats. As Udayaditya's men fired their guns from their boats, the two Mughal *kusas* closed in on his Mahalgiri boat. As they touch the rear of the Mahalgiri, Udayaditya was rescued by the boatmen of a faster moving *kusa* and he successfully escaped with his wives. Nathan says that the Mughal *kusas* did not pursue him any further because they had already boarded the Mahalgiri leaving their *kusas* behind when Udayaditya fled and also as they busied themselves looting that boat. Mirza Nathan, having no boat at hand while the enemy escaped, was left despondent.[26]

Logistics of river warfare

Bridges

The most usual way of crossing the streams and torrents that Babur's war band would encounter in Transoxiana was to ride across them. If the stream was too deep to be ridden across directly, it would be forded. The Indus was the first massive river system that Babur's war band came across. During his several expeditions on the Punjab plains between his occupation of Kabul in 1504 and the occupation of Delhi and Agra in 1526, he crossed the Indus and its tributaries several times. These crossings constituted the first phase of Mughal negotiation with the huge rivers of the Indian subcontinent and a quick look at them would be helpful in order to understand how from this time onwards the Mughals came to negotiate with the big South Asian rivers. For example, in 1505, on his way back to Kabul from his raiding expedition, Babur's band came to a torrent called Ab-i istada from the plain Kattawaz. Unable to locate a ford, horses and camels were made to swim across the stream, probably carrying the men on their backs. Some of the baggage was carried on their backs, while some of it was hauled over the stream by ropes.[27] Thus, at this stage, the Mughals were still not equipped enough to tackle huge water bodies. Gradually, during their raids into the Punjab region over the next two decades, Babur's war band gradually became more and more familiar with the Indus and its tributaries. In case it would not be possible to ford these rivers, boats would be procured for the men and horses to be shipped across to the other side. This was the case while crossing the Indus in 1519 when Babur's army crossed it in six boats.[28]

Under the reign of the successive two *padshahs*, the Mughal Army traversed enormous areas of the South Asian subcontinent and was compelled to negotiate aquatic spaces of diverse sizes. In the course of the sixteenth century, they learnt to build different sorts of bridges over different types of rivers and also picked up the art of riverine warfare. The most elaborate description of Mughal bridge building activities in the sixteenth century comes not from any Mughal court chronicle, but in fact from Father Monserrate, who accompanied Akbar on his expedition towards Kabul against the latter's political rival, Mirza Muhammad Hakim 1581. On the way, the Mughal army had to cross the Indus and its tributaries on several occasions. A detailed look at how the army crossed these rivers in this expedition, as narrated by Monserrate, may not be out of context here. It fact, it can be useful in giving an idea about how the Mughals went about their business under Akbar.

Monserrate says that marching from Delhi, the army first crossed the Yamuna at Karnal. However, it is difficult to imagine why an army marching from Delhi towards the north-west would have to cross the Yamuna. Hence, we can assume that this was some other water body, perhaps a canal from the Yamuna. The infantry, says Monserrate, crossed over a stone bridge, which probably was a permanent one as there is no mention of the bridge being constructed then and there, as was the case for most of the later bridges. It was probably the close proximity of the place to the Delhi–Agra heartland that made the Mughals build a permanent stone bridge there. Elephants, camels and cavalry crossed the river at a nearby ford.[29]

After some days, the Mughal army arrived on the banks of the Beas. The army marched along the river for two days in search of a ford where the elephants could cross and for a narrow place where a wooden bridge could be built. Scouts were sent ahead of the main army for the purpose and once they located a spot suitable for crossing near Kalanaur, the army camped there and subsequently crossed the river. The elephants were made to ford the river while the horse, infantry, along with draught animals like mules and camels crossed over the bridge.[30]

Proceeding from Kalanaur, the army crossed the Ravi at Ramgarh by means of a similar bridge built for the army. Advancing further, the Mughal force reached the Chenab. The Mughal army faced great difficulty in crossing this tributary of the Indus. No bridge could be built on it, probably owing to the great width of the river. A number of men who tried to ford it were drowned. There was a great scarcity of boats, in spite of Akbar's order to local villagers to bring their boats to the Mughal camp. Mughal and local intelligence, which was the main source of information for Monserrate, held that when Mirza Hakim fled across the river, he broke and burnt all the boats in which his army crossed, in order to delay the Mughal pursuit and thus buy some time. His purpose was served, as the Mughals were delayed by three days in trying to cross the river. This was a type of scorched earth tactics that Mirza Hakim had adopted with respect to boats and was evidently quite successful in delaying Akbar's force. Around 400 Mughal soldiers of different ranks perished while trying to cross the river. Probably the entire army had to cross in a small number of boats that they ultimately succeeded in gathering. This must have delayed the operations further.[31]

Marching away from Chenab, the next tributary of the Indus, the Jhelum was reached soon. In order to avoid any recurrence of the disaster on the Chenab, the army took a longer route in search of a place suitable for crossing. Then finding a satisfactory location at Rasulpur, the army camped there. Monserrate describes the river as broad and deep at this place, making it impossible for men, horses and even elephants, to swim across. Consequently, a wooden bridge was constructed on the river in eight days. The army crossed over this and marched ahead.[32]

After several days, the army reached the Indus. The river was so fast flowing and wide that only elephants could ford it, and with difficulty. Realizing that building a bridge on such a huge river would not be possible, Akbar tried to gather boats for crossing by establishing good terms with the local Dilzak Afghans and obtained 40 boats and much timber from them. Another 40 boats were constructed from this supply of timber. Once the boats were made and gathered, Akbar sent his son Murad ahead with an advance party in pursuit of Mirza Hakim. Then, he himself crossed the Indus along with the main army in boats. From the description of Monserrate, it is evident that this crossing too was not very comfortable, as the number of boats being inadequate for the large army.[33]

The Persian chronicles, however, give a slightly different account of the crossing of the Indus. According to them, the Mughals, having reached the Indus, decided on the construction of a bridge of boats. Officers therefore set about collecting them. The place of confluence of the Kabul and the Indus rivers was selected for crossing the river. Qasim Khan was given the charge of building the bridge. As there was a paucity of boats in the area, the *mansabdar*s and soldiers were assigned different *chaukis* or localities and were ordered to procure boats. Evidently, this bridge could not ultimately be constructed owing to the great width of the river. And when the Persian sources say that a part of the army was made to cross the river and sent off towards Kabul, they probably indicate, in agreement with Monserrate, that they crossed the river in boats. After a few days, Akbar crossed the Indus.[34]

During the return journey, Muhammad Qasim Khan was sent ahead of the royal camp to build a bridge on the Indus. Much to the relief of the army, he found that the water of the river had greatly decreased and this time they crossed the river by building a bridge on it. He built a bridge of boats and the army crossed over swiftly in just a day. On this journey to Fatehpur, Akbar crossed the Beas and Sutlej and, in all probability, the other tributaries too, over similar bridges.[35]

Monserrate's account of this expedition shows us the various means adopted by the Mughals to cross the rivers of the Indus Basin. They crossed over bridges, on the back of elephants, on boats etc. The first instinct of the commanders was always to try and locate a ford where the army could cross. For this purpose, scouts were sent ahead of the main army. In case the river was too deep for riding across, elephants were used to transport the troops, especially in times of urgency. This precluded any expenditure of money or time on bridge building activities. In case the river was a big one, bridges were built. Writing on Mughal roads and systems of communications, Abul Khair Muhammad Farooque categorizes Mughal bridges into four groups: wooden bridges, boat bridges, permanent stone bridges and rope bridges.[36] Of these, the first two types were the two most widely used varieties. These were meant especially for the horse, infantry, along with draught animals like mules, camels, etc. Elephants were preferably made to ford the river, lest their weight brought the bridge down. In case the river was too wide or turbulent for any such bridge to be built, the usual procedure was to cross over in boats.[37]

Various measures were adopted to ensure a safe crossing of the river without any material loss. In order to prevent chaos and commotion while crossing the bridge and enforce some amount of discipline on the troops, a trench would often be built in front of the bridge, from where troops would be disbursed systematically during the crossing. Father Monserrate points out such an instance of Mughal ingenuity:

> the King [Akbar] overcame the difficulty and danger of constructing bridges; for if these were built over a broad river-bed they are apt to be swept away by the force of the current, and hence disaster on the army crossing them. It is the custom in India to make temporary bridges of boats, which are tied together only by grass ropes. Over these boats is laid a roadway made of branches of trees, bushes and hay. The King, however, gave orders that care should be taken to see that only one type of troops or transport should approach the bridges at a time: and that the cavalry, the infantry, the camels, the other baggage animals, the flocks and the herds, should pass over both separately and in a single file, so that if the bridge parted, the river should take no great toll of men and supplies. Therefore on nearing a river, a small blockhouse was set up and occupied by the King's officers, who took care that a large number should not carelessly crowd the bridge at one and the same time, and so sink the boats. Moreover, elephants were not allowed to cross such bridges, lest they would sink them by their weight.[38]

Bridges made of boats became very popular with the Mughals and during their campaigns in the Ganga–Brahmaputra Delta, these would be quickly constructed on the rivers and rivulets for the army to cross over. The easy availability of boats in this part of the subcontinent made this task easy and prompt. The two following events during Mughal campaigns against Musa Khan during 1609–1610, described by the *Baharistan*, may be taken to be representative of this trend: 'Muhammad Khan Pani was posted at the *Mohana* of Khizrpur which is the confluence of the rivers Dulay and Lakhya with five hundred horsemen, and a bridge was constructed at the mouth of the river with *katari* and *maniki* boats of the artillery.'[39] Mirza Nathan then ordered the boats left behind by the enemy and lying about in the canal, to be gathered together; these were arranged like a bridge and the soldiers were ordered to carry their saddles on their heads and shoulders and make their horses swim to the other side by holding their reins. The infantry also crossed that way.[40]

Boats[41]

The Mughal army did not care for boats much before the 1570s. The *Baburnama* tells us how Babur used to have wild drinking parties in his boats on the Indus in his pre-Panipat days while he was still the lord of Kabul; but the military necessity to build a riverine fleet first arose in the 1570s, when they started campaigning in the Indus and Ganga basins. The first large contingent of boats was prepared when Akbar decided to sail downstream along the Ganga to supervise the siege of Patna in 1574. While the main army would march along the banks, the emperor decided to advance by the river with the ladies of his harem and his select courtiers. A large contingent of boats was procured for the purpose. From the description provided in contemporary sources, these boats were massive, ornate and diverse in shape. There were several types of boats. Some were residential, used for carrying Akbar and his family members, while others carried his political elite and their families and offices. Some boats carried military and logistical equipment, treasury, wardrobes, carpets, animals to be used in battles and hunting. Animals transported by boats included around 500 elephants. It must have required a very large number of boats to carry them, as a single boat could carry three elephants at most. Apart from these, there were large boats, called *ghurabs*, fitted with cannons and fully equipped for war. A fascinated Abul Fazl describes how the sterns of the boats were built in the shape of animals and some bigger vessels even had gardens on them. The main body of the crew was known as *kharwaha*, who are described as a social group linked with riverine transportation. While the Mughal Navy had an official Mir Bahr, or commandant of the fleet, these boatmen were recruited on an ad hoc basis.

Eager to expand their resources of boats, the Mughals welcomed any potential ally who could strengthen them in this respect. In the Ganga–Brahmaputra Delta in 1586, the Arakan ruler of northern Burma sent the Mughal governor of Bengal presents, including elephants, which were held in high value both valuable militarily and symbolically in both Bengal and Burma, as a diplomatic gesture. Abul Fazl's comment, quoted earlier, in the context of the description of this incident[42] betrays the relief at the settlement of peace with a militarily and logistically superior power. This attitude of healthy respect of the Mughals towards another contending political power was not very typical of them. The Gangetic Delta, however, was an entirely different terrain (literally so), as the Mughal cavalry that dominated the plains of the northern parts of the subcontinent were practically crippled here owing to the aquatic body dominated topography. Again, in 1596, Lakshminarayan, the Raja of Kuch Bihar surrendered to the Mughals. Abul Fazl mentions him having 4,000 cavalry, 2,000,000 infantry,

700 elephants and, more importantly, 1,000 war boats. The Mughals found in him an important ally to the north of Ghoraghat and also against the Isa Khan clique. His war fleet provided much needed support to the meagre Mughal resources in this respect. Boats formed a regular and prized form of war-booty and apart from the boats built or hired by the Mughals, boats captured from the enemy or obtained as tribute from a subordinate ally formed an important part of the Mughal fleet.[43]

Mughal resources in terms of boats increased dramatically once they established themselves in the eastern parts of the Ganga–Brahmaputra Delta. It was probably in this period, when the Mughals penetrated the Bengal Delta far beyond their base in Dhaka and engaged and ultimately defeated their enemies following a policy of all out military aggression, that a proper Mughal war fleet was formed for the first time under the initiative of Islam Khan, the then *subahdar* of Bengal. *Baharistan* mentions a whole range of indigenous boats used by the Mughal Army during their campaigns of 1608–1612. These include *kusa*, *khelna*, *katari*, *maniki*, *bachila*, *jaliya*, *dhura*, *sundara*, *bajra*, *piara*, *balia*, *pal*, *ghurab*, *machua*, *pashta* etc. Some of these, like *ghurab*, *maniki* and *katari* were used as war boats and fitted with artillery pieces of different range and weights. Others were used for transporting men, animals and supplies. Merchant boats, loaded with supplies, also probably accompanied the Mughal war fleet, keeping the army supplied.[44] The ones among the *bara bhuiyan* (12 warlords) who were defeated and absorbed into the rubric of Mughal service joined the Mughals with their own war fleets during military expeditions.

Manpower and war animals

The campaigns in the Bengal Delta involved a massive number of skilled and unskilled labourers for the execution of military operations. The largest among these was the corps of boatmen and crews of the Mughal riverine fleet. In times of need, they also doubled as labourers and were deployed by the Mughals to execute a variety of actions. One of these was digging up silted canals and facilitating the passage of boats. The *Baharistan* provides numerous examples of such occasions. On one occasion, the Mughal detachment besieging Musa Khan's fort at Dakchara found it extremely difficult to take this location. The Mughal commanders were then advised by Raja Raghunath to enter the Ichhamati river with their fleet through a canal leading up to the river from their trenches. The only problem was that the canal had dried up and its mouth was blocked by a large heap of sand. The Mughal commander, Islam Khan, intent on executing this manoeuvre, which, as Raghunath suggested, would help them occupy not only the fort of Dakchara, but also Musa Khan's base at Jatrapur, requested Mirza Nathan take up this task. Nathan got on to the task immediately. The *Baharistan* says, out of the 12,000 boatmen and crews of the fleet, he left 2,000 to man the boats and employed the remaining 10,000 in the excavation of the canal. To boost the morale of the workers, Nathan supervised the work personally and distributed among them ample quantities of copper coins, rice, *bhang* (Indian hemp) and opium. The entire canal was dug up and its mouth cleared within a span of a week and the Mughal flotilla sailed through it to enter the Ichhamati.[45]

The crews were also occasionally employed as unskilled labour during siege operations. During the siege of Musa Khan's fort at Dakchara, the Mughals at one point found it extremely difficult to approach the fort, due to bamboo spikes fixed all around it as impediments. The ingenious Mirza Nathan then ordered the captains of his fleet to divide the corps of their boatmen into two groups – one group to be ready with bundles of straw and the other with basketfuls of earth, so that the bamboo spikes, and eventually the ditches around the fort, could be covered with straw and earth and facilitate the advance of the Mughal Army. The captains instructed their boatmen accordingly and kept 5,000 bundles of straw and 5,000 baskets of earth ready for use. Under cover of fire by the Mughal soldiers, the boatmen worked throughout the night to cover up the bamboo spikes and level the ditches. Once this had been done, the Mughals stormed the fort with their war elephants.[46]

The Mughals also required their boatmen to erect barricades and mud forts. As I have already mentioned, the marine soldiers of Bengal were famous for their famous mud forts. By the first decade of the seventeenth century, the Mughals had learnt the usefulness of the mud fort from their opponents. Consequently, from this time onwards, wherever they would go, they would make their boatmen set up temporary block houses for them to defend their positions. In times of combat, the Mughals would defend their key positions by erecting mud forts and stationing their artillery, archers and musketeers behind them.

Unlike the campaigns in the Lower Indus Basin, those in the Ganga–Brahmaputra Delta demanded the deployment of a large number of elephants. The nearby forests of Orissa served as one of the main regions from where the Mughal Army would procure these animals. The *Baharistan* mentions one incident where some Mughal officers had gone to purchase elephants in Orissa and on their way back, were robbed by a group of Afghans, who took the elephants as well. Soon, a Mughal detachment advancing against them, defeated the Afghans in an engagement and took possession of, among other things, the elephants that the previous Mughal party had been robbed of. This event is illustrative of several important points. In the first place, it underlines the logistical and military importance of elephants in the Mughal force that operated in the Bengal Delta in particular, and elsewhere at large. Second, it shows that Orissa was the main supplier of these animals to the Mughals during the opening decade of the seventeenth century. We know that the forests of Orissa have always been very rich in elephants, so much so that the *rajas* of the Mahanadi Basin would traditionally style themselves as *Gajapatis* (lords of elephants). And for this reason, these forests had always been lucrative for political powers in the northern Indian subcontinent, anxious to recruit more and more elephants for combat and logistical branches. In the fourteenth century, the sultans of Delhi also procured their elephants from these parts. Mughal presence in the Mahanadi Basin was secured under Akbar during the 1570s. Probably, from this period onwards, the

Mughals acquired elephants from the forests of this region. Before this, elephants must have been principally supplied from the jungles of the Middle Ganga Basin. A passage in *Baburnama* says:

> The elephant, which the Hindustanis call *hat(h)i*, is one of the wild animals peculiar to Hindustan. It inhabits the borders of the Kalpi country, and becomes more numerous in its wild state the further east one goes. From this tract it is that captured elephants are brought; in Karrah and Manikpur elephant-catching is the work of 30 or 40 villages.[47]

But, as this region presumably had a lower elephant population than Orissa, we may argue that after Orissa came under direct Mughal control in the 1570s, there was a considerable increase in the number of elephants in the Mughal Army. Quite naturally, the case of a Mughal party being robbed of their elephants while returning from Orissa left the Mughals worried. Even apart from the political implications of the incident, the robbery signalled an endangering of the supply line of elephants from Orissa. For the Mughal Army that had grown increasingly dependent on elephants in terms of military deployment and logistical purposes, this was alarming. This explains the instant despatch of a Mughal force for chastising the Afghan insurgents and to ensure the continued Mughal control over this crucial supply area, without which their ongoing campaigns in the Ganga–Brahmaputra Delta would suffer heavily.

However, as the Mughals penetrated deep into the Bengal Delta, they found a new mine of elephants in northern and north-eastern Bengal. *Baharistan* narrates several occasions where we find Mirza Nathan and his associates engaged in *kheda*, a type of hunt to capture elephants. From the description of one such *kheda* operation in Nazirpur, we are informed that its usual format was the same as the usual *qamargah* hunt, a legacy that the Mughals brought into the subcontinent from Central Asia. A large area of the forest would be selected and then the elephants of that area be driven into a smaller enclosure for them to be captured. A passage from the *Baharistan* explains:

> Iftikhar Khan, Mubariz Khan and Raja Satrajit wrote that they had brought a large number of elephants to the 'Qamargah' (ring hunt). Therefore, Islam Khan marched quickly and reached early in the morning a place at a distance of five *kos* from Nazirpur where the elephants were decoyed to the hunting enclosure. They began the game of *kheda* and captured one hundred and thirty-five elephants by trapping them in the *khad* (ditch) and tied them to the noose of trained hunting elephants.[48]

Some of the elephants procured were be absorbed into the personal retinues of the *mansabdars* and the rest were sent to the *Padshah* as tribute.

Conclusion

The historiography of Mughal state formation and warfare has tended to highlight either the cavalry or the gunpowder artillery as the chief driving force behind the establishment and consolidation of the Mughal state. While on the one hand, thorough studies of the *mansabdari* institutions highlight the centrality of the cavalry in Mughal state and psyche; on the other hand, works on Mughal technology focus heavily on gunpowder technology. As a result, the role of other forms of military technology in the history of Mughal military expansion has tended to become marginalized. It is only when we shift our attention to other forms of military enterprise, like sieges, amphibious engagements and logistical arrangements that this picture begins to get more complicated. This chapter is an attempt in looking at one such form of military activity, amphibious wars, and thereby trying to highlight the different factors, other than cavalry and gunpowder artillery, that went into the complex process of Mughal military expansion. On the other hand, this chapter tries to explore how the Mughal post-nomadic experience was conditioned by the specific geographic conditions of the Indian subcontinent and how this contributed to the transformation of the nature of Mughal war making. One hopes that the appreciation of this evolutionary dimension of Mughal warfare in particular and the Mughal experience in general will make the analytical category of post-nomadism more nuanced and better equipped to analyse the South Asian case.

Notes

1 Jos J.L. Gommans, 'Warhorse and Post-Nomadic Empire in Asia, *c.*1000–1800', *Journal of Global History*, vol. 2 (2007), pp. 1–21.
2 Jos J.L. Gommans, 'The Silent Frontier of South Asia, c. 1200–1800', *Journal of World History*, vol. 9 (1998), pp. 1–23.
3 Jos J.L. Gommans, *Mughal Warfare: Indian Frontiers and High Roads to Empire, 1500–1700* (London: Routledge, 2002).
4 Anatoly Khazanov and Andre Wink (eds), *Nomads in the Sedentary World* (Richmond: Curzon Press, 2001); Thomas J. Barfield, *The Perilous Frontier: Nomadic Empires and China, 221 BC to AD 1757* (Cambridge, MA, and Oxford: Blackwell, 1992); Russel J. Smith, 'Grassland and Farmland as Factors in the Cyclical Development of Eurasian History', *Annals of the Association of American Geographers*, vol. 33, no. 3 (1943), pp. 135–161.
5 Jos J.L. Gommans, 'Warhorse', p. 14; Jos J.L. Gommans, *Mughal Warfare: Indian Frontiers and High Roads to Empire, 1500–1700* (London: Routledge, 2002), pp. 100–111.
6 Jos J.L. Gommans, 'Warhorse', p. 24.
7 Pratyay Nath, 'Siege Warfare in Mughal India, 1519–1538', in Kaushik Roy (ed.), *Warfare and Politics in South Asia from Ancient to Modern Times* (New Delhi: Manohar, 2011), pp. 121–144.
8 For Burma, see Michael W. Charney, 'Shallow-Draft Boats, Guns, and the Aye-ra-wa-ti: Continuity and Change in Ship Structure and River Warfare in Precolonial Myanmar', *Oriens Extremus*, vol. 40, no. 1 (1997), pp. 16–63. For a general overview of warfare in South-East Asia, see Michal W. Charney, *Southeast Asian Warfare, 1300–1900* (Leiden and Boston, MA: Brill, 2004).

9 See, for example, Aniruddha Ray, *Adventurers, Landowners and Rebels: Bengal, c.1575–1715* (New Delhi: Munshiram Manoharlal, 1998), pp. 63–97; Atul Chandra Roy, *History of Bengal: Mughal Period, 1526–1765* (Calcutta: Nababharat Publishers, 1968).

10 Jean Deloche calls the Ganga river 'the royal route of the northern plains'. For details regarding the riverine transport in the Ganga Basin, see Jean Deloche, *Transport and Communications in India: Prior to Steam Locomotion*, 2 vols (Delhi: Oxford University Press, 1994), *Water Transport*, vol. 2, pp. 18–23.

11 Jean Deloche, *Transport and Communications in India prior to Steam Locomotion*, 2 vols (Delhi: Oxford University Press, 1993), *Land Transport*, vol. 1, pp. 24–30.

12 Abul Fazl Alami, *Akbarnama*, Persian text (ed. Maulawi Abdur Rahim), 3 vols, Bibliotheca Indica, New Series nos 352, 353, vol. 3 (Calcutta: Asiatic Society of Bengal, 1876), pp. 584–585, 601; Abul Fazl Alami, *Akbarnama* (English tr. H. Beveridge), 3 vols (Delhi: Low Price Publication, 1989), vol. 2, pp. 886–887, 917–918.

13 Jean Deloche, *Transport and Communications*, vol. 3, pp. 14–18.

14 O.H.K. Spate, A.T.A. Learmonth and B.H. Farmer, with the collaboration of A.M. Learmonth, *India, Pakistan and Ceylon: The Regions* (New Delhi: B.I. Publications, 1972), pp. 534–538.

15 Nizamuddin Ahmad, *Tabaqat-i-Akbari* (English tr. Brajendranath De, ed. Baini Prasad), 3 vols (reprint, Delhi: Low Price Publications, 1992), vol. 2, pp. 637–641; Abdul Qadir Badaoni, *Muntakhabu-t-Tawarikh* (English tr. W.H. Lowe, rev'd B.P. Ambashthya), 3 vols (reprint, Delhi: Renaissance Publishing House, 1986), vol. 2, p. 392.

16 Jean Deloche, *Transport and Communications*, vol. 1, pp. 37–43, 45–46.

17 *Akbarnama*, Persian text (ed. Abdur Rahim), vol. 2, pp. 361–364; *Akbarnama* (tr. H. Beveridge), vol. 2, pp. 526–528.

18 *Akbarnama*, Persian text (ed. Abdur Rahim), vol. 3, pp. 601–603, 606, 608–609, 613–615, 633–634; *Akbarnama* (tr. H. Beveridge), vol. 3, pp. 917–920, 925, 929–931, 938–940, 971–973; *History of the Arghuns and Tarkhans of Sind (1507–1593), An Annotated Translation of the Relevant Parts of Mir Masum's Tarikh-i Sind* (English tr. Mahmudul Hasan Siddiqi) (Sind: Sind University Press, 1972), pp. 194–206.

19 William Foster (ed.), *Early Travels in India, 1583–1619* (Delhi: Low Price Publications, 2007), p. 28.

20 *Akbarnama*, Persian text (ed. Abdur Rahim), vol. 3, pp. 479–480; *Akbarnama* (tr. H. Beveridge), vol. 3, p. 722.

21 Mahinder S. Kingra, 'The Trace Italienne and the Military Revolution during the Eighty Years' War, 1567–1648', *Journal of Military History*, vol. 57, no. 3 (1993), pp. 431–446; Weston F. Cook Jr, 'The Cannon Conquest of Nasrid Spain and the End of Reconquista', *Journal of Military History*, vol. 57, no. 1 (1993), pp. 43–70.

22 There are some mentions of temporary forts or fortified locations being built out of mud and sand in the Lower Indus Basin during the Mughal campaigns against Thatta in the 1590s. However unlike the Bengal Delta, the incidence of building such temporary forts was scare. Both the parties tried to make use of sand dunes lying on the bank of the Indus for their own strategic advantage in defending these fortified locations. *Tarikh-i Sind* (tr. M.H. Siddiqi), pp. 195–200, 202–204.

23 Ayesha Begum, 'Mughal Fort Architecture in Bengal with an Introduction to Some Important River Forts', *Journal of Asiatic Society of Bangladesh*, vol. 47, no. 1 (2002), pp. 1–24.

24 Mirza Nathan, *Baharistan-i Ghaybi: A History of the Mughal Wars in Assam, Cooch Behar, Bengal, Bihar and Orissa during the Reigns of Jahangir and Shah Jahan* (English tr. M.I. Borah), 2 vols (Guwahati: Government of Assam: 1992), vol. 1, pp. 126–127.

25 Mirza Nathan, *Baharistan-i Ghaybi*, vol. 1, pp. 127–128.

26 Mirza Nathan, *Baharistan-i Ghaybi*, vol. 1, pp. 128–130.

27 Zahiruddin Muhammad Babur, *Baburnama* (English tr. A.S. Beveridge), 2 vols (New Delhi: Munshiram Manoharlal, 1998), vol. 1, p. 240.
28 Zahiruddin Muhammad Babur, *Baburnama* (tr. A.S. Beveridge), vol. 1, pp. 378–392.
29 *The Commentary of Father Monserrate on His Journey to the Court of Akbar, 1580–1582* (English tr. J.S. Hoyland) (New Delhi and Chennai: Asian Educational Services, 2003), p. 98.
30 *The Commentary of Father Monserrate*, p. 104.
31 *The Commentary of Father Monserrate*, p. 109.
32 *The Commentary of Father Monserrate*, pp. 109–110.
33 *The Commentary of Father Monserrate*, pp. 121–135.
34 *Akbarnama*, Persian text (ed. Abdur Rahim), vol. 3, pp. 353–354; *Akbarnama* (tr. H. Beveridge), vol. 3, p. 519; *Tabaqat-i Akbari* (tr. B. De), vol. 2, pp. 548–549.
35 *The Commentary of Father Monserrate*, pp. 156–157; *Akbarnama*, Persian text (ed. Abdur Rahim), vol. 3, pp. 372–373; *Akbarnama* (tr. H. Beveridge), vol. 3, p. 546; *Tabaqat-i Akbari* (trans. B. De), vol. 2, pp. 551–552.
36 Abul Khair Muhammad Farooque, *Roads and Communication in Mughal India* (Delhi: Idarah-i Adabiyat-i Delli, 1977), pp. 40–54.
37 Such was the case when the Mughal Army attempted to cross the Chenab in 1581. Abul Khair Muhammad Farooque, *Roads and Communication in Mughal India*, p. 109.
38 *The Commentary of Father Monserrate*, p. 81.
39 Mirza Nathan, *Baharistan-i Ghaybi*, vol. 1, p. 78.
40 Mirza Nathan, *Baharistan-i Ghaybi*, vol. 1, p. 84.
41 For a general treatment of boats and other means of riverine transportation in the Indo-Gangetic Basin in the pre-colonial period, see Jean Deloche, *Transport and Communications*, vol. 2, pp. 143–170.
42 *Akbarnama*, Persian text (ed. Abdur Rahim), vol. 3, pp. 479–480; *Akbarnama* (tr. H. Beveridge), vol. 3, p. 722.
43 *Akbarnama*, Persian text (ed. Abdur Rahim), vol. 3, pp. 716–717; *Akbarnama* (tr. H. Beveridge), vol. 3, pp. 1066–1068.
44 Mirza Nathan, *Baharistan-i Ghaybi*, vol. 1, p. 48.
45 Mirza Nathan, *Baharistan-i Ghaybi*, vol. 1, pp. 60–64.
46 Mirza Nathan, *Baharistan-i Ghaybi*, vol. 1, pp. 66–69.
47 Zahiruddin Muhammad Babur, *Baburnama* (tr. A.S. Beveridge), vol. 2, p. 488.
48 Mirza Nathan, *Baharistan-i Ghaybi*, vol. 1, p. 23.

9 Indo-Portuguese naval battles in the Indian Ocean during the early sixteenth century

K.S. Mathew

The discovery of the direct sea-route connecting the Indian Ocean regions with the Atlantic ports by Vasco da Gama, the undaunted admiral of the Indian Ocean, and the subsequent establishment of Portuguese factories and fortresses on the coastal regions of the subcontinent of India are considered important milestones in world history. The sixteenth century witnessed a significant encounter between the West and the East in the socio-economic realms. The Westerners observed with great interest whatever they came across in the East and reduced to writing most of the interesting aspects of life in the East to which probably an Indian observer paid only scant attention. The rich corpus of information available in the Portuguese archives and libraries throws considerable light on the naval activities of Indians and especially the Indian encounters with the Portuguese. The present study is an attempt to highlight some of the aspects of the Indo-Portuguese naval battles during the early decades of the sixteenth century. Contemporary European as well as 'native' sources are made use of in the preparation of this chapter. This period is chosen chiefly for the sake of brevity and also because of the fact that the Portuguese during this point of time had confrontations with the Indians on the eastern as well as on the western coasts of India and by this time they had established a sort of naval supremacy in the Indian Ocean regions.

Indian ships were found in the far-flung regions of the Indian Ocean before the arrival of the indefatigable mariners of Portugal. When Vasco da Gama reached the East African coast, there were Indian merchants in Moçambique, presumably Gujaratis, selling cloths, cloves, pepper, ginger, pearls, silver rings and rubies, most of which was taken to the East African coast by vessels from India. Huge vessels sewn with coir and propelled with the help of veils made of palm trees, owned by Indian merchants used to leave for Soffala.[1] The Gujarati merchants in large number lived in Mombasa and Melinde in the last decade of the fifteenth century.

Apart from the Gujarati merchants, Indian Christians, probably those from Cranganore, were found engaged in trade in Melinde on the East African coast when the first successful voyage of Vasco da Gama reached this port. Christians from India presumably those of St Thomas were found trading in Mombasa. They purchased gold, amber and ivory against copper, quicksilver and cotton cloths. On being asked about the place from where these merchants came to Melinde, they told Vasco da Gama that they were from Cranganore and followers of St Thomas. In fact it was a Gujarati pilot living on the East African coast who led Vasco da Gama to Calicut.[2] The Gujarati merchants had very long-standing trade relations with the Persian Gulf and Red Sea regions. Aden, Ormuz and Mocha were frequently visited by them. They took gold, quicksilver, vermillion, copper, rosewater, camels, scarlet-in-grain, coloured woollen cloth, glass beads and weapons which were brought by the merchants from Cairo to Aden. The above-mentioned items were collected by merchants from Italy, Greece and Damascus. Horses from various parts of Arabia and Persia especially from Ormuz were brought by the Gujarati merchants to India.[3]

The merchants from Gujarat held Malacca as their chief centre of sea-borne trade. Over 1,000 merchants were found settled there in the fifteenth century. Moreover, about 4,000–5,000 Gujarati merchants visited Malacca every year for the sake of trade. It was reported by a Portuguese observer of the first quarter of the sixteenth century that the interdependence of Malacca and the Kingdom of Gujarat was such that the former could not survive without the latter and vice versa. The Gujarati ships visited places like Bengal, Pegu, Siam, Pedir, Pase, Kedha, Java, Sunda and Grisee. They had efficient pilots who were expert in navigational skill. Moreover, they kept a number of men-at-arms to defend their ships. As they were known as good navigators, merchants from Ormuz, Kilwa, Malindi (Melinde), Mogadishu, Mombasa, Cairo, Mocha, Aden, Abyssinia, Persia, Turkey, Armenia, Khorasan and Shiraj accompanied them to Malacca. They were compared to the Venetians and Genoese on account of their skill in navigation and maritime trade. Similarly, merchants from Chaul, Dabul, Goa, Malabar, Bengal, Coromandel especially Pulicat, Nagore, Kayal, Tranquebar, Tirmel-wassel, Kalapet and Pondicherry frequented Malacca in connection with trade and commerce.[4] Malacca was reckoned as the most important emporium of the East–West axis of international maritime trade connecting the 'Orient' with Venice through Cairo and Alexandria during the fifteenth century.[5]

Portuguese writers while emphasizing the importance of navigation and sea-borne trade conducted by the Gujaratis spoke of their trade network with the West as well as the East. It is reported by Tomé Pires: 'Cambay [kingdom of Cambay] chiefly stretches out two arms, with her right arm she reaches out towards Aden and with the other towards Malacca.'[6]

The international maritime trade conducted with the port towns on the East African coast, the important centres of the Persian Gulf and the Red Sea regions as well as the most important oriental emporium of Malacca, by the Indian merchants in the century prior to the opening of the direct sea-route connecting the Indian Ocean regions with the ports of the Atlantic regions speaks aloud for the Indian navigation. Similarly, the fact that the Indian navigators were not highly impressed by the navigational equipment shown to them by the Portuguese admiral substantiates the view that the Indians had a well-developed technique of navigation before the dawn of the sixteenth century. This makes one conclude that the Indian navigators who operated in the various important regions of the

Indian Ocean were not lagging behind the Europeans in terms of sailing technology.

The naval confrontations of the Indians with the Portuguese right from the inception of the mutual contacts provide a picture slightly different from what is seen in the case of peaceful navigation and trade. We shall address ourselves to discuss the details of these confrontations taking into account that the Indian ships were not always armed with guns, while the non-Indian ships were armed with guns made of iron.

One of the earliest naval encounters between the Portuguese and the Indians under the leadership of the Zamorin of Calicut took place in September 1498. Since the first meeting between Vasco da Gama and the Zamorin of Calicut could not end on a happy note, about 70 vessels equipped with arms followed the Portuguese fleet, south of Calicut. They attacked the three ships under Vasco da Gama. The Portuguese fleet proceeded in the direction of Angediva. Eight of the ships sent by the Zamorin confronted the caravel *Bérrio* of the fleet of Vasco da Gama. The Portuguese ship, because of it being a caravel, had the advantage of sailing to windward and facing the Indian ships. Moreover, it had guns and cannons made of bronze which were quite powerful to cause harm to the Indian ships which were put to flight by a single caravel. The Portuguese took over one of the abandoned ships (*nau*)[7] wherein the personnel of *Bérrio* found a number of bows, arrows, swords, lances, javelins and other weapons. There was no trace of any gun. It was ascertained that these ships were sent by the Zamorin with a view to capturing the Portuguese and not permitting them to leave for Portugal with the claim that they discovered the direct sea-route connecting India and Portugal. Another day, seven vessels (pinnace[8] and fusta[9]) belonging to the famous 'pirate' Timoja confronted the Portuguese ships S. *Gabriel* and S. *Rafael* and opened fire. As they could not do any harm, and *Bérrio* started chasing them; finally the Indian ships left the sea for safety.[10]

We find that a single vessel belonging to the Portuguese was able to defeat a fleet of eight Indian ships. Similarly, the seven ships of Timoja were faced by a single Portuguese vessel and were put to flight. It is to be noted that the Portuguese ship was equipped with bronze cannons and guns while the Indian ships were devoid of any artillery. Moreover, it is worth noting that the Portuguese vessel with which the Indians were confronted was a caravel which by its nature had certain advantages. Another technical aspect that should be underlined is that the Portuguese ships were manufactured with the use of iron nails while the Indian ships were sewn with coir.

Another important naval encounter between the Indians and the Portuguese took place in 1500. The Portuguese fleet consisting of 13 ships with 15,000 persons on board left Lisbon on 6 March 1500 under Pedro Álvares Cabral for India. Only six of them reached the western coast of the subcontinent. After peaceful negotiations, a factory was established at Calicut for conducting trade in spices. Since the initial relations were friendly, the Zamorin asked Pedro Álvares Cabral a favour. The former was interested in one of the seven elephants carried in a ship belonging to a merchant from Cochin which was passing by

Calicut. As a token of friendship, Álvares Cabral was requested to capture the ship and get the elephants on which the Zamorin's eyes were fixed. Though Cabral did not want to run the risk of offending the King of Cochin, he had to come forward to show a good gesture to the Zamorin. He put two noble men and 60 soldiers in charge of a *nau* and ordered them to capture the elephants along with the ship of the Cochin merchant. Pêro de Ataide was put in command of the Portuguese vessel which was supposed to overpower the ship of the above-mentioned merchant well armed with 300 fighters on board. Pêro de Ataide confronted the Indian ship in the vicinity of Cannanore. The Indian ship sent a host of arrows and shots of cannons from its guns towards the Portuguese ship which responded promptly with all its artillery. As desired by the Zamorin, the coveted elephants were delivered to him by Pêro de Ataide after capturing the ship. This boosted the military prestige of the Portuguese. Serious damage was inflicted on the Indian ship. There is no doubt that the Indian ship was quite large having the capacity of transporting seven elephants and 300 soldiers.[11]

Besides, Pêro de Ataide managed to destroy four ships of the Muslims near Cannanore and a few *paraus*.[12] On another occasion, five ships were put to flight by Pêro de Ataide. As the prestige of the naval power of the Portuguese went on increasing day by day, the Zamorin himself began to fear that the Portuguese might destroy the Kingdom of Calicut. This conviction was confirmed by the report given by his own men. As a result, the Zamorin permitted the Muslims to attack the Portuguese factory at Calicut who killed Aires Correa and 50 Portuguese men in the factory. In fact Pedro Álvares Cabral expected the Zamorin to apologize for the casualties. Since this was not forthcoming, Cabral attacked ten ships in the port of Calicut and killed 600 men in the ships. Subsequently, the cargoes of the ships were appropriated by the Portuguese. Cabral bombarded the city from dawn to dusk. The naval superiority of the Portuguese spread like a wild fire and the kings of Cannanore, Cochin and Quilon invited Pedro Álvares Cabral to their ports. Cabral took required cargo from the port of Cochin and was about to proceed to Lisbon when he got the information that a fleet of the Zamorin consisting of 25 vessels and *paraus* with 15,000 men was approaching the Portuguese ships for a naval confrontation. On account of the fear of the Portuguese artillery instilled in the hearts of the Indian soldiers, they did not dare face the Portuguese. So the Portuguese squadron returned to Lisbon safely. On his way, Pedro Álvares Cabral captured a ship of the kingdom of Gujarat laden with valuable merchandise. After taking over the cargo, he permitted the ship to go in peace.[13]

As Dom Manuel I, the King of Portugal, was convinced that trade and commerce had to be established not by force of arms but by peaceful means, he sent four ships in 1501, out of which three belonged to private persons, to India. Of the 350 persons in the fleet, only eight were soldiers under João da Nova. On the way back, he had to face near Cannanore, a fleet organized by the Zamorin of Calicut which consisted of 40 ships and approximately 180 *paraus* and *zambuks*[14] with over 7,000 men. Though the King of Cannanore counselled restraint and requested João da Nova not to proceed towards the high seas, the latter, however, decided to face the enemy on the high seas only as that is where he could show the superiority of his artillery and better quality of the ships. The naval engagement took place on 31 December 1501 and lasted until 2 January 1502. The artillery of the Portuguese ships resulted in a significant victory over the fleet of the Zamorin. About 400 Indian soldiers were put to death and a host of them were seriously injured. João da Nova captured two other ships coming from Mocha. The naval battle took place between Mount Deli and the Bay of Cannanore. The naval superiority of the Portuguese based on artillery and probably also on better technique of shipbuilding was clearly manifested in this encounter.

A naval confrontation between the Portuguese and the forces of Cairo as well as those of Indians took place in the Arabian Sea in 1502. Thirsting to take revenge on the Zamorin for the damage caused to the Portuguese establishment in Calicut in 1500, Vasco da Gama was once again sent to Calicut. A fleet consisting of 15 big ships which were supposed to return to Portugal with cargo after the naval operation against the Zamorin and five other smaller ships to give protection to the Portuguese on the western coast of India were sent in 1502 under the command of Vasco da Gama. One of them was lost on the way and the remaining 19 ships came together near the Island of Angediva and proceeded down south.

A ship, *Meri*, belonging to the Sultan of Cairo, carrying many rich Muslims from Calicut with their families bound for Mocha came face to face with the Portuguese fleet near Mount Deli. She had about 260 soldiers and mariners and was laden with a very valuable cargo. A few vessels in the fleet under Vasco da Gama fired at *Meri* which was equipped with iron cannons. The Portuguese ships equipped with bronze cannons defeated *Meri* and captured the ship and took over the valuable cargo.[15]

Proceeding further south, Vasco da Gama with the fleet reached Calicut and tried to negotiate. After three days of fruitless labour to get compensation for the damage caused to the Portuguese in 1500, Vasco da Gama decided to take up arms against the ships of the Zamorin. Being afraid of the possible destruction of the city of Calicut, the Zamorin once again tried to start negotiations with Vasco da Gama. But, the latter, not having any trust in the Zamorin, remained firm in his decision. The Zamorin ordered 29 ships of the Moors berthed in Calicut to attack the fleet of Vasco da Gama. In the ensuing encounter 300 Moors were killed by the Portuguese. The Portuguese fleet consisting of 13 ships laden with spices left for Lisbon.[16]

After the departure of Vasco da Gama for Portugal, the Zamorin of Calicut concentrated his attention on Ponnani and prepared an army of 50,000 men to chastise the King of Cochin, his own vassal, for having given asylum to the Portuguese. Being aware of the attempts on the part of the Zamorin, the Portuguese factor at Cochin insisted upon Vicente Sodré to remain in India with the fleet of five ships and a caravel with the objective of providing protection to Cochin. But, unmindful of the request and assuming that the Zamorin would not take any step against the Portuguese during the monsoon season, Vicente Sodré went to the Red Sea area for chasing the ships from Mocha. In the meantime, by the close of the monsoon season a fleet under Francisco de Albuquerque reached the Western coast of India giving confidence to the Portuguese factor at Cochin. The Zamorin had already invaded the Kingdom of Cochin as soon as Vicente Sodré had left for the Red Sea area. The King of Cochin took refuge in neighbouring island along with the Portuguese officials of the factory. It was at this juncture that Francisco de Albuquerque reached Cochin to the great jubilation of the Portuguese as well as that of the people of Cochin. In the face of the Portuguese naval threat, the forces of the Zamorin retreated from Cochin and Francisco de Albuquerque reinstated the local king with the assistance of the military power of the place. The Portuguese made good use of their vessels equipped with artillery. With the arrival of Alfonso de Albuquerque, the construction of the Portuguese fortress at Cochin begun by Francisco de Albuquerque was speeded up. The defence put up by Duarte Pacheco Pereira with the assistance of four Portuguese vessels and the *paraus* of Cochin, in which approximately 150 Portuguese and 500 men from Malabar Coast took part, formed an integral part of the saga of the Portuguese in the East. He withstood the attack of the numerically superior army of the Zamorin.

During the Portuguese encounter with the forces of the Zamorin, two Italians who came to India from Venice in the Portuguese ships passed to the side of the Zamorin with the intention of teaching the men of Malabar how to produce artillery. They opted to work with the Zamorin against the Portuguese. On the other side, the Turks had supplied a large quantity of cannons and guns to the Zamorin. Therefore, by 1504, the Zamorin was well equipped to fight against the Portuguese and the kingdom of Cochin. Moreover, Francisco de Albuquerque had also left for Lisbon. The encounter took place in April 1504 near Cumbalam. The army of the Zamorin consisted of more than 84,000 men and the fleet had about 100 *paraus* – each of which was equipped with two short cannons for throwing stones, and five guns – besides 100 other small vessels (*tonees*)[17] with one small cannon each. Duarte Pacheco who had to face such a formidable force of the Zamorin had only one ship with a captain and 24 men, two caravels with 25 men each and two *batels*[18] each of which had 20 soldiers. Duarte Pacheco tried to give confidence to the King of Cochin who was disheartened at the sight of the formidable force of the Zamorin. It is calculated that over 1,300 men of the Zamorin lost their lives in the first confrontation at Cumbalam. The Zamorin put up a fight again and attacked the Portuguese ships and began to bombard the city. Duarte Pacheco ordered his men not to react until the forces of the Zamorin came closer. The tactic followed by him was quite successful and on being attacked by the Portuguese on close quarters, the forces of the Zamorin fled away.

As the Zamorin failed miserably three times, in the battles near Cumbalam, in inflicting any serious defeat on the Portuguese, he decided not to attempt any more attacks. But, his subjects as well as his vassals prompted him to take further offensive action in order to retrieve his lost prestige as it would be shameful to retreat with the formidable fleet in front of the Portuguese squadron consisting of just four vessels. So, the Zamorin made the preparations to attack Cochin by crossing the shallow portion of the sea. Being informed of this new strategy, Duarte Pacheco deployed two of his caravels at the Palluruthy pass to obstruct the movement of the forces of Calicut. The Zamorin started the attack on Cochin from the sea and land on 1 May 1503 with the help of 250 vessels. The Portuguese from their position at Palluruthy went on firing at the vessels of the Zamorin incessantly. A lot of people were put to death. The battle was resumed on 7 May 1504 after a short interval. But, there ensued a large number of casualties on the side of the Zamorin, who had decided to have recourse to strategies other than face to face naval combat. The Zamorin took steps to attack the town of Cochin, poison the wells and victuals sold in the town and thus persuade Pacheco to abandon the watch over the Palluruthy pass and go over to the town for its defence. This snare laid by the Zamorin could not meet with any success since Pacheco was fully aware of the purport of the new move on the side of the Zamorin. The attack started by the Zamorin in May/June 1504 was vigorously resisted by the Portuguese with extensive deployment of artillery. A few vessels belonging to the Zamorin and some small guns were captured by Pacheco. This victory was celebrated with great jubilation in the town of Cochin. The conflicts between the Zamorin and Duarte Pacheco ended on 3 July 1504 after seven successive victories for Pacheco and assassination of five local rulers supporting the Zamorin.

The fleet that reached India in September 1504, under Lopo Soares de Albergaria, had to face the five ships and 80 *paraus* of the Zamorin near Palliport in the vicinity of Cranganore. The Portuguese bombardment put to flight the forces of the Zamorin. At the request of the ruler of Tanur who was at loggerheads with the Zamorin, Lopo Soares de Albergaria sent a caravel with 100 soldiers to Tanur to extend help in the fight against the Zamorin. The king of Tanur courted victory and became a vassal of the Portuguese king. Similarly, at Pantalayani the Portuguese ships laden with spices and proceeding to Lisbon fought against 16 ships from Mocha that were loading commodities for the Red Sea regions. These were bombarded by the Portuguese and overpowered. The role played by the Portuguese artillery in the naval confrontations was highly appreciated by the writers of the sixteenth century. In fact, the large number of Indian vessels that confronted the small number of Portuguese vessels could not bring about any substantial advantage for the Zamorin.[19]

The confrontation of the Indian and Arab ships with the Portuguese fleet under Francisco de Almeida at Honavar in October 1505 reflected the weakness of the Indian war ships. The ruler of the region of Honavar made preparations to attack the Portuguese ships that were anchored off Angediva where a fortress was being constructed under the orders given by the Portuguese king. There were 14 ships from Mocha that were anchored off Honavar besides a large number of ships belonging to the pirate Timoja and those of the local ruler that attacked the Portuguese ships. The Arab ships were set on fire and the rest of them were destroyed. Finally, Timoja apologized for the mistakes on his part and those of the local ruler.[20]

Further fear was instilled in the minds of the Indians with the defeat of a fleet of 27 ships at Quilon by 11 *naus* and two caravels of the Portuguese in November 1505.[21] Similarly, the confrontation of the fleet of the Zamorin of Calicut with the Portuguese near Cannanore in March 1506 exposed the weakness of the Indian naval power. It was learnt by the Portuguese through the Italian who spent some time in Calicut disguised as a Muslim, Ludovico di Varthema, that the Zamorin got 200 vessels constructed all ready to put up a stiff fight against the Portuguese. The Zamorin was helped by the two Italians who taught the men of Calicut how to manufacture a better quality of artillery. The three Portuguese *naus* and a caravel bombarded the 60 *naus* and hundreds of *paraus* as well as *zambuqes* of the Zamorin. The Portuguese artillery played an important role especially the big guns installed on the wall of the fortress at Cannanore. About 3,000 men of the Zamorin were killed in this naval battle. This was a decisive victory for the Portuguese over the Zamorin and brought great prestige to them. The entire operation was executed by Lourenço de Almeida, the son of Francisco de Almeida who was put in charge of the naval surveillance in the area between Cambay and Cape Comorin.[22]

Three Portuguese *naus* under Pêro de Anaia which were found at Angediva were destroyed by Sabaio of Goa who besieged the Portuguese fortress at Angediva in 1506 during the post-monsoon season. Subsequently, orders were issued by Francisco de Almeida to dismantle the fortress since it did not serve any important purpose. Lourenço de Almeida after having demolished it proceeded towards Dabul where a number of vessels from Cochin and Cannanore were found, having been chased by the vessels of the Zamorin. Despite the earnest request made by the people of the ships from Cochin and Cannanore the Portuguese did not dare attack the ships of the Zamorin and subsequently Zamorin's men ransacked and destroyed these ships. This incident upset Francisco de Almeida who reprimanded his son and the other Portuguese officials. This was considered a great blow to the prestige of the Portuguese naval power since they did not dare confront the forces of the Zamorin at Dabul.

Taking advantage of the low profile of the Portuguese evinced in the demolition of the fortress at Angediva and the failure to confront the naval forces of the Indians at Dabul, the Zamorin decided to encounter them once again on the Malabar Coast. The fact that no fleet from Portugal reached Indian waters in 1506 emboldened the Zamorin to take this step. He convinced the ruler of Cannanore of the possible victory of the proposed attack on the Portuguese during the monsoon of 1507. The Portuguese fleet that had left Lisbon in 1506 under Tristão da Cunha after spending the winter in Moçambique reached the Malabar Coast by the close of the monsoon of 1507. Therefore, the Portuguese with the arrival of the new fleet at Cannanore, succeeded in defeating the combined forces of the Zamorin and of the king of Cannanore who had besieged the Portuguese fortress at Cannanore. The Portuguese viceroy, Dom Francisco de Almeida, being informed of the fact that a large number of vessels had taken cargo from the port of Ponnani in the Kingdom of Calicut and were about to leave for the Red Sea areas, decided to attack them with a view to retrieving the prestige lost on account of the two incidents of Angediva and Dabul. He assumed the command of the fleet consisting of 12 vessels: four *naus*, six caravels[23] and two gales[24] and proceeded to Ponnani.

The port town of Ponnani was defended by a strong battery of artillery and a number of well-armed ships under the famous captain of the Zamorin Kuttiali with 7,000 men who had taken the oath to die if needed for the defence of the city. The Portuguese fleet had about 500 soldiers in the vessels *Pero Barreto*, *Diogo Pires*, *Lourenço de Almeida*, *Nuno da Cunha*, and *Francisco de Almeida*. Some of the important Portuguese noblemen participated in this confrontation. The 13 ships laden with spices from Ponnani were burnt and forces of the Zamorin were totally defeated in November 1507 to the great jubilation of the Portuguese.[25]

The confrontation between the combined forces of the Indians and the Ottoman Turks (including the *Mamluk* Sultanate of Egypt) assisted by the Venetians on one side and the Portuguese on the other, which took place in March 1508 in Chaul and the defeat as well as the subsequent murder of Dom Lourenço de Almeida, the son of Francisco de Almeida is considered a landmark in the history of naval confrontations in the Indian Ocean region. The Turks had made use of the same technology as the Portuguese. They too used artillery with bronze cannons and employed several European renegades. The Venetians provided the required carpenters and experts for artillery for the fleet that was prepared by the Sultan of Cairo to oust the Portuguese from the Indian Ocean regions. Encouraged by the request made by the Zamorin of Calicut and the Sultan of Gujarat, the Turks made the preparation for the naval confrontation. According to the contemporary Portuguese historian, Castanheda, and other writers of the sixteenth century, the Venetians took part in the battle waged in Chaul and later in Diu. Two of the warships were of the Venetians.[26]

The Turkish fleet consisting of 2,000 men under *Amir* Hussain proceeded from the Red Sea with the destination of Diu to join forces with the men of Malik Ayaz of Diu and of the other rulers of the Indian coast. There were 12 ships well equipped with bronze cannons and experienced soldiers including European renegades. Maymame, the envoy of the Zamorin to the Sultan of Cairo was also in the fleet under *Amir* Hussain. On the Indian side, Malik Ayaz led the fleet composed of the men of Gujarat, Bijapur, Ahmednagar and Calicut. The combined forces proceeded to Chaul where the Portuguese fleet under Lourenço de Almeida was found. Malik Ayaz sent 34 fustas from Diu to Chaul on seeing that the combined forces were making headway to confront the Portuguese who had only three *naus*, one round caravel, two caravels with lateen sails and two gales with 500 men. About 20 ships from Cochin laden with cargo were in the creek of Chaul which the Portuguese tried to attack. The chief of the artillery of the anti-Portuguese combined fleet was a German.[27] The Zamorin had sent about 40 vessels to the north to take part in the battle.[28] The Portuguese lost one *nau* and 150 men including the son of Francisco de Almeida. In total, 20 persons were taken as captives by Malik Ayaz. The 20 ships from Cochin were, however, set free by the Portuguese. When the news of the victory of the combined forces over those of the Portuguese reached the Venetians, their joy knew no bounds.[29]

The Portuguese viceroy, Dom Francisco de Almeida, decided to attack the combined fleet again for the death of his son. Though Alfonso de Albuquerque, the new Governor, showed willingness to take revenge upon the victorious fleet of *Amir* Hussain and Malik Ayaz, Almeida himself took up the challenge. The latter proceeded to the north from Cannanore on 12 December 1508 with 19 vessels and 1,200 men among whom were 400 men from Malabar.[30] The Zamorin in the meantime had sent necessary communication to *Amir* Hussain in Diu regarding the preparations going on in Cannanore and Cochin against the combined contingent of Egypt and Gujarat. About 6,000 soldiers of the united front fought against the Portuguese in this battle.

The Portuguese viceroy after having avenged Dabul and Chaul for the help extended to *Amir* Hussain and Malik Ayaz in March 1508, proceeded to Diu and reached the destination on 2 February 1509.[31] The contingent under *Amir* Hussain consisting of Egyptians, Venetians and others in cooperation with the Indian fleet composed of the soldiers of Malik Ayaz and of the Zamorin faced the Portuguese fleet in Diu and fought desperately. The Zamorin had sent 100 *paraus* to Diu in spite of the vigilant blockade of the Portuguese.[32] The Portuguese fleet routed the combined forces completely in the confrontation at Diu and consequently Malik Ayaz sought for peace through the mediation of Cidy (Siddi) Ali and the terms of peace were agreed upon on 3 February 1509.[33] About 1,500 soldiers of the combined forces died during this confrontation. When the news of defeat percolated to the Venetians, they accused the Muslims for their inability to use the armaments and to adjust themselves to the Italian way of fighting.[34] The Portuguese writers consider this victory the greatest in the history of the Portuguese maritime encounters and compare it with the battle of Lepanto (1571), Trafalgar (1805) or that of Tsushima (1905). This in fact marked the beginning of the Portuguese ascendancy in the Indian Ocean region.[35]

Adil Shah of Bijapur took the help of the Turks and made ready about 20 *naus* and several *paraus* at Goa to confront the Portuguese. As the sultan was on bad terms with the ruler of Vijayanagar, the Portuguese with the aid of the latter captured Goa in February 1510, without any serious resistance from Adil Shah. But, after a period of three months, the sultan came to terms with the ruler of Vijayanagar and took steps to get Goa back from the hands of the Portuguese. He organized an army of about 40,000 soldiers, a good number of whom were Turks, and confronted the Portuguese. A Turkish leader Sufolarim by name led the *paraus* of Adil Shah. The subsequent battle waged in Mondovi during the monsoon season was won by the Portuguese under Alfonso de Albuquerque. Later, on 25 November 1510, with the assistance of 1,200 men in 30 vessels, Alfonso de Albuquerque captured Goa and constructed naval installations for the repair and manufacture of ships.

Caravels were used always by the Portuguese as warships in the early years of confrontations with the Indians. They realized the importance of gales in the naval encounters in the Indian Ocean regions especially on the western coast of India and so Alfonso de Albuquerque briefed the Portuguese regarding the necessity of manufacturing chiefly gales for the Indian waters. Since the time of Albuquerque, gales, galeotas[36] and fustas were preferred for naval battles. Gales and fustas had certain advantages over vessels of other types. First, they could be more easily and quickly constructed with local materials during the time of monsoons. And when they were not needed, they could be easily hauled aground without any damage of expenses for maintenance. Second, not being painted thickly and due to their low draught, they could easily ply in rivers. They were propelled both by oars and sails. Therefore, they could sail even when there was no wind or the wind was in the opposite direction. The advantages of the caravel over the vessels propelled by oars were their ability to use guns of greater calibre, better speed with fresh wind, greater autonomy, ability to withstand the storm and, above all, few crew requirements due to its smaller size.

In 1517, 15 fustas of Diu encountered the Portuguese fleet under João de Monroi in the vicinity of Chaul and then combat started with bombardment on both sides. Probably on account of the better quality of artillery, the Portuguese defeated the ships sent from Diu at the request of the ruler of Mahim against them. Similarly, another fleet of 14 fustas put up a fight against the fleet of Monroi near Dabul in the same year. The Portuguese fleet consisting of seven fustas defeated the squadron of 14 fustas with the help of superior artillery.

The Portuguese faced the Indian ships near Chittagong, one of the important ports of Bengal in the sixteenth century. The ships of that region did not have cannons. They fought only with bows and arrows against the Portuguese during the naval encounter in 1518. The Portugese were commanded by Lopo Soares de Albergaria. The vessels from Chittagong were defeated with the help of artillery carried by the Portuguese on their ships.[37] The fustas of Dabul destroyed a Portuguese ship laden with cargo in 1518 on its way to Cambay from Cannanore. The people of Dabul made use of artillery and so they were successful in their confrontation with the Portuguese *nau.* Later, in 1521, the fustas of Diu led by Aga Muhammad put up a stiff fight against the Portuguese ships. As the fustas were well equipped with artillery in the same way as those of the Portuguese, they did a lot of harm to the Portuguese vessels.[38]

On account of the importance of Chaul in the trade in horses and the necessity of having a well-fortified centre from where they could fight against the ruler of Diu, the Portuguese began the construction of a fortress in Chaul with the consent of Nizam Shah of Ahmadnagar. Malik Issack of Diu had his fears about the Portuguese establishment in Chaul. So, he asked Aga Mohammed to proceed to Chaul with 40 fustas to obstruct the Portuguese building activities there. Diogo Lopes de Sequeira organized a fleet of three galeões,[39] a caraval and three gales under the command of Dom Aleixo de Meneses to confront the 50 fustas from Diu. Each of the fustas was equipped with cannon that could be charged in the breech and could be moved in any direction. These artillery pieces were of small calibre sending shots of stone to short ranges while the Portuguese guns were charged though the mouth and the shots were heavier than those of the

Indian guns. This confrontation took place in November 1521. After a long-drawn combat of artillery, the fleet of Aga Muhammad retreated in disgrace.

The fact that Indian merchant marines were found in various ports of the Indian Ocean regions proves that they were used to peaceful commerce and they maintained simple weapons capable of warding off the common pirates. Of course, they possessed necessary technology for navigating freely in the Indian Ocean regions and were never behind the Europeans in the matters of techniques of navigation. This shows the high achievements of Indian navigation.

The weakness of the Indian vessels in their confrontation with the Portuguese consists chiefly in the fact that they were not as well equipped as the Portuguese vessels. For example, the ships captured by the Portuguese in 1498 did not contain any cannons. The traditional bows and arrows and spears were the only items of equipment that were stored in the ships. It was the same case with the ships that confronted the Portuguese ships near Chittagong. Further, the vessels manufactured in India did not have sufficient strength to withstand the shock administered by the Portuguese cannon balls. Even those vessels of the Arabs and those of the Indian equipped with guns and cannons were not capable of resisting the aftermath of a bombardment especially the ones from the European ships equipped with bronze cannons. Sewing with coir did not give the strength to withstand the shock. Moreover, the ships of the Indian Ocean regions were in a much more disadvantageous situation compared to the European ships which were manufactured using iron nails, giving more strength to the body of vessel.

Some of the Portuguese writers do argue that they did not come to establish trade and commerce by force of arms. The use of force by the Portuguese according to these writers was chiefly because of the attitude taken by the Zamorin. They note the instance of Pedro Álvares Cabral leaving a Gujarati ship in peace after taking out all the cargo in 1501 as an example. Similarly, the sending of four ships under João de Nova in 1501 with only 80 men-at-arms on board is also shown as a case substantiating their view.

The superiority of European artillery in naval battles against the backdrop of Indian confrontations becomes clear from the outcome of the Battle of Chaul fought in 1508. Lourenço de Almeida, the valiant son of Francisco de Almeida died and a number of Portuguese vessels were destroyed by the combined forces of the Zamorin of Calicut, Sultan of Gujarat, under Malik Ayaz, and Admiral Amir Hussain representing the Sultan of Cairo. The victory for the combined forces was made possible chiefly because the arms and ammunitions sent by the Venetians, another European power, and which were 'properly' used by the naval crews and the mariners. This would mean that either qualitatively or quantitatively, the Venetian artillery was superior to that of the Portuguese. As we have seen above, the well-armed ship belonging to the Sultan of Cairo, *Meri*, was easily defeated by the Portuguese because the arms and ammunitions used in *Meri* were not a match for those of the Portuguese.

The defeat of the combined forces of the Indians as well as those of the Arabs by the Portuguese in 1509 at Diu was bemoaned by the Venetians who asserted that the Arabs did not know how to use the weapons and artillery supplied by them. This would mean that those who used the type of artillery furnished by the Venetians for the Battle of Chaul in 1508 did not survive or were not present at the Battle of Diu in 1509. At any rate both the victory of the combined forces at Chaul in 1508 and that of the Portuguese at Diu in 1509 testify to the strength of the European artillery in comparison to that of India in particular and Asia in general.

In conclusion, it may be stated that the naval battles waged in the Indian Ocean by the Indians and the Europeans in the period before 1700 were won by those who had better artillery. Indian ships were not usually equipped with arms and ammunitions. Even those equipped with 'native' arms could not withstand the heavy artillery with which the European ships were furnished. The technique of shipbuilding also had a lot of impact on the victory or defeat in the naval battles. Use of iron nails in the construction of European ships had an edge over the use of coir in sewing planks for shipbuilding in Indian waters. Therefore such ships made in India could not withstand the impact of bombardment from a European ship. Similarly, the Europeans used ships with special architectural styles suited to fight on high waters easily damaged the Indian ships. Therefore, the Portuguese had several advantages in the naval battles waged against the Indians during the period of our study.

Notes

1 Fernão Lopes de Castanheda, *Historia do Descobrimento & Conquista da India pelos Portugueses*, Livro I (Coimbra: Imprensa da Universidade de Coimbra, 1924), pp. 17–31.
2 Fernão Lopes de Castanheda, *Historia do Descobrimento*, pp. 31–33.
3 Tomé Pires, *The Suma Oriental of Tomé Pires* (reprint, New Delhi: Asian Educational Service, 1990), vol. 1, pp. 41–44; Duarte Barbosa, *The Book of Duarte Barbosa*, vol. 1 (reprint, Nendeln/Liechtenstein: Kraus, 1967), pp. 128–131.
4 Tomé Pires, *The Suma Oriental*, vol. 1, pp. 45–46, vol. 2, pp. 268–273.
5 K.N. Chauhuri, *Trade and Civilization in the Indian Ocean* (Cambridge: Cambridge University Press, 1985), p. 110; K.N. Chaudhuri, *Asia before Europe: Economy and Ciivilization of the Indian Ocean from the Rise of Islam to 1750* (Cambridge: Cambridge University Press, 1990), p. 367.
6 Tomé Pires, *The Suma Oriental*, vol. 1, p. 42.
7 *Nau* had a tonnage of 100 to 120 in the early phase of the Portuguese contacts with India. It had two decks, the first one extending from the rear to the front containing the cargo hold, store-room for water and provisions, cables, cloths and gunpowder etc. The second deck in the rear had the captain's quarters. It had three masts. K.M. Mathew, *History of the Portuguese Navigation in India* (New Delhi: Mittal Publications, 1988), pp. 288–292.
8 Pinnace was a narrow and light vessel with oar and sails and with three masts and a square stern. It was used for reconnaissance and disembarkation of people on land. Mathew, *History of the Portuguese Navigation in India*, p. 292.
9 Fusta was a long and flat ship propelled by means of lateen type sails and oars and with one or two masts. It has a capacity of 300 tons. It was used in Asia for warfare. It had ten to 20 oars on the two boards and the mast in the middle with lateen sails.
10 Gaspar Correa, *Lendas da India*, tomo I, Part I (Coimbra: Imprensa da Universidade de Coimbra, 1922), pp. 123–129.

11 Fernão Lopes de Castanheda, *Historia do Descobrimento*, pp. 83–84; João de Barros, *Da Asia, Decada* I (Lisboa: Livraria Sam Carlos, 1778), p. 425.

12 *Parau* was a small warship. It was used also to carry merchandise. This was compared by the European writers with fusta and galeotas. Sebastião Rodolfo Dalgado, *Glossario Luso-Asiatico*, vol. 2 (Coimbra: Imprensa da Universidade de Coimbra, 1921), pp. 170–171.

13 Gaspar Correa, *Lendas da India*, p. 226.

14 *Zambuk* (*Zambuque*), *Sambook*, *Sanbuk*, *Sanbuka* was a kind of a small vessel formerly used in western India and still in use on the Arabian Coast. It was smaller than the *bagala* and was chiefly used to communicate between roads and the shore or to go inside the reefs. Henry Yule and A.C. Burnell, *Hobson Jobson: A Glossary of Colloquial Anglo-Indian Words and Phrases and of Kindred Term, Etymological and Historical, Geographical and Discursive* (reprint, Delhi: Munshiram Manoharlal, 1968), p. 788.

15 João de Barros, *Da Asia, Decada* I, Part II (Lisboa: Livraria Sam Carlos, 1777), pp. 29–38.

16 João de Barros, *Da Asia, Decada* I, Part II, pp. 55–76.

17 *Tonee* was small vessel with single mast and oars used in rivers.

18 *Batel* was a sort of boat used in western India, Sind and Bengal.

19 Gaspar Correa, *Lendas da India*, pp. 424ff.

20 Gaspar Correa, *Lendas da India*, pp. 233–239.

21 Gaspar Correa, *Lendas da India*, pp. 489–492; Fernão Lopes de Castanheda, *Historia do Descobrimento*, pp. 253–255.

22 Fernão Lopes de Castanheda, *Historia do Descobrimento*, pp. 263–270.

23 Caravel: the word is alleged not to be oriental, but Celtic, and was connected in its origin with the old British *Coracle.* The Portuguese caravel is described by Bluteau as a round vessel that was not long and sharp like a galley with lateen sails ordinarily of 200 tons. It had triangular type sails and could sail with beam-wind even when it was coming from off the bow and thus was able to face the natural calamities of the sea. There were two types of caravels, namely, lateen caravels and round caravels. The lateen caravel had lateen mast and the main mast of the round caravel was round. Henry Yule and A.C. Burnell, *Hobson Jobson*, p. 162.

24 Gale was a warship of small tonnage with triangular lateen sails. About 25 to 30 oars on each side were used in a gale. It had 25 to 30 benches with five to six rowers on each bench. A gale usually was about 25 palms long and 30 palms wide having two masts and two lateen sails.

25 João de Barros, *Da Asia, Decada* II, Part I, pp. 51–81.

26 *Cronica do Descobrimento e Conquista da India pelos Portugueses* (Coimbra: Imprensa da Universidade de Coimbra, 1974), pp. 170–171; Fernão Lopes de Castanheda, *Historia do Descobrimento*, p. 384.

27 João de Barros, *Da Asia, Decada* II, Part I, p. 181.

28 Shaykh Zaynud Din, *Tuhfat-al Mujahidin* (tr. David Lopes as *História dos Portugueses no Malabar*) (Lisboa: Sociedade de Georgraphia de Lisboa, 1898), p. 40.

29 Rinaldo Fulin (ed.), 'I Portoghesi nell India e I Veneziani in Egito' dai *Diarii di Girolamo Priuli* in *Archivio Veneto*, vol. 22, Part 1, Venice, p. 205ff. For a detailed study of the battle see K.S. Mathew, 'The Mercantile Battle in the Indian Ocean: The Afro-Asian Front against the Portuguese (1508–1509)', in *II Seminario Internacional de História Indo-Portuguesa-Actas* (Lisboa: International Seminar on Indo-Portuguese History, 1985), pp. 65–79.

30 João de Barros, *Da Asia, Decada* II, Part I, pp. 231, 256; Fernão Lopes de Castanheda, *Historia do Descobrimento*, p. 424; Damião de Gois, *Cronica de Felicissimo Rei D. Manuel*, Part 2 (Coimbra: Imprensa da Universidade de Coimbra, 1953), p. 128.

31 João de Barros, *Da Asia*, *Decada II*, Part I, p. 2990; Damião de Gois, *Cronica*, p. 183.

32 Antonio Brasio, *Uma Carta inedita de Valentim Fernandes* (Coimbra, 1959), p. 20.
33 Shaykh Zaynud Din, *Tuhfat-al Mujahidin*, p. 41; Fernão Lopes de Castanheda, *Historia do Descobrimento*, p. 439.
34 Rinaldo Fulin (ed.), 'I Portoghesi nell India e I Veneziani in Egito' dai *Diarii di Girolamo Priuli*, pp. 206–207.
35 Saturnino Monteiro, *Batalhas e Combates da Mrinha Portuguesa*, vol. I, *1139–1521* (Lisboa: Livraria Sá Costa, 1989), p. 192.
36 Galeota was a warship having 15 to 20 benches on each side with one man at each oar. Its dimension was less than that of a gale.
37 Fernão Lopes de Castanheda, *Historia do Descobrimento*, Livor 4 (Coimbra: Imprensa da Universidade de Coimbra, 1928), pp. 439–47.
38 Gaspar Correa, *Lendas da India*, tomo II (Coimbra, Imprensa da Universidade de Coimbra, 1923), pp. 609–625.
39 It was basically a warship, lighter and more heavily equipped with guns. The tonnage varied from 100 to 1,000. A few other vessels used in the naval confrontations of this period was the *almadia*. The *almadia* was introduced into Portuguese order of battle from Moorish *al-m-diya.* It indicates a raft. But is generally used by the writers to describe a canoe or small 'native' boat.

10 The politics of military control in the west coast

Marathas, Mughals and the Europeans, 1650–1730

Amarendra Kumar

Indians have a fairly rich tradition of maritime activities. With the Himalayas in the north, they, for centuries have depended on sea routes for trade and communication with rest of the world. Vital sea links, therefore, emerged over a period of time facilitating trade, commerce and exchange of culture. The Indians not only constructed seaworthy vessels as shipbuilders, but also showed great aptitude as sailors. On the west coast of India, overseas transactions had been a crucial component of commercial activities for a long period of time. The ports of Gujarat, Konkan and Malabar transacted international trade in a huge volume. Many of the Konkan ports such as Byzantym, Syemulla etc., were known to the ancient seafarers and had always attracted brisk overseas commerce. Interestingly enough, the very idea of the control of the sea was not there as the sea was regarded as a wide common passageway. In Indian waters, thus, trade and navigation were open to all without discrimination. Such a situation continued until the time the Europeans entered Indian waters. Thereafter, the idea of control of the sea – largely based on the use of military/naval prowess – replaced the notion of equality on the sea.

It has been argued that politics was introduced to Indian waters with the coming of the Portuguese in India. The politics that the Portuguese introduced in the water bodies around South Asia was mainly confined to obtaining monopolistic rights of certain commodities they preferred to trade in and earn huge profit. Since, the exclusion of potential 'competitors' was also a pre-requisite to this process, the theory of 'sovereignty over the sea' was astutely floated by the Portuguese trading company. Subsequently the Portuguese King Dom Manuel I designated himself as 'Lord of Navigation, Trade and Commerce of India' in 1501. This paved the way for the foundation of the Portuguese *Estado da India*, the basis of which was to ensure the Portuguese 'exclusivity' in Indian waters by supervising naval operations through the various bases/forts established by them all along the coast from the Red Sea to the Arabian Sea. It was decreed that (in India) 'all the ships should collect passes (*cartaz*) issued by the Portuguese captain or factor, failing which they would be attacked and confiscated'.[1] But, the native Indian rulers (especially those settled along the west coast) did not readily buy this 'Portuguese diktat' since the concept of control of sea water never existed in the surroundings of the Indian Ocean/Arabian Sea. The likes of the Zamorins of Calicut, the Adil Shah of Bijapur, the Kunjali Marakkars and the Sultans of Gujarat, therefore, ferociously reacted and forged military (naval) coalition with outside powers in order to thwart the Portuguese attempts of controlling Indian waters.[2]

The Portuguese, however, succeeded in their endeavour. This happened because of the following factors: (a) the failure of anti-Portuguese military (naval) alliances; (b) mutual (selfish) differences among the native powers; (c) the indifference of the Indian rulers towards the concept of the 'control of sea'; (d) the determined attempts of the Portuguese to enforce their military control over Indian waters.[3] For over a century, the Portuguese *Estado* reigned supreme over the coastal waters of India. However, the seventeenth century saw a gradual decline in the power and prestige of the Portuguese in India. One of the reasons, arguably the most important one, behind the decline of the *Estado* was the rampant corruption in its administrative set up – the supervision of which was difficult from Portugal. Another reason, no less important by any means, was the entry of other European trading companies from Britain, Holland and France in the Indian Ocean. They were too modern and competitive in their outlook to have allowed the Portuguese to corner the benefits of the Indian trade without hard toil. The monopoly of the *Estado*, thus, was seriously challenged by these new entrants. The Portuguese, who tried to jealously guard their privileged position, had no option but to fight with them to retain their privileged position.[4] The tussle for the west coast further intensified with the emergence of the Marathas in the Deccan under the dynamic leadership of Shivaji, who never ceased to look for a possibility of extending his control beyond the Konkan Coast. The Mughals too, had a large stretch of the west coast under their control with the port of Surat as its main outlet. The geopolitics of the west coast of India in the mid-seventeenth century, thus, seems to be quite competitive and complex with the political and commercial landscape being 'marked by the overlapping conflicts engaging the Marathas, Mughals, other indigenous powers, and the Europeans'.[5]

It would not be out of context here to briefly discuss the rise of the Marathas as a political power in the territory adjoining the west coast of India in the second half of the seventeenth century. It has been viewed as a 'phenomenon', which was made possible by Shivaji who struck a fine balance between the unique geographical situations of the Deccan country on the one hand and the aspirations of the Maratha people for their *Swarajya*, on the other. Choosing the path of relentless military campaigns and conquests, he expanded his domains up to the west coast.[6] It was here that the dynamics of the sea-front attracted his interest. This shrewd statesman, therefore, immediately assimilated naval considerations in his policy matters. He embarked upon the project of building a navy and gave the Maratha *Swarajya* a fully fledged and competent naval establishment (as per the yardstick of the pre-modern era) which could also be scrupulously used as a tool in response to various political compulsions. Shivaji could not have risked overlooking or undermining the economic prospects of the Konkan region. His *Swarajya* too, was in need of finances to stand up on its own, and a national navy would have definitely helped to accomplish his ambition.[7] In the coastal surroundings, it must have been the perceived threat from the potent quarters (particularly the Europeans) that impelled the great visionary to set the keel afloat at Kalyan-Bhiwandi immediately following its conquest in 1658.[8] In the matters of navigation, the helplessness of the native rulers (with the exception of the Mughals) against the merciless exploitations of the European naval fleets could not escape the attention of this genius. Hence, Shivaji showed utmost urgency in organizing a naval establishment as he was convinced that a planned use of navy would not only help him scrupulously guard his political future, but also add to the economic fortunes of his nascent *Swarajya*.

Though the contemporary Marathi chronicles do not subscribe any particular date to the beginning of the Maratha naval enterprise, it is assumed that the building of vessels commenced immediately after Kalyan-Bhiwandi was captured by Shivaji in 1657–1658. The first ever definite information in this regard comes from a Portuguese document of the year 1659.[9] It seems that Shivaji's naval establishment also benefited from the services of the Europeans.[10] For example, some Portuguese artisans designed and equipped the first 20 armed Maratha ships.[11] It was not just about raising a fighting fit fleet, but Shivaji was also concurrently strengthening his coastal defences. In fact, before his military rendezvous with *Maharaja* Jai Singh, the Mughal commander of the Deccan, Shivaji had laid the foundations of his first marine fort at Sindhudurg in Malwan, which eventually became the naval headquarters of the Marathas. James Grant Duff also confirms that Shivaji had built Sindhudurg (Malvan) fort and taken possession of many other coastal forts and refurbished them.[12] It is possible that Shivaji anticipated a military action of such a magnitude from the Mughal side; hence he wanted to secure an outlet towards the sea in case the situation became more critical on the land.

It is really interesting to study how the Marathas placed themselves in matters related to trade and navigation with respect to the other players on the western coast. A contemporary Marathi source *Rajniti* of Ramchandrapant Amatya outlines: 'Those merchants who reside in the foreign countries should be persuaded to come and settle … by sending an assurance of safety to seafaring merchants at various ports, they should be given the freedom of intercourse in trade.'[13] Shivaji's desire to accommodate the Europeans in trade and other commercial activities has been well reflected in the accounts of a contemporary French traveller Abbe Carre, who writes:

> In fact along the stretch of the sea where he (Shivaji) was the master, there never passed a ship of Europe to which the governors (of those coastal areas under Shivaji) did not send refreshments with all the good offices that could hardly be expected by an allied prince. I passed that way in 1668 with two ships of the company and we were treated in a manner which was beyond our expectation. It was an act of his policy, but it was also due to the preference he felt for the people of Europe.[14]

The nature of hospitality extended to the Europeans, thus, testifies the desire of the great Maratha ruler to promote the commerce of his realm and ensure that his

Swarajya prospered. At the same time, however, there was a certain element of caution and anxiety as regards the Maratha policy towards the Europeans. The *Rajniti* continues:

> Their [the European traders'] intercourse should be restricted to the extent of only their coming and going [for trade] . . . therefore, if any place is at all given to them, it should be given in the midst of two or four famous great towns distant about eight to 16 miles from the mouth of the sea . . . thus by fixing their place of habitation factories should be permitted to be built.[15]

V.B. Kulkarni, a leading biographer of Shivaji, has summarized the Maratha approach towards the Europeans in a brilliant manner, 'Shivaji's dealings with the European traders in India were marked by caution and kindness and, at the same time, by a determination not to allow them to encroach upon his *sovereign* rights.'[16] It was indeed impossible for him or any of his contemporaries to foresee that one of these trading communities was destined to impose its sway over the length and breadth of the country.

It is true that the Portuguese power in India was on the path of virtual decline in the second half of the seventeenth century. The Dutch and the English were aggressively expanding their respective spheres of influence in India at the cost of the Portuguese. But, the fleet of the '*Estado*' still was good enough for any of the native Indian rulers including the Marathas. But on land their weakness was too obvious to be overlooked, hence, they sounded politically more moderate and diplomatically polished while interacting with the native powers. With Shivaji, for example, though they professed friendship, they used deceitful methods in chorus with the purport of causing impairment to his naval endeavours. One of the Portuguese letters read, 'to ensure that these men-of-war do not set sail, we have ordered the Portuguese Captain not to let them come out of these ports'.[17] On the other hand, despite the threat potential of the Maratha Navy, the Portuguese, at times, preferred to help Shivaji (though in a covert manner) in his struggle against the Mughals. In a letter dated 31 March 1665, addressed to Jai Singh, the Portuguese viceroy of Goa assured him that no help would be rendered to Shivaji during the course of Mughal action against him. However, on 18 April 1665, the viceroy confidentially wrote to the captain of Bassein directing him:

> you should order [to your people] that nothing should be done from which they [the Mughals] could have suspicion [that the Portuguese show favour to Shivaji]. However, if without the risk you could secretly give any aid with munitions or food-stuffs to Shivaji you should do it for money, because it is not desirable that if he is driven from his lands, the Mughals should remain the lords of them. But this should be done with such great caution that never should he (Jai Singh) be able to guess, much less to verify it. To Shivaji you will write how much better it is for him and for us that his retreat, in case it should be necessary to do so, should not be Chaul, but

> rather to Goa, where he would be more safe and we would not have to break with the Mughal.[18]

Jai Singh, however, had a different perception of Deccan geopolitics. He was convinced that Shivaji's political edifice would not survive for long;[19] hence, he planned to use him as a lever against Bijapur in consonance with an eternal principle of diplomacy. In June 1665, he wrote to Aurangzeb:

> Concerning the territories of Bijapur – of which Bijapuri Tal-Konkan [on the west coast] is actually in his [Shivaji's] possession and some other tracts of Bijapuri Balaghat are desired by him – if your majesty is planning to punish him-, in view of his insincerity and alliance until recently with the enemies of the Emperor (i.e. Bijapur), and his consent to accompany the imperial army in this very expedition, now that the time is favourable (to us), what can be better than this that we first overthrow Bijapur with the help of Shiva [ji]?[20]

The strategic importance of the island fortress of Janjira in the affair of things on the west coast could not escape the awareness of Jai Singh who wanted to extend Mughal control over it. But, rather than applying direct military pressure, he wanted Shivaji to accomplish it for him. He, therefore, proposed to allow Shivaji to continue having his control over the 'environs and suburbs of Janjira', yielding 10,000 *hun*.[21] At the same time Jai Sing was also courting the *Sidis* of Janjira to join the Mughal military campaign against Shivaji.[22]

The *Sidis* of Janjira, it is pertinent to bring out here, remained the most hated enemy of the Maratha *Swarajya* for over a century. The *Sidis* first gained recognition as admirals of the Nizams of Ahmednagar.[23] After the dissolution of the Ahmadnagar Kingdom, the *Sidis* were taken into service by the Adil Shah of Bijapur. The Marathi chronicles describe the *Sidis* as rats in the house[24] and despise them as trouble makers who must necessarily be eliminated. Shivaji, though succeeded in defeating the *Sidis* on land and deprived them of all their landed possession on the west coast, the Maratha Navy failed to emulate this feat on the coastal waters. The *Sidis* survived not just because of the weakness of Shivaji's nascent navy (mainly the lack of effective artillery), but also because of the alliance(s) of convenience they could successfully forge with the other players in the Konkan waters. There are numerous references of the *Sidi* fleet wintering in the Bombay/Mazagaon dockyard of the EIC despite serious objections from Shivaji.[25] Even the Portuguese were found willing to support the *Sidis* in the struggle against Shivaji on the condition that the *Sidis* transfer their allegiance to the *Estado*.[26]

Shivaji could not have done much under such circumstances, but he had not run out of options either. Even though the Marathas depended on European help to augment their artillery department, Shivaji instituted his own ways to express unhappiness. References of Maratha military actions in the form of naval raids involving the capture of European ships/vessels/people as a prize, strangulating supply of provisions from the landed side etc., could be repeatedly found in the contemporary records. On one occasion Shivaji threatened to ally with the Dutch in the latter's endeavour to take Bombay to keep the British away from helping the *Sidis*.[27] The *Sidis*, on the other hand, could not handle the incessant Maratha pressure and decided to join the Mughals in 1670.[28] As a result, the balance of power further tilted in favour of Aurangzeb, who got a ready-made ally against Shivaji. Consequently, Maratha–*Sidi* rivalry got intensified and became increasingly lethal. An English letter of 1672 records:

> King Aurangzeb having sent a fleet of 30 Sails of small frigates from Surat to assist the *Sidi* of Danda-Rajapuri against him [Shivaji] by sea, which fleet hath done him much mischief, burning and destroying also about 50 of his vessels.[29]

Khafi Khan notes:

> The [Mughal] fleet sailed down [from Surat, under the command of Khan Jehan] to Janjira. One night it attacked the vessels of Shivaji and captured them with two hundred men. One hundred Maratha sailors were killed by tying stones to their feet and throwing them into the sea. From that day the animosity between the *Sidi* and Shivaji grew more violent. The Raja built a new fleet and there were frequent naval fights [between them].[30]

Success (in capturing Janjira) still eluded Shivaji. But, the network of marine forts constructed by him along the west coast prevented the *Sidis* from making any further dent in his coastal possessions. In response to the *Sidi* of Janjira's actions, Shivaji could console himself by occupying and fortifying an off-shore rock – Khanderi – strategically located at the entrance to Bombay harbour.[31] Khanderi indeed offered some leverage to the Marathas with respect to the naval/commercial operations of the British EIC originating from its base in Bombay. The British could never reconcile to the presence of Maratha naval strongholds (namely, Kulaba, Khanderi) so near to their headquarters at Bombay. Hence, they clashed with the Maratha Navy head-on in the early eighteenth century and tried to dominate the geopolitics of the west coast.

Sambhaji, the son and successor of Shivaji, needed to logically complete the unfinished assignments of his predecessor. He began by occupying and simultaneously fortifying an island near Elephanta.[32] For Janjira, he had an interestingly innovative idea of filling up the sea channel roughly measuring 800 × 30 yards between the coast and the fort. He planned to raise a bridge to the island by filling the channel with stones, timber and bags of cotton[33] so as to take his cavalry and guns/cannons there and subdue it. Such an ambitious project required, in addition to time, phenomenal efforts and resources. However, with the arrival of Mughal forces in the Deccan,[34] Sambhaji had to hurriedly abandon this plan. Aurangzeb had started a 'terminal war' against the Marathas. Sambhaji did not want the Europeans to take advantage of the situation; hence he continued his offensives against them and also entered into a naval alliance with the Arabs 'directed against the combined enemies of the Marathas and the Arabs'.[35]

Interestingly, Sambhaji could sense his chance against the *Sidi* after Richard Keigwin assumed the command of Bombay in 1683. This was the time when Bombay had openly refused any kind of help to the *Sidi*. Keigwin also courted friendship with Sambhaji keeping in mind the Maratha predominance in the area around Bombay. He sent Captain Henry Gary and Lieutenant Wilkins as British envoys to Sambhaji. Apparently, an agreement was signed between them regarding mutual assurance of help.[36] The short reign of Keigwin (even though illegal from the point of view of the British EIC's administrative set-up in India),[37] could be seen as a special period for both the sides as it was characterized by overall peace. This enabled Sambhaji to undertake military adventures against the Portuguese too, as their possessions on the west coast were attacked. The Persian *akhbarat* dated 20 June 1682 reports that the Marathas had surrounded the fortress of Chaul.[38] The *Jedhe* chronology confirms that Sambhaji had besieged the *Firangan* (indicative of the area under the Portuguese control).[39] The island of Karanja was occupied,[40] threatening the safety and security of Thane – an important Portuguese settlement. In April 1683, the Marathas attacked the area between Daman and Bassein. Villages were burnt, ships were taken away and even the priests were captured, as per the Portuguese records.[41] About 4,000 people fled from Chaul to Bombay.[42] The Portuguese, sensing real danger for Chaul, pressed the panic button and sent for Mughal help. The *Sidis* grabbed this opportunity (to act against the Marathas) by immediately sending 400 men to the aid of the governor of Chaul. According to Portuguese documents, *Sidi* Yakut Khan's ships ferried a specially constituted military contingent by Manuel Lobo De Sylveria, the General of Bassein, to relieve the garrison of Chaul.[43] It is also said that Yakut Khan, displaying his tactical acumen, anchored few of his fighting vessels in Karanja river and prevented Sambhaji's naval fleet, stationed at Nagaon, from entering the open sea. By doing so, he cautiously avoided picking up a direct conflict with the Marathas. Meanwhile, the Portuguese governor of Goa devised another strategy to divert the attention of Sambhaji from Chaul. He deployed a strong military contingent to reduce the Maratha fort of Phonda, just a couple of miles away from Goa.[44] In this endeavour, he was joined by a few 'promising' allies like the Desais and Sawants of south Konkan, who were always waiting for any opportunity to break free from Maratha control.[45] As expected, Sambhaji in his eagerness to retain a military base close to Goa, came charging down with a big military contingent that comprised 8,000 horses and 1,500 infantry.[46] Phonda was relieved and the Maratha troops, in the heat of the moment, invaded Goa.[47] The Portuguese situation was critical as the river separating Goa and Phonda was fordable under normal circumstances, and the adjoining area was held by the Marathas. Success, however, eluded the Marathas as a high tide occurred during their onslaught giving Goa a providential escape. Sambhaji could save himself from drowning (along with his horse) with great difficulty.[48] The Portuguese restrengthened the defences of Goa by deploying naval vessels in the surroundings of Goa, as they could not afford taking any further risk against the Marathas. Sambhaji, on the other hand, retreated to his capital because of the advancing Mughal forces in his pursuit.

After the capture and execution of Sambhaji by Aurangzeb, the *Sidis* got another opportunity to expand on the west coast. Aurangzeb issued them a *sanad*, granting even those territories which they had held prior to the rise of Shivaji. The *Sidis* immediately occupied Suvarnadurg and Anjanwel and wrested the forts of Rajpuri and Raigad from the Marathas.[49] But, before they could consolidate further, Aurangzeb ordered an attack on Bombay. Bombay, it should be remembered, was recently upgraded as EIC's headquarters in India. Aurangzeb was indeed inspired by his 'final' success against the Marathas, but there were other pretexts too for the attack on Bombay.[50] The role played by the EIC in the Mughal–Maratha conflict was never viewed beyond suspicion by both the sides, and Aurangzeb utilized this opportunity to 'teach the British a lesson'. The British, however, got an inkling beforehand and shifted their governor's office to Swally. Bombay also appeared determined for a 'show-down' with the Mughals as per Governor Child's letter to the Company's head office in Britain. He wrote, 'Better to fight it out than purchase a peace that will not only be dishonourable and base and mean, but bring with it great evils and such continual uncertainty to your trade that it may be insupportable.'[51] However, the final outcome of this episode proved that Sir John Child was not as selective and meticulous with his actions as he was with his words, even after committing almost the entire resources at the disposal of Bombay. The *Sidis* outnumbered the Bombay garrison by landing a large number of mercenaries at numerous points on the island of Bombay which lay unguarded and undefended. The garrison, on the other hand, suffered from the action of deserters, who proved to be 'more formidable foes to the defenders than 100 black enemies', since they gave their new associates the advantage of their military skill, 'teaching them the art of mining and sheltering themselves in trenches and basket works', with such effect, says Ovington, that they were 'able in time to bombard the fort with massive stones instead of iron balls, while our shots from thence was scarce able to annoy them'.[52] The British realized that mere possession of a well-equipped navy was not enough to hold ground against a strong, land based force – that, too, with a workable flotilla. Terms of peace, therefore, were discussed and were quickly agreed to by the Mughals, who were aware that until the time they improved their means to effectively tackle the British/European navies it wouldn't be easy to coerce them thoroughly.

Aurangzeb, however, did not restrict his offensives against the British alone. With a view to establish Mughal control far and wide on the west coast, he targeted the Portuguese too. The most 'pressing' pretext for the Mughal action, interestingly, was Portuguese sympathy and their 'secret helps' to the newly acclaimed Maratha King Rajaram.[53] Aurangzeb had asked the viceroy of Goa to be vigilant in view of Rajaram's flight/escape through Portuguese territory. The viceroy wrote back that Desai Khem Sawant had been alerted in this regard.[54] Another Portuguese letter dated 3 September 1689 reads,

> you [the Mughal authority] ask me to post vigil on land as well as on sea to catch *Ramo Raze* [Rajaram] who wants to flee to Karnataka with a treasure which is also to be captured.... On this I had posted the necessary vigil ... so that he might not have a chance to escape.[55]

Despite such assurances, the Mughal wrath could not be avoided. Bruce mentions that the *Sidis* attacked Bassein and threatened Salsette in the year 1692 and, for the next couple of years, continued to ravage the Portuguese territories.[56] Rajaram, at the same time, was courting the Portuguese for good relations.[57] The Portuguese were quick to learn their lesson as they started sounding diplomatically correct in the Indian affairs. One of their letters read,

> of late he [*Sambhaji*] had been repeating this request [to renew the alliance of peace and friendship which this state formerly maintained with Shivaji]. We have responded to it with the same demonstration of friendship (as we did in the case of the Mughals) in such a manner and with so great a care that both the vassals of the Mughal and those of Rajaram believe that this government is inclined to the interest of either.[58]

The character of the politico-military equations on the west coast started changing quickly with the rise of Kanhoji Angre – the daring Maratha seafarer – early in the eighteenth century. His father Tukoji Sankpal[59] was *Sarnaubat* or the Deputy Commander of a Maratha marine fort Suvarnadurg.[60] Kanhoji's upbringings were thus conducive to 'bound future generations of Angres to ship and to the sea'.[61] He was appointed to the post of the deputy to the Maratha Admiral (*Sarkhel*) Sidhoji Gujar and granted the *Subah* of Konkan by Rajaram in 1696.[62] In due course of time Kanhoji Angre, by 'royal command', got an independent charge of 150 miles of wide open coastline.[63] Armed with royal sanction he began projecting his 'legal' authority over the adjoining sea by forcing everyone to purchase his passes if they wanted to avoid their vessels being taken as legitimate prizes by his men.[64] He knew that it would not be easy for him to do so, but he took his chances and succeeded too. S.N. Sen writes,

> his prowess was soon rewarded with rich spoils and his neighbours copiously, though involuntarily, contributed to his coffers ... but the very process involved hostilities with his neighbours among whom the English and the Portuguese were the most prosperous and powerful. The *Siddi* of Janjira was the avowed enemy of his King and therefore his legitimate prey.[65]

He had shifted his base from Suvarnadurg to Kolaba (Kulaba) for obvious strategic consideration by the time he was elevated to the position of the Maratha *Sarkhel*.[66] Sen notes,

> With the headquarters at Kulaba, Kanhoji was in the happy position to command and control the high road of sea borne trade that literally passed

> by his door. With Bombay, Bassein and Surat on the north, and Chaul, Janjira and Goa on the south, the Kulaba fleet could keep a vigilant eye on all vessels sailing from the north to the south and vice-versa while the intermediate ports offered safe asylums in case of emergency and danger.[67]

On the coast, Kanhoji tried to recover the area occupied by the *Sidis* in the interregnum following the fall of Sambhaji and flight of Rajaram to Jinji. The *Sidis*, thus, felt the heat both on land and sea (Kolaba was too close to Janjira). They therefore, enforced a naval blockade around Kolaba[68] so as to starve the garrison and coerce it to surrender. Being confident, the *Sidis* put this plan into action once the rains were over.[69] However, Kanhoji Angre had more surprises in store for the *Sidi* to frustrate his plans.

Kanhoji Angre joined the mainstream of the Maratha politics as he was persuaded by Balaji Visvanath, the *Peshwa* or Prime Minister of the 'legitimate' successor to the Maratha throne Shahuji in 1713.[70] It is interesting to note that out of the territory awarded to Kanhoji, a large portion was under the control of the *Sidis*.[71] No sooner than this treaty was signed, a joint military expedition was undertaken by the *Peshwa* (Prime Minister of the Maratha Empire) and Kanhoji Angre against the *Sidis*. With no immediate help coming from any quarters, the *Sidis* applied for peace. In 1715, the *Sidis* signed a treaty with the Marathas relinquishing control over a few forts including some on the west coast.[72] The Europeans now remained to be dealt with. They did not disappoint either! In the decade following Maratha–*Sidi* peace, they made four determined attempts on various strongholds of Kanhoji.[73]

Early in the eighteenth century, the Portuguese had approached the British at Bombay for the purpose of a joint attack on Angre. Bombay Governor Aislabie, however, not only rejected the Portuguese proposal but, in fact, signed peace with Kanhoji when approached by the latter.[74] But, Bombay's understanding with Kanhoji lapsed after Charles Boone took charge as governor in 1715. With the agenda to improve the EIC's administration and trading fortunes, Boone worked upon the improvement of the defences of Bombay and simultaneously attempted to dislodge Kanhoji Angre from his advantageous position in the Konkan waters. The first of such attempts was made on Vijaydurg in 1718. The pretext(s) for the English action were too many; namely, (a) Kanohji's making prize capture of English trading ships/official vessels,[75] (b) Kanhoji's refusal to pay harbour dues at Bombay,[76] (c) Kanhoji's vessels forcibly removing timber pieces from British ships,[77] etc. Boone had approached the Portuguese to join him in this expedition but was politely refused by them.[78] For the demolition of Vijaydurg, Boone relied on his 'war ships', some of which – like the *Britannia* (18 guns), the *Fame* (16 guns), *Revenge* (16 guns) and the *Victory* (24 guns) – were 'purpose built' at Karwar, Surat and Bombay.[79] Boone was deliberate in choosing the month of May/June for his expedition as the monsoon set in by that time and sea-borne activities were largely avoided by the coastal powers.[80] He planned to set the Maratha ships (hauled ashore during monsoon) on fire; and land British troops and storm the fort before any help reached Kanhoji from

Kolaba or Sindhudurg. Boone, however, was totally outwitted by Kanhoji's preparations[81] which prevented the British 'bomb' ships from reaching the Maratha ships and consigning them to flames (because of the erection of a boom in the channel leading to the site where the ships were hauled up). As a result, neither could Boone land his troops ashore nor could he inflict any damage to the walls of the fort through bombardment. Totally outwitted, he called off the mission, but secretly worked on another plan to corner the Maratha *Sarkhel*. After the rains subsided, in November 1718, another naval expedition (which could be justifiably called an extension to the Vijaydurg expedition) targeting Khanderi was launched. This time the King of Kolhapur joined hands with the English.[82] The Bombay fleet initiated the action virtually enforcing a blockade of Khanderi as they anchored off the fort at a safe distance. The British plan was to batter the walls of the fort by a barrage of bombardment, strangle the supply route to Khanderi, starve the garrison and force them to surrender. But, the bombardment from the British vessels failed to make any impact on the walls of the fort as they could not come close enough to the fort for the fear of submerged rocks. The Khanderi garrison, on the other hand, briskly returned fire, making it difficult for the EIC's soldiers to land and implement an assault to overpower the garrison. Boone's efforts to motivate his soldiers for a second landing also failed despite announcement of incentives for the 'volunteers'.[83] In sheer frustration, therefore, Boone ordered the Bombay fleet to return back. This was the second 'significant failure' of the determined British against an 'insignificant enemy'; hence responsibility had to be fixed. The British records oblige by attributing treachery and cowardice of the officers and soldiers as the main causes for the failure of the Khanderi expedition. The names of Manuel de Castro, a naval officer, and Rama Kamathi, a rich merchant of Bombay, cropped up as the main conspirators against the British.[84]

Realizing the gravity of the situation (as appraised by his admiral), the Maratha king opened channels of negotiations with the British early in 1719. Boone, however, did not pay any heed to the Maratha requests and kept working on his plans to wipe out the Maratha power from the west coast. Another attack on Vijaydurg was launched in 1721. This time the Sawant of Wadi, the old foe of the Maratha king, approached Brown, the commander of the British expedition, with a proposal of 'joint and simultaneous attack' on the landed possessions and another marine fort *Jaigarh* of the Marathas. Boone was confident of achieving success as he had meticulously worked on the project. However, one of his 'purpose-built' bomb vessels named *Phram* (also referred to as 'floating castle'), failed to deliver because of a 'design fault' (its holes were cut too low) – resulting in the shots falling in the water nearby. It was ordered to be rectified in the midst of the action which retarded the momentum of the British attack considerably, thus, easing pressure on the besieged Maratha garrison. The logbook of the *London*, one of the English ships participating in the action, notes, 'The boats [of enemy] full of armed men were constantly crossing the river', but nothing was done to intercept them. In other words, the defences of the Maratha fort were restrengthened and the retaliatory fire from the fort prevented the Company soldiers from landing ashore. Even though the defect of *Phram* was rectified, due to a 'chance accident' (in the course of return fire from the fort) leading to its ammunition box catching fire that wounded and scorched some British soldiers, it had to be withdrawn from action. To add to the miseries of the British, the Sawant, despite his assurances, did not turn up in the thick of action. Boone called off the expedition citing 'swiftness and manoeuvrability of Angre's ships' as the reason for the British debacle. Boone immediately placed orders for the construction of three fast, light ships, each of about 160 tons, for service under the Bombay council.[85] Happily for Boone, the Company administrators (at London) appeared to be convinced with his logic and had assured that a squadron of the British Navy would be sent out to assist the EIC's marine establishment at Bombay. Boone was becoming more impatient with each failure against the Marathas. In December 1721, with the Portuguese on his side, he launched another attack on Kanhoji Angre's possessions. The target, this time was Kolaba, the headquarters of the Maratha Navy. In addition to their naval squadrons, the European allies committed their armies as well, to achieve their objective. To divert the attention of Kanhoji Angre on two fronts, a naval blockade was enforced on the Kolaba Fort and, on the coast, a large European force waited at some distance from the fort. The idea was to strangle the supply to Kolaba Fort and prevent any possible help from reaching there from either sea or land.[86] Though Kanhoji was besieged at Kolaba with 1,000 infantry and 700 cavalry, a military contingent under the command of Pilaji Jadhav was immediately dispatched by King Shahu for his rescue. Upon its arrival, this contingent, comprising 1,500 cavalry and 1,000 infantry, was deployed near the northern side of the Kolaba Fort. On the other hand, even after cornering Kanhoji Angre initially, the Europeans did not really commence any noteworthy action from the sea side. They, instead, began their offensive from the land side and tried to reach and force their entry into the fort (during low tide). The fort's garrison fought bravely and frustrated the British attempt to gain entry. In the meanwhile, Pilaji Jadhav's contingent ferociously charged and routed the Portuguese contingent which had, so far, kept aloof from the action. Elated by the proceedings, Kanhoji sent a detachment from the fort's garrison to deliver another blow, this time on the British contingent involved in the attack on the fort. Without much difficulty, the Maratha detachment was able to convert its assigned task into brilliant tactical success. As a result, the Europeans left the scene of action in the darkness of night. Kanhoji cleverly approached the Portuguese with generous terms of peace. The Portuguese viceroy, in turn, responded positively to the Maratha overture, and signed an 'offensive–defensive' treaty with them. Both the sides agreed to help one another against enemies who were not at peace with either of the signatories.[87] The Portuguese also agreed to supply and sell ammunition and artillery to the *Peshwa* at fair price.

The Maratha accord with the Portuguese, and the departure of Governor Boone from Bombay, necessarily slowed down the spate of European hostilities on Angre, but such a situation did not last long. The new governor of Bombay, William Phipps, found a new ally in the French Company. The two companies quickly resolved their mutual differences and not only agreed to honour the passes issued by each other to the vessels plying on the Malabar Coast, but also promised 'not to assist the enemies of each other with ammunitions, guns or provisions directly or indirectly, nor give convoy or protection to each others' enemies'.[88] At the same time, it seems, the Sawant of Wadi was trying to induce the Dutch to launch an attack on Vijaydurg. Under such circumstances, Kanhoji Angre thought it prudent to improve relations with Governor William Phipps. He sent a letter to Bombay offering peace and it was well reciprocated by the Bombay authorities.[89] In this way, though the situation on the west coast became peaceful for the time being, the causes for future conflict could not be avoided altogether. The circumstances after the death of Kanhoji Angre in 1729 compelled his successors to compete for the west coast and, in the following decade, with varying success, 'the Angrian fleet remained a terror which the sea farers of the eighteenth century were glad to avoid'.[90]

Notes

1 K.S. Mathew, 'Taxation in the Coastal Towns of Western India and the Portuguese in the Sixteenth Century', in K.S. Mathew (ed.), *Mariners, Merchants and Oceans: Studies in Maritime History* (New Delhi: Manohar, 1995), p. 143.

2 See for example, O.K. Nambiar, *The Kunjalis: Admirals of Calicut*, 2nd rev'd edn (Bombay: Asia Publishing House, 1963), pp. 60–64, describes the alliance between Calicut and Egyptian fleet commanded by one Mir Hussain. Malik Ayaz, the Governor of the Sultan of Cambay also joined the fray. It also mentions that several Arab ships were destroyed by Don Lorenzo, the commander of the Portuguese fleet.

3 In 1534, when Bahadur Shah of Gujarat was preoccupied fighting with Chitor, he surrendered Bassein (in lieu of Diu) and the seven islands of Bombay to the Portuguese and also gave them the right to collect taxes on land and sea at and around Bassein. Accordingly, all the vessels bound for the ports of Red Sea or the Persian Gulf regions were obliged to collect passes from the Portuguese captains at Bassein; and on their return were required to visit Bassein and pay customs. In 1535, Bahadur Shah permitted the Portuguese to occupy Diu on a purely temporary basis and gave permission to the Portuguese to construct a fortress at any place of their choice in Diu. They were also entitled to collect customs on horses brought from Ormuz and Arabia. The right of issuing *cartaz* and collection of duties was thus shifted from Bassein to Diu. Thereafter, in 1539 a new treaty was signed between Sultan Muhammad Shah and the Portuguese regarding revenue sharing at Diu. However, the Sultan of Gujarat lost Diu forever to the Portuguese in 1546, when the troops of Khwaja Safar, who had been trying to retake Diu for the Sultan, were routed. The Portuguese thus collected all the customs revenue as also revenue from local taxes at Diu. After Gujarat was annexed by Akbar in 1572, friendly relations were established with the Portuguese at Diu. It was agreed that Akbar would be given every year a free *cartaz* to send his ship from Surat to Mocha. This implied that the ship on its return trip would not be taxed. Elsewhere, the King of Cochin put up a vigorous fight against the Zamorin for giving asylum to the Portuguese, losing a few of the members of his royal family in the process. He accepted the position of a vassal and took oath of fealty and vassalage to the King of Portugal at the time of his coronation in 1505. See, Mathew, 'Taxation in the Coastal Towns of Western India and the Portuguese in the Sixteenth Century', pp. 143–147.

4 In 1610, Middleton brought ships of the English Trading Company on the west coast

of India, but the Portuguese hastily dispatched some big galleons from Diu to warn him off. The local native governor also refused to have any dealings with the British for fear of the Portuguese, and the British departed. But in 1612, the London ships appeared again under Captain Thomas Best determined to 'assert their right to trade'. The Portuguese challenged them, and a naval war was fought between them at Swally, near Surat, resulting in the defeat of the Portuguese. G.A. Ballard, *Rulers of the Indian Ocean* (1927, reprint, Delhi: Neeraj Publishing House, 1984), pp. 160–161.

5 See, Ruby Maloni, 'European Enclaves in the Konkan (16th–18th Centuries)', in D.S. Gaikwad (ed.), *Some Aspects of Medieval India* (Bombay: Samant Publications, 2006), p. 153.

6 Shivaji was expanding his dominions largely at the expense of the Bijapur state, which, it was obvious, was not in a position to subdue him. As a result, negotiations for permanent peace were opened (by Bijapur) in 1662 and Shahji Bhonsle, Shivaji's father (who was an employee of the Bijapur state), was chosen as the envoy. All the demands of Shivaji were met. His independence was conceded and he was recognized as the ruler of the coast from Bombay to Goa. David Kincaied, *Shivaji the Founder of Maratha Empire (The Grand Rebel)* (reprint, Delhi: Discovery Publishing House, 1984), pp. 128–130.

7 U.M. Pathan (ed.), *Krishnaji Anant Sabhasad Virachit Chhatrapati Sri Shivaji Raje Yanchi Bhakhar* (Poona: Snehwardhan Publishing House, 2000) (hereinafter *Sabhasad Bakhar*), p. 94. Sabhasad mentions that the *Raje*'s (Shivaji's) ships fought against numerous sea powers and used to bring him grain and other provisions.

8 Jadunath Sarkar, *Shivaji and His Times* (1919, reprint, Delhi: Orient Longman, 1973), p. 245.

9 Portuguese letter dated 28 November 1659, quoted by P. Pissurlencar, *Portuguese Marathe Sambandha (arthat Portugezanchya Daftaratil Marathayancha Itihash)* (Poona: Pune University Press, 1967), p. 41. A contemporary British letter also suggests that Kalyan-Bhiwandi was under the control of Shivaji. B.G. Paranjape, *English Records on Shivaji, (1659–1682)*, 2 vols (Poona: Shiva Charitra Karyalaya, 1931) (hereafter, *ERS*), vol. 1, letter no. 160.

10 G.V. Scammell, 'European Exiles, Renegades and Outlaws and the Maritime Economy of Asia *c.*1500–1750', in K.S. Mathew (ed.), *Mariners, Merchants and Oceans: Studies in Maritime History* (New Delhi: Manohar, 1995), pp. 121–142. See also G.S. Sardesai, *New History of Marathas* (1946, reprint, Delhi: Munshiram Manoharlal, 1986), vol. 1, p. 219. Sardesai writes that Shivaji borrowed the services of a few naval and artillery experts from Goa and established his own shipbuilding yards and arsenals at Malwan.

11 It is stated that two Portuguese artisans namely Roe Leitao Viegas and his brother Fernao Leitao Viegas supervised some 340 workmen, Portuguese and others, and helped Shivaji to construct the fighting vessels. Cited by P. Pissurlencar, *Portuguese Marathe Sambandha*, pp. 41–42.

12 Grant Duff writes, 'He (Shivaji) built Sindeedroog or Malwan.... He rebuilt and strengthened Kolabah; repaired Severndroog and Viziadroog and prepared vessel at all these places.' See, Grant Duff, *History of the Marathas*, 3 vols (1863, reprint, Jaipur: Avishkar Publishers, 1986), vol. 1, p. 137.

13 Srikrishna Venkatesh Puntambekar, *A ROYAL EDICT on The Principles of State Policy and Organization*, otherwise called Ramchandrapant Amatya's *Rajniti* (reprint from *Journal of Indian History*, vol. 8, Parts I and II, Madras: Diocesan Press, 1929) (hereafter, *Rajniti*), p. 31.

14 S.N. Sen, *Foreign Biographies of Shivaji*, rev'd 2nd edn (Calcutta: K.P. Bagchi & Company, 1977), p. 155.

15 *Rajniti*, p. 32.

16 V.B. Kulkarni, *Shivaji: The Portrait of A Patriot* (New Delhi: Orient Longmans Limited, 1963), p. 163. Italics mine.

17 P. Pissurlencar, *Portuguese Marathe Sambandha*, p. 42. It is also revealed in Portuguese correspondence that they secretly wanted to set some Maratha vessels on fire.
18 *Indian Historical Records Commission Proceedings*, vol. 9, pp. 112–113, cited by Bal Krishna, *Shivaji the Great*, 4 vols (Bombay: D.B. Taraporewala Sons & Co., 1938), vol. 2, pp. 499–501.
19 Jai Singh wrote to Emperor Aurangzeb, 'that Shiva(ji) is a *Zamindar* and the pillars of his *Zamindari* would not endure beyond a period of seven or eight years'. See, Jagadish Narayan Sarkar (ed. and tr.), *The Military Dispatches of a Seventeenth Century Indian General* (Calcutta: Scientific Book Agency, 1969), p. 15.
20 Jagadish Narayan Sarkar (ed. and tr.), *The Military Dispatches of a Seventeenth Century Indian General*, p. 29.
21 Jagadish Narayan Sarkar (ed. and tr.), *The Military Dispatches of a Seventeenth Century Indian General*, p. 94.
22 On the eve of the invasion of Bijapur, Jai Singh got to know that the salaries of Bijapuri *mansabdars* (including the *Sidi* of Janjira) and artillerymen were in arrears for six months. So he tried to use money to win the *Sidi* for his cause. Jagadish Narayan Sarkar (ed. and tr.), *The Military Dispatches of a Seventeenth Century Indian General*, p. 30.
23 For a general discussion on the rise of *Sidis*, see, Santi Sadiq Ali, *The African Dispersal in the Deccan* (New Delhi: Orient Longman, 1996), pp. 157–192.
24 *Sabhasad Bakhar*, p. 91.
25 See, for example, *ERS*, vol. 1, nos 167, 168, etc.
26 Many of the Portuguese documents of that period point towards the fact that the Portuguese were favourably disposed towards the *Sidis* and helped them in various ways against Shivaji. Many a times they secretly supplied grains to the *Sidis* in case Janjira was blockaded by the Maratha Navy. According to a Portuguese letter the *Sidis* were induced to accept Portuguese overlordship in the year in 1667. Quoted by Pisurlencar, *Portuguese Marathe Sambandha*, p. 47.
27 Owen C. Kail, *The Dutch in India* (Delhi: Macmillan, 1981), p. 77; see also, *ERS*, vol. 1, p. 345.
28 The *Sidis* seem to have approached the British as well as Aurangzeb for help and that Aurangzeb had asked Shivaji to withdraw his forces from the surroundings of Janjira. On the ground, however, the *Sidis* did not receive any practical assistance. Hence, the *Sidis* decided to make terms with Shivaji. But, the Sidi of Janjira's subordinates prevented him from doing so by staging a coup and finally took up service under the Mughals. See William Foster (ed.), *English Factories, 1668–1669* (London: Oxford University Press, 1921), pp. 242–243; *ERS*, vol. 1, no. 171; Jadunath Sarkar, *Shivaji*, pp. 261–262.
29 W.S. Desai, *Bombay and the Marathas upto 1774* (New Delhi: Munshiram Manoharlal, 1970), p. 13.
30 *Muntakhab-ul-Lubab* of Khafi Khan (tr. H.M. Elliot and J. Dowson), *History of India as told by its own Historians*, 8 vols (1867–1877, reprint, New Delhi: Low Price Publications, 2001), vol. 7, p. 258. Khafi Khan writes, 'Shivaji had secured several islands in the sea by means of a fleet which he had formed.... He had also built many forts in those parts.'
31 John Fryer, *A New Account of East India And Persia* (ed. with notes and introduction, William Crooke), 3 vols (Nendeln/Liechtenstein: Kraus Reprint Limited, 1967), vol. 1, pp. 159–160. Fryer calls it 'the great rock or barren islet of Henry-Kenry', and describes Bombay, Trombay, Elephanta, Karanja and Rock of Henry-Kenry as 'so many mountains arising out of sea'.
32 Quoted by Kamal Gokhale, *Chhatrapati Sambhaji* (Poona: Navkamal Publications, 1978), p. 82. This fort was named Kamsa or Padamdurg; however it could not be completed the way Sambhaji would have liked.
33 Robert Orme, *Historical Fragments of the Mughal Empire, of the Morattoes, and of*

the English Concerns in Indostan (ed. and introduction, J.P. Guha) (1782, reprint, New Delhi: Associated Publishing House, 1974) p. 75; See also Jadunath Sarkar, *History of Aurangzeb*, 5 vols (1919, reprint, New Delhi: Orient Longman, 1973), vol. 4, p. 312.

34 Jadunath Sarkar, *House of Shivaji*, 3rd edn (Calcutta: Orient Longmans, 1955), p. 223. Aurangzeb was also furious in view of the rebellion of his son Akbar who, as a fugitive, collaborated with Sambhaji.

35 The Portuguese were apparently uncomfortable with the Arab fleet. Contemporary Portuguese sources occasionally refer to this 'unholy alliance' between the two which was fearsome with respect to the Portuguese interests. See V.S. Bendre, *Sri Chhatrapati Sambhaji Maharaj (Yanche Vichikitsaka Charitra)*, 2nd edn (Bombay: P.P.H. Bookstall, 1971), pp. 228–229, 235. We have the reference of one English ship, the *President*, being engaged and attacked by the Maratha vessels early in January 1683. The British records allege that the Maratha vessels were aided by the Arab fleet in the attack on the *President*. The British casualties were stated as 11 dead and 35 wounded whereas some Maratha galvats were also reportedly sunk due to the British counter-offensive. See also Robert Orme, *Historical Fragments of the Mughal Empire*, pp. 82–83; Kamal Gokhale, *Sambhaji*, pp. 87–88.

36 It seems that Sambhaji not only confirmed his friendship, but also promised to pay the British 12,000 *pagodas* on account of the losses suffered by them due to Maratha military action at Rajapur and Hubli in the past. He also granted permission for the establishment of British factories at Cuddalore and Thevanapatnam, and allowed them exemption from customs duty in the Carnatic. See W.S. Desai, *Bombay*, p. 33.

37 A.M. Philip Anderson, *The English in Western India being History of the Factory at Surat, of Bombay and the Subordinate Factories on the Western Coast*, 2nd edn (London: Smith Elder & Co., 1856), p. 184. Anderson writes, 'It is an episode of which we are not ashamed. Keigwin's moderation and shining qualities, his sound internal and external policy, and his honesty of purpose, more than atoned for his unconstitutional conduct.'

38 Quoted by Kamal Gokhale *Sambhaji*, p. 112; V.S. Bendre, *Sri Chhatrapati Sambhaji Maharaj*, pp. 232–233. The main cause of the Maratha grievance was refusal of the Portuguese viceroy to allot a strip of land to them at Chaul. Sambhaji's contention was that since similar permission was granted to the *Sidis* in the past, it was unfair to deny the 'friendly' Marathas a similar privilege.

39 A.R. Kulkarni (ed.), *Jedhe Shakawali-Karina* (Poona: Maan Samman Prakashan, 1999), p. 315. The *Shakawali* records that it was captured in the month of *Margashirsha* (i.e. November to December), 1683. For a detailed description of the siege of Chaul, see also, Jadunath Sarkar, *History of Aurangzeb*, vol. 4, pp. 317–335.

40 Robert Orme, *Historical Fragments of the Mughal Empire*, p. 87.

41 P. Pissurlencar, *Portuguese Marathe Sambandha*, pp. 80–81; V.S. Bendre, *Sri Chhatrapati Sambhaji Maharaj*, pp. 232, 235.

42 P.R. Kakodkar, *The Portuguese and the Marathas* (Bombay: State Board of Literature and Culture, 1975), pp. 92–95. It should be kept in mind that, for the Portuguese, the only possible option to make good their escape from Chaul to Bombay was via the sea.

43 See P. Pissurlencar, *Portuguese Marathe Sambandha*, p. 82 (fn. 220). Yakut was the same person who had transported provisions for the Mughal troops in the Deccan when they were sent against Sambhaji. The Portuguese record also alleges that the *Sidis* 'lent' 400 men for the same purpose.

44 Bombardment on Phonda commenced in November 1683. See V.S. Bendre, *Sri Chhatrapati Sambhaji Maharaj*, pp. 237–239.

45 See, A.R. Kulkarni, 'Portuguese in the Deccan Politics: A Study of New Marathi Documents from Lisbon', in T.R. D'Souza (ed.), *Indo Portuguese: Old Issues, New Questions* (Delhi: Concept Publishing Company, 1985), pp. 114–122.

46 D.V. Apte and S.M. Divekar (eds), *Shiva Charitra Pradeep* (Poona: Bharat Itihash Samshodhak Mandal, 1925), p. 32.

47 Manucci also provides us with valuable details of Sambhaji's Goa campaign. See N. Manucci, *Storia da Mogor* (tr. William Irvine) (New Delhi: Atlantic Publishers, 1989), pp. 272–275.

48 It is said that because of the rising tide, the horse Sambhaji was riding was being forced down by the current in the stream. Seeing this, Khandoji Ballal jumped into the water without any thought of his safety and rescued his master. He was generously rewarded by Sambhaji for this brave deed. See, P.V. Mawjee and D.B. Parasnis (eds), *Sanads and Letters* (Published with the Permission of the Government of Bombay by P.V. Mawjee, Poona: 1913), Letter no. 61.

49 P.V. Mawjee and D.B. Parasnis (eds), *Kaifiyat, Yadi* (Published with the Permission of the Government of Bombay by P.V. Mawjee, Poona: 1908), p. 3; Grant Duff, *Marathas*, vol. 1, p. 231.

50 On the east coast, the EIC had picked a fight with the Mughals, apparently because of illegal exactions (from the Company establishments) by the Mughal officials. The British subsequently sacked Hugli in 1686 and committed atrocities at Balasore in 1688, before finally making peace with the Mughals. Probably, to take revenge, Aurangzeb ordered the *Sidis* to attack the British establishment at Bombay. See, Jadunath Sarkar, *History of Aurangzeb*, vol. 5, pp. 246–255.

51 Arnold Wright, *Annesley of Surat and His Times: The True Story of the Mythical Wesley Fortune* (London: Andrew Melrose Ltd, 1918), p. 128.

52 J.A. Ovington, *Voyage to Surat in the year 1689* (ed. H.G. Rawlinson) (London: Oxford University Press, 1929), p. 94.

53 Y.S. Paranjpe, 'Military Traditions of Maharastra', in A.K. Bhagwat (ed.), *Maharastra: A Profile* (Kolhapur: Amrit Mahotsava Satkar Samiti, 1977), pp. 412–413.

54 P. Pissurlencar, 'Rajaram and the Portuguese', *Indian Historical Records Commission*, vol. 17 (1940), p. 223.

55 P.R. Kakodkar, *Portuguese and the Marathas*, pp. 155–157.

56 J. Bruce, *Annals of the Honorable East India Company* (London: Black, Parry & Kingsbury, 1810), vol. 3, p. 124.

57 Rajaram appealed to the Portuguese to continue with good relations with the Maratha Kingdom as it had existed during the time of his predecessors. It seems that the Portuguese were forcing the Maratha navigators in the Konkan waters to purchase their *cartaz* after the fall of Raigad in 1689. See S.N. Sen, 'Half a Century of Maratha Navy', *Journal of Indian History*, vol. 10, Part I–II (1933), p. 256.

58 P. Pissurlencar, 'Rajaram', p. 224.

59 P.V. Mawjee and D.B. Parasnis (eds), *Kaifiyat, Yadi*, p. 2. The surname Angre is derived from the name of the family's hereditary village Angarwadi near Pune.

60 P.V. Mawjee and D.B. Parasnis (eds), *Kaifiyat, Yadi*, p. 2. Tukoji was subsequently elevated and made a commander of 200.

61 Manohar Malgaonkar, *Kanhoji Angre: The Maratha Admiral* (Bombay: Asia Publishing House, 1959), p. 12.

62 Manohar Malgaonkar, *Kanhoji Angre*, p. 60; D.G. Dhabu, *Kulabkar Angre*, 1st edn (Alibag: Privately published by D.G. Dhabu, 1939), p. 11. In fact there were two deputies to the Maratha *sarkhel* with a defined sphere of influence. See, P.V. Mawjee and D.B. Parasnis (eds), *Kaifiyat, Yadi*, p. 3.

63 Subsequently, Kolaba province, which was held by one Bhawanji Gujjar Maratha, was also transferred to Kanhoji Angre to discharge his duties effectively. See P.V. Mawjee and D.B. Parasnis (eds), *Kaifiyat, Yadi*, p. 4.

64 B.K. Apte, *A History of the Maratha Navy and Merchant Ships* (Bombay: State Board for Literature and Culture, 1973), pp. 201–203. Apte gives details about two types of passes – *dastak* and *kaul*. While a *dastak* was a pass permitting the ship (possessing it) to sail within the jurisdiction of the issuing power, a *kaul*, on the other hand,

allowed the holder to move on all the seas, on the authority of the issuing government. The *kaul* enabled to identify the nationality of the ship. It seems that a ship possessing a *kaul* was not necessarily exempt from purchasing a *dastak* of a foreign government.

65 S.N. Sen, 'Half a Century', p. 258.

66 Rajaram had made a comeback to the Deccan in 1698, and offices/responsibilities were redistributed; D.G. Dhabu, *Kulabar Angre*, pp. 11–12.

67 S.N. Sen, 'Half a Century', p. 258.

68 The English factors record, 'The *Sidi* lies before Caulaby [Kolaba] and Cundry [Khanderi], though hath grand force yet Marathas find a way to galbats [galvats] from one place to another.' Quoted by V.G. Khobrekar, *Konkan: From the Earliest to 1818 A.D. (A Study in Political and Socio-Economic Aspects)*, (Poona: Snehvardhan Publishing House, 2002), p. 127; see also Manohar Malgaonkar, *Kanhoji Angre*, pp. 84–85.

69 It seems probable that the *Sidi* fleet must have charged from Mazagaon where it wintered during the monsoon. See D.R. Banaji, *Bombay and the Sidis* (New Delhi: Macmillan, 1932), pp. 56–57.

70 See, D.G. Dhabu, *Kulabkar Angre*, p. 10.

71 It yielded an annual income worth Rs34 lakh. See D.R. Ketkar, *Sarkhel Kanhoji Angre* (Bombay: Mrunmayi Rugvedi Prakshan, 1997), p. 31.

72 V.G. Khobrekar, *Konkan: From the Earliest to 1818 A.D.*, p. 139.

73 The British attacked Vijaydurg (Gheria) twice; Khanderi was also attacked by them; a joint Anglo-Portuguese expedition was undertaken to reduce the Angre stronghold of Kolaba.

74 This proposal was made in the light of Kanhoji's bold activities on the sea. In 1712, he raided the armed yacht of the governor of Bombay, together with the *Anne* Ketch from Karwar. A month later two British ships (on voyage from London to India) were attacked near the coast of Goa by a ghurab and a galvat of Kanhoji's fleet. The Portuguese were also not spared; it is reported that the fleet of Kanhoji Angre had captured the greater part of a Portuguese '*armado*', and disabled a 30-gun man-of-war that was conveying it. John Biddulph, *The Pirates of Malabar and an Englishwoman in India Two Hundred Years Ago* (reprint, New Delhi: Lancer International, 1992), p. 81.

75 Manohar Malgaonkar, *Kanhoji Angre*, p. 199; W.S. Desai, *Bombay*, p. 65; John Biddulph, *The Pirates of Malabar*, p. 99.

76 See John Biddulph, *The Pirates of Malabar*, p. 99.

77 Manohar Malgaonkar, *Kanhoji Angre*, p. 191.

78 The Portuguese governor seemed to be pre-occupied with the threat of the pirates from Muscat. See, W.S. Desai, *Bombay*, p. 66.

79 John Biddulph, *The Pirates of Malabar*, pp. 87–88. Biddulph's sarcasm, at the same time, could not be concealed when he commented on the preparations made by Boone to tackle Kanhoji Angre and others 'pirates'. He wrote,

> It was easier to build ships than to get sailors to man them, in view of the miserable pay given by the Company, and the attractions of service under native chiefs. Many of the crews were foreigners, who were ready enough to take service under the Angre, if the inclinations took them, and the bulk of the crews were Indian lascars.

80 As per traditions followed in Maharastra, when the sea becomes quite rough at the onset of the monsoon, the sea-going vessels are pulled ashore for repairs etc. Moreover, because of the heavy and incessant rains, it is not possible to undertake any sea voyage/operation of fruitful nature. Once the rains subside (by the month of September/October), the Marathas venture again into the sea preferably on the auspicious day of *Narli Purnima* (still celebrated in Maharashtra). Even J.A. Ovington, *Voyage to Surat*, pp. 81–82, takes note of this festival.

81 Details of the English action could be found in Clement Downing, *History of Indian*

Wars (Introduction and Notes, William Foster) (London: Oxford University Press, 1924), pp. 64–67.

82 P. Pissurlencar, *Portuguese Marathe Sambandha*, p. 141.

83 Boone reportedly asked his sailors to 'volunteer' for landing. Rich rewards, namely, 40 rupees in cash (nearly equal to five pounds at that time) compensation for the loss of a leg or arm at 30 pounds; employment for life under the company; the married men were promised, if killed, that their widows should receive 30 pounds, ten pounds extra for each child. See, John Biddulph, *The Pirates of Malabar*, pp. 124–125.

84 G.W. Forrest, *Selections from the Letters, Dispatches and other State Papers preserved in the Bombay Secretariat, (Maratha Series)* (Bombay: Bombay Central Press, 1885), vol. 2, pp. 12, 15. For a discussion on the Rama Kamathi episode, refer, Ravindranath Vaman Ramdas, 'Rama-Kamati: The Man and His Achievements', in V.S. Bendre (ed.), *Nibandha Samgraha* (Proceedings of the Fourth Session of the Maharastra History Congress, organized by the Shivaji University Kolhapur, 25–26 January 1969, Bombay), pp. 249–255.

85 R.A. Wadia, *The Bombay Dockyard and the Wadia Master Builders Wadia* (Bombay: Privately published by R.A. Wadia, 1957), pp. 102–131. As early as 1673, two new vessels built at Bombay were taken by Governor Aungier to boost the strength of the Bombay Marine. References of repair works being carried out at Bombay are also numerous in English records.

86 The timing of the assault (full credit must be given to the Europeans) could not have been more suitable as the troops of the Maratha King, led by the *Peshwa*, were busy fighting the Nizam of Hyderabad. That fight was over the interpretation of the Mughal *firman* to the Maratha Monarch Shahuji, which virtually made him the de-facto sovereign of the area earlier held by Shivaji. It also gave Marathas the right to collect *Chauth* and *Sardeshmukhi* of the Deccan *Subah.*

87 It has been argued that the viceroy's considerations for the 'prospects of peace and amity with the Marathas' were more a result of news of the arrival of a strong Maratha force under the command of the *Peshwa* for the rescue rather than sheer generosity or for the sake of the 'treaties of the past'. See P. Pissurlencar, *Portuguese Marathe Sambandha*, p. 142.

88 G.W. Forrest, *Maratha Series*, vol. 2, p. 36; see, articles 3 and 4 of this treaty.

89 G.W. Forrest, *Maratha Series*, vol. 2, pp. 37–41.

90 See S.N. Sen, 'Half Century of Maratha Navy', *Journal of Indian History*, vol. 11, Part III (1934), p. 27.

Part III

Military culture, state and society

11 Command, control and castration

Eunuch supervisors in the armies of the Tang dynasty

David A. Graff

One of the perennial characteristics of military organization in imperial China was the use of a variety of institutional mechanisms to check the power of the military and prevent it from posing a threat to the ruler. Although the details varied over time, most dynastic regimes instituted a more or less elaborate structure of divided command that served to fracture the military establishment. Another common control mechanism was to give command or supervisory authority to trusted individuals who were not part of the regular military establishment. Depending on the period, these might be high-ranking civil officials, censors, imperial favourites or palace eunuchs. The later part of the Tang dynasty, in particular, was distinguished from other periods by the routine and systematic employment of eunuchs in the command and control of imperial armies. This phenomenon was most evident in the imperial palace armies, which were placed under the control of trusted eunuchs by the emperor Suzong during the An Lushan rebellion. With the exception of a brief interlude between 770 and 783, the famous Shence (or 'Divine Strategy') Army and the other palace forces were controlled by senior eunuchs almost until the end of the Tang dynasty. With a total strength of about 180,000 men by the early 820s, the Palace armies were the mainstay of the court's military power, and eunuch control of these forces was institutionalized in the office of palace commandant-protector (*hujun zhongwei*), normally held by the two most powerful of the palace eunuchs.[1] Although the commandant-protectors rarely exercised direct command over troops in the field, they had firm control of army administration and personnel management and cemented their hold over the 'professional' leadership cadre by forming fictive kinship ties with many of their generals.

A second capacity in which eunuchs served to control the late Tang military was as army supervisors (*jianjun* or *jianjun shi*), an office that Samuel B. Griffith once likened to that of political commissar in the Leninist armies of the twentieth century.[2] From the An Lushan rebellion onward, the Tang court routinely and systematically assigned eunuchs to keep watch both on armies in the field and on the provincial garrison forces commanded by the military governors (*jiedushi*). Although the supervisors were present primarily as observers, to ensure that imperial orders were obeyed and any signs of incipient disloyalty promptly reported to the capital, their direct channel to the centre of power often enabled them to dominate military governors and usurp their command authority. From Tang times onward, generations of traditional Chinese scholars denounced the system of eunuch supervisors as fundamentally dysfunctional. The eunuchs were accused of taking bribes to turn a blind eye towards disloyal activity, and of submitting secret reports that slandered loyal generals; their interference in matters of operational and tactical command, and their overall military incompetence, was blamed for numerous defeats. According to this critique, the system of eunuch army supervisors was not a support for imperial authority but rather a serious liability that contributed to the fall of the dynasty.[3]

We must, however, be sensitive to the fact that the scholar-officials who controlled the production and interpretation of the historical record took an almost uniformly hostile view of eunuchs, who were often their rivals at court and also stood out as monsters and misfits in the family-oriented Confucian moral order. To what extent is the indictment of the eunuch army supervisors an accurate reflection of Tang military realities, and to what extent is it a product of the general tradition of literati hostility towards the court eunuchs and all of their works? This study examines both the development and functioning of the system of army supervisors and the specific accusations levelled against it by literati critics. It finds that while abuses were widespread and many criticisms justified, the system also had its positive aspects and on balance probably contributed to the longevity of the Tang dynasty rather than hastening its downfall.

There are numerous examples from throughout China's imperial history of rulers appointing trusted individuals to accompany armies and keep an eye on the activities of the generals. Under the Tang, what had in former times been an irregular and informal practice was regularized as a 'system' during the period of Wu Zetian's ascendancy. An edict issued in the autumn of 684 made it clear that the Censorate was the agency responsible for supervising the armies, and a few weeks later the first known army supervisor of the Tang received his appointment. This was the palace censor Wei Yuanzhong, who was sent with the military commander Li Xiaoyi to put down Xu Jingye's uprising at Yangzhou.[4] Up until the fourth decade of the eighth century, the men chosen to act as army supervisors were usually censors – attendant censors (*shi yushi*) of the sixth rank, palace censors (*dian zhong shi yushi*) of the seventh rank and investigating censors (*jiancha yushi*) of the eighth rank.[5] Since censors were responsible for uncovering and denouncing official malfeasance, their supervision of generals in the field was a logical extension of their usual duties. Beginning in the early 730s, however, censors were replaced in the role of army supervisors by trusted eunuchs sent from the imperial palace. This was in keeping with a general trend towards greater eunuch authority at Xuanzong's court, which also saw the debut of the first eunuch-controlled military force, the Flying Dragon Palace Army. The supervision of generals soon became the exclusive province of the eunuchs, and remained so until the last days of the dynasty.[6]

Like the general to whom he was assigned, the army supervisor held an extraordinary, ad hoc position that was not a regular part of the imperial bureaucracy, but his rank was usually much lower than that of the general. Supervisors did not have the authority to issue orders to generals, though some (such as the censor Wei Yuanzhong) did offer strategic advice and participate in headquarters deliberations. Generals were apparently free to reject whatever advice was offered. When the frontier commander Gao Xianzhi launched his epic campaign against the pro-Tibetan Himalayan kingdom of Little Balûr in 747, and his eunuch supervisor Bian Lingcheng objected to advancing so deep into hostile territory, Gao simply left the eunuch behind at a base camp with several thousand 'weak' troops while he himself pressed forward to victory.[7] The supervisors' basic duties were to 'oversee reward and punishment', uncover falsehoods and report any illegal or treasonable activities. They could not arrest, dismiss or punish generals on their own authority, but a negative report submitted to the emperor might result in the dismissal or even execution of the offending commander.[8] After the Little Balûr campaign, when Gao Xianzhi had difficulties with his superior officer, Fumeng Lingcha (who felt that Gao had grabbed too much of the credit for himself), Bian Lingcheng submitted a memorial defending Gao.[9] A few years later, however, during the early days of the An Lushan rebellion, Bian took the opposite position. When Gao and another imperial general, Feng Changqing, were routed by the rebel forces near Luoyang, he was able to secure the prompt execution of both of the defeated generals.[10]

Before the An Lushan rebellion, it was not the Tang practice to appoint army supervisors to accompany every general in garrison or in the field. The government made these decisions on a case-by-case basis, and it seems that in quite a few instances no supervisor was appointed.[11] It has been suggested, quite plausibly, that the institution of the army supervisor was introduced and developed in response to the demise of the early Tang system of ad hoc expeditionary armies and the emergence of potentially dangerous standing armies on the frontiers of the empire.[12] However, the failure to make consistent and systematic use of the army supervisors guaranteed that they would be ineffective as a first line of defence against the rebellion launched by the powerful frontier general An Lushan in the waning days of 755. This horrific experience seems to have led the later Tang emperors, not unreasonably, to distrust their generals and place greater authority in the hands of the throne's closest and most dependent servitors, the palace eunuchs. During the reigns of Suzong, Daizong and Dezong (covering the half century from 756 to 805), the supervision system was expanded and regularized to become the standard mechanism for army control. Eunuch commissioners continued to oversee imperial armies in the field, but a striking new development was the regular assignment of eunuch supervisors (*jianjun shi*) to the provincial military commands.[13] These supervisors were supposed to serve for terms of three years; they had their own administrative and clerical staff, and it seems that they installed subordinate eunuch supervisors (*jianjun*) in the prefectures and outlying garrisons of the province. Many of them had armed followings that could be as large as several thousand strong. In some cases these were detachments from the Imperial Palace armies, and in other cases bands of private retainers.[14] The military governor and the army supervisor were supposed to be equal in status. While the governor exercised command authority, the supervisor's job was to keep the court informed of developments in the garrison and to act as necessary to prevent rebellion and put down mutinies. The collapse of the eunuch-controlled Imperial Palace army during the Huang Chao rebellion in the late ninth century undermined the authority of the eunuch supervisors, but they remained in evidence in the provincial garrisons until they were swept away in the wake of Zhu Wen's massacre of the court eunuchs in 903.[15]

Over the centuries, Chinese scholar-officials took a consistently dim view of the Tang army supervisor system. The criticism began in the days when the system was still in effect. It can be seen in memorials submitted by late Tang officials such as Xiao Fu, Bai Juyi and Li Deyu, and is reflected in the work of the official historians who compiled the *Jiu Tang shu* (Old Tang History), completed in the middle of the tenth century. It is repeated in the *Xin Tang shu* (New Tang History) completed in 1060 by Ouyang Xiu and his collaborators, in the other great masterpiece of Northern Song historiography, the *Zizhi tongjian* (Comprehensive Mirror for Aid in Government) of Sima Guang and in the strongly didactic work of Sima Guang's collaborator Fan Zuyu, the *Tang jian* (Mirror of Tang). The same historical examples were repeated again and again, and finally brought together by the great eighteenth-century scholar Zhao Yi in a fairly comprehensive list amounting to a bill of indictment against the eunuch supervisors. The particulars include corruption, slander and improper interference in command decisions (coupled with military incompetence), all of which led to defeats and setbacks for the Tang court.

Venality is a fundamental element of the eunuch character as depicted in traditional Chinese historiography, perhaps with the underlying implication that the eunuch's lust for gain should be understood as a way of seeking compensation for what he had lost. In the military context, venality comes into play when eunuchs are sent as commissioners to report on conditions in an army or garrison. The historical literature contains several examples of eunuchs bribed to overlook clear signs of treason, or present false reports that serve the interests of mutineers and rebels.[16] The failure of a general to deliver the expected bribe might result in a false denunciation, and fear of denunciation might push a general into rebellion (as happened in 764 when Pugu Huai'en failed to provide the eunuch commissioner Luo Fengxian with the sumptuous banquet he anticipated).[17]

The sources contain numerous examples of army supervisors submitting false denunciations of military governors.[18] The most colourful case comes from the year 800, when the eunuch supervisor of the Yicheng Army, Xue Xingzhen, sent a subordinate to the capital bearing a memorial falsely accusing the Yicheng military governor Yao Nanzhong of serious offences. One of Yao's followers, apparently acting on his own initiative, pursued the messenger, caught up with him at post station en route, cut him down and disposed of the memorial in a latrine. The officer then wrote a memorial of his own exonerating Yao and committed suicide. This prompted the emperor to summon the eunuch Xue to court for questioning. Fearing that this audience would provide his opponent a further opportunity for slander, governor Yao also rushed to the capital to advance his own cause before the emperor. The upshot was that Xue was removed from his position as supervisor of the Yicheng Army, but suffered no punishment and was actually promoted to an even more powerful office in the capital.[19] In this case, as in most others, there is no mention of bribery or extortion. It seems that the root cause of the trouble at Yicheng was not money but power. Xue had been meddling in military administration and trying to usurp the command authority of the military governor; when Yao resisted his efforts, Xue began to slander him in his secret memorials to the emperor.

The sources offer several other examples of eunuch supervisors usurping the command authority of military governors. The Hedong province with headquarters at Taiyuan seems to have been particularly troubled. When Li Yue was serving as acting military governor there near the end of Dezong's reign, all personnel appointments and decisions of military administration were made exclusively by the army supervisor Wang Dingyuan. Although Wang eventually overreached himself, one of Li's successors at Taiyuan, Yan Shou, was also dominated by his eunuch supervisor.[20] In addition to throwing their weight around in the garrisons, eunuch supervisors also accompanied armies on campaign and often meddled in operational and tactical decision making. In the spring of 761, when a Tang army was trying to retake Luoyang from the rebel forces under Shi Siming, the supervisor Yu Chao'en secured an edict from the emperor compelling the commanding general, Li Guangbi, to advance upon the enemy. Li moved forward against his better judgment and suffered a severe defeat at the battle of Mangshan.[21] In 800 when general Han Quanyi led a composite army drawn from 17 provinces against the rebellious military governor of Huaixi, more than a score of eunuch supervisors sat in his council of war and bickered with one another, making it impossible to arrive at any command decisions.[22] Military commanders operating against the rebellious Hebei garrisons in 821 were also hamstrung by interference from their supervisors, resulting in the failure of the campaign.[23]

The eunuchs were blamed for the failure of the 761, 800 and 821 campaigns, and the absence of eunuchs came to be regarded – at least by scholar-officials – as an important harbinger of military success. When general Gao Chongwen was sent to put down a revolt in Sichuan in 806, the chief minister Du Huangshang memorialized that Gao be allowed to operate without a supervisor.[24] Du's request was turned down, but in 817 the chief minister Pei Du, who was then leading another campaign against Huaxi, did succeed in getting the supervisors withdrawn so that his generals could have exclusive control over their troops and all orders would emanate from a single source.[25] The chief minister Li Deyu pulled off another such coup in the early 840s, when he negotiated an agreement with the most powerful court eunuchs that the supervisors would not be allowed to meddle in military affairs during a campaign against the rebellious military governor of Zhaoyi.[26] Later commentators attributed the success of both the Huaixi and Zhaoyi campaigns to the generals' freedom from eunuch intervention.

Even when they were not actively interfering in command decisions, the eunuch supervisors still seem to have been a liability on the battlefield. They were accused of requisitioning the best troops in the provincial armies to serve in their own bodyguards, leaving only the 'weak and decrepit' soldiers to fight in the battle line.[27] And Li Deyu claimed that when a supervisor was present on the battlefield, he would display his flag prominently to show his position. Upon the slightest setback, however, he would furl his banner and depart from the field – often prompting the rest of the army to join him in flight.[28]

As bad as the army supervisors appear to have been, the scholar-officials sometimes faced an even worse scenario: the formal appointment of a eunuch as generalissimo commanding a major imperial field army. The first move in this direction was taken in 758, when the Tang court was about to launch a major campaign against An Lushan's son and successor, An Qingxu. Nine military governors and their forces were gathered for the offensive, but the two most senior of them, Guo Ziyi and Li Guangbi, were of equivalent rank and merit, making it difficult to place either man above the other. In this situation, the emperor Suzong chose to name neither as commander-in-chief but instead appointed the favoured court eunuch Yu Chao'en to be 'commissioner to arrange and dispose of matters concerning conditions in the armies and manifestations of imperial conciliation' (*guan junrong xuanwei chuzhi shi*).[29] It is unclear to what extent Yu's authority exceeded that of an ordinary army supervisor, but he must surely have been expected to participate in command decisions that called for coordination of effort among the various provincial armies.

A much more alarming development came in 809, when the emperor Xianzong appointed his favourite eunuch, Tutu Chengcui, to a position that implied direct, operational control over the armies that were being gathered for an invasion of the rebellious Chengde province in Hebei. (Tutu's title was 'troop commissioner of the mobile encampment of the Left and Right Shence and the Hezhong, Heyang, Zhexi, and Xuanshe armies, and commissioner for arranging and disposing of matters concerning bandit suppression' – *zuo you Shence, Hezhong, Heyang, Zhexi, Xuanshe xingying bingma shi, zhaotao chuzi shi*.) This move aroused intense opposition from the censors and many of the most senior officials in the capital, and elicited an eloquent memorial of protest from the Hanlin scholar Bai Juyi, who argued that the appointment was unprecedented, shameful and likely to lead to defeat because the generals would be so humiliated to have to serve under a eunuch that they would simply not be able to perform their duties properly and well.[30] The emperor had no choice but to downgrade Tutu's title to something more suggestive of a supervisor than a generalissimo, but it seems that the eunuch did have considerable de facto authority for operational decision making during the Chengde campaign of 809–810. When the offensive failed, scholar-officials attributed it to Tutu's lassitude and military incompetence, and called for his execution.[31]

With all of the damage that was seen to flow from the supervision of armies by eunuchs, scholars were unequivocal in their condemnation of the *jianjun* system. In 784 Xiao Fu, who served as minister of personnel under Dezong, submitted a memorial arguing that eunuchs were suited only to menial tasks in the palace and calling for their removal from all positions of military authority.[32] Later, in the second half of the eleventh century, Fan Zuyu would comment,

> Since ancient times, when eunuchs interfere with military administration, there has never been a case in which the state was not defeated and the army was not lost. And the Tang was the worst of all. It would be better to select generals carefully, give them the authority they need, and not suspect them. Moreover, if generals are loyal and worthy, there is no need to supervise them. At the very least generals will be concerned about their families, which is more than can be said for eunuchs. I've seen their harm, but never any benefit from them.[33]

Fan's conclusion was repeated by Zhao Yi in the eighteenth century: 'That eunuchs supervising armies is all harm and no benefit can be seen very clearly.'[34]

The judgment delivered by the literati was unequivocal, and the evidence supporting it might at first glance appear to be overwhelming. Yet closer inspection of the sources and reflection on the nature of the most serious military challenges faced by the later Tang emperors suggests that there was another side to the story.

Despite their overall tone of hostility towards the eunuchs, the Tang dynastic histories and the *Zizhi tongjian* also contain discordant elements hinting at a more complex and multivalent reality. The sources sometimes contradict themselves and one another, and an obscure snippet of information found in one chapter may cast doubt on the accuracy of a sweeping assertion made in another chapter. For example, the claim made in the *Jiu Tang shu* biography of the chief eunuch Gao Lishi that Bian Lingcheng and other eunuch supervisors of Xuanzong's time wielded greater command authority than the generals to whom they were assigned is thrown into doubt by the same book's biography of the general Gao Xianzhi who, we are told, dealt with Bian's objections to the Little Balûr campaign simply by leaving the eunuch behind at base camp.[35] The sources also mention a few episodes that are not consistent with the general image of the eunuch supervisors as military incompetents and trouble makers, though these successes are not treated with quite the same loving detail accorded to examples of eunuch incompetence and failure. In 799, for example, Ju Wenzhen, the supervisor of the Xuanwu Army at Kaifeng, played a key role in restoring order to the garrison after the military governor Lu Changyuan was murdered by mutineers, and in 844 the Hedong army supervisor Lü Yizhong summoned loyal troops from an outlying garrison to crush a mutiny at Taiyuan.[36] To be sure, such examples are far less numerous than the cases of eunuch-spawned disasters – but this may be more a reflection of the anti-eunuch bias of the historians than of Tang realities.

The most outstanding – and unusual – example of a successful eunuch army supervisor is surely Yang Fuguang (842–883), who received the most laudatory biographies accorded to any eunuch in the Tang dynastic histories.[37] Yang spent most of his career serving the Tang rulers as an army supervisor, in which role he performed with exemplary effectiveness and scored success after success during the years of the Huang Chao rebellion. Interestingly, Yang is praised for the same sorts of actions that led the historians to condemn so many other eunuchs. When the general he was supervising was defeated by the rebels, Yang pushed him aside, took personal command of the troops and advanced to attack the rebels at Hongzhou and capture one of their leaders. Yang repeated this exercise in 881, this time executing the dilatory commander Wang Shu, taking control of his troops and advancing to defeat the rebels at Nanyang and recover Dengzhou. In addition to being a strict disciplinarian and a formidable campaigner, Yang also had impressive powers of persuasion. On one occasion, we are told, he almost persuaded Huang Chao himself to surrender, and he was later able to bring about the defection of Zhu Wen (the future founder of the Liang dynasty) from the rebel camp. Yang played a key role in recruiting the Shatuo Türk leader Li Keyong to participate in the 883 campaign to recover Chang'an, and appears to have once again gone beyond the merely supervisory role to hold an important troop command during the advance on the capital.[38] In Yang's case, the historians deploy the sorts of descriptive tags and topoi that are more often encountered in the biographies of exemplary officials and loyalist heroes; we are told that he was generous, restrained and righteous, that he was capable of making strategic plans, that he called off one of his campaigns to observe mourning for his mother and the soldiers were moved to tears when he himself passed away at the age of only 42 *sui*.

What are we to make of this exemplary figure in the context of the otherwise highly negative representations of eunuch supervisors in the Tang histories and later literati writings? It is tempting to argue that Yang Fuguang is the exception that proves the rule; that is, the willingness of the historians to praise Yang for his accomplishments shows that their anti-eunuch bias did not prevent them giving credit where it was due (thus confirming their otherwise dismal portrayal of eunuch supervision). We need to bear in mind, however, that Yang Fuguang was in many ways not a typical eunuch. Educated, literate and cultivated, he was the scion (by adoption, of course) of an old and distinguished eunuch lineage. In the fractious palace politics of the late ninth century, he belonged to an 'aristocratic' eunuch faction that allied with prominent scholar-officials to resist the machinations of eunuch parvenus such as Tian Lingzi.[39] Yang was, in effect, an honorary scholar-official, and the usual conventions regarding the representation of eunuchs were suspended in his case. Far from being the rule-proving exception, the example of Yang Fuguang demonstrates that the system of eunuch supervisors was by no means consistently or uniformly dysfunctional, and that impressive results could be attained by precisely the sorts of usurpations and interventions that the historians were usually at pains to condemn. It also suggests that there must have been other competent and effective eunuch supervisors whose successes went unrecorded because they lacked Yang's very special qualifications.

There are other reasons for believing that the eunuch-supervisor system did not, on the whole, do nearly as much damage as claimed by the scholar-officials. While almost all late Tang armies had eunuch supervisors assigned to them, not all campaigns ended in disaster. Despite the chief minister Du Huangshang's efforts to the contrary, army supervisor Liu Zhenliang accompanied Gao Chongwen's army during the campaign to put down Liu Pi's rebellion in Sichuan – and the eunuch's presence did not prevent the expeditious military suppression of the revolt.[40] It is also interesting that the emperors rarely punished eunuch supervisors very seriously, even in the wake of quite serious defeats. Yu Chao'en, for example, suffered no punishment at all and continued to serve in a supervisory capacity with the armies in Henan even after the 759 disaster at Xiangzhou had occurred on his watch.[41] Cases of this sort may perhaps be attributed to blind favouritism, but it is also possible that the eunuchs got off fairly easily because the fault was not entirely theirs. Quite apart from the treatment of individual eunuchs, it is noteworthy that the Tang emperors – even strong emperors such as Dezong and Xianzong – made no attempt to weaken or abolish the eunuch-supervisor *system*. As Charles Peterson has observed, 'In spite of its defects, inherent or otherwise, and the opposition that it raised, the system proved effective from the throne's point of view and was retained, with some subsequent modifications, throughout the remainder of the dynasty.'[42]

It was effective because, during most of the period when the system was in operation, the greatest threat faced by the Tang court was not foreign invasions or popular uprisings but revolts and mutinies arising from within its own armies. The two-week occupation of Chang'an by the Tibetans at the end of 763 was certainly a disaster and an embarrassment, but as a threat to the survival of the Li family's dynastic state it paled in comparison with the mutiny of the Jingyuan Army in 783 which put Dezong's life in jeopardy and the capital in the hands of the mutineers for more than six months. Even if we set aside such spectacular episodes, it is clear that the long-term survival of the dynasty depended upon maintaining discipline in the provincial garrisons and resisting the efforts of military governors to gain greater autonomy for themselves. In this day-to-day struggle the eunuch army supervisors were the throne's first and best line of defence, given their ability to provide intelligence and to intervene as necessary in the politics of the garrisons. In exchange for these advantages, a slight diminution of military effectiveness may have seemed a small price to pay – and the discomfiture of the generals a positive benefit. Scholar-officials might deplore the phenomenon of eunuchs bullying and dominating military commanders, but from the perspective of the throne this may have been a sign that the system was working precisely as intended.[43]

Notes

1 For this figure, see Song Minqiu (compiler), *Tang da zhao ling ji* (Beijing: Shangwu yinshu guan, 1959), ch. 2, p. 12. For more information on eunuch control of the Shence and other palace armies, see: He Yongcheng, *Tang dai Shence jun yanjiu* (Taibei: Taiwan Shangwu yinshu guan, 1990); Zhang Guogang, *Tang dai zhengzhi*

zhidu yanjiu lunji (Taibei: Wenjin chubanshe, 1994), pp. 113–156; Liu Yat-wing, 'The Shen-ts'e Armies and the Palace Commissions in China, AD 755–875' (PhD dissertation, University of London, 1970).

2 Sun Tzu, *The Art of War* (tr. Samuel B. Griffith) (Oxford: Oxford University Press, 1963), p. 82, n. 1.

3 Ouyang Xiu and Song Qi, *Xin Tang shu* (Beijing: Zhonghua shuju, 1975) (hereafter cited as *XTS*), ch. 207, p. 5856.

4 Zhang Guogang, 'Tang dai jianjun zhidu kaolun', *Zhongguo shi yanjiu*, no. 2 (1981), pp. 120–122; Dong Hao (comp.), *Quan Tang wen* (Taibei: Jingwei shuju, 1965) (hereafter cited as *QTW*), ch. 96, p. 14a; Liu Xu, *Jiu Tang shu* (Beijing: Zhonghua shuju, 1975) (hereafter cited as *JTS*), ch. 92, p. 2951. Du You claims that censors were used as army supervisors from the very beginning of the dynasty, but there are no examples prior to 684; Du You (comp.), *Tong dian* (Beijing: Zhonghua shuju, 1988) (hereafter cited as *TD*), ch. 29, p. 605.

5 Zhang Guogang, 'Tang dai jianjun zhidu kaolun', p. 122.

6 Zhang Guogang, 'Tang dai jianjun zhidu kaolun', p. 123; *TD*, ch. 29, p. 605. According to Wang Pu (comp.), *Tang hui yao* (Beijing: Zhonghua shuju, 1955), ch. 65, p. 1131, eunuchs were appointed as early as 705. The same source says that the appointment of censors as army supervisors was halted at the end of 687, but there are examples of later appointments (see *JTS*, ch. 185B, pp. 4807–4908). For the Flying Dragon Palace Army and the rise of the eunuchs under Xuanzong, see Edwin G. Pulleyblank, *The Background of the Rebellion of An Lu-shan* (London: Oxford University Press, 1955), pp. 66–67 and p. 144, n. 24; J.K. Rideout, 'The Rise of the Eunuchs during the Tang Dynasty', *Asia Major* (New Series), vol. 1 (1949–1950), pp. 53–72.

7 Sima Guang, *Zizhi tongjian* (Beijing: Zhonghua shuju, 1987) (hereafter cited as *ZZTJ*), ch. 215, pp. 6685–6686.

8 Zhang Guogang, 'Tang dai jianjun zhidu kaolun', pp. 122, 127; *JTS*, ch. 92, p. 2951 (Wei Yuanzhong).

9 *ZZTJ*, ch. 215, p. 6886. I am indebted to Professor Jonathan Skaff for bringing this episode to my attention.

10 *JTS*, ch. 104, pp. 3206, 3209–3211.

11 Zhang Guogang, 'Tang dai jianjun zhidu kaolun', p. 122; *ZZTJ*, ch. 204, pp. 6446–6447. Edwin G. Pulleyblank, *The Background of the Rebellion of An Lu-shan*, p. 74.

12 Zhang Guogang, 'Tang dai jianjun zhidu kaolun', p. 128.

13 Due to the limitations of available evidence, it is not possible to determine whether eunuch supervisors were assigned to *all* of the provinces. What is clear is that all of the heavily garrisoned provinces did receive supervisors. See Charles A. Peterson, 'The Restoration Completed: Emperor Hsien-tsung and the Provinces', in Arthur F. Wright and Denis Twitchett (eds), *Perspectives on the T'ang*, (New Haven, CT, and London: Yale University Press, 1973), p. 175, n. 107.

14 Zhang Guogang, 'Tang dai jianjun zhidu kaolun', pp. 123–125.

15 *ZZTJ*, ch. 263, pp. 8594–8595; Wang Gungwu, *The Structure of Power in North China during the Five Dynasties* (Kuala Lumpur: University of Malaya Press, 1963), pp. 33, 83.

16 Zhao Yi, *Nian er shi zha ji* (Taibei: Hong shi chubanshe, 1974), ch. 20, pp. 264–265.

17 Zhao Yi, *Nian er shi zha ji*, ch. 20, p. 264; *XTS*, ch. 207, p. 5862.

18 See, for example, Zhao Yi, *Nian er shi zha ji*, ch. 20, p. 265; *ZZTJ*, ch. 237, p. 7678; *JTS*, ch. 184, p. 4763; *XTS*, ch. 207, pp. 5861, 5863, 5875.

19 Fan Zuyu, *Tang jian* (Shanghai: Shanghai guji chubanshe, 1984), ch. 8, pp. 9b–10a; *ZZTJ*, ch. 235, pp. 7586–7587.

20 Zhao Yi, *Nian er shi zha ji*, ch. 20, pp. 265–266; *JTS*, ch. 146, pp. 3958–3959.

21 *ZZTJ*, ch. 222, pp. 7105–7106; Zhao Yi, *Nian er shi zha ji*, ch. 20, p. 265.

22 Fan Zuyu, *Tang jian*, ch. 8, pp. 10a–b; *ZZTJ*, ch. 235, pp. 7587–7588.

23 Fan Zuyu, *Tang jian*, ch. 10, p. 4b.
24 Zhao Yi, *Nian er shi zha ji*, ch. 20, p. 265.
25 Zhao Yi, *Nian er shi zha ji*, ch. 20, p. 265; *JTS*, ch. 170, p. 4418.
26 Zhao Yi, *Nian er shi zha ji*, ch. 20, p. 265; Fan Zuyu, *Tang jian*, ch. 10, p. 13a–b.
27 Fan Zuyu, *Tang jian*, ch. 10, pp. 4b, 13a; Zhao Yi, *Nian er shi zha ji*, ch. 20, p. 265. Li Deyu is also credited with negotiating an agreement with the senior eunuchs to limit the troops siphoned off for bodyguard duty to only ten per thousand.
28 Fan Zuyu, *Tang jian*, ch. 10, p. 13a; Zhao Yi, *Nian er shi zha ji*, ch. 20, p. 265.
29 *JTS*, ch. 184, p. 4763; *ZZTJ*, ch. 220, p. 7061. The translation of the title is borrowed from Charles O. Hucker, *A Dictionary of Official Titles in Imperial China* (Stanford, CA: Stanford University Press, 1985), p. 284.
30 *ZZTJ*, ch. 238, pp. 7666–7668; *JTS*, ch. 184, p. 4768; *XTS*, ch. 207, p. 5869. For the text of Bai Juyi's memorial, see *QTW*, ch. 668, pp. 2a–3a.
31 *JTS*, ch. 184, p. 4768; *ZZTJ*, ch. 238, p. 7679; *QTW*, ch. 666, pp. 18b–19a.
32 *JTS*, ch. 125, p. 3551; *ZZTJ*, ch. 229, p. 7397.
33 Fan Zuyu, *Tang jian*, ch. 8, p. 10b.
34 Zhao Yi, *Nian er shi zha ji*, ch. 20, p. 265.
35 Compare *JTS*, ch. 184, p. 4757 with *JTS*, ch. 104, p. 3204 (and *ZZTJ*, ch. 215, pp. 6685–6686).
36 *ZZTJ*, ch. 235, p. 7582; *JTS*, ch. 174, p. 4527.
37 For Yang's biographies, see *JTS*, ch. 184, pp. 4772–4774, and *XTS*, ch. 207, pp. 5875–5877. The information in this paragraph is taken mainly from the *Jiu Tang shu* biography.
38 The titles that Yang held in this period are indicative of supervisory status (*JTS*, ch. 184, p. 4773; *XTS*, ch. 207, p. 5877; *ZZTJ*, ch. 255, p. 8279), but there is a passage in the *Xin Tang shu* biography of Tian Lingzi that mentions several 'subordinate generals' of Yang Fuguang commanding a total of 20,000 men (*XTS*, ch. 208, p. 5887). It is also noteworthy that it was Yang Fuguang who submitted the 'announcement of victory' (*lubu*) reporting the recovery of Chang'an; this was normally the prerogative of the principal field commander. For the text of Yang's *lubu*, see *QTW*, ch. 998, pp. 16a–17b.
39 Robert M. Somers, 'The End of the T'ang', in Denis Twitchett (ed.), *The Cambridge History of China*, vol. 3: *Sui and T'ang China, 589–906*, Pt 1 (Cambridge: Cambridge University Press, 1979), pp. 716–717.
40 Zhao Yi, *Nian er shi zha ji*, ch. 20, p. 265; *XTS*, ch. 207, p. 5868; Charles A. Peterson, 'Court and Province in Mid- and Late-T'ang', in *Cambridge History of China*, vol. 3 (1979), p. 324. For other examples of supervised armies meeting with success rather than failure, see *ZZTJ*, ch. 220, pp. 7062–7063, and *XTS*, ch. 207, p. 5863.
41 See Yu's biographies (*JTS*, ch. 184, p. 4763; *XTS*, ch. 207, p. 5863) and *ZZTJ*, ch. 221, pp. 7069–7070 and ch. 222, pp. 7105–7106. Tutu Chengcui also got off lightly after the 810 Chengde campaign; see *JTS*, ch. 184, p. 4768.
42 Charles A. Peterson, 'The Restoration Completed', p. 176.
43 Shih-shan Henry Tsai has reached similar conclusions regarding the role of eunuchs in the Ming dynasty; see his book, *The Eunuchs in the Ming Dynasty* (Albany, NY: State University of New York Press, 1996), p. 58.

12 The cultural construction of war in Tang frontier poetry

Ming Jiang

The Tang Dynasty (CE 618–907) has been seen at once as one of the high points of Chinese warfare and as the apogee of Chinese poetry. It is particularly appropriate to explore the cultural construction of war in the Tang by way of poetry, not only because poetry has always played a unique role in Chinese civilization as a highly favoured form of cultural expression, but especially because poetry was instituted as the most important literary and social genre by the Tang era. Of more than 48,900 poems by some 2,200 poets preserved in *The Complete Collection of Tang Poetry* (*Quan Tang shi*), a good number of them can be categorized as war poems.[1] A majority of these war poems are frontier poems, which were written by some of the most talented poetic geniuses and are accounted as being among the best Tang poems.[2] In this chapter, I intend to explore the literary aspect of Chinese military culture in the Tang. More specifically, I will: examine how wars, frontier wars in particular, were seen and represented in the High Tang (CE 712–756) frontier poems; analyse the dynamic interaction between *wen* (civil) and *wu* (military) in frontier poems, which reflected the general attitude of the society towards the military, and consequently influenced the literati's career choices and paths; and explore how the meaning of war was understood and constructed by the intellectual elite through their literary creations.

The shifting *wen–wu* dyad in Tang frontier poetry

The *wen–wu* dyad is a useful Chinese conceptual framework that can be deployed here to discuss the dynamics of the Tang military culture in terms of the poetic representation of war in the High Tang that valued the military, praised military campaigns and raised the status of a military career. In the *wen–wu* dichotomy, *wen* refers to civil, mental, literate, or distinct literary and cultural attainments, while *wu* refers to military, physical, martial, or training in the skills and ethical codes of the warrior. In the long interrelation of the dyad of *wen* and *wu* throughout Chinese history, *wen* generally ruled supreme, not only in politics and society, but also in aesthetic and literary tradition. In the High Tang, and especially in High Tang frontier poetry, however, the balance of the dyad tipped strongly in favour of *wu*.

'Give up pens to follow your ambition,/Pick up swords to drive out barbarian flying generals' (*QTS*, 8838).[3] The line from a High Tang frontier poem by Dai Xiuting epitomized the spirit shared by many Tang poets who favoured, instead of scholarly achievements and literary talent, *wu* both in their real lives and in their literary creations, especially in their frontier poems. A lot of them travelled to the remote frontiers, either to join the military campaigns to serve in the military headquarters, or simply to experience the frontier regions. Ren Wenjing counted 127 poets who not only visited the frontiers, but also composed frontier poems that were included in various poetry collections.[4] A good number of poets served in the frontier military headquarters. According to the distinguished Tang poetry scholar of the Ming, Hu Zhenheng, 'It was a prevailing practice for Tang poets to serve as military governors' staff or in frontier military headquarters.'[5] In their frontier poems, some Tang poets resolutely expressed, just like Dai Xiuting, their desire to give up their pens in favour of swords. Abandoning the traditional path to officialdom by excelling in study, poet Ceng Shen, for example, became convinced that 'Official ranks can be earned only on horse-back' (*QTS*, 2061). In the same vein, another poet Yang Jiong proclaimed that he would 'Rather serve as an insignificant warrior,/than live as an illustrious scholar' (*QTS*, 615). Other frontier poets even went so far as to trivialize and denigrate themselves and other Confucian scholars in order to praise warriors. While admiring 'heroic generals', Gao Shi, one of the most famous frontier poets, derided the literati in 'Song at the Frontier', 'Laugh at those literati,/Who waste their whole lives on study of the classics' (*QTS*, 2189–2190). Wang Wei concurred with his poet friend Gao when composing a poem to send off a friend with a military career, 'Don't emulate those scholars,/Who waste their whole lives studying the classics in studies' (*QTS*, 1271).With verses like these, the Tang frontier poets equated the military (*wu*) with the heroic, the extraordinary and living a worthy life, while associating the scholar (*wen*) with the quotidian, the trivial and idling away one's precious time. No wonder the poet Li He would suggest that instead of making a living with pen, 'A man should bear swords' (*QTS*, 4414).

The Tang poets' growing partiality towards *wu* was telling when one considers that in the Tang era, *wen* was actually strengthened in some very important aspects. For example, the Civil Service Examinations, which selected administrative officials for the state bureaucracy based on scholarly understanding and exposition of the Confucian classics, began to develop systematically and mature in the Tang. Even skill in composing poetry was tested in the *jinshi* examination, the most prestigious type of imperial examination, to qualify a person for holding public office, which was seen as the main objective of traditional Chinese literati.[6] This enhanced the political leverage and high social prestige of *wen*. Indeed, in the Kaiyuan era (CE 713–741), which coincided largely with the High Tang, many excellent poets served as government officials, and some even at the highest level, such as Song Jing, Zhang Jiulin and Zhang Yue, who all became chancellors, the highest-ranking officials at the royal court.[7] This raises the question as to why the balance of the *wen–wu* dyad tilted towards *wu* at this time.

As the Tang was constantly at war with neighbouring countries and regions for either expansion or survival, it is not surprising that *wu* (the wars and warriors) was highly valued by people in all walks of life, including the intellectual elite and the poets, because of their patriotic passion, imperialistic arrogance or simply their thirst for excitement, exoticism or power. And more importantly in our case, the great dependence of the Tang on its military effectiveness raised the status of the military, increased the power of military leaders on the frontiers and created more career opportunities for both the military and literati, whose chances of holding public office through the Civil Service Examinations was, while markedly improving, still relatively limited in the Tang. Military generals, for example, often sought talented poets and scholars to compose, in elegant style, official documents, especially memorials to the emperor in order to impress him. And the commanding generals had the prerogative to appoint his staff. These and many other opportunities created vacancies, and thus an alternative path, for poets and other literati to officialdom through the military. Thus, military service became a powerful incentive for the literati's career advancement, which drove the Tang poets from studies to battlefields on the frontiers and consequently helped change the dynamics of the *wen–wu* dyad.

A saying in *The Analects* of Confucius famously goes, 'One who excels in study should follow an official career' (*Analects*, 19.13). This adage guided the career paths of intellectuals in traditional China and proved true in the Tang as well. For the Tang literati, *wu* was, in fact, a different means to the same end, and that end was, as hinted in the above-cited Gao Shi line, 'official ranks' or holding public office, which induced the literati to study (*wen*) before and drove them to military (*wu*) now. Gao himself was a good example for those poets who eventually received high posts in the central government through the detour of serving in military headquarters. His misfortune in failing to achieve office through normal academic avenues drove him to serve in general Geshu Han's headquarters, which paved the way for his political success. He was later awarded a number of high positions, including that of regional commander and prefect.

To be sure, military service was not equally appealing to everyone. The Ming scholar, Hu Zhenheng, talked about two groups of literati to whom the military service was particularly attractive:

> According to the Tang system, candidates who had successfully completed the *jinshi* examination often went to serve ranking generals. Among poor but talented literati, even more people relied on their services in military headquarters in order to enter the state bureaucracy and to advance their careers.[8]

The first group of scholars mentioned by Hu consisted of the highest academic achievers who passed the highest level of the Civil Service Examinations. After being awarded the *jinshi* title, however, they had to wait for their appointments by the Ministry of Official Personal Affairs according to the Tang system, and in some cases the waiting could last for years.[9] During this waiting period, some of them would choose to serve in military headquarters, where some could be appointed as staff right away, but many of them were there simply to gain military experience and administrative skills on the frontiers, which would increase their chances of getting appointments faster or getting better jobs later if they performed well. The second group referred to those literati who lived in meagre circumstances and had little prospect for official careers. Thus, serving in the military might be their only chance to gain access to public office. In addition to these two groups, there were other scholars who sought military service to advance their careers. Some of them were educated provincials who failed to gain civil office through the central government; others were ambitious scholar-bureaucrats who were frustrated by the lack of opportunities to advance their careers. As successful military service or the recommendations of military generals were highly appreciated and valued by the royal court, serving under military generals could pave the way for those literati who were low- or middle-ranking officials to advance their careers in the central government.

These High Tang literati mostly served with pens on the frontiers, often in the headquarters of commanding generals and other military institutions, rather than fighting with swords on battlefields. Nevertheless, the fact that a large number of talented poets and writers served in the military on the frontiers or visited the frontier regions did significantly contribute to the flourishing of frontier/war poetry, which in turn helped engender a changing dynamic of *wen* and *wu* in favour of the latter among intellectuals and in society as a whole. This change was characterized by a new masculine, heroic, at times even belligerent, ethos, which was particularly evident in the Tang frontier poetry. The masculine and vigorous tone in the Tang frontier poems differed unmistakably from a poetic tradition that had seen melancholy as its dominant mode.[10] Critics of contemporary and later ages often described the Tang frontier poetry with words such as 'precipitous' (*qijun*), 'resolute and manly pathos' (*beizhuang*), 'energy and sturdiness' (*qigu*) and 'powerful, masculine emotions' (*xiong yiyu*). These terms reflect a literary style and poetic tone that endow the Tang frontier poetry with a quality of typical *wu*.

The High Tang frontier poets considered war and warriors to be the apotheosis of masculinity and constructed, based on this perception, a masculinity that was closely associated with courage, heroism, endurance and ultimate sacrifice. In this spirit, Gao Shi praised heroic Tang soldiers 'Steadfast to death,/and never a thought of glory or reward' and proclaimed, 'True men by their very nature/love the long military campaign' (*QTS*, 2218).[11] Such a masculine, heroic feeling towards war even drove a few frontier poets to go beyond the eruption of sensation and euphoria to assenting to war's violent grandeurs. A case in point was Gao Shi's poem, composed when general Geshu Han defeated the Tibetans, 'The streams spurt blood of the many tribes,/Winds drive the souls of dead nomads,/Heads fly as they gather thousands of pikes,/Hands tied behind, massed at the general's gate' (*QTS*, 2234). Clearly brutal violence on the battlefields was celebrated as the manifestation of the superior military power of the Tang Empire.

Such 'manly pathos' and 'masculine emotions' not only existed in the depiction of military actions, but were also expressively reflected in the descriptions of landscape and nature on the frontiers, which was the most popular topic in frontier poetry. While previous frontier poetry was dominated by bleak and barren landscapes and a dreary and desolate mood, the Tang frontier poetry, although containing many elements of its predecessor, captured some startling and unique aspects of the frontier landscape and rare sensations, conveyed an uplifting spirit and demonstrated a fascinating mixture of exoticism and powerful feelings towards war. To give but a couple of examples: 'A white sun emerges from the ground,/The Yellow River comes from the sky;/The sand rolls, its ridges like waves' by Zhang Bin (*QTS*, 8144), and 'Great desert: one column of smoke stands straight;/Long river: the setting sun hangs round' (*QTS*, 1279) by Wang Wei. Even when writing about the piercing cold and severe weather conditions on the frontiers, another popular topic in frontier poetry, Tang frontier poems often strike a tone of heroic exuberance by commending the soldiers' endurance and sacrifices, as demonstrated in Cen Shen's lines, 'At midnight they march, lances ram together./The headlong wind like dagger, men's faces seem slashed./Snow hangs upon the horses' hair, sweat steams upwards.... Hearing of this, the tartar cavalry is surely filled with terror' (*QTS*, 2059). As we see from the above examples, the High Tang frontier poets adopted many of the most frequently used stock images concerning nature and military life such as limitless sand and desert, the severe cold, beacon fires, dagger, lances, horse, cavalry, suns and the Yellow river. However, what made the Tang frontier poetry differ prominently from its predecessors was the fact that it charged the old images and themes with new energy, dynamic vitality and heroic gaiety.[12]

It should be noted that in spite of the fact that a good number of Tang frontier poets explicitly praised *wu* at the cost of *wen*, they, in general, did not think of *wen* and *wu* in terms of a polarity. They saw them, instead, as competitive, complementary concepts, rather than opposing especially when it came to the pathways to political power and public office. What they admired most, both rhetorically and in real life, were those who were at once brilliant court officials with excellent academic training and talented commanders-in-the-field who accomplished martial exploits, such as Guo Zhen and Zhang Yue. Indeed, 'to be a really powerful man', as Kam Louie pointed out, 'it is essential to have both *wen* and *wu*.'[13]

Constructing the meaning of war

For the High Tang poets, their frontier poems were not only a way of relating war, but also a means of constructing the meaning of it. It can be often observed that the same Tang poets who in some poems enthusiastically *praised* some military campaigns and frontier wars, in other poems resolutely opposed other campaigns and wars, especially those caused by the emperor and the military generals' indulgence in wars of aggression. In other words, their poems revealed a rather complex stance towards war, which, depending on how they understood and interpreted the meaning of war, ranged from euphoric embrace to despairing condemnation. Many Tang poets willingly embraced frontier wars because they readily bought into the belief that many of those wars were necessary and meaningful for various political, ethical, ethnic and personal reasons.

Like society in general, Tang poets believed in a just war concept, while their perspectives of 'just' seemed to be both Confucian and ethnocentric.[14] The 'righteous army' (*yibing*), the first Chinese terminological equivalent to the term of the 'just war', was believed to appear in *The Spring and Autumn Annals of Mister Lü* (*Lüshi chunqiu*),[15] but the most influential notion of a 'just war' in Chinese history might be, however, Mencius' view on the issue. Mencius stated:

> *The Spring and Autumn Annals* acknowledges no just wars. There are only cases of one war being not as bad as another. A punitive expedition is a war waged by one in authority against his subordinates. It is not for peers to punish one another by war.[16]

In this statement, Mencius made a clear distinction between a campaign of justified punishment (*zhengfa*) waged by the authority (the emperor) and the unjustified acts of inter-state aggression. Mencius' view here was drawn from Confucius' idea of a 'just war' in *Analects*, expressed in this statement, 'When the Way prevails in the Empire, the rites and music and punitive expeditions are initiated by the Emperor' (*Analects*, 16.2).[17]

Such a take on a 'just war' was clearly reflected in the Tang poets' conception of and responses to war. It was in this spirit that the renowned poet Wang Wei, for example, tried to justify the Tang's military campaigns against neighbouring communities on the frontiers. In a preface accompanying a collection of poetry written by the participants in a farewell party for a poet friend who set off to the frontier, Wang Wei emphasized the moral justification and the 'authority' of Chinese frontier wars with these opening sentences: 'Following the Way the emperor controls vast remote and uncivilized regions, and employing military power he rules the world.'[18] Following Confucius' lead, Wang Wei vindicated Chinese frontier wars by establishing the moral justification when he claimed that the Chinese emperor ruled by following 'the Way'. Just as important for the Confucius–Mencius understanding of 'just war', Wang made it clear that these wars were led by the 'authority', that is, the emperor, or 'proceeded from the son of Heaven', to use Confucian words. In the preface, Wang Wei clearly assumed a hierarchical relationship between the authority (the Chinese emperor) and his subordinates that also included all neighbouring foreign countries and communities. He once suggested in one of his poems, 'All barbarian countries should be advised,/To learn from Kunya and subordinate themselves to the Chinese emperor' (*QTS*, 1273). If 'barbarians do not understand this, and subsequently act in a conceited and arrogant way', Wang Wei warned in the preface, the Chinese emperor and military forces would have the right to 'trample the barbarian territories, and capture all barbarians'. He saw it as justified for the Chinese emperor to wage wars against barbarians in those vast remote and uncivilized regions because those wars were waged by 'the one in authority' as 'a punitive expedition against his subordinates', which was defined by Mencius as a 'just war'.

When Mencius stated that 'rival states do not punish one another by resorting to war', he was actually talking about the Chinese feudal states only. In other words, he did not believe that the same definition of 'just war' should be applied to conflicts between the Chinese and the non-Chinese. While he believed that feudal Chinese states with the same language and cultural practices should not resort to war to solve their conflicts, he unambiguously agreed with what the Duke of Zhou did to barbarians as described in the *Book of Songs*, which he cited in *Teng Wen Gong I*, 'It was the barbarians that he attacked;/It was Ching and Shu [non-Chinese states] that he punished.'[19] Here we can see an ancient Chinese concept, the Hua-Yi distinction, at work. The Hua-Yi distinction differentiates a culturally defined 'China' (called *hua* or *huaxia*) from cultural or ethnic outsiders or 'barbarians' (*yi*). The concept developed through othering an unequal relationship between Chinese and non-Chinese and served to justify the Chinese domination in the peripheral communities and to facilitate the subordination of 'barbarians'. This concept was precisely what poet Wang Wei advocated in his preface and was shared by some Tang poets, frontier poets included. Based on this notion, some Tang poets considered it justified for the Tang, as a culturally superior empire, to wage wars against 'other' groups or peoples who were considered culturally inferior. It was in this sense that the Hua-Yi distinction helped shape the Tang poets' thinking concerning the meaning of war.[20]

When Gao Shi, for example, enthusiastically supported a clearly unjust Chinese military aggression against Nanzhao, a neighbouring non-Chinese country in the south-west, the Hua-Yi distinction seemed to be the major factor for him to judge whether the war was meaningful. In lines such as 'Our Sage glowed wrathful in his rage,/Ordered attack on the south west barbarians' (*QTS*, 2209), he left no doubt that the Chinese 'sage', a man supposedly with superior moral and spiritual authority, had the right to order an attack against 'barbarians', even though they posed no threat to the Chinese at all. Even in Li Bai's well-known frontier poem 'Fighting South of the Wall', which has an unmistakable antiwar overtone, one can still see, although in a much more subtle way, this Hua-Yi distinction at work. In striking contrast to the Chinese civilization based on agriculture and on cultivating farm fields, Li Bai wrote, 'Xionglu's cultivation is nothing but killing,/They always sowed their fields – the yellow desert with bones of the dead' (*QTS*, 1684). As a result, as the ensuing lines hinted, the Chinese had to 'build the Great Wall to defend China against the barbarians', and this was also why 'wars at the frontier can never end'. Thus the Hua-Yi distinction caused military conflicts between the Chinese and nomads, and the Chinese frontier wars against Xionglu could be justified both as a war of civilization against barbarians and as a defensive act. The latter was a critical assumption pertaining to the conception of the 'just war', which was shared by Li Bai and many other Tang poets. In other words, a war could be justified if it was waged for survival or as a legitimate defensive act. Nevertheless, Li Bai warned in the poem against any war because the Tang had already seen too many wars and too much suffering by both soldiers and civilians. He concluded the poem by essentially citing Laozi, 'Therefore arms are an instrument of evil/A sage uses them only on the compulsion of necessity.'

For the High Tang poets, a war would be meaningful if it was ethically, politically, ethnically or racially appropriate and just, and it could also make sense if it gave meaning to them as individuals or satisfied their personal needs and desires. As discussed before, the frontier wars provided an alternative career path for the Tang poets, and especially for those who felt deeply disappointed in their official carriers, and thus saw little of value in their lives. As their primary object in life was to hold public office, the frontier wars helped them achieve this objective and thereby gave meaning to their lives. For some Tang poets, wars allowed them to be noble and to get the feeling of some kind of heroism, and thus helped them, through experiencing excitement, overcome everyday trivia and ennui in their scholarly lives, which were seen by them as wasted 'on the study of the classics', as Wang Wei and Gao Shi wrote in their poems. The patriotic poets who searched for meaning in living for a higher purpose also found meaning in war, as they believed that sacrificing one's life for the survival of the country or ensuring safety for the people would serve a higher purpose, and thus make their lives meaningful, as stated in Yuan Qianyao's verse, 'Sacrifice for the country makes one's life worthwhile' (*QTS*, 1109). A good number of poets were ready, at least in their poems, to sacrifice their lives and die on the battlefield for their country. While Dai Shulun would 'Wish to sacrifice my life for the country,/I do not expect to return alive to the Yumen Pass' (*QTS*, 3097), Li Xizhong claimed that 'One should sacrifice for loyalty and justice/And die to repay the country' (*QTS*, 1620).

It is worth noting that Li Xizhong's line here stressed both the moral and political aspects of the Tang poets' thinking about sacrifice, namely to sacrifice for both moral values – 'loyalty and justice' (*zhongyi*) – and for the political entity 'country' (*guo*). Joseph Levenson calls these two aspects the 'time-honored co-ordinates in Chinese political thinking',[21] which he discusses when analysing Chinese 'culturalism' and nationalism, expressed respectively by the terms *tianxia* (the world, all under heaven or the Chinese empire), and *guo* (country, state). Levenson believes that the term *guo* in traditional China needs to be understood in reference to *tianxia* and makes a clear distinction between *guo*, 'a political unit held by de facto power' and protected by military force, and *tianxia*, which refers to a 'regime of value' and 'a conception of civilized society'.[22] Chinese scholars were not only worried about the country as a political entity, but also, often even more, were concerned for Chinese moral superiority based on Confucian ideology. In Wang Wei's preface cited above, he clearly had the Chinese state *guo* as a political entity in mind, which was a restricted notion of the Han community and the fatherland in which the 'barbarians' had no place. At the same time, he explicitly emphasized that the Chinese emperor ruled the world (*tianxia*) 'following the Way', which legitimated a concentric universal empire and Chinese superiority. To this world that was defined by the Chinese culture and values also belong those non-Chinese who accepted Chinese culture and values. This would explain why the poet, while expressing seemingly ethnocentric viewpoints, devoted almost half of the same preface to enthusiastically glorifying Geshu Han, a famous Chinese general who himself was ethnically a non-Han Chinese. In Li Zhongxi's line 'One should sacrifice for loyalty and justice/And die to repay the country', the poet demonstrated not only his passionate patriotism towards the *guo*, but also his devotion to the typical Confucian moral values tied to the state and, thus, in a broader sense, to the 'culturalism' in relation to *tianxia*, which refers to the Chinese literati's conviction of Chinese cultural superiority and universality.

Notes

1 Hong Zan counted about 2,900 war poems written by about 360 poets in *The Complete Collection of Tang Poetry*, the most comprehensive anthology of Tang poems. See Hong Zan, *On War Poetry in the Tang Dynasty* (*Tang dai zhanzheng shi yanjiu*) (Taibei: Wenshi chuban she, 1988), p. 403.

2 The famous Song literary critic and poetry theorist, Yan Yu (*c.*1180 to *c.*1235), commented, for example, 'Good Tang poems often wrote about being enlisted on the frontiers, displacement, embarking on the road, and separation, which stimulate emotion and excitement.' Yan Yu, *Cenaglang's Remarks on Poetry* (*Canglang shihua*), in *Shi hua cong kan* (Taibei: Hong dao wen hua shi ye you xian gong si, 1971) , vol. 1, p. 636.

3 Abbreviation: *QTS*: *Quan Tang shi* (*The Complete Collection of Tang Poetry*) (Beijing, Zhonghua shuju, 1999).

4 Ren Wenjing: *The Cultural Interpretation of the Tang Frontier Poetry* (*Tang dai biansai shi de wenhua chanshi*) (Beijing: Renmin chuban she, 2005), pp. 111–119. The actual number of poets who travelled to the frontier regions was obviously significantly larger than 127, because this excluded those poets who went but did not leave frontier poems behind, or those who wrote poems that were not included in any poem collection or anthology.

5 Hu Zhenheng, *The Compendium of Tang Poetry* (*Tang yin guiqian*) (Shanghai: Shanghai guji chuban she, 1981), vol. 27, p. 285. The literati's frequent visits to the frontier regions and especially them serving in military headquarters and other military institutions in large numbers in the High Tang are an important social and literary phenomenon, which demonstrated the literati's interest in the military and reflected its improved social prestige. There have been some lively discussions and debates on this topic, focusing on what caused this phenomenon, the social mobility of Tang intellectuals, how many poets actually went to frontier regions and served in military institutions, what impact this phenomenon had on literary creation, the military and society as a whole, as well as more concrete issues such as who were those literati who went to frontier regions and served in the military, what positions they held in the military and what duties they performed, what was the military recruiting policy of the literati and the military structure, etc. See, for example, Dai Weihua, *Tang dai shifu yu wenxue yanjiu* (Guilin: Guangxi shifan daxue chuban she, 2007); Dai Weihua, *Tang fangzhen wenzhi liaozuo kao* (Guilin: Guangxi shifan daxue chuban she, 2007); Chen Tieming, 'Lun wenren chusai yu shengtang biansaishi de fanrong', *Wenxue yichan*, no. 3 (2002), pp. 23–38; Ran Xu, 'Lun sheng Tang shire rumu de shikuang he dui biansaishi de yingxiang: cong mingren "llijiu pi waimu" shuo qi', *Journal of Shanghai Teachers University (Social Science)*, vol. 29, no. 4 (November 2000), pp. 90–97.

6 According to Wang Dingbao, poetic composition was introduced into the *jinshi* examination in CE 680. See Wang Dingbao, *Tang zhiyan* (Shanghai: Gudian wenxue chuban she, 1957), p. 9.

7 Hu Yinling, a renowned Ming literary scholar, counted 12 literati who served as chancellors in the Kauyuan era, and saw this period of time as the golden era for poets serving as high-ranking officials. See Hu Yinling, *Thickets of Poetic Criticism* (*Shisou*) (Shanghai: Guji chuban she, 1958), Waibian, vol. 3, p. 175.

8 Hu Zhenheng, *The Compendium of Tang Poetry* (*Tang yin kuiqian*), no. 10, vol. 27.

9 Sima Guang, *Zizhi tongjian* (Shanghai: Zhonghua chuju, 1956) (orig. comp. 1084), vol. 14, p. 6789.

10 Marie Chan and Ts'en Shen discussed such a style change in detail, especially in Cen Shen's poems. See Marie Chan and Ts'en Shen, 'The Frontier Poems of Ts'en Shen', *Journal of the American Oriental Society*, vol. 98, no. 4 (1978), pp. 420–437.

11 I used Stephen Owen's poem translations of Gao Shi and Wang Wei (*QTS*, 126–176), as well as his translations of the terms *qiyun, beizhuang, qigu, xiong yiyu*. See Stephen Owen, *The Great Age of Chinese Poetry: The High Tang* (New Haven: Yale University Press, 1981). For Zhang Bin's and Cen Shen's (*QTS*, 135–106) poems, I used Marie Chan and Ts'en Shen's translations (see Note 10).

12 It is worth noting here that the Tang wars not only changed the overall tone and the literary style of frontier poetry, but also, along with the social upheavals caused by wars, lent a strong thrust to the emergence and flowering of new narrative poems (*xin yuefu*) represented first by poet Du Fu, and continued by Yuan Ji, Gu Kuang, Bai Juyi, Zhang Ji, etc.

13 Kam Louie, *Theorizing Chinese Masculinity: Society and Gender in China* (Cambridge: Cambridge University Press, 2003), p. 83.

14 Although expressions such as 'just swords' (*yijian*) can be found in Tang poems, Tang poets generally did not explicitly use the term 'just war' (*yizhan*).

15 For a discussion of the Chinese term of 'just war', see Don J. Wyatt, 'In Pursuit of the Great Peace: Wang Dan and the Early Song Evasion of the "Just War" Doctrine', in Don J. Wyatt, *Battlefronts Real and Imagined: War, Border, and Identity in the Chinese Middle Period* (New York: Palgrave Macmillan, 2008), pp. 75–109.

16 Mencius, *Mencius* (tr. D.C. Lau) (Hong Kong: Chinese University Press, 1984), vol. 2, p. 287.

17 Confucius, *Analects* (tr. D.C. Lau) (New York: Penguin Books, 1979), p. 139.

18 Dong Gao *et al.*, Qin *ding Quan Tang Wen* (Taiwan: Jinwei shuju, 1965), vol. 7, p. 4168.

19 Mengzi, *Teng weng gong* (Beijing: Zhonghua shuju, 1956), p. 345.

20 When talking about the Hua-Yi distinction in the Tang, it should be noted that the Tang was a cosmopolitan empire and it absorbed significant influences of foreign cultures, including the ones from the nomads or 'barbarians'. However, when it came to frontier wars, such a cosmopolitan spirit was much less visible in frontier poems.

21 Joseph R. Levenson, *Confucian China and its Modern Fate: The Problem of Intellectual Continuity* (Berkeley, CA: University of California Press, 1958), p. 99.

22 Joseph R. Levenson, *Confucian China and its Modern Fate*, pp. 99, 101.

13 Martial values in painting

Chinese bannerman painters at the Qing Court[1]

Nixi Cura

During the Qing (1644–1911), the complementary relationship between *wu* and *wen*, the military and the civil or cultural, was often represented in ethnic terms, with the allied Manchus and Mongols epitomizing *wu* and the Han Chinese subjects representing *wen*. Idealized notions of rulership and personal cultivation exemplified a balance between *wu* and *wen*, as Qing emperors self-consciously aimed to conquer on both military and cultural fronts. The Eight Banners, the Qing military and administrative organization, developed as the proving ground for policies assimilating military qualities necessary for conquest and cultural sensitivity necessary for peacetime rule.

This chapter explores the position of the *Hanjun* bannerman,[2] ethnically Chinese but ostensibly part of the Manchu military system, within the ideological construct of a harmonious multi-ethnic polity under enlightened Qing rule. In what Mark C. Elliott calls an 'ethnic sovereignty', this ideological unity paradoxically relied on the maintenance of bounded identities.[3] I focus here on one particular Chinese bannerman, the poet-painter Zhu Lunhan (1680–1760). Examined contextually, his military rank, poetry and painting reveals the inherent tensions and external pressures that ultimately undermined the hybrid ideal of military prowess and cultural refinement.

Imperially sponsored publications praised and promoted bannerman achievements in the Chinese arts of poetry and painting. So in the late eighteenth-century *Xichao yasong ji* (Anthology of elegant songs in our splendid dynasty), edited by Tiebao, a Manchu literatus par excellence, we read lyrical landscape poems by Manchu generals and romantic verses on hunting by Chinese bannermen. The *Baqi hua lu* (Record of painting in the Eight Banners), written by a Qing loyalist soon after the fall of the dynasty, nostalgically reconstructs a school of painting defined not by style or geography, but rather by participation in this military and cultural hybrid.[4] An entry on Zhu Lunhan appears in both texts; his reputation as a poet and painter, when juxtaposed with his military office, satisfied the criteria for the exemplary Chinese bannerman.

Zhu Lunhan's portrait (Figure 13.1) appears in his poetry collection, *Xianqingtang shiji*, published on or after 1778. It shows him at the height of his career, stroking his beard in self-satisfaction. His lips are pressed closed, but barely suppressing a smile, reinforced by the laugh lines radiating from his eyes.

Figure 13.1 Portrait of Zhu Lunhan (source: *Xianqingtang shiji* in 10 *juan*, latest preface dated 1778, Shanghai Library *guji* #357201-06).

The individual hairs just growing back on his head show without doubt that he is a subject of the Qing, which required all subjects of the realm to shave the front of the head and grow hair long in the back, braided in a queue. Its absence here is not unusual in such standardized frontal portraits, but as Slavoj Žižek points out, 'The truth is out there', that is, the external manifestations of and by the subject, and 'not the depth of the subjects' inner convictions and desires, is the true *locus* of the fantasy which sustains an ideological edifice'.[5] The submersion of the queue, brought to the surface by the unusual prominence of the stubble, exposes the underlying importance of the Zhu Lunhan's Qing subjecthood.

Zhu Lunhan attained his official emoluments through the military examination system. The *Baqi wenjing* (Classic of literature in the Eight Banners), notes that he captured second place in the metropolitan military *jinshi* examination of 1712, after which he served in the Kangxi emperor's Imperial Guard.[6] From the rigour of the military examinations – which involved competitions of mounted archery at full gallop, archery on foot, swordsmanship, demonstration of strength by lifting boulders and, finally, recitation from the military classics – we can fully appreciate the extent of Zhu's military skills. Only one text, the imperially commissioned *Xichao yasong ji* (Anthology of elegant songs in our splendid dynasty), provides anecdotal evidence of his military abilities. I paraphrase:

> His reputed strength with the sword and halberd surpassed that of other men. He was skilled in both left- and right-handed archery and adept at all techniques. At age nine he bravely travelled on the Shandong road and encountered bandits along the way. Without dismounting, he grasped his bayonet and stabbed at them. The bandits, taken by surprise, fled. During peacetime he hunted birds with his long willow bow, but outside of this, there were no further displays of his military prowess. Henceforth he served as a civil official in the Inner Court.[7]

Indeed, after leaving the Imperial Guard, he served as Prefect in Zhejiang province (where he also led successful campaigns against bandits),[8] was subsequently appointed to the Grain Tax and General Surveillance Circuits there, then received a series of promotions under the Censorate. The prefaces and epitaphs attached to the *Xianqingtang shiji* attest to his diligence and forthrightness – to the extent of insulting superiors, including the emperor.[9] He attained his highest rank as vice commander-in-chief of the elite Chinese-martial Yellow Banner in 1735 and remained in that position until his death in 1760 at age 80. Despite having returned to an ostensibly military position, his duties were primarily administrative, with little opportunity to re-exercise his military talents.

Zhu Lunhan's military exploits appear only in imperially commissioned compendia extolling bannermen achievements in poetry and painting. Not a single privately printed work on his poetry or painting brings up his military background or even his bannerman status, starting with his poetry collection. In this compilation, the 1757 preface by Xu Yan (active eighteenth century), from Donghai, Zhu Lunhan's hometown, limits itself to boundless praise of Zhu's poetry, painting and picture-poems, calling forth comparisons to the greatest poet-painters of the Tang (618–907), Song (960–1279) and Yuan (1271–1368) dynasties: Wang Wei (699–759), Mi Fu (1051–1107), Su Shi (1037–1101), Zhao Mengfu (1254–1322), Ni Zan (1301–1374). The 1764 preface by Yao Nai (1731–1815), renowned essayist and editor of the imperial compendium *Siku quanshu* (The complete library of the Four Treasuries), focuses on Zhu's painting and civil career, hardly even mentioning his poetry. Likewise, the last preface, written in 1778 by the dramatist and poet Jiang Shiquan (1721–1799), limits itself to generalized praise of Zhu Lunhan's character and artistic achievements.

The 300 or so poems in Zhu Lunhan's poetry collection accord with established lyric themes: drinking, touring, landscape, friendship, with the occasional painting inscription and funerary record thrown in. Verses with martial imagery are rare. Even the imperial compilation *Xichao yasongji* carries only one poem with military overtones out of 22 selections. In Zhu's poem 'Crossing the Luo River at the Ruins of a Military Camp', dancing trees bear mute witness to the discarded accoutrements of battle – a corner of a drum, an overturned horse trough. The poem expresses more about the ephemerality and futility of warfare, rather than the heady expectation of victory or the exultation of combat.[10]

Zhu Lunhan's poetry does not appear in modern compendia of Qing poetry; nor did the emperors under whom he served – Kangxi (r. 1661–1772),

Yongzheng (r. 1722–1735) and Qianlong (r. 1736–1795) – ask for his poems as they did his paintings. Judging by the prefaces to his poetry collection, Zhu was well respected, even well liked, but the avoidance of direct critical commentary raises doubts about the potency and originality of his poetry. The most effusive praise for his poems comes again from the *Xichao yasongji*, and only in the context of his painting: 'his roots were in the Song poets, and the sentiment in his paintings and the feeling in his poetry mutually interact and surge up, so that each [poem] captivates [the reader] with its inchoate sighs and diffusions'.[11] As one of many representations of Qing multi-ethnicity, this collection of bannerman poetry proves most efficacious in accentuating the literary aspects of its chosen exemplars, both by delimiting poetic range to its overtly Chinese features and by amplifying the quality of the works in question.

Zhu Lunhan's middling reputation as a poet stands in stark contrast to his painting career, which was marked with and shaped by imperial favour, starting from his tenure as Imperial Guard, when his paintings hung in the Forbidden City, Wuyingdian (the Hall of Martial Brilliance) and the Yangxindian, where imperial audiences and conferences with high officials took place. At one point, the Kangxi emperor conferred his calligraphy on a fan painting by Zhu and bestowed it to the ruler of the state of Gaoli, present-day Korea, who liked the painting so much that he pleaded for more, even offering the Chinese court a large amount of money.[12] Zhu received commissions from both Yongzheng and Qianlong emperors, with entries of his paintings in the Imperial Household Department records numbering close to 50.[13] For the pleasure of the court, he submitted brush and finger paintings in silk and paper eventually mounted as fans, scrolls, on walls, on screens and even as lampshades, and meticulous ledgers indicate the specific locations of these works in both Beijing and the hunting retreat in Jehol (Rehe). Some of these were produced in collaboration with other artists at court, such as the Manchu imperial tutor Tangdai (1673–1754) and Zhang Zhao (1691–1745), president of the Board of Punishments. The Qianlong emperor appreciated how Zhu Lunhan 'brightened the paper' to such an extent that he continually commanded his eunuchs to restore and rehang works that had suffered wear. To this day his numerous extant works remain largely uncatalogued and unpublished.

In Zhu Lunhan's portrait, the unusual prominence given his long nails points to his renown in finger-painting, in which long curved fingernails acted as an inkwell. Cheng Jinfang (1718–1784), an eighteenth-century bibliophile, relates the origin myth of his nails and his initiation as a painter:

> Hanzhai's painting technique is conferred by heaven. At the age of four, he smeared a wall with coal, painting images of ghosts and spirits, birds and beasts. Those who witnessed [this] were astonished, not daring to request [more], [but compelled to] watch. One day, [while] grasping a coal cart to obtain more coal, [he] fell to the ground and injured the middle finger of his right hand. After [it] healed this fingernail [became] unusually thick and sharp, with [a] slight indentation, [which] could hold ink and [thus] substitute the finger for a brush.[14]

Finger-painting plays a significant role in the mythology of the exemplary Chinese bannerman. Although artists outside the orbit of Beijing or the Eight Banners, such as Luo Ping (1733–1799) or Gao Fenghan (1683–1749), also painted with their fingers, the majority of practitioners came from the Chinese banners. The first Qing emperor, Shunzhi (r. 1664–1661), credited as the founder of the bannerman painting school in the *Baqi hualu*, is noted by the Qing poet Wang Shizhen (1634–1711) as having painted water buffalo along a riverbank by using the pads of his fingers.[15] The finger-painter utilized the full range of effects created by using different parts of the fingers, fingernails, knuckles, hands and forearms. Speed and surety in execution was of the essence, for any hesitation caused ungainly smears and spotting, owing to the awkward angle of the fingernail and its relative inability to hold and deliver ink. The sharpened tip of the pinky and pointer fingernails supplied finer lines for figures and details of drapery, and fingers, together or singly, applied broad washes and contours.

In painting style and subject matter, Zhu Lunhan is greatly indebted to Gao Qipei (1660–1734), often noted as Zhu's maternal uncle as well as teacher. Zhu humbly corrects this misperception in his colophon to a finger painting by Gao, and suggests that many have confused him with Li Shizhuo (1690–1770), another Chinese bannerman finger-painter of renown.[16] Since Gao and Zhu both circulated among *Hanjun* circles from Shandong and in the capital Beijing, it is likely that Zhu first encountered Gao's works and/or Gao himself as a youth. In their official capacities, these *Hanjun* participated in the ideological apparatus promoting a multi-ethnic state. Tellingly, their bannerman status does not arise in either visual or textual sources identified with the sphere of *wen*, which includes all their paintings, the *Treatise on Finger Painting* by Gao Qipei's grandson Gao Bing or Zhu Lunhan's poetry collection. Their achievements in the sphere of *wen* attains a discursive purity, untainted by their military background.

With Gao as his model, Zhu excelled at finger-painting, and various texts give special notice to his portrayal of tigers and his large-format landscapes. When painting tigers, the abbreviated but powerful slashes constituting the tiger's stripes show how the finger painter overcomes limitations of the flesh in holding and manipulating ink. A work by Zhu dated 1758 (Figure 13.2), formerly in a Japanese collection, presents the blurred edges of each stroke conveying the depth and softness of the tiger's fur. These large muscular finger-painted tigers bear a striking resemblance to contemporaneous Korean *minhwa* folk paintings of tigers, and hints at only one of the many cultural exchanges occurring along those borders in north-east China. Perhaps the Korean king mentioned above received paintings of tigers by Zhu Lunhan.

It is tempting to speculate that the performative bodily gestures involved in directly applying ink without the mediation of a brush explains why Chinese bannermen in the orbit of the Manchu court were the primary proponents of finger painting. But no connection between finger-painting and martial arts, explicit or rhetorical, appears on the paintings or in discussions of finger-painting. Again focusing on the external appearance of an object as a *locus* of an

Figure 13.2 Zhu Lunhan (1680–1760). *Tiger*, dated 1758 (source: *Ming ren shu hua ji* (Collection of Painting Calligraphy and Painting by Renowned [Artists]) (Shanghai: Shang wu yin shu guan, 1920–1922), p. 1757).

Note
Detail of hanging scroll, ink on paper. Dimensions n.a. Current location unknown.

underlying conflict, I propose that in Zhu's case the density and encrustation of finger marks betrays an anxiety about a perceived inferiority of finger-painting to brush-painting. C.C. Wang (Wang Jiqian, 1907–2003), the renowned modern painter and connoisseur, dismisses finger-painting as a trick, a passing fad irrelevant in an art history that fetishizes the traces of the brush as indicative of the artist's individual character.[17] Wang echoes the sentiments of Zhu Lunhan's contemporary, the painter Zou Yigui, who in his treatise on painting dismissed finger-painters as incapable of brush-painting – 'vulgar painters who do not know how to wield their brushes, longing to imitate the great [Gao Qipei], only waste ink and paper'.[18]

In theory, however, discarding the brush allows a more direct expression and cultivation of this character. Furthermore, these excruciatingly detailed finger-painted landscapes may be seen as an attempt to incorporate the upstart technique of finger-painting within the venerated genre of Chinese landscape. In terms of size, composition and execution, Zhu's landscapes resemble those produced at the Qing court, though imbued with an atmospheric quality reminiscent of Ming court painting. The large scale of Zhu Lunhan's *Mountain Landscape with Architecture* (Figure 13.3), wholly painted with the fingers, would satisfy in surfeit a patron's desire for a work by an artist who defied the hegemony of the brush. At the bottom foreground, the style seems inseparable from other

Figure 13.3 Zhu Lunhan: *Mountain landscape with architecture* (source: Klaas Ruitenbeek, *Discarding the Brush*, p. 70 (see Note 15)).

Note

Hanging scroll, finger painting in ink and pale colour on silk, 169.1 × 64.2 cm. Liaoning Provincial Museum, Shenyang.

eighteenth-century precision brush landscapes produced at court; but then the eye floats upwards and gets lost in the muted and misty imprints of leaves and rocks. As the connoisseur Zhang Geng (1685–1760) succinctly remarked: 'Each hill and ravine, though strange, comes from the orthodox (*zheng*).'[19] It seems perfectly plausible that a Chinese bannerman already caught up in multiple self-representations in daily life should come up with a singular metier melding orthodox painting practices at the Manchu court and the so-called eccentric painting by Han Chinese artists in the south. In the wider context as a signifier of *wen*, or Han Chinese-ness, finger-painting tested and went beyond the very limits of the brush. The very physicality of the act of finger-painting incorporated both the martial and the painterly gesture – not merely an empty gesture within the operative Qing discourses of balance between *wen* and *wu* embodied in the Chinese bannerman, but a bold assertion of difference from ordinary Chinese.

The discourse of hybridity in the Qing social structure has a material weight (as seen in texts, imperial proclamations, paintings, etc.), but the maintenance of this hybrid fantasy requires that its constituent parts retain its essential properties. For the Chinese bannerman, this meant – no, *required* – no real participation in a military theatre, but rather the semblance of military capability. Try to imagine a scenario in which these Chinese bannermen actually did follow through with the proffered option of actually leading an army into battle. As Etienne Zi pointed out in his study of military examinations, such attempts led to a farce on the battlefield, primarily because graduates of military examinations had taken that route in order to forward private interests without expectation of combat duty.[20] Zhu Lunhan, along with Chinese bannermen generally, chose to fashion himself as a hybridized symbol in the Qing multi-ethnic polity, just as many other non-banner Chinese chose to reject its advantages. For no doubt, those who played the game enjoyed the benefits of high social status, including the leisure time for poetry and painting. Part of their assigned role in this tableau demanded the downplaying of martial exploits in their literary affairs, thus showcasing their symbolic capital as *wenren*.

Just as Chinese bannermen were portrayed as strongly as possible as symbols of *wen*, so did the stability of the fantasy depend on maintaining the *wu* essence as embodied in the Manchus. With more and more Manchus living among the Chinese populace, this meant that as early as the Yongzheng reign, re-instilling the military skills associated with Manchu-ness became a priority. In the Qianlong reign, a torrent of translation projects were designed to bolster Manchu literacy, and the *Manzhou yuanliu kao* (Researches on Manchu origins) established Manchu lineages propagating the origin myth of Manchu martial superiority.[21] The Qianlong emperor spoke forcibly on this issue:

> If the bannermen concentrate on and are allowed to be successfully examined on the Confucian classics, they will despise the bow and the horse, and that will not enhance our military preparedness; on the contrary they will contravene the very purpose for which the nation established garrisons.[22]

In the case of the Chinese bannermen, renewed military adventurism during the mid-eighteenth century necessitated shifting imperial attention from maintaining a balance between *wen* and *wu* towards re-inculcation of a military mindset among bannermen. This involved increased Manchu set-asides in hiring, the inquisition and persecution of Chinese writers during the *Siku quanshu* project and, finally, the gradual ejection of the majority of *Hanjun* from the Eight Banners between 1742 and 1779.

The efflorescence of a distinctly hybrid Chinese bannerman culture, nourished by the initial optimism of the Qing imperial ascendancy, then banished by the anxiety of Sinicization, lasted only a brief period, roughly from the suppression of loyalist rebellions in the 1680s until the drastic reduction in the numbers of Han-martial bannermen in the mid-1700s – in effect, Zhu Lunhan's lifetime. This historical correspondence allows us to view the commingling of military and civil personae in full articulation, as necessary but essentially disparate practices in the construction of an idealized hybrid Qing subject, one that went beyond the generic essentials of *wu* and *wen* to forge real innovation in painting. This hybrid being, like many a biological or alchemical mutation, proved short lived and ephemeral, once the conditions nurturing its survival changed drastically.[23] Through Zhu Lunhan and other transitory figures, we may still capture the ghostly imprints of a forgotten art and life.

Chinese names and terms

Hanjun 漢軍
Kangxi 康熙
Qianlong 乾隆
Shunzhi 顺治
Wang Shizhen 王士禎
wen 文人
wenren 文人
wu 武
Yongzheng 雍正
Zheng 正

Notes

1 This chapter was presented in two versions, first as 'A Chinese Bannerman's Artful Self-Representation', delivered at the Association for Asian Studies Annual Meeting, 2004, in the panel 'Martial Art: Strategic Deployment of Wen and Wu in Chinese Painting Practice', sponsored by the Chinese Military History Group; and in 2009 under the present title for the conference 'War and Devastation in the Qing and Ottoman Empires', Bogazici University, Istanbul, Turkey.

2 Writers of the 'new Qing history' have debated the English-language translation of *Hanjun*, to differentiate ethnic Chinese inside and outside the Eight Banners. To simplify matters, 'Chinese bannerman' will be used interchangeably with the more literal 'Han-martial' or 'Chinese-martial'. On the 'new Qing history', see Joanna Waley-Cohen, 'The New Qing History', *Radical History Review*, vol. 88 (Winter 2004), pp. 193–206.

3 Mark C. Elliott, *The Manchu Way* (Stanford, CA: Stanford University Press, 2001), p. 4.
4 Li Fang李放, *Baqi hua lu* 八旗畫錄 (Record of painting in the Eight Banners) (Tianjin: preface dated 1919).
5 Slavoj Žižek, *The Plague of Fantasies* (London: Verso, 1997), p. 6.
6 Shengyu 盛昱 (1850–1900) and Yang Zhongxi 楊鍾羲 (1865–1940), *Baqi wenjing* 八旗文經 (Classic of literature in the Eight Banners) (*Zhonghua wenshi congshu*中華文史叢) (Taipei: Huawen shuju, 1969), vol. 90, p. 58/2b.
7 Tiebao 鐵保 (1752–1804), *Xichao yasong ji* 熙朝雅頌集 (Anthology of elegant songs in our splendid dynasty) (Shenyang: Liaoning daxue chubanshe, 1992), pp. 24/1a–7b.
8 Jiang Shiquan preface dated 1778 in Zhu Lunhan 朱倫瀚 (1680–1760), *Xianqingtang shiji* 閒青堂詩集 (Collected poems of Zhu Lunhan) (China: between 1778 and 1795).
9 Yao Nai preface written in 1764 in Zhu Lunhan, *Xianqingtang shiji*.
10 Tiebao, *Xichao yasong ji*, pp. 24/1a–7b.
11 Tiebao, *Xichao yasong ji*, pp. 24/1a–7b.
12 Yu Jianhua 俞劍華 (1895–1979), *Zhongguo meishujia renming cidian* 中國美術家人名辭典 (Dictionary of Chinese artists), 6th edn (Shanghai: Shanghai renmin meishu chubanshe, 1992), p. 211.
13 References in this paragraph to Zhu Lunhan's paintings appear in the *Neiwufu zaobanchu huojidang* 內務府造辦處活計檔 (Household registry files of the Imperial Household workshops), Number One Historical Archives, Beijing, ranging throughout the Yongzheng era (1729) and that of Qianlong (1736, 1741–1745, 1747–1749, 1751, 1758, 1760–1761). Kangxi-period records have yet to be translated into Chinese from the original Manchu.
14 Li Fang, *Baqi hua lu*, p. 8a.
15 Li Fang, *Baqi hua lu*, p. 1, citing Wang Shizhen, *Chibei outan* (Occasional chats north of the pond), published 1691. No such work by the Shunzhi emperor has survived, but fingerprinted buffaloes do appear in a singular work by Gao Qipei (1660–1734), *Three buffaloes on the shore, with willows*, dated 1682 (album leaf 1/12, ink and colour on paper, 21.8 × 23 cm), in the Liaoning Provincial Museum, Shanghai. Klaas Ruitenbeek, *Discarding the Brush: Gao Qipei (1660–1734) and the Art of Finger Painting* (Amsterdam: Rijksmuseum; Gent: Snoeck-Ducaju & Zoon, 1992), p. 2.
16 Klaas Ruitenbeek, *Discarding the Brush*, p. 274.
17 Author's conversation with C.C. Wang on Gao Qipei and finger painting, 18 December 1999.
18 Zou Yigui 鄒一桂 (1686–1772), *Xiaoshan hua pu* 小山畫譜 (Zou Yigui's painting manual), *c.*1753 (*Meishu congshu* 美術叢書 5) (Taipei: Yiwen shuju, 1975), vol. 2/2, p. 137. Wang Yao-ting 'Gao Qipei *Horse and Groom*', Seattle Art Museum, http://chinesepainting.seattleartmuseum.org/OSCI/view/objects/asimages/search@?t:state:flow=695a24c5-c580-43aa-a28d-cf335b3705ba (accessed 28 April 2014).
19 Li Fang, *Baqi hua lu*, p. 8, citing Zhang Geng, *Guochao hua zheng lu* (Painters of the Qing dynasty), published 1739.
20 Étienne Zi, *Pratique des examens militaires en Chine* (The practice of military examinations in China) (*Variétés Sinologiques*) (Shanghai: Imprimerie de la Mission catholique a l'orphelinat de T'ou-sè-wèi, 1896), vol. 9, pp. 141–142.
21 Pamela Kyle Crossley, '*Manzhou yuanliu kao* and the Formalization of the Manchu Heritage', *Journal of Asian Studies*, vol. 46, no. 4 (November 1987), p. 774 ff.
22 Pamela Kyle Crossley, *A Translucent Mirror: History and Identity in Qing Imperial Ideology* (Berkeley and Los Angeles, CA: University of California Press, 1999), p. 268.
23 'Une forme donnée ne peut être indéfiniment soumise à des étirements, elle doit être abandonnée. L'évolution adoptera le véhicule d'une nouvelle forme.' Pierre Mabille, *Égrégores ou la vie des civilisations* (Egregores, or the life of civilizations) (Paris: Jean Flory, 1938), pp. 42–45.

14 Disorder in the general staff

A corruption case during the First Jinchuan War (1747–1749)

Ulrich Theobald

Speaking of corruption one usually thinks of one of the following two scenarios: first, the 'unlawful exaction of money for property through intimidation', originally being a 'complement of bribery, both crimes involving interference with or by public officials'. Second, a 'fraudulent misappropriation of goods of another by a servant, an agent, or another person to whom possession of the goods has been entrusted. [...] Typically, [this] embezzlement occurs when a person gains possession of goods lawfully and subsequently misappropriates them.'[1] More generally, corruption is defined as 'misuse of entrusted power for private gain'.[2]

As for the second definition, such a case happens to have occurred during the first Jinchuan 金川 war (1747–1749), which took a remarkable form: General Zhang Guangsi 張廣泗 did not divert state funds into his own pockets (at least not to such an extent that it was critically noticed), but, instead, he squandered the money allocated for war to blacken his disliked superior, Grand-Minister commander (*jinglüe dachen* 經略大臣) Nocin 訥親, a Manchu. The attacking 'weapon' of Zhang backfired on him and he was executed. This special case of corruption is worth close scrutiny in order to enlighten a new facet in the colourful construction of miraculous possibilities, which had enabled Chinese officials to misuse the power trustfully assigned to them.

The case of Zhang Guangsi was not only in the mind of contemporary scholars and historians, but preoccupied generations of later historians and historiographical novelists, who raised the question of how a commander in a position of such high earnings, with such military success could fall for the worst practices of corruption. A novelist of early Republican times (1912–1949) thus tried to shift the blame from Zhang to his rival Nocin, who had fallen for the temptations of a Tibetan princess: the 'Seductress' (*yaoji* 妖姬) of Jinchuan.

However, both kinds of sources – the historical documents and the short novel – show that the emperor, as representant of the law, would never put up with private intrigues which would under all circumstances be interpreted as corruption or high treason.

I will first narrate the historical facts surviving in documents about the first Jinchuan war and then discuss the matter in how far the case can be termed one of corruption and in how far not. After a short view on the novelist interpretation of the case from the early 1920s, I will come back to the question of whether the case was one of corruption or not.

Zhang Guangsi as a commander

In a well-known article entitled 'Corruption in Eighteenth-Century China' the author writes that the military commander Zhang Guangsi (d. 1749) was charged posthumously with corruption and that an investigation 'substantiated these allegations', with the respective biography of Hummel's *Eminent Chinese of the Ch'ing Period* as a footnote.[3] However, neither Hummel[4] nor his sources or other primary sources like the official chronicle of the first Jinchuan war (*Pingding Jinchuan fanglüe* 平定金川方略, Military annals of the pacification of Jinchuan)[5] or documents in the *Qingshilu* 清實錄 (Veritable records of the Qing)[6] speak of corruption in the sense of misappropriation of state-owned money. The 'posthumous charge' and the 'ensuing investigation' cited in this article cannot be proven by documentary sources.

In order to gain an overview of what really happened in Jinchuan, when the two highest leaders of the campaign started to defame each other, we will look at the official biographies of the three persons involved as compiled by the Historiography Institute (*guoshiguan* 國使館): Zhang Guangsi, Nocin 訥親 (d. 1749) and Yue Zhongqi 岳鍾琪 (1686–1754). Biographies of all three can be found in the collection *Qingshi liezhuan* 清史列傳 (Biographies of Qing history).[7] They are identical to those in the voluminous *Qingguoshi* 清國史 ([Biographically arranged] history of the Qing dynasty).[8] A biography of Yue Zhongqi can be found in the biography collection with the impressive title *Guochao qixian leizheng* 國朝耆獻類徵 (Topically arranged [biographies of] the honoured of Our Dynasty),[9] which does not include the biographies of Zhang Guangsi and Nocin. This is a first hint at how the imperial court's historiographers saw these two characters: in a political sense both were 'non-persons' who did not deserve to be commemorated officially with a biography that was sanctified by the emperor. Yet even in the other two biography collections it becomes evident that the Qianlong 乾隆 emperor (r. 1736–1795) was 'not amused' by the affairs that occurred in the distant mountain area of outer Sichuan.

Zhang Guangsi, so it is written in his biographies, had earned greatest merits during the Dzungar campaign (1732–1735) and the suppression of the Miao rebellions in south-western China (1726–1728).

During the Dzungar campaign Zhang Guangsi became acquainted with Yue Zhongqi when Zhang was made second commander only after Yue. Zhang wrote a memorial in which he harshly criticized his superior for the bad tactics of the latter. Yue Zhongqi was recalled from office, degraded and replaced by Zhang Guangsi. Yue Zhongqi was reprimanded by the emperor for his poor leadership performance and his harshness against his subordinates. The Grand-Ministers decided that he should be decapitated, but he was never executed and later pardoned by the Qianlong emperor.

Zhang Guangsi earned new laurels in numerous campaigns which he led against the rebelling Miao tribes (1735–1736) and other native tribes in his domain as governor-general of Hu-Guang. He was able to bring these campaigns to a successful end within shortest time. For this reason the highly honoured Zhang Guangsi was appointed supreme commander in the war against the Tibetan petty kings of Greater Jinchuan (Da Jinchuan 大金川).[10] The emperor expected that Zhang Guangsi would instantly pacify the Jinchuan rebels.

Jinchuan is located in the western part of present-day Sichuan. At that time it was administered indirectly in a 'loose rein' (*jimi* 羈縻) system by hereditary native kings, the so-called *tusi* 土司, who obtained their seal of office from the emperor. But the native kings of that region permanently staged raids on the villages of their neighbours and thus brought unrest to the region.

The Jinchuan war was an attempt to bring the permanent territorial conflicts of the region to an end. In the eyes of the Qing government the quarrels along the border were a sign of pure disrespect towards the emperor who had officially installed each native king as an autonomous governmental official. This autonomy was granted as long as no territory of China proper was endangered, which would always be the case, when one of the native kings would become too mighty and would disturb the balance of power in the border region. This was exactly what had happened when the king-priest (in Chinese transcription *shaluoben* 沙羅奔) of Greater Jinchuan had occupied Lesser Jinchuan (Xiao Jinchuan 小金川).

The two Jinchuan wars were no singular expeditions fought in the eastern outskirts of Tibet. They were only two cases in a long series of colonial wars which was waged in order to dominate Tibet. In 1700 Qing troops occupied the important place of Dartsedo (Chinese transcription Dajianlu 打箭爐) to safeguard the highway from Sichuan to Tibet. In 1718–1720 Central Tibet was invaded and liberated from the domination of the Ölöd Mongols. At the same time the rulers of China installed their own candidate as Dalai Lama for the first time ever. Between 1721 and 1724 Kokonor and Amdo were pacified and assigned the name Xining 西寧 region (later Qinghai), in 1727 a last rebellion in Lhasa was quelled; in 1745 the native king of 'Zhandui' 瞻對 started a rebellion, in 1750 the Tibetans rebelled against the Qing garrison in Lhasa and in 1752 the king of 'Zagu' 雜谷 rebelled not far from Jinchuan. From 1771 to 1776 the second Jinchuan war ended the independent rule to the native kings. The final incorporation of Tibet into the Chinese orbit came after the expulsion of the Nepalese Gurkha that had invaded Lhasa in 1788. From the year 1793 on all important matters of the administration of Tibet (border defence, tax levies, contacts with foreign countries and appointment of ecclesiastical dignitaries) were within the responsibility of Chinese officials.[11]

Military failure and the 'corruption' case

Zhang Guangsi, sent out to castigate the disobedient Jinchuan 'barbarians' (*fan* 番) in 1748, had made strategic mistakes. His experiences in the Dzungar desert and the subtropical hills of Guizhou did not help him to subdue the *montagnards* in their war-towers (*diao* 碉) that were erected on steep mountain slopes and could neither be circumvented nor brought down by sappers, climbers or by cannonade. The emperor, worried about the long duration of the campaign and the loss of some 40,000 troops, sent Duke Nocin, a Manchu civil official of great merit, to take a look at what happened in the field camps of Jinchuan. At the same time he sent out fresh generals. Among them was the already well-known Yue Zhongqi who at that time occupied the post of provincial military commander (*tidu* 提督) of Sichuan. The emperor had sent him to the Jinchuan area especially because of the experience Yue had made with the Tibetan feuds from an earlier date (1722, 1724) when he had taken part in the invasions in Tibet and Qinghai.

Nocin, who was appointed Grand-Minister Commander only in order to investigate how matters were going,[12] tried to interfere in the command of the armies instead. Zhang Guangsi on the other hand – an experienced military commander – was meanwhile accustomed to the topographical conditions in Jinchuan. He despised Nocin, considering him an imbecile civilian who would try with improper tactics to conquer a territory dominated by guerrilla groups.

Once and again the emperor wrote that he urgently expected news from the advancing front but Zhang Guangsi reported minor successes only. Nocin for his part tried to take over command and to change tactics: he suggested that the Qing troops should also build war-towers, like their enemies, in order to conquer the territory. This proposal enraged the emperor who was surprised about the passive stance of Zhang Guangsi obviously just watching what would happen with Nocin as new commander. Zhang's own strategy was to surround the hostile dens by advancing on 12 different routes, a method that weakened each single battalion. Thus Nocin accused Zhang Guangsi of 'sapping his mens' energy and wasting their supplies' (*lao shi mi xiang* 勞師靡餉),[13] and of 'squandering [state funds] instead of making them useful' (*mi fei bu zi* 糜費不貲).[14] Zhang was also notorious for his unfair treatment of subordinates, especially of troops that had failed to achieve victories, and he would often refuse to send auxiliary troops. Both commanders thus accused each other of inappropriate military methods and ceased to talk to each other. Nocin, totally frustrated by his military failure, left the work to Zhang Guangsi and preferred to stay in his tent. He even let the emperor wait for weeks before answering his letters. Zhang Guangsi, angry about his civilian surveillant, neglected his military obligations with the following reasoning: in the case of a victory he only would be ranked the second triumphant; in the case of continuing defeat everything could be blamed on Nocin.[15]

Yue Zhongqi meanwhile discovered something that substantially altered the situation: he found out about spies among the local scouts and the native allies that Zhang Guangsi made use of. Among the staff there was a Chinese called Wang Qiu 王秋, who seemed to be in contact with the Tibetan enemy and to transmit all military plans to the Tibetan kings. 'Liang'erji' 良爾吉, a Tibetan baron (*tushe* 土舍) from Lesser Jinchuan, who was a son-in-law of the head of the Tibetan rebels, the king-priest of Greater Jinchuan, personally led a contingent of native auxiliary troops supporting the Qing. These not only ran away in front of the war-towers but also handed over weapons and food to their cousins, the rebels.

Yue sent a secret letter to the emperor in which he informed him about Zhang Guangsi's betrayal. The emperor immediately dismissed Zhang and summoned him to Beijing for personal trial on the Yingtai Terrace 瀛臺 in the Forbidden City. The downfall of Zhang Guangsi, however, also required the dismissal of Nocin. Both 'had been involved in outrageous private affairs and had totally neglected their duty to serve the cause of the state' (*zi cheng qi si, wang xu guo shi* 自逞其私，罔恤國事),[16] they were even charged with the offence of high treason (*wu guo zhi zui* 誤國之罪).[17] While Zhang Guangsi was termed 'a mean man of recalcitrant and selfish character' (*gangbi ziyong* 剛愎自用),[18] Nocin was considered a mean man because he was a real 'coward' (*qienuo* 怯懦) or 'a wimp' (*yinrou* 陰柔).[19] This becomes evident when Nocin asked the emperor to be allowed to return to the capital after his rival had been executed, with the pretext to personally present a memorial (*mianzou* 面奏),[20] but without success: the emperor believed that Nocin only tried to leave his uncomfortable post in Jinchuan. He even ironically suggested to him to use Manchu for writing his memorials in case he would think his Chinese not being good enough to express what he really wanted to say.[21]

Zhang Guangsi had blamed his own troops and Nocin for military failure. Likewise Nocin, as a military 'layman' endangered in his authority, totally blamed Zhang for all military disasters: thus the two were 'incriminating each other with ever-changing absurd arguments' (*zhanzhuan guaimiu* 輾轉乖謬).[22]

The emperor saw this case as a betrayal of the benevolence which he had displayed against the two officials, who had already served under his father. In a final edict he stated that Zhang Guangsi had seen only the weak goodwill of his emperor and not the martial side of his sovereign's character.[23] There was an important fact which the two commanders had totally ignored: it was not their war but the Son of Heaven's war. Thus, the Qianlong emperor, who did not personally conduct his wars, like his father and grandfather had done, also desired to know each minuscule detail of what was happening at the front and was very upset that the generals did not report the actual circumstances and the actual reasons for them having so many problems with the Tibetan 'bandits' (*zei* 賊).

Zhang Guangsi was executed on imperial order. Nocin for his part was forced to commit suicide (*cisi* 賜死) with the sword of his own grandfather Obilun 遏必隆 in the presence of the whole army.[24] The downfall of Zhang was a pity, but the ruin of Nocin was a shame.[25]

When using the term 'corruption' as an argument, Zhang Guangsi's fault was him wasting material, money and manpower not against the enemy but with the awareness that he was merely playing a game with Nocin's fate. The troops lost their motivation in a war without reasonable tactics and their lives against an enemy who knew better what was happening than the Qing troops themselves. Possibly the pure hatred against the inexperienced civilian Nocin who was now

Zhang Guangsi's superior instigated him to intrigues that even risked the sacrifice of the lives of his own troops, the sole objective being to discredit Nocin.

In the field of military performance it must be mentioned that when the experienced military Zhang Guangsi adopted the old and time-proven principle of 'fighting the barbarians with barbarians' (*yi fan gong fan* 以番攻番), he was far more the victim of the enemy's spies than the official historigraphy admits. It was well known to everyone that all the Tibetan kings were interrelated by marriage, a bond which made it very easy to suddenly change sides because of 'old acquaintance' and outstanding liabilities or simply because some of the kings were afraid that Greater Jinchuan would take revenge later on for them siding with the Qing army. Zhang, after spending a year in the mountains of Jinchuan, could – or rather should – have known that. It was a very lackadaisical care for the safety of the troops and the whole enterprise as a such, that virtually everyone, from money changers, sutlers, rice porters, storytellers, barbers, monks, etc., had access to the camps.

Zhang Guangsi had cooperated with Liang'erji, a native baron, and 'Akou' 阿扣, a niece (daughter?) of the main scoundrel, the king-priest of Greater Jinchuan. Especially this woman or girl was, according to historiography, a really vile woman described with a lot of derogatory terms. She was a member of a 'clique of bandits' (*zeidang* 賊黨), a 'wing of the faction [of the rebels]' (*dangyu* 黨羽), the 'clique of the relatives [of the Shaloben]' (*yindang* 姻黨) or the 'traitor-woman' (*waijian* 外奸) out of a 'traitorous clique' (*jiandang* 奸黨) or 'bad gang' (*edang* 惡黨) that served as 'the eyes and ears of the bandit [leader]' (*wei zei er mu* 為賊耳目).[26] Yet the actual role of Akou within the 'gang' is not described in the historical documents.

Whatever the circumstances might have been, the case of Zhang Guangsi is clearly not a case of corruption in the commonly accepted definition (entrusted money being misappropriated) but one of private intrigues (in order not to say 'squabble', *juyu* 齟齬),[27] which influenced the progress of the war negatively ('forfeiting all chances for successful combat', *yiwu junji* 貽誤軍機)[28] and thereby – of course – also squandered state finances.

With Fuheng 傅恆 (d. 1770) as the new commander, assisted by Yue Zhongqi (who was finally avenged for his elimination by Zhang Guangsi in Dzungaria), the war soon came to an end, and therefore it comes as no surprise that all the failures of the past could be attributed to Zhang Guangsi and Nocin, with the pretence that both had been 'corrupt'.

If Zhang's and Nocin's behaviour was no corruption in the usual sense, why then were both doomed to die? If money were the issue, it would probably have been a better solution to disrate the two generals and to claim back the squandered money from their private purses by deducting it from their salaries. This was a common method to penalize officials who had misappropriated state finances. Private assets were also confiscated in cases where a state official or a private person was executed. In such a case, the government would be able to compensate any losses made by embezzlement. Although the word 'corruption' is not used in the Zhang/Nocin case, it might well have been that both also had embezzled funds and lined their own pockets.[29] Two circumstances support this conjecture: first, it was found out that a retainer (*jiaren* 家人) of Zhang Guangsi with the name Xue Er 薛二 was involved in a case of extortion. Together with Wang Qiu he had seized valuables from the belongings of King 'Zewang' 澤旺 of Lesser Jinchuan.[30] Wang Qiu was apparently a very intimate collegue of the household of Zhang and intrigued against anyone involved in the war, seeking his profit from both sides, betraying the Qing commanders and blackmailing the native lords. Zhang Guangsi had known all these things and protected Wang Qiu in his foraging activities. For his subordinates, Zhang seemed to have been a bad example to be imitated. Reversing the idom 'like master, like servant', it might be concluded that Zhang, too, misappropriated and embezzled the funds of the state. Effectively, Zhang is called 'obstinate, covetous, and disrespectful' (*guaizhang*, *tanzong*, *bu su* 乖張貪縱不肅).[31] Supposing Zhang had in fact embezzled state funds or extorted the pay of his subordinates, even this would have been a peccadillo in an environment where corruption was ubiquitous and only casually punished.

Second, the first Jinchuan campaign cost 7.6 million silver *liang*.[32] It is unknown how much money was spent for particular items since no final account was drawn up, either because everything was all right and no expenditure to be dicussed in a kind of auditing, or because the whole campaign was a military and financial disaster so that the government wanted to forget about it as soon as possible, and to sweep bad thoughts of it under the carpet.

There is a whole set of reasons why the emperor decided to have Zhang and Nocin executed. Instead of fulfilling their duties in office, they fought against each other, mixing private interests with official duties. The second reason was the espionage problem. Yet even this matter did not outweigh the two reasons, why the emperor repeatedly reprimanded Zhang and especially Nocin: instead of developing better tactics, they waited for something to happen at the front and, even worse, they declined to answer the imperial letters of command. They refused to obey the imperial orders and even disregarded the person of the emperor. This – and not corruption – was the real issue. In an edict, the emperor even puts Nocin's faults in opposition to bribery: 'It would have been a mere petty matter if he had taken bribes for his own profits' (*ji ling na hui xing si, ci yi xiao guo* 即令納賄行私，此亦小過).[33]

A novelist reinterpretation of the case

The successes which were achieved by Zhang before him being appointed highest General of the Jinchuan campaign, make the sudden failure of such a highly decorated general hard to understand. Even the harshness of the conditions under which the Jinchuan army was fighting did not outweigh this disproportion when looking for the origins of his failure. In the eyes of some naive people a hero like Zhang must have been the victim of intrigues. Thus an attempt to present him as not guilty and to blame his downfall on someone else, was made in a novel in the early Republican Era whose arguments shall now be examined.

A rather unknown early Republican writer with the name of Xu Zhiyan 許指嚴 (1875–1923/25)[34] has written an unofficial 'wild' history (*yeshi* 野史) of the first Jinchuan war. He explains that the driving force behind all the intrigues and misadventures in Jinchuan was nobody else than a woman that seduced all leading characters of the story, namely Akou. His unofficial history is therefore called *The Seductress of Jinchuan* (*Jinchuan yaoji zhi* 金川妖姬志).

While Xu's longer novels might have attracted a wider audience – visible in the large number of republications[35] – it is not very probable that the *Seductress* attracted a wide readership, although title and content might have had some marketing appeals. Xu Zhiyan was personally interested in the genre of anecdotic historical novels (*zhanggu* 掌故), but his own economic situation also forced him to write what the public wanted – and that did not always correspond to historical documents.

In the short novel, Xu thus has altered one of the main reasons why Zhang and Nocin were executed: the permanent problem of the espionage, which must have been serious, and which would eventually lead to the great disaster of the Qing army during the second Jinchuan war, is played down. The novelistic Zhang understands the situation in Jinchuan and knows how to solve the problems among the native kings, yet he is caught by the seductress and cannot get rid of her, although he pretends to deal with her and the enemy in a strategic manner. The intriguer Nocin on the other hand 'lost his head over her inside his tent' (*hun mi yu zhang xia* 昏迷於帳下).[36]

Zhang is furthermore described as a victim of Nocin's and Yue's jealousy, while the actual delicts leading to his execution are not even mentioned: in the short novel, it is Nocin refusing to send relief troops, and it is Nocin who does not respond to the emperor's commands. Xu even involved Yue Zhongqi in the intrigue, who in reality had nothing to do with the squabble between the two men.

One can even detect a clearly political aspect in the novel – too clear to be overlooked. Zhang is being portrayed as a hero who is waging an effective war. He is overthrown for the only reason that his incapable superior, the Manchu Nocin, has fallen for the charms of the seductress Akou and for the fact that his envious rival Yue Zhongqi misinterprets Zhang's promising strategies as collaboration with the enemy. Even the emperor – a Manchu, too – is not spared and is accused of rather trusting intriguers than those telling the truth.

A further aspect of great interest is that Zhang, like a hero in an ancient Greek tragedy, is depicted as a character suffering the fate of destruction not because of his own wrongdoings but because of a tragic destiny that cannot be escaped. He is not guilty whatever he has done wrong: Zhang can be exculpated because he became a victim of Nocin, Yue and Akou.

Was there corruption?

To come back to the question about possible corruption and the kind of corruption used, we have to conclude that Zhang and Nocin, in both the historical documents and in the short novel, fought a battle for power costing the state much money and many people's lives. This is where the accusation of corruption sets in (squandering state finances). Instead of fighting the enemy, they engaged in intrigues against each other, and thus protracted the war. Yet the problem with the term corruption is that they did not funnel larger amounts of state-owned money into their own purses, nor did they use the money to directly achieve a certain purpose (say, the forgery of palace memorials reporting victory instead of 'all quiet on the front'). And, third, there is no precise sum of money which was misappropriated and could thus be repaid by the culprits. In a real case of corruption, when money 'disappears', a censor would have been sent out to check the ledgers and to interview the persons involved.

Instead, three other reasons lead to the decision to have Zhang and Nocin executed, namely that, first, a very intensive contact with the enemy led to the conjecture that military secrets were carelessly transmitted; second, a thorough misconception of how to wage war against the Jinchuan rebels, the two generals not being able to develop proper tactics, which is nothing else than incompetence;[37] and, finally, disobedience, in coincidence with disrespect for the Son of Heaven. Behind the crime of high treason, the charge of corruption – which can not even be substantiated – is diminishing. When using the word 'corruption' in this context, it has the meaning of mixing private matters with public affairs. Simple 'corruption' in dealing with money might be pardoned, but the emperor could not excuse such careless, disobedient and treacherous officials and therefore had to deal with them in the same manner as he did the Jinchuan rebels, even if Zhang and Nocin in the past had achieved greatest merits.

Nancy Park has put out two very important things about corruption in late imperial China, namely, that it was 'a continuing dilemma of how to conceptualize, define and compare the concept of corruption', and that a charge of corruption (in any shape) was an 'all-purpose tool' to eliminate political rivals.[38] Both points could also be observed in the Zhang/Nocin case. Although neither Zhang nor Nocin were officially charged with corruption, their behaviour had an effect that brought it in the very vicinity of embezzlement, of collaboration, of high treason, in one word, of 'moral corruption' in the widest sense. This was all the more possible because the Qing code (*Da-Qing lüli* 大清律例)[39] does not include a proper definition for 'misuse' of state funds, like the squandering of funds earmarked for the suppression of the Jinchuan enemies, and not even for the private use of state funds (embezzlement or malversation). Apart from 'their downfall out of their own responsibility' (*qi zi xian zuili* 其自陷罪戾),[40] it seemed that Yue Zhongqi's memorial initiated the ruin of Zhang and Nocin, with the result that Yue was, according to one source, temporalily entrusted with the post of the Grand-Minister commander (*she jinglüe* 攝經略), as Nocin's official biography holds.[41] The downfall of Zhang and Nocin lead to the rehabilitation of Yue. After the war he was first granted the title of Junior Guardian of the Heir Apparent (*taizi shaobao* 太子少保) and then the brevet title of Minister of War (*bingbu shangshu* 兵部尚書) and was enfeoffed as a duke of the third rank (*sandeng gong* 三等公). He was furthermore allowed to ride a horse inside the Forbidden City. His debts of 70,000 *liang* that he owed to the government from the Dzungar campaign (because he had been charged with embezzlement of state funds during that time and was thus, unlike his collegues, really charged with corruption) were dispensed with, his sons were granted the ranks of Junior Guardsmen (*lanling shiwei* 藍翎侍衛) and, as a brilliant capstone of this pyramid of honours, the emperor wrote a poem lauding Yue's excellent military performance.[42] Although Yue Zhongqi highly profited of Zhang's dismission and execution he had certainly not machinated against his superior with the intention to eliminate him as a political rival. Yet Yue was intelligent enough to perceive the chance offered to him by the enourmous defects that Zhang's negligent style of administration showed, and so was at the same time able to take revenge and to emerge from the ashes with laurels around his head.

The case Zhang/Nocin is characterized by a lot of criteria that bring it into the vicinity of corruption. Yet the charge, for which Zhang and Nocin were executed, was military failure (*yi yi shiwu junji lü ying zhanjue* 議以失誤軍機律應斬決 'it was deliberated that [Zhang] was for his malfeasance be punished with execution according to military law'),[43] and not corruption, at least not in the narrower sense, but probably in a wider sense of 'misuse of entrusted power for private gain.'

Notes

1 *The New Encyclopaedia Britannica, Micropedia*, 15th edn (Chicago etc.: Encyclodaedia Britannica, Inc., 1991), vol. 4, pp. 469, 637.

2 See http://transparency.org/news_room/faq/corruption_faq#faqcorr1 (accessed 1 November 2011).

3 Nancy E. Park, 'Corruption in Eighteenth-Century China', *Journal of Asian Studies*, vol. 56, no. 4 (1997), pp. 967–1005, see p. 997. The involvement of Nocin into the case is not mentioned by the author. Vice versa, Mark C. Elliott, *Emperor Qianlong: Son of Heaven, Man of the World* (New York: Longman, 2009), pp. 62–63, deals a lot with the Qianlong emperor's relation to Nocin and does not mention the name Zhang Guangsi.

4 Arthur W. Hummel, *Eminent Chinese of the Ch'ing Period* (Washington, DC: United States Government Printing Office, 1943), pp. 43–45 (Zhang Guangsi), pp. 957–959 (Yue Zhongqi).

5 *Pingding Jinchuan fanglüe* 平定金川方略, comp. Laibao 來保 *et al.*, *Siku quanshu* 四庫全書 edition.

6 *Qingshilu* 清實錄, *Da-Qing Gaozong Chun huangdi (Qianlong) shilu* 大清高宗純皇帝（乾隆）實錄 (short: *Gaozong shilu* 高宗實錄) (Taibei: Taiwan huawen chubanshe, 1964).

7 *Qingshi liezhuan* 清史列傳 (Beijing: Zhonghua shuju, 1987), vol. 5, ch. 17, pp. 1263–1270 (for Zhang), pp. 1249–1261 (for Yue), and vol. 6, ch. 22, pp. 1632–1648 (for Nocin), respectively.

8 *Qingguoshi* 清國史 (Beijing: Zhonghua shuju, 1993), part *Dachen zhengbian* 大臣正編, ch. 122 and 146 (vol. 6, pp. 512–519, 520–524 and 902–912).

9 *Guochao qixian leizheng* 國朝耆獻類証 (Yangzhou: Jiangsu guangling guji keyinshe, 1990), vol. 13, ch. 280, *Jiangshuai* 將帥 20, fol. 1-a-4a (pp. 333–335).

10 The inhabitants of the Jinchuan area were called rGyalrong in their own language. Most scholars identify their language as being an archaic form of Tibetan. Today, the

Chinese government defines them as a part of the Qiang ethnicity, distant relatives to the proper Tibetans. Compare Guillaume Jacques, 'Phonologie et morphologie du Japhug (rGyalrong)', doctoral thesis, Paris, 2004. An excellent introduction into the background of the war can be found in Roger Greatrex, 'A Brief Introduction to the First Jinchuan War (1747–1749)', in Per Kværne (ed.), *Tibetan Studies: Proceedings of the 6th Seminar of the International Association for Tibetan Studies, Fagernes 1992* (Oslo: Institute for Comparative Research in Human Culture, 1994) , pp. 247–263.

11 W.W. Rockhill, 'The Dalai Lamas of Lhasa and their Relations with the Manchu Emperors of China, 1644–1908', *T'oung Pao*, vol. XI (1910), pp. 1–104, see pp. 27–51.

12 *Qingshi liezhuan*, ch. 22, p. 1636.

13 *Gaozong shilu*, ch. 330, fol. 2a (Qianlong [QL] 13/12/xinsi=1). The word *lao* 勞 'to make suffering' is sometimes also written 老 'to distress', and the expression also appears as *mixiang laoshi*, like in *Pingding Jinchuan fanglüe*, ch. 12, fol. 22a (QL 13/9/jiwei=8).

14 *Pingding Jinchuan fanglüe*, ch. 19, fol. 4b (QL 13/12/renchen=12).

15 *Qingshi liezhuan*, ch. 17, p. 1269.

16 *Qingshi liezhuan*, ch. 17, p. 1269.

17 *Qingshi liezhuan*, ch. 22, p. 1648.

18 *Qingshi liezhuan*, ch. 17, p. 1269.

19 *Qingshi liezhuan*, ch. 22, pp. 1643, 1647.

20 The emperor ironically suggested to him to use Manchu for writing his memorials in case he would think his Chinese was not good enough to express what he really wanted to say. *Qingshi liezhuan*, ch. 22, p. 1642.

21 *Qingshi liezhuan*, ch. 22, p. 1642.

22 *Gaozong shilu*, ch. 330, fol. 27a (QL 13/12/xinmao=11).

23 *Qingshi liezhuan*, ch. 17, p. 1270.

24 So goes the official statement in Wei Yuan 魏源, *Shengwuji* 聖武記, *Sibu beiyao* 四部備要 edition, ch. 7: *Qianlong chu ding Jinchuan tusi ji* 乾隆初定金川土司, fol. 15a–20a. However, not all sources speak of suicide. Some use the term *zhengfa* 正法 instead ('law was enforced', like *Pingding Jinchuan fanglüe*, ch. 20, fol. 7a [QL 13/12/guimao=23], or *Gaozong shilu*, ch. 331, fol. 36b [QL13/12/xinchou=21]), or *zhan* 斬 ('he was beheaded', like in Jaoliyan 昭槤, *Xiaoting zalu* 嘯亭雜錄 (Beijing: Zhonghua shuju, 1980), ch. 1, *Sha Nocin* 殺訥親, p. 14). The most detailed description of the execution is to be found in Nocin's biography in *Qingshi liezhuan*, ch. 22, p. 1648, where it is said that he was in the end spared the humiliating suicide in front of the whole army and was in the first month of QL 14 (i.e. February/March 1749) executed (*fu zhu* 伏誅) in the camp at Mt Banlan 斑爛山 in Jinchuan.

25 During the first years of the Qianlong reign Nocin was a highly favoured official in the central government and held several important posts concurrently. Some officials therefore criticized him for accumulation of offices and titles. *Qingshi liezhuan*, ch. 22, p. 1633.

26 *Pingding Jinchuan fanglüe*, ch. 11, fol. 16a (QL 13/8/yichou=7); ch. 13, fol. 13b (QL 13/9/gengwu=19), fol. 19b (QL 13/9/dingchou=26), fol. 30a (QL 13/9/gengchen=29); ch. 15, fol. 30b (QL 13/11/wuwu=8); ch. 21, fol. 15a (QL 14/1/jiayin=5); ch. 28, fol. 3a, *Pingding Jinchuan ya* 平定金川雅 (a poem by Minister of War Liang Shizheng 梁詩正); ch. 31, fol. 3a, *Pingding Jinchuan ge* 平定金川歌 (a poem by Minister of personnel Chen Dashou 陳大受).

27 Xu Guoying 許國英 [i.e. Xu Zhiyan 許指嚴], *Jinchuan yaoji zhi* 金川妖姬志, originally published in *Man-Qing yeshi sanbian* 滿清野史三編, no. 19 (Chengdu: Changfu, 1925), republished in *Zhongguo yeshi jicheng* 中國野史集成 (Chengdu: Ba-Shu shushe, 1993), vol. 40, pp. 106–111; see fol. 1a.

28 *Qingshi liezhuan*, ch. 17, p. 1269.

29 Dai Yingcong has written several important contributions to explain how much the military lived on extra funds paid out during war, see Dai, *The Sichuan Frontier and Tibet: Imperial Strategy in the Early Qing* (Seattle, WA: University of Washington Press, 2009), esp. ch. 6; Dai, 'Military Finance in the High Qing Period: An Overview', in Nicola di Cosmo (ed.), *Military Culture in Imperial China* (Cambridge, MA: Harvard University Press, 2009), pp. 269–316, 380–382; or Dai, 'To Nourish a Strong Military: Kangxi's Preferential Treatment of His Military Officials', *War and Society*, vol. 18, no. 2 (2000), pp. 71–91.

30 *Gaozong shilu*, ch. 328, fol. 28a (QL 13/11/wuwu=8); ch. 334, fol. 4a–4b (QL 14/2/xinsi=3).

31 *Pingding Jinchuan fanglüe*, ch. 25, fol. 18b (QL 14/3/wuwu=10).

32 *Pingding Liang Jinchuan junxu li'an* 平定兩金川軍需例案 (ed. Zheng Qishan 鄭棲山), republished in the series *Xizangxue Hanwen wenxian huike* 西藏學漢文文獻會刻 2 (Beijing: Quanguo tushuguan wenxian suowei fuzhi zhongxin, 1989), ch. 1, fol. 1b. A silver *liang* is an ingot with a standard weight of *c.*37.3 g, cf. Luo Zhufeng 羅竹風 (ed), *Hanyu da cidian* 漢語大辭典 (Xianggang: Sanlian shudian, 1995), vol. *Fulu suoyin* 附錄索引, p. 19.

33 *Pingding Jinchuan fanglüe*, ch. 14, fol. 8b (QL 13/10/guiwei=2).

34 For a short biography of Xu see e.g. Ma Liangchun 馬良春 and Li Futian 李福田 (eds), *Zhongguo wenxue da cidian* 中國文學大辭典 (Tianjin: Tianjin daxue chubanshe, 1991), vol. 4, p. 2361.

35 For instance, *Nanxun miji* 南巡秘記 (Shanghai: Guohua shuju, 1915) (republished Shanghai: Shanghai shudian, 1997 and Taiyuan: Shanxi guji chubanshe/Shanxi jiaoyu chubanshe, 1999); or *Shiye yewen* 十葉野聞 (Shanghai: Guohua shuju, 1917) (republished Taibei: Wenhai chubanshe, 1972 ; Zhengzhou: Henan daxue chubanshe, 1991; Shenyang: Liao-Shen shushe, 1994; Taiyuan: Shanxi guji chubanshe, 1995 and 1992; Chengdu: Ba-Shu shushe, 2000); or *Xinhua miji* 新華秘記 (Shanghai: Qinghua shuju, 1918) (republished Taiyuan: Shanxi guji chubanshe/Shanxi jiaoyu chubanshe, 1999).

36 *Jinchuan yaoji zhi*, fol. 10b.

37 Nocin is later often quoted not because of his insolence towards the emperor, but because of his strange tactics of waging war against the war-towers by building war-towers. Governor-general Artai 阿爾泰 was so likewise charged with military inactivity. *Pingding Liang Jinchuan fanglüe* 平定兩金川方略, *Siku quanshu* 四庫全書, ch. 8, fol. 8b (QL 36/10/xinsi=14); ch. 18, fol. 16a (QL 37/2/dingmao=2). For Artai's execution, see ch. 48, fol. 9a (38/1/jiachen=14). In the same book, Zhang Guangsi is not mentioned at all.

38 Nancy E. Park, 'Corruption in Eighteenth-Century China', pp. 968, 998.

39 *Da-Qing lüli* 大清律例 (ed. Gugong bowu yuan 故宮博物院), published in the series *Gugong zhenben congkan* 故宮珍本叢刊 (Haikou: Hainan chubanshe, 2000), vols 331–332.

40 *Pingding Liang Jinchuan fanglüe*, ch. 28, fol. 11b (QL/5/renyin=2).

41 *Qingshigao* 清史稿 (Beijing: Zhonghua shuju, 1979), ch. 301, p. 10444. No such statement can be found in Yue's own biography, or in other sources. In fact, the seal of the Grand-Minister Commander was handed over to the imperial guardsman (*shiwei* 侍衛) Oši 鄂實 and not to Yue Zhongqi, see *Pingding Jingchuan fanglüe*, ch. 13, fol. 32a (QL 13/9/gengchen=29).

42 *Qingshi liezhuan*, ch. 17, p. 1258.

43 *Qingshi liezhuan*, ch. 17, p. 1270.

15 Fortified walls and social ordering in China's late eighteenth-century revolts

Daniel McMahon

Significant changes in Qing China's administrative planning accompanied the great revolts of the early Jiaqing 嘉慶 (r. 1796–1820) period: the 1795–1797 Miao uprising in west Hunan and 1796–1804 'White Lotus' rebellion in China's central highlands. Among the most important responses of the day was a strategy of military pacification that combined roving army campaigns with new 'local defence' community mobilization. This shift towards greater civilian involvement, epitomized by state-supervised militia and fortified walls, is understood to have been instrumental in restoring imperial order.[1]

Historians have noted the military role of local defence, particularly *tuanlian* 團練 militia. As recent research on Chinese 'military culture' suggests, however, this topic might be further examined from cultural and social perspectives. The scholar Peter Perdue offers an intriguing starting point. As Perdue argues, examination of China's martial past requires an effort

> to understand the relationship between war, society, and thought beyond the empirical level and to recognize the ways in which intellectual, civilian, and literary developments intervened to shape the nature of military institutions, military theory, and the culture of war.[2]

Such a perspective, considering the linkages shaping military and social thought, resonates with local defence work at the turn of the nineteenth century. Consider that its architects and implementers were largely *not* soldiers, and that its plans consistently intertwined both defence and reconstruction initiatives. One might ask what relationship civilian mobilization, particularly fortified walls (*bao* 堡), had to Qing administrative planning. Were walls simply an expedient to shield people and supplies? To what extent did they serve a larger purpose?

This chapter argues that fortified walls functioned as a vital tool for not just defence, but also social ordering. *Bao*, that is, served concurrently to isolate disruptive rebels and regiment imperial subjects, working much like an instrument for disease control. In conception and implementation, that is, the essential focus of these structures was *quarantine* – to shut the 'ill' out (or in) while shielding, constraining and strengthening the 'healthy'. Walls provided the physical barrier of separation, setting and shielding spaces within which corrective imperial ordering would take place. In this way, they fit seamlessly with other state reformist measures that extended beyond defence to designs of imperial settlement, control and inculcation.

Border revolt and local defence initiative

Late eighteenth-century insurgency erupted along the inner peripheries of the Qing empire, if in different social milieu. The Miao revolt, raging on the mountainous Hunan–Guizhou provincial border, was an ethnic conflict in which native Miao clans united to reclaim lands lost to intrusive Han Chinese settlers.[3] In contrast, the White Lotus revolt – spread along the highland border linking Hubei, Shaanxi and Sichuan provinces – was a conflict not of natives, but of scattered, rootless and predatory Han Chinese. In scope and numbers, it was several times larger than the Miao uprising.[4]

In both conflicts, the Qing armies encountered difficulty when utilizing established military methods. In west Hunan, campaigns to apprehend insurgents and solicit surrender shattered the native alliance, but left lingering insurgency. On the 'three-province border', initial efforts failed to yield even this scant success. Rebels remained elusive and the rapaciousness of local soldiers and officials exacerbated, rather than curtailed, the conflict. Rural communities were vulnerable, homes were destroyed, families were uprooted, fields went to seed, people seethed in bitterness and (later reports lamented) imperial 'subjects were turned into bandits'.[5]

As disorder reigned on these borderlands, local people turned to their own defence. This began as early as 1796 as communities were organized by gentry, in some cases with the aid of concerned field officials such as Fu Nai 傅鼐 (d. 1812), Fang Ji 方積 (1765–1815) and Gong Jinghan 龔景翰 (1747–1802). Militia formed, walls were girded around towns and forts arose along roads and strategic passes – mobilization funded by merchant donations, led by local elite and manned by local stalwarts. The justification advanced for this unsanctioned action was that residents would protect their homes more fiercely than soldiers, communities could provide necessary mutual aid, rural fortification would slow roving rebels and there would be chaos without it.[6]

Reports of these grassroots efforts were communicated to superiors in the Qing bureaucracy, but were coolly received. In 1797, pacification general Delengtai 德楞泰 (1745–1809) acknowledged a need to 'succor the refugees and impede the rebels'. He recommended that locals 'be encouraged to repair and build earthen fortifications [*tubao* 土堡]', as had already begun. In this way, subjects would be systematically sheltered under state supervision and 'to kill one rebel is to have one rebel less'.[7] The notion of arming civilians, however, was anathema to the elder (retired, but ruling) Emperor Qianlong 乾隆 (r. 1736–1795), echoed by many others in government who feared the border peoples' potential to aid the enemy.[8]

Change came in 1799, with Qianlong's death and the consolidation of power under his successor, Emperor Jiaqing. In the context of the 'Jiaqing Reforms' –

Confucian promotion of enhanced fiscal responsibility, public morality and governmental integrity – pacification policy also shifted. Speaking of the White Lotus conflict, Jiaqing proclaimed that 'officials have forced subjects to rebel' and 'in all places good subjects have been pillaged, their villages burned … they roil in confusion and cannot but follow the rebels'.[9] Given such circumstances, 'good subjects' – now indicating compliant regional populations – would be permitted to participate in imperial defence. Indeed, such participation was imperative if they were to avoid 'confusion', coercion and collusion with state foes. Accordingly, Jiaqing praised a case in which locals had 'dredged deep moats and erected ramparts of earth' to defy rebels and deprived them of resources. 'This is a good strategy', the emperor proclaimed, and he commanded that 'village fortification' (*cun bao* 村堡) be extended to Hubei, Sichuan and Shaanxi.[10]

Local defence and imperial objectives

Civil fortification of China's war-torn borderlands had a clear military purpose. It sheltered subjects from attack, protecting homes and families while mitigating a crippling terror that left highlanders scattered and prone to flight. As important, new walls deprived the enemy of resources such that insurgents had 'nothing to which to return and nothing they can loot in the wilds'.[11] 'If the subjects have one more day of grain', one strategist wrote, 'the bandits have one day less of food.'[12] The larger regional designs for militia and fortified walls, however, differed. In Miao areas, the objective was an interconnected line of fortification (forts, blockhouses and sentry posts) that would divide settlers from natives. This partition was to be organized and financed using 'military farms' (*tuntian* 屯田), in which men were granted land to serve as both farmers and soldiers.[13] In the White Lotus area, where an extended dividing wall was not feasible, the goal was concentration of borderland populations in secure locations: a strategic hamlet policy of 'strengthening the walls and clearing the countryside' (*jianbi qingye* 堅壁清野).

What should be observed in Jiaqing's support of these local defence initiatives, however, was not just their military necessity, but also the manner that they spoke to Confucian social concerns. Fortification was the backbone of resettlement and all that it implied. 'Subjects desire to return home', the emperor wrote, 'but have nothing to which to return. If they return, they have no food.'[14] With walls and militias, refugees would now have security to rebuild, restore livelihoods, start families, learn proper teachings and be thus vindicated in their contribution to the Qing cause. This path – based on an established template of dynastic reconstruction planning – suggested restoration not just of essential order, but of imperial civilization. Enabling this process was an array of interlocking and mutually supporting plans that accompanied fortification: *baojia* household surveillance, village policing, granaries, lectures on Confucian ethics and promotion of agriculture and trade. Defence and rebuilding measures were thereby advanced in tandem in order to 'transform good subjects' such that 'far and near would be one body and above and below would be one heart'.[15]

A common metaphor applied to new defensive strategizing was one of Chinese medicine, in which society was depicted as a body, dissent as sickness, and compliance as health. The 'rebels are like a disease', literatus Zhou Yangpu 周錫溥 lamented during the White Lotus revolt.[16] 'Defending against the Miao is like defending against an illness', the scholar Wei Yuan 魏源 (1795–1826) said of the Miao uprising.[17] Such portrayal of rebellion as affliction mandated accommodating forms of curative treatment. As the scholar Yan Ruyi 嚴如熤 (1759–1826) advised the Hunan governor, 'since ancient times, when healing an illness, one looks first at its root'. He then recommended two strategic 'cures' indicative of prevailing Jiaqing reform views on pacification: isolating illness such that 'it will not manifest any symptoms' and nourishing the body such that its health overcomes sickness.[18] Specific prescriptions to effect this restoration were then required 'modified based on [the doctor's] own insight into the sickness'.[19]

As such medical metaphors suggest, new local defence was never regarded as just a matter of plans. When Emperor Jiaqing advanced reforms, he concurrently patronized Men of Ability: his 'doctors'. Promoted to border posts and mandated to speak out were able field officials such as Liu Qing 劉清 (1742–1827), Gong Jinghan, Fang Ji, Yan Ruyi and Fu Nai.[20] These were not soldiers, but officials versed in both civil (*wen* 文) and martial (*wu* 武) traditions, whose early service had led to successful innovation of community fortification concepts. Wary of the Qing armies, if coordinating with them, these administrators elaborated on the methods they believed worked best, notably Fu Nai's proposal for 'rectification of the border' (*xiubian* 修邊) in Hunan and Gong Jinghan's call to 'strengthen the walls and clear the countryside' in Sichuan.[21]

Isolating illness: *bao* fortification and the Miao revolt

Turning to the case of the Miao conflict, how did local defence ideas play out in west Hunan, particularly in relation to *bao* fortified walls? What root problems were identified and what curative 'prescriptions' were offered? As will be seen, solutions were sought in a separation of Miao and Han peoples capable of isolating and eliminating disorder-generating 'symptoms'. *Tubao* (earthen fortifications) and *diaobao* 碉堡 (blockhouses) served as the fundamental structure of this separation.

A better definition of *bao* is needed. This term has typically referred to fortified walled enclosures positioned in strategic locations to enhance defence. As Yan Ruyi wrote of *tubao* in 1795, 'some rely on mountains and rivers and some rely on town walls'.[22] In highland areas, these were ideally built on level ground possessing springs and arable land sufficient to support several hundred families, prepared with a stone foundation and a circular wall of stamped earth some four metres high and 50 metres around. Dirt would be drawn from just beyond the walls to create moats, with thorny vines planted beyond. Thus prepared, they had 'strength comparable to a small city': 'arrows cannot pierce them, fire cannot burn them, and rebels cannot cross them'.[23]

Use of such walled fortification in China extends back over 4,000 years, but came to particular prominence during the North Song 北宋 dynasty (960–1127), in the course of a 'stockades and forts' (*zhaibao* 寨堡) policy against the Xixia 西夏 kingdom. Organized in the context of a military farm (*tuntian*) network, border populations were concentrated in secure sites, with some 500 men per fort, providing fields and martial training. This militarized coordination of self-sustaining stockades allowed the Song to claim Xixia lands via gradual extension of protected settlements.[24]

During the Ming 明 dynasty (1368–1644), a system of 24 *bao* was built on the unruly Hunan 'Miao Frontier' (*Miaojiang* 苗疆), but was later dismantled.[25] From 1795, Qing strategists called for its revival. As the frontier opened in the eighteenth century, permitting Han to migrate in and Miao to expand out, natives became scattered, impoverished and bitter – grown desperate in a mountain environment ideally suited for guerrilla resistance. Restored walls, strategists claimed, would not just defend against attack, but also better regulate the region's seething border society.[26]

As the Miao uprising broke out, Yan Ruyi presented plans for military forts to simultaneously reclaim strategic locations and encircle native insurgents. Soldiers were to start at the edges of the highlands and then push in along roads used by the rebels, building stockades at intervals. With mutual coordination and gradual application, the 'fierce Miao' would be naturally fenced in'.[27] Refugees and settlers could then follow the advancing line, settling around forts, sheltering within them and contributing to their sustenance in the form of manpower, food and supplies. This method, the scholar argued, 'establish[ed] proper conditions for reconstruction'.[28]

Subsequent planning reflected much the same principles, albeit advanced on a larger scale. As Qing troops were moved north to fight the White Lotus revolt from 1797, local Han had necessarily to see to their own protection. In response the magistrate Fu Nai, supported by the Qing court from 1799, established a border-wide system of some 1,000 military- and civilian-manned fortifications: 'watch towers for observation, cannon towers to block the enemy, forts to gather families into a home'.[29]

This expanded fortification functioned not just to enhance military defence, but also to provide ethnic separation. Fu Nai, as Yan Ruyi, held that the source of local disorder lay in the unregulated mixing of socially and culturally dissimilar Miao and Han peoples. Accordingly, the restive natives had not only to be better controlled, but also physically contained to curb inter-ethnic alliances and Han Chinese interference. Fu Nai's 'border wall to restrict the frontier' thus focused on situating Miao in designated territory and then encircling this territory with a 'Miao Pale' that demarcated, defended and sustained Han/Miao division. Doing so served to 'isolate the root illness'.[30]

Fortified walls also structured an agenda of social correction and imperial 'health' on both sides of the divide. As Fu Nai argued, '[Han] people can hide in the *bao*'.[31] If given land nearby, they could then rely on forts to settle, farm and trade. 'Call back the refugees and build military farms', Fu advocated, 'disseminate weapons and tools, then command them to restore their livelihood.'[32] In this way, 'for several hundred *li* 里 fathers and elders, sons and younger brothers, would be sheltered to produce and be safe'.[33] Confucian logic suggested that they would then start families and naturally embrace imperial authority, ethics and practices. The Miao, in contrast, were subject to a sterner correctional focus. New fortifications housed a system of subsistence and supervision in which natives were employed as tenant farmers, regulated by 'Miao officers' and educated by better-acculturated natives. In this way, authorities sought to 'use Miao to rule Miao and ... instruct Miao' such that 'their customs change, families stabilize, minds become correct', and the natives 'transform into good subjects'.[34]

By 1812, the fortified network connected to military farm administration (*tunzheng* 屯政) had linked settlements, controlled roads and waterways, and systematically segmented Miao and Han territories. These *bao* walls were supported by some 23,000 acres of appropriated arable land, manned by 8,000 irregular soldier-farmers and 5,000 allied Miao troops. The militarized line, exemplified by the 40 *li* 'Hunan Great Wall', grounded a new order on the Miao Frontier that lasted to the early twentieth century.[35] The result was a segregated and stratified mountain society, albeit one that largely lacked the security, clean divisions and cultural transformation originally envisioned.[36]

Nourishing the body: *bao* fortification and the White Lotus revolt

As rebellion ravaged the highland border of Hubei, Sichuan and Shaanxi provinces, Qing planners again turned to the solution of *bao* walls. Mirroring implementation of earthen fortifications in west Hunan, newly built 'stockades and forts' (*zhaibao*) interlocked with an array of measures to promote subsistence, order, education and mutual defence. These structures served the common function – at once, social and martial – of dividing, defining and strengthening a chaotic borderland society. In this case, however, the region in question was far larger and populated predominately by a single ethnic group. Loyal Han subjects in these hills were immersed in a sea of seditious, coercive and drifting Han insurgents. Accordingly, defence required a different kind of quarantine: one in which the 'good' (rather than rebellious) were to be sealed within walls.

In 1799, Emperor Jiaqing embraced local defence initiatives and called for plans to enable community fortification. Responses came from men such as Yan Ruyi (now an official), Liu Qing and Gong Jinghan.[37] The ideas they offered showed much the same principles, logic and technology seen in Miao frontier fortification. Walls were to be of stamped earth surrounded by moats: a design simple enough, with materials accessible, to be built in large numbers. Positioning in this case, however, was to be more radical. Not only were 'walls strengthened', but the 'countryside was cleared' via forcible concentration of highland populations into centralized stockades – here planned as temporary relocation rather than lasting resettlement.[38] These fortifications shielded locals, refugees, food, livestock and tools. With expanded coordination between communities, they concurrently provided defenders means to harass and delay the enemy such that the rebels, exhausted and afraid, could be destroyed by Qing troops.[39]

In light of regional conditions, such defensive measures were deemed a military necessity. The enemy here was not the Miao: conspicuously different people with clan affiliations and home territories. Rather, the adversary was other Han Chinese: drifting and predatory insurgents submerged within a vast pool of fellow Han mountaineers, a disorderly population viewed with almost as much anxiety as the rebels themselves. The guerilla bands lacked homes or independent resources. Their survival depended upon extraction of materials and assistance from collaborators. Accordingly, the destruction of the insurgents lay in ensuring that extraction and collaboration did not occur. If the rebels lacked supplies they would starve; if they lacked supporters they would fade.[40] *Bao* (the walls strengthened and the area within which the countryside was cleared) were intended to provide the space where this interference would occur.

New walls, however, functioned for more than just military defence. On the three-province border, as in west Hunan, fortifications also served to 'establish clear dividing lines in rural society'.[41] Such imperial separation was critical in this case, as authorities had little other means (whether appearance, language or culture) to distinguish friend from foe. Insurgents looked the same as subjects, mixed with them, posed as them and had success in coercing and subverting them. *Bao* physically divided – and thereby distinguished – the loyal from disloyal, 'good' from 'chaotic', healthy from ill. Walls, that is, clarified a social morass, providing a tangible inner and outer partition that could be ascertained, defended, reported on and gradually refined.

Given that virtually the entire highland population was viewed as a diffusion of common traits and suspected rebel sympathies, *bao* concurrently functioned not just to keep the enemy out, but also to ensure the loyalty of those within. That is, local defence initiative, bounded by fortified walls, was 'as much for internal control as external defense', giving aggressive attention to the 'unreliable', 'vagrant' and 'treacherous' folk within its purview.[42] Measures designed to strengthen internal observation and regulation, 'nipping disloyalty in the bud', included hierarchies of fort officers, militia training, *baojia* surveillance and careful oversight by regional officials.[43]

Under these circumstances, notions of quarantine became particularly significant. Victory was contingent not just on isolating 'illness', sheltering loyalty and regulating deviance, but also on nourishing more dynamic forms of corrective 'health'. As also seen on the Hunan Miao frontier, defensive walls girded state efforts to inculcate and shape imperial subjects, evidenced by renewed emphasis on schools, Confucian classics, Sacred Edicts and civic responsibilities. Gong Jinghan argued that only a combination of imperial education and control would ensure that collaborators were 'afraid and not dare to be unrestrained and so can be gradually transformed into good subjects'. In this way, all within the walls would 'unite together as one family' to support the Qing.[44]

Arguably, the ambitions of this effort advanced a step further. *Bao* walls, and the ordering practices they aided, likewise reflected a strategy of using 'health' to overcome 'illness' in ways that aggressively pushed the scope of imperial influence. Recall that loyal and disloyal on this borderland were of one folk and, unlike on the Miao Frontier, planners saw a clear zero-sum relationship between them. As Gong Jinghan observed, 'if there is one more subject, there is one less rebel'.[45] Transforming Qing subjects behind fortified walls reduced the insurgent pool. But this was only the first step. Expansion of defence networks also progressively *increased* the loyal population. Herein lay a vision for comprehensive cure. Given time, the entire war zone could be encompassed by networks of sheltering and transformative walls. The inner would thereby replace the outer, the rebels would disappear, all would become 'healthy' and there would cease to be need for fortified divisions at all.

The martial, if not social, effectiveness of this fortification initiative has been noted. Susan Mann Jones and Philip Kuhn, for example, wrote that the White Lotus armies 'constantly in flight and cut off by a spreading strategic hamlet strategy ... were hunted and destroyed'.[46] The immediate impact of new community walls, however, may have been overstated. Contemporary critics argued that rebel forces, numbering over 100,000 at their height, were powerful and forts were weak, thus *bao* 'helped concentrate people so that they might all be destroyed at once!'[47] Strategist such as Yan Ruyi rejected this view, but recognized that the potency of rural defence declined as rebel forces became smaller and more determined: a key reason why the revolt endured an additional five years. By 1804, however, most of the three province border region had been 'cleansed', although some fortification remained to guard against 'scattered bandits'.[48]

Conclusion

As Philip Kuhn observes in his classic study of Qing militarization, the aim of local defence at the turn of the nineteenth century 'was never simply a matter of keeping armed rebels out'.[49] It was, rather, a system of external defence and internal control inextricably linked to the designs and institutions of the imperial state. This brief discussion of *bao* fortification in China's late-eighteenth-century-Miao and White Lotus revolts supports Kuhn's insight, but suggests that Qing administrative plans for state-sponsored community mobilization had even greater social ramifications.

In the context of the 1799–1805 Jiaqing reforms, *bao* – a seemingly simple expedient to defend villages and guard strategic sites – played a vital role in new bureaucratic projects for sustained state ordering. Conceived and constructed in the context of overarching military farm and strategic hamlet initiatives, locally manned ramparts formed the stage for the realization of imperial rectification. Walls not just protected and policed, but concurrently strengthened state authority, secured livelihood and girded orthodox education. Towards these ends, *bao* were mobilized in specific response to borderland ecology, social conditions, national political change and court patronage of a new and more martial breed of civil borderland administrator.

Contemporary pacification metaphors of imperial 'illness' and 'health' were not incidental to this development. Rather, they lay at the heart of a larger social vision for fortified walls. With the Miao revolt, fortification not only separated Han from Miao, it quarantined natives – suppressing the 'symptoms' of illness – in the context of a self-sustaining line that bifurcated highland society and strove for gradual acculturation across a policed divide. In the White Lotus conflict, in contrast, *bao* walls quarantined Han subjects (from Han insurgents), serving as a foundation to isolate dissent, strengthen control and secure compliance. Here, the focus on 'nourishing health' proved even more aggressive, with an agenda of mass transformation that would ideally eliminate the entire rebel population. *Bao* walls, thus, were more than just physical barriers to keep Miao from crossing or rebels from pillaging. They were the material lines that re-inscribed the Qing's tumultuous borderlands: tangible tools to divide, define, shelter and encompass in pursuit of restored imperial civilization and social order.

Such circumstances, if accurate, suggest at least three points of scholarly significance. First, if the Miao and White Lotus revolts presaged 'dynastic decline' – as has often been observed – the Jiaqing regime's use of fortified walls provide signs of continuing administrative flexibility and vigour, in evident continuity with eighteenth-century precedent.[50] Although China was in crisis, its government yet retained sophisticated means to confront this challenge.

Second, the early nineteenth-century proliferation of *bao* walls (and related *bao* discourse) suggests a further 'militarization' of Qing political culture. This was a regional and popular phenomenon, and thus differed from the Manchu martial aesthetics of eighteenth-century imperial promotion identified by Joanna Waley-Cohen. Consistent with that overarching elevation of military achievement, however, this turn towards civilian mobilization likewise aimed 'to forge a new and distinctively Qing culture that would generate a shared sense of community'.[51] Such linkage suggests that the militarization of China's culture not only continued past the 1799 mark set by Waley-Cohen, penetrating more deeply into society, but also shifted towards regional strategists, Han military traditions and the maintenance of Qing territorial integrity.

Third, the planning of *bao* walls, evidenced by metaphors of health and designs of social structuring, also indicate an opposing line of influence: civil concerns shaping military ideas, institutions and practices. With local defence, martial methods of drilling and fortification were redeployed beyond the Qing armies towards designs of mass mobilization, social order and regional reconstruction. This redirection – as, indeed, the larger cross-fertilization of Qing civil and martial spheres – presaged forms of state planning seen throughout China's subsequent nineteenth century. Innovated methods of community organization and defence were brought to bear on a larger scale in projects such as the 1851–1864 Taiping pacification and 1862–1874 Tongzhi Restoration. Imperially sanctioned community walls continued to be more than simple defensive expedients. In intent, if not always fact, they focused both civil and martial planning towards greater goals of imperial rehabilitation.

Notes

1 See, for example, Philip Kuhn, *Rebellion and Its Enemies in Late Imperial China: Militarization and Social Structure, 1796–1864* (Cambridge, MA: Harvard University Press, 1970), pp. 37–50; Susan Mann Jones and Philip Kuhn, 'Dynastic Decline and the Roots of Rebellion', in *Cambridge History of China*, vol. 10 (Cambridge: Cambridge University Press, 1978), pp. 107–612; Daniel McMahon, 'Restoring the Garden: Yan Ruyi and the Civilizing of China's Internal Frontier, 1795–1805' (Davis dissertation, University of California, 1999); Donald S. Sutton, 'Ethnic Revolt in the Qing Empire: The "Miao Uprising" of 1795–1797 Reexamined', *Asia Major*, vol. 16, no. 2 (2003), pp. 105–152; Daniel McMahon, 'New Order on China's Hunan Miao Frontier, 1796–1812', *Journal of Colonialism and Colonial History*, vol. 9, no. 1 (2008), pp. 1–26.

2 Nicola Di Cosmo (ed.), *Military Culture in Imperial China* (Cambridge, MA and London: Harvard University Press, 2009), p. 4.

3 Donald S. Sutton, 'Ethnic Revolt'.

4 Susan Mann Jones and Philip Kuhn, 'Dynastic Decline', pp. 136–144; Daniel McMahon, 'Restoring the Garden', ch. 4.

5 See essays in He Changling 賀長齡 (ed.), *Huangchao jingshi wenbian* 皇朝經世文編 (1826, 1827) (hereafter, *HCJSWB*), 88:1a–89:19b. For related discussion, see Philip Kuhn, *Rebellion*, pp. 37–50; Susan Mann Jones and Philip Kuhn, 'Dynastic Decline', pp. 136–144; Daniel McMahon, 'Restoring the Garden', chs 2, 4; Daniel McMahon, 'New Order'.

6 *HCJSWB*, ch. (*juan* 卷) 89.

7 *HCJSWB*, 89:9a–10a.

8 Philip Kuhn, *Rebellion*, pp. 44–45; Susan Mann Jones and Philip Kuhn, 'Dynastic Decline', pp. 140–143.

9 *Da Qing Renzong rui (Jiaqing) huangdi shilu* 大清仁宗睿（嘉慶）皇帝實錄 (Taibei: Huawen shuju reprint, 1964) (hereafter, *DQRRHS*), 38:16a, 39:3a.

10 *DQRRHS*, 40:4b, 8b–9a.

11 *DQRRHS*, 39:3a.

12 *HCJSWB*, 89:11a.

13 Daniel McMahon, 'New Order', pp. 11–16; Wu Xinfu 伍新福 (ed.), *Zhongguo Miaozu tongshi* 中國苗族通史, vol. 1 (Guiyang: Guizhou renmin chubanshe, 1999), pp. 404–413.

14 *DQRRHS*, 39:3a.

15 *HCJSWB*, 89:12a. For similar views in regard to west Hunan see 88:1b and Yan Ruyi, *Leyuan wenchao* 樂園文鈔 (preface 1844) (hereafter, *LYWC*), 6:21b–22a.

16 *HCJSWB*, 89:19b.

17 *HCJSWB*, 88:1a.

18 *LYWC*, 5:23b–24a.

19 *LYWC*, 6:7b; Yan Ruyi (ed.), *Yangfang jiyao* 洋防輯要 (1838) *xu* 序, 2b.

20 Daniel McMahon, 'Southern Shaanxi Border Officials in Early Nineteenth Century China', *T'oung Pao*, vol. 95, nos 1–3 (2009), pp. 120–166.

21 *HCJSWB*, 88:1b–2a; 89:10a–11b.

22 *LYWC*, 6:21b.

23 *LYWC*, 5:5b, 19a–1b; HCJSWB, 88:1a; Daniel McMahon, 'Restoring the Garden', p. 112.

24 *Zhongguo junshi shi* 中國軍事史 (Beijing: Jiefang jun chubanshe, 1991) vol. 6, pp. 6–9, 209–210.

25 *LYWC*, 5:16b–17a.

26 See, for example, *LYWC*, 5:16b–20a.

27 *LYWC*, 5:23b.

28 *LYWC*, 5:18b–19a.

29 *HCJSWB*, 88:1a.
30 *HCJSWB*, 88:1a; *HMFTZK*, 11:5a–5b; Yan Ruyi (ed.), *Miaofang beilan* 苗防備覽 (preface 1820), 13:5b. The term 'Miao Pale' was coined by Donald Sutton.
31 *HCJSWB*, 88:1b, 10b.
32 Dan Xiangliang 但湘良 (ed.), *Hunan Miaofang tunzheng kao* 湖南苗防屯政考 (1884), 11:4b.
33 *LYWC*, 5:20a. *Li* is a unit of distance, approximately equal to half a kilometre.
34 *HCJSWB*, 88:2a, 2b; *LYWC*, 5:7a–7b.
35 Daniel McMahon, 'New Order', pp. 11–20; *HCJSWB*, 88:1a–2a, 10b, 12b: Dan Xiangli, *Hunan Miaofang,* ch. 11.
36 Daniel McMahon, 'New Order', pp. 16–18.
37 *HCJSWB*, 89:9b; Philip Kuhn, *Rebellion*, pp. 42–50; Daniel McMahon, 'Southern Shaanxi', pp. 131–142. For accounts, see Yan Ruyi (ed.), *Sansheng bianfang beilan* 三省邊防備覽 (preface 1822), 18:25b–30b, 34b–35a.
38 *LYWC*, 6:22a–22b.
39 *HCJSWB*, 89:9b, 11a–11b, 13a–13b.
40 For contemporary reports discussing these matters, see *HCJSWB*, 89:10a–17b; *LYWC*, 6:4a–7a (and ch. 6 generally).
41 Philip Kuhn, *Rebellion*, p. 49.
42 Philip Kuhn, *Rebellion*, p. 49.
43 *HCJSWB*, 82:17b, 89:11a; Philip Kuhn, *Rebellion*, pp. 48–49; McMahon, Restoring the Garden', ch. 5.
44 *HCJSWB*, 89:10a–12a. See also, *LYWC*, 6:8b, 10a; Daniel McMahon, 'Restoring the Garden', pp. 202–204.
45 *HCJSWB*, 89:11a.
46 Susan Mann Jones and Philip Kuhn, 'Dynastic Decline', p. 143.
47 *LYWC*, 6:22a–22b.
48 Daniel McMahon, 'Qing Reconstruction in the Southern Shaanxi Highlands: State Perceptions and Plans, 1799–1820', *Late Imperial China*, vol. 30, no. 1 (2009), pp. 100–104.
49 Daniel Kuhn, *Rebellion*, p. 49.
50 See, for example, Daniel Kuhn, *Rebellion*; Daniel McMahon, 'New Order' and 'Qing Reconstruction.'
51 Concerning militarized Qing culture, see Joanna Waley-Cohen, 'Militarization of Culture in Eighteenth-Century China', in Nicola Di Cosmo (ed.), *Military Culture in Imperial China* (Cambridge, MA and London: Harvard University Press, 2009), pp. 278–295, see p. 278 for the quotation.

16 Total war

Military supply and civilian resources during China's era of rebellions

Elisabeth Kaske

Accommodation and devastation

'十戶九空少積倉，年年採買供軍糧，我侯痛切民間苦，平價平收平斛量' 'When nine out of ten have emptied their stores, while year after year the army procures, our magistrate knows what the people deplore, and pays a fair price for a fair peck of corn.'[1] This poem was published in the 1873 gazetteer of Sanshui 三水 county to celebrate acting magistrate Jiang Tonggang 姜同岡, who also commissioned the edition, for his impeccable and (for the time) unusual behaviour of paying a fair price for the grain procured for the army. Sanshui (now called Xunyi 旬邑) is located in the north-west of Shaanxi, close to its border with Gansu province. Being connected by the Jinghe 涇河 river to the provincial capital Xi'an, Sanshui was the start of the 167 km long land route to Qingyang 慶陽 where the logistic headquarter for the eastern Gansu operations was located.

Sanshui county had seen its share of destruction and famine during the civil war between the Muslim (Hui 回) and non-Muslim (Han 漢) Chinese that ravaged Shaanxi and Gansu since 1862. Contemporary population historians estimate that the county lost 30 per cent of its population of almost 60,000.[2] In 1873, the gazetteer listed about 1,800 men and 300 women who lost their lives when the Muslim rebels ransacked Han villages in 1862/1863 and 1867/1868.[3] It does not list the Muslim losses, which may have been as high as 90 per cent of pre-war numbers,[4] or the number of victims of the famine of 1876/1877 that was yet to happen.

In accordance with custom, the civil administration of Sanshui was made responsible to protect the county (or at least the walled county seat) and provide supplies and smooth transport through its territory for outside troops. Since the statutory peace time bureaucracy was ill equipped to handle wartime tasks, the magistrate headed four temporary bureaus established for this purpose, two of them concerned with defence – a local militia bureau (*tuanlian ju* 團練局) and a town defence bureau (*chengfang ju* 城防局) – the two others for transport (*zhuanyun ju* 轉運局) and grain procurement (*caimai ju* 採買局).

Jiang Tonggang, the magistrate who had himself celebrated in the poem, had been appointed in 1869, and his arrival was depicted as a turning point. A note attached to the poem points the finger at government troops – invariably irregular mercenaries hired to supplement the insufficient regular forces of the Qing – which had at times directly exacted grain from the people in excess of the tax quotas and short-changed farmers in army purchases by using oversized pecks (or by not paying at all). With tax and grain purchases back in the hands of a benevolent civil administrator, not only was justice done, but the purchase of grain was also more efficient and the granary well stocked.

The county elites were unanimous in their feeling of relief over the new magistrate, if not for the military corvée quota of 600 donkeys conscripted for the ongoing grain transports to Qingyang:

> 忽來羽檄催軍需，轉運艱難費拮据，北路鴻溝多負袋，西坡鳥道不通車，俄警飢勇三千虎，飛調窮鄉六百驢，記得去年聞警日，正當穜麥露涼出。
>
> There came an urgent call for military supplies, to be carried on arduous roads for sky high costs. Deep gullies on the northern route – only pack animals can pass. Hanging paths on the western route – no cart will go through. Frightening news told of starving troops, all three thousand men. An urgent order deployed from our poor county six hundred donkeys. We still remember the day last year when the order came. It was just when the sowing of wheat began and the donkeys were needed to plough.[5]

The poem refers to events in the spring of 1869, when Qingyang had just been recovered from the hands of the Muslim rebels. The prefectural town was located in the fertile plain of Dongzhi 董志原 in eastern Gansu. When the Muslim rebels were driven out of Shaanxi in 1863, they poured into neighbouring Gansu. Qing government troops erected the headquarter in Qingyang to supply the frontline at Jingzhou about 100 km to the south-west. By 1866, however, the rebels were roaming the Dongzhi plain. The Qing managed to protect the walled towns of the county and prefectural seats, but the transport lines became extremely vulnerable. Remaining government troops brutally tortured the local population to find their last hidden grain reserves until, two years later, hunger and neglect brought the defences down. The result was total devastation. Hui Dengjia, a survivor, described apocalyptic scenes with rats as big as cats, packs of wolves hunting down their human prey weakened by hunger in broad daylight and robbers luring refugees with promises of grain to slaughter them for food.[6] Cao Shuji, who studied the population history of the war years, estimates that Qingyang prefecture lost 91 per cent of its 1.4 million population during this carnage.[7]

These two scenarios describe the best and worst cases of the Qing military efforts to solicit civilian resources during the civil wars, a civil administration that mitigates between the needs of the military and the plight of the people versus the total breakdown of order with marauding mercenaries foraging for food. However, the authors of the gazetteer of Sanshui county were coy to mention one event that made civil–military relations look a bit less bleak. In late 1864, Lei Zhengwan 雷正綰, one of the commanders whose troops were to be supplied by the Qingyang logistic headquarter, circumvented this unreliable supply line and directly sent commissioners to various counties in Shaanxi and Sichuan to solicit contributions from the wealthy landowner class. So-called contributions were the main way by which the Qing government exacted funds from local elites without alienating them too much. The trick was to exchange the silver or grain donated for official titles and rank, a practice authorized by the imperial government for the sake of alleviating the social fallout from constant warfare. Lei's commissaries approached the richest families of the county, especially the Tangs and the Zhangs. Both seemed reluctant to contribute initially, but the agents hit success with Tang Hongxu 唐鴻序, who donated two-thirds of the 30,000 taels of silver.[8] Tang Hongxu was one of the co-authors of the Sanshui gazetteer, which lists him as a department head in one of the Six Boards of the central government in Beijing, a typical rank obtained by contributions. Eleven more members of the Tang family are listed in the gazetteer sporting ranks also typically venal, in a clan already rich in members who made their way through the meritocratic civil service examinations.[9] No complaints about contributions are recorded in the gazetteer.

Thus, the troubled relationship between government armies and the civilian population can be described as oscillating between the two poles: on the one hand, there was accommodation of the elites through office selling and fair treatment of the commoners mitigated by a functioning civil administration, which was put in the service of the war effort. On the other hand, there was total destruction and victory achieved by exhaustion and mass slaughter. This chapter will attempt an overview of the logistic machines of the Qing armies and their changes from the 1850s to the 1870s. Special emphasis will be given to the military–civil nexus in the management of military logistics and to the question of how resources (mostly grain and silver) were obtained from the civilian populations. There has not been much research on this aspect of nineteenth-century military history in China,[10] and much of what will follow is based on original research of archival and reprinted sources.

A total war

The war fought by the Qing government to survive the flood of rebellions can be well described as a total war. There was warfare on multiple fronts. There was incredible carnage and loss of life. And there was a total mobilization of resources for the war on the part of the dynasty and China's rural gentry society in their joint struggle for survival.

The period in question, namely the 1850s and 1860s, is well known as the era of the Taiping Rebellion which managed to establish a nemesis of the Qing empire with its capital in Nanjing. But the Taiping were only one of a contagious wave of rebellions that overran almost the whole empire. Those whose names have become known in history are the following. The Taiping rebels, 1850–1864, whose well organized nemesis of the Qing emperor Hong Xiuquan 洪秀全 started to build up an army in Guangxi in the late 1840s, overran several southern Chinese provinces and finally established his capital in eastern Chinese Nanjing in 1856. In 1860, with the collapse of the Qing forces' siege of Nanjing, Taiping forces seized Jiangnan (southern Jiangsu and northern Zhejiang), the economic centre and rice basket of the empire.[11] The Nian Rebellion, 1853–1868, was active in most northern provinces of the Chinese heartland, from Zhili in the north to northern Jiangsu in the south, from Shandong in the east to Shaanxi in the west.[12] The Miao Rebellion, 1854–1873, started as a tax revolt and eventually involved not only Miao people, but also Han, Muslims and other minorities in poor and backward Guizhou province in the south-west.[13] The Yunnan Muslim Rebellion, also called Panthay Rebellion, 1856–1873, remained mostly confined to Yunnan in the south-western borderlands.[14] And, finally, there was the Muslim Rebellion of Shaanxi and Gansu, 1862–1873, described in the beginning of this chapter. The latter was particularly bloody, because it was an ethno-religious civil war between two local populations almost equal in size (especially in Gansu). It not only devastated these two north-western provinces but also cut China off from Xinjiang and central Asia (the recovery of Xinjiang took until 1878).[15] Looking at a Chinese map in the year 1860, we can find hardly any province that was not affected by some kind of warfare. Notwithstanding, the throne in Beijing simultaneously engaged in the Arrow War with the British Empire (1856–1860). In the autumn of 1860, Anglo-French forces entered the capital and burned down the imperial summer palace on its outskirts.[16]

Second, not only were these wars extensive, they were also one of the last examples of large-scale warfare fought largely with pre-industrial means. The use of modern imported weaponry was negligible and most of the arsenal in use looked like the list delivered to the Shaanxi armies fighting in Gansu in 1863–1864, which included iron cannon balls, gunpowder, fuses, shoulder matchlock muskets (*taiqiang* 抬槍), single matchlock muskets (*niaoqiang* 鳥槍), bows, arrows, rocket-propelled arrows, gunpowder flame throwers and swords.[17] It was also a war of Malthusian dimensions. Cao Shuji and other population historians in China today have been sifting through the available records and gazetteers county by county and arrived at estimates of population losses that by far exceed traditional body counts. According to the new numbers, the Taiping wars alone would have cost 73 million lives (up from the traditional 20–30 million). Including the repeated flooding of the Yellow river in 1851 and 1855 (when it changed its course), as well as 1867–1868 and 1887, and the North China Famine of 1876–1879, mortality may have been as high as 118 million people, or one-quarter of the total population of China during the three deadly decades of the mid-nineteenth century.[18] Most provinces saw their populations decline during this era by numbers ranging from 10 to 20 per cent (Shandong, Hubei) to a staggering 70 per cent (Gansu). Many provinces that formed the economic core of the Chinese empire had population losses in the 40 to 50 per cent range (Anhui, Zhejiang, Jiangxi, Shanxi and Shaanxi). Jiangsu, China's richest province, lost 34 per cent according to this estimate.[19]

Third, there was an enormous mobilization of fighting forces. Before the war, the Qing dynasty had two standing armies – the Chinese provincial forces known as Green Standard and the ruling Manchu dynasty's garrison forces, the Eight Banner Army. Thomas Wade estimated their strength at slightly under 620,000 and 300,000 respectively, but his numbers were only estimates of statutory numbers (no complete statistics were ever given in Chinese sources). Moreover, they were already outdated at the time they were published in 1851.[20] Studies of the era generally allege that both armies suffered from corruption of their offices, low fighting spirit and lack of readiness due to mostly hereditary military employment and a largely sedentary lifestyle. This probably being true, it is also evident that with Green Standard forces thinly spread through each of the 1,300-odd counties of inner China and Banner garrisons concentrated in Beijing, the north-east and a few major cities, these armies were hardly sufficient to deal with the scale of the conflict.

Thus, another layer of military mobilization was added by hiring volunteer irregular forces known as 'braves' (*yong* 勇) as opposed to regular 'soldiers' (*bing* 兵). Their strength has been estimated at 300,000 men, although there were probably much more of them and numbers greatly fluctuated. Every province had irregular troops, and they could be hired or disbanded whenever enough funds were available to pay the salaries.[21] One commander usually was in charge of several battalions of 500 to 680 men each, forming an army that often bore the commander's name. The Hunan Army (*xiangjun* 湘軍) of Zeng Guofan established the fame of being the most regulated of these armies with unified pay scales, organizational charts and laws. It became the model for other such forces such as the Hubei Army (*chujun* 楚軍) founded by Hu Linyi and the Huai Army (*huaijun* 淮軍) under Li Hongzhang. Gradually, these armies operated outside the area of their inception, and their names denominated their organizational style more than their origin. Much of the war in Shaanxi and Gansu was fought for example by Hubei Army forces since several of the governors appointed after the beginning of the war had risen as leaders of such armies in the war against the Taiping, including the governor of Shaanxi Liu Rong (1862–1865) and successive governors-general of Gansu and Shaanxi Yang Yuebin (1865–1866) and Zuo Zongtang (1869–1876).

The final layer of military mobilization consisted of local militia forces (*tuanlian* 團練) as the ones mentioned in the Sanshui gazetteer and studied intensively by Philip Kuhn.[22] Nobody has ever seriously attempted to estimate their numbers, but local militias were ubiquitous. Even though the local government was formally in charge, they were in fact hired and organized by the landowners and gentry (local elites with some sort of official rank). Here, too, the imperial government used official rank as an incentive for the elites to engage in militia organizations. Between 1852 and 1862, organizers of local militia could report their expenses for reward and receive the same preferential treatment as contributors to government armies.[23] Only later were rewards for contributions to the militias reduced substantially in order to channel more resources to the army.[24]

The Qing dynasty's fiscal system suffered immensely during the era of rebellion. For most of the time it switched into what we may call a 'survival mode', being totally devoted to financing the war, and a serious recovery did not began before the mid-1860s, especially after 1868 when all of the eastern provinces were pacified. Even during peace times, the pre-industrial agrarian empire spent 70 per cent of central government expenditures for its standing armies, not counting the funds provided by local governments for arms and equipment, military construction and the upkeep of the courier and relay system.[25] Wartime expenses by far exceeded available resources. This is, of course, true for most wars in world history, but in the 1850s China government borrowing as a funding source was virtually non-existent. Central government reserves – once a major source of war funding – were depleted even before the Taiping wars. Moreover, the capture of Nanjing by the Taiping forces in 1853 and their northern expedition along the Grand Canal had disrupted the fiscal operations of the imperial government, which was heavily dependent on silver and grain shipments from the wealthy Yangzi provinces. Beijing effectively lost the ability to provide war funding.

The provinces became the major source of funding for the various war efforts. Traditionally, the provinces held the power of taxation, but land taxes were raised by fixed quotas and most of the tax income of the provinces was earmarked for expenditures determined by the Board of Revenue in Beijing, including for interprovincial assistance, i.e. bullion to be shipped to poorer provinces along the western borderlands to pay for their administration and defence. Only a small percentage was left to the provinces to cover administrative expenses – never sufficient save for some creative financing measures. After 1853, the Board apportioned fixed quotas to be shipped to Beijing for the survival of the capital. Apart from this, provinces had to transfer war funds to wherever the throne determined was in a military emergency. These war funds had to be shipped in bullion on a monthly basis. As there were military emergencies almost everywhere, this brought the interprovincial assistance system to the brink of collapse, as I have shown in more detail elsewhere.[26]

With government finance in jeopardy, living off the land became again common for late Qing armies. During the prosperous eighteenth century, Qing forces were not forced to live off the land. Grain was purchased by the government and often shipped over vast distances to protect local residents from rising grain prices due to bulk army purchases.[27] After 1853, not only did provinces explore new sources of revenue to provide for their armies, armies themselves also started to directly forage for revenue. Logistic headquarters attached to the armies often gained the authority to collect local land taxes, but quotas were fixed (and not very high), and with much of the countryside in rebellion and many areas devastated, increasing land taxes – a 'sacred cow' of Qing fiscal policy, anyway – did not provide a solution. The answer was to raise taxes on trade[28] but since armies were deployed in areas where warfare had disrupted the economy, the famous Lijin tax (also known as Likin) on goods in transit did not provide the anticipated revenues. The alternative was contributions, the sale of rank and office mentioned above. They were quite effective in alleviating the problems of war funding, not only by raising new revenues, but also by allowing the government to partly default on the statutory salaries of their troops without instigating a mutiny.

Finally, there were the services that counties had to provide to the government armies. These reached from assisting to solicit ever more contributions from the local wealthy, sending grain from the county-operated granaries, organizing grain procurement for the troops and transport corvée, which not only included the sending of pack animals and carts but also the maintenance of the government relay stations used both for government messaging and travel as well as for the movement of troops. Counties further hired porters, and even though the pay of these porters could be reimbursed, much of the management going into the hiring could not.

The management of army supplies

The supply system of the Chinese armies was almost entirely the business of civil officials, although the civil–military distinction of the Qing bureaucracy was often blurred during the era of rebellion, since many of the most famous military leaders originated from the civil bureaucracy or made a civil career later on. Different from the eighteenth century when war funds came from central government reserves and current tax revenue silver shipped from the provinces, so that military supply was largely a function of disbursement and distribution, Taiping era supply institutions had to combine both revenue generating and more narrowly defined supply functions.

Theoretically, two types of administrations were authorized to raise revenues to supply the armies. The first were provincial governors and governor-generals, the latter having jurisdiction over one (Sichuan, Zhili), two or three provinces (for example, Shaanxi-Gansu or Jiangsu-Jiangxi-Anhui) or quasi-territorial administrative units (e.g. the governor-general of tribute grain transport). Governors and, even more so, governor-generals, incorporated both military and civil functions, even though they were formally heading the civil bureaucracy.[29]

The second type was an imperially commissioned director of logistics (*duban liangtai* 督辦糧臺) in charge of the logistic headquarter (*liangtai*) of a large-scale campaign fought under an imperially commissioned commander-in-chief (*duban junwu* 督辦軍務). In this constellation, the director of logistics was a high official (mostly of the second rank and with experience as provincial treasurer) who memorialized to the throne independently and was not subordinate to the military command. In practice, commanders-in-chief often circumvented a formal LHQ or operated without it, instead relying on a service department (*yingwuchu* 營務處) and a paymaster bureau (*zhiyingju* 支應局). Like Lei Zhengwan (assistant commander-in-chief of the eastern Gansu operations) in the example cited above they would independently solicit contributions or even collect taxes and, even though these army supply institutions were also staffed with civilians, these civilians were under the direct command of the military.

The question of whether logistic and financial operations should be independent from military operations and how provincial command and military command should be coordinated was a serious bone of contention since the beginning of the government campaigns to suppress the various rebellions ravaging the empire. During the 1860s this question was decided in favour of giving governorships to successful military leaders and having them leading the campaigns to pacify their provinces autonomously with full command over both military and logistic operations. The culmination of this development was Zuo Zongtang's pacification of Gansu after 1869 and his subsequent recovery of Xinjiang.

The focal point of this struggle was the institution of the independent logistic headquarter. The name '*liangtai* 糧臺' used to denominate this institution translates as 'grain station' and is a generic term that does not completely reflect the actual position of the institution within the ranks of the civil–military bureaucracy. Therefore, I will be speaking of an independent *liangtai* or formal LHQ if both an official of the second rank was in charge and if this official was imperially commissioned and made decisions independently from the military command or the provincial authorities. According to Gao Genghao, the name of '*liangtai*' and this particular institutional setting did not appear before the mid-eighteenth century. It was especially useful when a campaign was fought across provincial borders so that several civil administrations had to be coordinated and silver bullion paid by multiple provinces as part of their interprovincial assistance obligation could not be shipped to a provincial treasury.[30] In the reality of the mid-nineteenth-century wars, however, different institutional arrangements were possible, and I will speak of a provincial *liangtai*, if the *liangtai* was subordinate to provincial authorities, or an army *liangtai* (also called 'field *liangtai*', *xingying liangtai* 行營糧臺) if it was under a military command.

As already stated above, revenue generation was one of the new tasks of the mid-nineteenth-century logistic apparatus. While services provided by the county administrations never showed up in the accounts, three sources of revenue appear as the most important ones: interprovincial transfers of silver bullion, local taxes (both traditional land and salt taxes as well as new commercial taxes and wartime levies) and contributions. Taxes would be levied either through the county bureaucracy or by agents commissioned by the provincial governor or the *liangtai*. The same was true for contributions. Once regional wars became entrenched and troops were no longer on the move, the *liangtai* could become a quasi-territorial unit which took over the tax income of a given area.

Interprovincial assistance, which was assigned to both provinces in need as well military headquarters via an independent *liangtai*, was more difficult to enforce. Donor provinces were frequently at war themselves. Their revenue was insufficient to meet the three big demands of provincial expenditure: paying for the alimentation of the capital Beijing, providing internal security and supporting other provinces in their security efforts. Therefore, one of the main tasks of the director of logistics was to write letters, send commissioners or complain to the throne in order to remind governors of their payment obligations. This role was the great advantage military leaders saw in having a formal LHQ. Its head was high enough in the hierarchy to directly address both the emperor and provincial governors. On the other hand, formal LHQs were relatively expensive to maintain and the director of logistics was appointed by Beijing. Provinces and some military leaders increasingly preferred army *liangtai* and ad-hoc bureaus staffed by commissioned agents, because they were easier to control and cheaper to maintain.

Another problem was that donor provinces often replaced their payment obligations by sending troops instead, so-called 'guest armies' (*kebing* 客兵), into the theatre of war. In theory these 'guest armies' were to be financed and supplied by the sending province alone, i.e. were neither within the jurisdiction of the LHQ nor in possession of a campaign fund financed by multiple provinces. One province simultaneously could house a hodgepodge of arrangements including a formal LHQ, several independent 'guest armies', armies directly under the provincial treasury and troops of a new governor-general on his (sometimes protracted) way to office. Eastern Gansu, described in the beginning of this chapter, was a typical example of this chaos.

The independent *liangtai*

To make this general overview more tangible, two war theatres and their revenues are compared here. The first war theatre is Jiangsu province, which for over a decade housed three logistic headquarters supplying government troops fighting an entrenched defensive war on two fronts, against the Taiping in the south and the Nian in the north. The second one is eastern Gansu during the Muslim Rebellion, where the attempt to establish a formal LHQ failed after only a year. The comparison shows that the war in the resource-poor and land-locked north-west could only be won by mobilizing the resources of the whole empire and finding new ways of managing their transport into the region.

The war in the lower Yangzi region was an entrenched operation with the government troops clearly on the defensive. Jiangsu province alone was carved up into what we may call 'protective cells' with various civil–military administrations lined up along the Grand Canal, a major strategic supply route for the capital Beijing, competing for local revenues and inter-provincial assistance transfers. Apart from the governor-general of the southern course of the Grand Canal (*Nanhe hedao zongdu* 南河河道總督) and the governor-general of tribute rice transports (*caoyun zongdu* 漕運總督), who jointly controlled the defences of the Grand Canal and resided in Huai'an 淮安 prefecture in north-central Jiangsu,[31] there were three logistic headquarters in Jiangsu: the Jiangnan and Jiangbei *liangtai* just south and north of the Yangzi river in the south and the Xuzhou *liangtai* in the north-western corner of the province.

The Jiangnan *liangtai* supplied the Jiangnan Army Group that had laid siege on Nanjing, which had been occupied by the Taiping since 1853. Originally located in the region known as Jiangnan (south of the Yangzi), it was routed when the Taiping lifted the siege and routed the Qing armies for the second and final time in May 1860. The *liangtai* officials relocated to Gaoyou north of the Yangzi into the territory of the Jiangbei *liangtai*.[32] They continued to supply the defence forces of Zhenjiang, the last bastion of the government troops on the southern side of the river. After much tussle with the Jiangbei *liangtai* over the revenue base and payment obligations the two finally merged two years later.[33]

The Jiangnan *liangtai* was directly subordinate to the Jiangsu provincial authorities. Different from the Jiangbei *liangtai*, which administered its own revenue, revenue generation for the Jiangnan *langtai* was in the hands of another provincial institution called the Jiangsu Revenue Bureau (Jiangsu Chouxiangju 江蘇籌餉局). Both institutions were headed by officials of the second rank, the latter by provincial treasurer Wang Youling 王有齡. Moreover, the governor-general of Jiangsu, Jiangxi and Anhui also put his weight behind the province's efforts at producing revenue. As a result, despite its dismal military performance, the Jiangnan *liangtai* was relatively well supplied, prompting the head of the Jiangbei *liangtai* to remark enviously:

> Because the governor-general heads the liangtai, the military is more cautious and do not dare to make unreasonable demands. The local magistrates have someone to fear and orderly deliver their tax revenues. And the merchants do not dare to look down upon and play cat and mouse with [the *liangtai*] to avoid contributions, instead they are eager express their gratitude. That's why [governor-general] He Guiqing can boast of his success with fundraising.[34]

Before the collapse of the Nanjing frontline, the *liangtai* received all its funding from the Revenue Bureau. After the relocation, its revenue shrunk to a quarter of the 1859 numbers, but it is clear that traditional land taxes, merchant contributions and trade taxes – these irregular taxes were also called 'contributions' (*juan* 捐), although they were not rewarded with office and rank – played an important role. Post-1860 accounts show that the *liangtai* also received interprovincial assistance from at least four provinces outside Jiangsu, namely Hubei, Guangdong, Shanxi and Henan.[35]

On the expenditure side, the Liangtai supplied a traditional army group which concentrated detachments from the widely scattered regular standing armies (both Eight Banner and Green Standard forces) and combined them with a large number of irregular forces. Before the collapse, irregular troops consumed more than half of the salary expenditures, Green Standard troops made up 20 per cent, Eight Banner Armies from as far as Jilin and Heilungjiang up to 15 per cent (the strength of the forces is not given in the source). After the end of the Nanjing siege, total troop numbers shrank to 20 per cent of previous strength (as measured by expenditures), but irregulars were reduced even more than regular troops, so that basically only a small garrison of regular troops (with a few irregular auxiliaries) was left to defend Zhenjiang.[36]

The Jiangbei LHQ was established in 1854. For over ten years, it was located east of Yangzhou north of the Yangzi river, catering to the Yangzhou Army Group that fought constant Taiping onslaughts from the south. The city of Yangzhou was overrun by the rebels three times in 1853, 1856 and 1858, but the *liangtai* – protected by the Grand Canal and Shaobo lake in the west and the Yangzi river in the south – remained intact and allowed government troops to recover the city each time. In the spring of 1859, the Yangzhou Army Group was made a subsidiary of the Jiangnan Army Group, but the logistic headquarters remained separate. Following the collapse of the Nanjing front, the Jiangbei *liangtai* had to subsidize the Jiangnan *liangtai*, but while the latter continued to supply the Zhenjiang defence forces south of the Yangzi, the Jiangbei *liangtai* came under a new military command under Duxinga 都興阿, the first military leader who had irregulars of the Hubei Army system (*chuyong* 楚勇) under his command.[37]

Different from the Jiangnan *liangtai*, the Jiangbei LHQ had no backing from provincial authorities and was responsible for both revenue generation and disbursement. It was a classic independent *liangtai*. In theory, half its revenue should have come from inter-provincial war funds from four provinces and one customs office, namely Shanxi, Shaanxi, Shandong, Henan and a foreign customs station, but transfers were notoriously unreliable and almost dried up after 1860.[38] The Jiangbei LHQ was fortunate to be located in a rich and fertile plain crisscrossed by rivers and channels. It had abundant agricultural and commercial resources even during war time. In 1859, local revenue was said to be 80,000 tael.[39] In 1864, a year after the merger with the Jiangnan *liangtai*, the monthly income from local sources was actually double that amount. Of these, only 22 per cent came from the newly levied Lijin tax on transit goods, even though the Jiangbei headquarter controlled important trade routes along the Yangzi river. Thirty-one per cent came from traditional agricultural taxes (land tax, grain tribute, reed tax), and 14 per cent from the salt tax. Contributions made up 14 per cent of the revenue.[40]

We do not have the complete accounts of the Jiangbei *liangtai* prior to its merger with the Jiangnan LHQ, but we may use Xuzhou in the north-western corner of the province bordering the provinces Shandong and Henan to show that such an abundance of local resources was the exception rather than the rule. The Xuzhou *liangtai* existed since 1853 and by 1860 supplied an army group under Fu Zhenbang 傅振邦,[41] which showed the usual mixture of regular armies (only Green Standard in this case) and a majority of irregular forces. It was in fact an independent *liangtai*, although it was classified as one of secondary order and headed only by a circuit intendant (fourth rank) instead of a higher-ranking official. The average monthly income of the *liangtai* between May 1860 and September 1864 was about 51,000 tael, but the sources of this income changed abruptly after 1862, when the main donor provinces for interprovincial assistance Shanxi, Henan and Shandong stopped payments, because they became embroiled in warfare themselves. Interprovincial assistance transfers fell from 75 to 20 per cent. Since the *liangtai* had only taxing authority in one (not very wealthy) prefecture, land tax and *lijin* revenue combined summed up to no more than 8 to 11 per cent. The gap was largely replaced by contributions, which made up over 60 per cent of the *liangtai*'s income after 1862.

What was called 'contribution' was in fact a default on the salaries and purchases made by the army. When silver from interprovincial assistance stopped flowing, the *liangtai* handed out salary tickets (*xiangpiao* 餉票) to the soldiers and merchants paying 20 to 30 per cent of the sums owed. These tickets were not meant to be redeemed in species, neither would they circulate as money as the government cash notes (*baochao* 寶鈔) which were also used to pay the troops. Instead, soldiers sold these debentures for a fraction of their face value to contributors who used them at face value to purchase offices and titles. This policy was to the disadvantage of the soldiers, but it benefited wealthy merchants and landholders who could submit these tickets at face value and thus obtain official status at discounted prices. And, in this way, paper notes became designated 'contributions' on the revenue side of the *liangtai* accounts.[42] During the periods studied, the silver income of the *liangtai* never exceeded 60 per cent (at times it was as low as 15 per cent), while it still needed large amounts of specie (both silver and copper cash) to pay the armies. To receive more silver from contributions, new regulations after 1863 required that half of the purchase price for office and rank had to be paid in specie.[43] The same happened at the Jiangbei *liangtai* in 1864 where only half of the contribution income came from salary tickets, while the rest was paid by office purchasers in specie or grain. At the same time, soldiers received between a third and half of their pay in salary tickets.[44]

The correspondence of Qiao Songnian 喬松年 who managed the affairs of the Jiangbei *liangtai* between 1859 and 1861 provide us with an unusual insight into the operations and difficulties of an independent *liangtai*. Qiao Songnian was the rather typical case of a civil administrator who was commissioned to concurrently manage logistic affairs. As a salt controller (*yanyunshi* 鹽運使) for one of China's most important salt producing regions, Qiao was put in charge of the LHQ in 1859.[45]

The organizational structure of the Jiangbei *liangtai* appears to be typical for most logistic headquarters of the era, and so were its problems. The *liangtai* consisted of the director and his staff, a war chest and a number of functional bureaus (*ju* 局). The war chest of the Jiangbei *liangtai* was originally located in Shaobozhen 邵伯鎮 on the eastern banks of the Shaobo lake. In 1860, when Taiping troops reached the western banks of the lake, Qiao relocated the war chest, which was stored in a temple and could not be secured against an attack, to the walled town of Taizhou 泰州, the seat of his regular office as salt controller and its treasury.[46] Those bureaus responsible for disbursement of salaries and supplies were located in Xiannüzhen 仙女鎮 closer to the troops in Yangzhou, including the Arms Manufacturing Bureau (*zhizao junxie ju* 製造軍械局), the Food Supply Bureau (*bingmiju* 兵米局) and the Paymaster Bureau (*zhiying ju* 支應局). The revenue of the *liangtai* was designated by the throne and included inter-provincial assistance and a local revenue base of ten counties in the region from which the *liangtai* was authorized to draw taxes and contributions.[47]

In his memorials to the throne, Qiao constantly complained about his difficulties in running the *liangtai*. Most of these troubles were in fact caused by the setup of the institution of *liangtai* within the civil and military bureaucracy itself, rather than by the chronic lack of funds (especially shortfalls in interprovincial assistance transfers) or military issues such as the infiltration of rebels into the area or the treatment of refugees. First of all, as a salt controller Qiao Songnian was only an official of the third rank, not only was this below the rank of a provincial treasurer, but it was also a functional office with no direct jurisdiction over the county magistrates, the actual source of tax extraction. As a result, so Qiao complained, counties only showed lacklustre performance in their delivery of taxes to the *liangtai*: 'The magistrates consider the salt controller not as their superior official. Only some of the magistrates cooperate in the interest of the greater situation, but many see it as a different turf.'[48]

His rank was also in the way when it came to contributions. Each of the civil–military administrations in the area was authorized to solicit contributions in the territories under their jurisdiction, but prospective purchasers of office and rank were free to deliver their silver or grain to wherever they wanted. Most, so Qiao claimed, chose the highest-ranking administration available, which was the governor-general of the Grand Canal who resided in Qingjiangpu 清江浦 near Huai'an 淮安 (today part of the city) to the north of the Jiangbei *liangtai*.[49] The Jiangbei *liangtai* had other issues with the Qingjiangpu administration as well. In 1860, governor-general Gengchang 庚長 not only opened Lijin tax stations on the territory of the Jiangbei *liangtai*,[50] he also intercepted interprovincial assistance funds send from Shanxi bound for the Jiangbei *liangtai* under the pretext that the salt controller's office – Qiao Songnian's regular bureaucratic position – was owing his office funds for Grand Canal repair. All funds being shipped in silver bullion before the government adopted bank remittance in the late 1860s, this incident had Qiao exclaim angrily:

> If such a practice becomes a precedent for others everybody could intercept military funds. I could as well intercept provincial funds delivered to Beijing [*jingxiang* 京餉] delivered by Guangdong, or the administrative fees delivered to the Board of Revenue [*bufan* 部飯] from the Lianghuai salt administration when they pass through Shaobozhen![51]

Finally, the relationship between the civilian *liangtai* and the army was continuously strained. Since the military and its logistic support system were parallel rather than hierarchically subordinate administrations, commanders recruited new troops, especially irregulars, without asking the logistic supply system which was overwhelmed by sudden new demands for provisions. They also did not report losses but continued to receive the full pay for dead soldiers.

The *liangtai*, or its paymaster bureau, did not hand out provisions directly to individual soldiers, but always to the commanders of the battalion or even to the commander-in-chief of a small army of several battalions. Theoretically, these commanders should produce a complete list of officers and soldiers currently on the rolls upon receipt of the rations, but they simply refused to do so and threatened (and committed) violence when the paymaster withheld the provisions.[52]

When Duxinga 都興阿 took over the military command in the region and brought in his Hubei Army irregulars (*chuyong* 楚勇), he demanded that all irregulars in the region were reorganized and paid according to the Hubei system pay scales. The famous Hubei Army was not stronger because of better soldiers or better training, but mostly because they were hired under a contract that promised them full pay, in specie and on any account. When this promise was broken they rioted as much as any other irregular army as will be shown for the Gansu case below. In contrast, Jiangbei soldiers in 1859 only received half of their pay in silver, the rest in almost worthless pay tickets. Later, their salaries were cut by a third, and the monetary payments were often replaced by rice handouts. Duxinga's demand to pay every soldier in full terrified the man actually in charge of providing the funds; so much so that he not only pleaded to the throne to prevent this from happening, but also requested to be relieved of his post citing an old eye disease (he actually remained in office as salt controller and later successively became governor of Anhui and Shaanxi).[53]

To sum up, the *liangtai* system of logistic support was an ad-hoc civilian administration originally construed to receive government funds (from the central government or interprovincial assistance) and manage their disbursement for the need of the army supplies. When government funds dried up they also gained revenue generating functions by being superposed over the local administration as the recipient of tax revenues and by directly administering new taxes. In the fertile and wealthy province of Jiangsu, the *liangtai* were able to sustain themselves by these local recourses, despite difficulties in dealing with competing administrations and an unruly and demanding army. This was not the case in poverty stricken Gansu.

The failure of the independent *liangtai*

Gansu differed considerably from Jiangsu. Much of the province consists of a narrow fertile corridor surrounded by the Gobi desert in the north and the Qinghai and Tibetan plateau in the south. Its arid west forms the gate into Xinjiang, while the centres of population are in the prefectures east of the capital Lanzhou. The province was a logistic nightmare for armies and transport officials. Except from one water route from Sichuan into southern Gansu up the *Bailongjiang* 白龍江 where rafts had to navigate dangerous rapids making transport forbiddingly expensive,[54] the province was completely landlocked. There were few good roads that could accommodate mule or ox carts; most transport was done by pack animals.

As a poor border province Gansu had insufficient tax revenue to maintain the province's own administration and military even in peace times and was thus a regular recipient of interprovincial assistance transfers. Local trade was underdeveloped, and the transit trade to Xinjiang relied on one main road through the Gansu corridor which was controlled by the Muslim rebels for most of the time. Moreover, while in Jiangsu a relatively wealthy and loyal population was attacked by external enemy forces of the Taiping, the Gansu rebellion was a civil war between the local Han and Muslim population. The Muslims were not a small minority. Before the war, they comprised up to half of the population, and, according to one observer, owned half of all grain stocks in Gansu. When civil strife broke out and the government decided to side with the Han, Muslim lands became either enemy territory or lay fallow because their owners had fled or were massacred. Tax revenues decreased and the outlook for army provisions was bleak.[55]

Although the war pervaded the whole province, the following discussion will focus on the eastern Gansu war theatre, which includes today's Qingyang, Pingliang and the Ningxia Hui Autonomous Region and was completely cut off from the provincial capital Lanzhou in central Gansu for most of the time between 1863 and 1869. As mentioned in the introduction, the region had one independent *liangtai* in Qingyang. But several armies in the same theatre of war were not under the supply of Qingyang including two so-called 'guest armies' supplied by their sending provinces Shaanxi and Hubei. Mutushan's army in Ningxia had its own field *liangtai*. But for many years the commitment of donor provinces to support the Gansu war was extremely low, and armies trying to live off the land in this poor province devastated by war soon found that they could not do so without casting off the mitigating influence of civilian controlled administrations and resort to extortion and looting. The Qingyang *liangtai* was abandoned. No independent *liangtai* was ever again established in Gansu. Only after 1869, seven years into the war, did a systematic effort of external supply and internal rehabilitation start under Zuo Zongtang who established a network of so-called rear *liangtai* (*houlu liangtai*), transport bureaus (*zhuanyun ju*) and procurement bureaus (*caimai ju*) outside the province which was gradually enlarged until it reached as far as Shanghai and Tianjin. The financial and logistic efforts of the whole empire were needed to recover the province.

The Muslim Rebellion originally started in neighbouring Shaanxi in the early summer of 1862. The government had been organizing local communities into militias to fend off an attack by a Taiping force that year, so that a simple brawl quickly evolved into a major civil war in which government armies sided with Han militias to fight well organized and fiercely combative Muslim militias. The initial response of the Shaanxi provincial government proved disastrous. With Muslim rebels roaming the fertile Guanzhong 關中 plain and repeatedly cutting off the supply lines of the capital Xi'an, provincial authorities lost access to tax revenues and stopped paying salaries to regular Green Standard garrisons throughout both Shaanxi and Gansu (traditionally paid for by Shaanxi). The payments had not resumed three years later.[56]

The ensuing war against the Taiping, Muslims and, later, Nian rebels in Shaanxi and against the Muslims in Gansu was mostly fought by irregular soldiers brought in from outside under commanders who were experienced in earlier battles against the Taiping and Nian rebels in the south and east. For almost eight years (1862–1869) this meant that wave after wave of new troops were recruited (mostly from Hunan and Hubei, but also from Henan and Anhui) and brought into the war theatre under a string of successive military and civil officials (imperially commissioned commanders-in-chief, governors of Shaanxi, governor-generals of Shaanxi and Gansu) with military disasters outnumbering military successes. Government armies spent almost two years to pacify Shaanxi – at least temporarily – and eventually drive the Muslim rebel forces out of the province into Gansu, where the Shaanxi Muslim leaders had already established a solid base of followers since shortly after the beginning of the uprising. By mid 1863, Muslims started to attack the walled cities of eastern Gansu. At the same time, Shaanxi was gradually disposing of its surplus of irregular armies by sending them into Gansu to continue to fight the Muslims there.[57]

Only a part of these troops, mainly those fighting under the above mentioned Lei Zhengwan, were taken under Gansu pay. In December 1863, when Lei Zhengwan's troops reached Gansu from Shaanxi,[58] an imperial edict ordered governor-general of Shaanxi and Gansu Xilin 熙麟 (–1864) to establish a *liangtai* in order to provide for the transport of funds and supplies to the front.[59] Qingyang 慶陽 was chosen as logistic headquarter rather by chance, because Xilin got stuck there on his way to Lanzhou after the main road into the capital was blocked by the rebels. It had robust city walls and the prefecture was one of the few in Gansu that did not have a Muslim population.[60] Because Qingyang was too far from the frontline, its designation was that of a rear *liangtai* (*houlu liangtai* 後路糧臺) whose main task was to receive and distribute interprovincial war funds bound for Gansu.

By 1865, the Qingyang *liangtai* was entitled to receive monthly war fund shipments of 78,000 taels from Sichuan, Shandong, Shanxi, Henan, although some of these funds never materialized. It supplied 52 battalions, of which ten alone were used to protect the *liangtai*. Thirty battalions (about 15,000 men, later said to have grown to 40 battalions) under Lei Zhengwan were bearing the brunt of fighting. A secondary *liangtai* had to be set up south-east of Qingyang in Jingzhou to be closer to the front line. The Qingyang *liangtai* also supplied a small cavalry force (1,000 men) under Duxinga 都興阿, formally commander-in-chief, that operated near Ningxia 寧夏 (today Yinchuan 銀川) in the north to relieve the Manchu garrison of the town, and, for a short period of time, local forces in Dingbian 定邊 in northern Shaanxi (160 km south-east of Ningxia).[61] The two subsequent directors of the Qingyang Liangtai, Wenyu 文煜 (–1884) and Yang Nengge 楊能格, were not local officials, because officials of appropriate rank were not resident in the area. Both had previous experience at both the Jianghan and Jiangbei *liangtai*. Both were made responsible for military defeat and dismissed from their previous positions. They were sent to Qingyang to accumulate sufficient merit for a reinstatement.[62] When Yang was appointed, the provincial judge Zhang Jixin on duty with Shaanxi troops nearby admitted in his diary that he was 'an old hand in logistics', but added mockingly that, 'without rice to cook, he will lack the long sleeves necessary for a good dance'.[63] Indeed Yang Nengge would fail as head of the *liangtai* only a few months later.

Available accounts of the Qingyang *liangtai* between 1864 and 1866 show that it had almost no local income, but was completely dependent on interprovincial assistance transfers. Of the statutory 78,000 taels per month, the *liangtai* actually received less than 43,000 taels. The accounts also show that the protection of the fortified *liangtai* itself consumed 40 per cent of all expenditures, even though only ten battalions were deployed there. Another 40 (under Wenyu) to 50 (under Yang Nengge) per cent reached the 30 to 40 battalions at the front line *liangtai* in Jingzhou, while a small amount was sent north to Ningxia.[64] The Jingzhou *liangtai* was directly responsible for provisioning Lei Zhengwan's troops.[65] Staff in the Jingzhou *liangtai* consisted of local officials who were considerably lower ranking than that of the central Qingyang *liangtai*, its head was acting *zhou* magistrate Lin Fashen 林發深.[66] In fact, however, the funds were managed not by the *liangtai*, but by Lei Zhengwan's quartermaster (*guan yingwu* 管營務) Zhou Xiancheng 周顯承, who was also commander of the battalions which guarded the supply station.[67]

To make things worse, not all troops in the eastern Gansu theatre of war fell under the jurisdiction of the Qingyang *liangtai*. Two 'guest armies' under Tao Maolin 陶茂林 and Cao Kezhong 曹克忠 (Zhang Jixin was accompanying the latter) – together 33 battalions or roughly 20,000 men – and two Manchu cavalry squadrons were under pay of the Shaanxi *liangtai* in Xi'an.[68] These two armies established independent supply lines, although they operated in the same war theatre as Lei Zhengwan. The Qingyang *liangtai* also flatly refused to also supply Mutushan's 穆圖善 20 battalions of Hubei Army irregulars, although they operated in the Ningxia theatre together with Duxinga. Thus Mutushan was forced to maintain his own field *liangtai*, which received interprovincial assistance from Shanxi and Shaanxi.[69]

In early 1864, Shaanxi spent more than its revenue from taxes and interprovincial assistance on paying for the war (the deficit was covered by paper money or by defaulting on the pay of the regular Green Standard army). Half of the expenditures were disbursed through the provincial *liangtai*, which financed the armies fighting outside the provincial capital and spent one-third of its funds for troops fighting in Gansu.[70] By the end of the year, Shaanxi supported a total of 53 battalions in the eastern Gansu theatre, most of them irregulars organized according to Hubei Army regulations. Their commanders and some of their troops had fought in Shaanxi before they were transferred to Gansu, but commanders also constantly recruited new soldiers, mostly on the spot in Shaanxi and Gansu.[71]

In fact, Shaanxi only paid a fraction of the Gansu armies' statutory monthly pay (8,000 of 47,500 tael according to Tao Maolin.)[72] Instead, governor Liu Rong in early summer 1864 sent his own Hunan troops, which had brought him from Sichuan into Xi'an one year earlier, as 'guest armies' into southern

Gansu.[73] To pay for them, he applied with the throne for the allocation of interprovincial assistance funds. Both southern and eastern Gansu forces received such allocations, Sichuan province and the foreign customs offices in Jiangsu and Fujian in Tao Maolin's case and from Zhili, Shandong and Shanxi in the case of the southern Gansu armies. But none of these actually arrived.[74] The historiographers of Zuo Zongtang's successful Gansu campaign later alleged that the sending of 'guest armies' was a common trick of cash strapped provinces to avoid responsibility. A governor could hope to hit three birds with one stone with this move: first, he could avoid demobilizing his mercenary armies by sending them abroad; second, he could gain the reputation of supporting a neighbouring province in its defence efforts; third, he could try to force wealthier provinces to increase their war fund transfers.[75]

The combination of this institutional set-up and the constant lack of funds and grain together proved lethal. In 1865 the situation escalated, and the hungry armies started rioting, extorting grain from the population, and pillaging villages and towns. First, May Tao Maolin's troops mutinied and many of the troops deserted, looting and killing on their way back to Shaanxi.[76] In July, government troops suffered a debilitating defeat at Jinjibao, the headquarter of the Muslim rebels. Lei Zhengwan was stripped of his title of deputy commander-in-chief, but in fact remained in charge of the eastern Gansu troops. In October, Lei Zhengwan's troops rioted and pillaged the towns Guyuan and Pingliang and several villages in the vicinity. Parts of the mutineers were stopped and reassembled by quartermaster Zhou Xiancheng in Jingzhou.[77]

Emotions ran high among the civil–military leadership about who was to blame. Two theories were brought forward. One faction suspected that Lei Zhengwan or some of his officers, disaffected by Lei's loss of command, colluded with the enemy to effectively abandon Guyuan. This version was supported by the magistrat Lin Fashen in charge of the Jingzhou *liangtai* and eventually by governor of Shaanxi Liu Rong and governor-general Yang Yuebin.[78] In fact, two of Lei Zhengwan's officers, one of them his brother Lei Heng, seem to have joined the rebels.[79] The other version, mainly Lei Zhengwan's own explanation, was that the Guyuan mutiny was a hunger riot that resulted from a collusion of Lin Fashen with quartermaster Zhou Xiancheng to withhold funds and supplies from the forces fighting at the front line. This version was strongly supported by Yang Nengge, manager of the central *liangtai* in Qingyang. Interestingly, the Qingyang *liangtai* was not blamed to withhold funds from Jingzhou.[80]

Whatever the truth was, the fallout from the riot was to reveal the deep cleavages between the various administrators involved which aggravated the current food crisis. Yang Nengge supported Lei Zhengwan and refused to supply the mutinying soldiers that Zhou Xiancheng had reassembled into an army.[81] Yang accused Zhou of wrestling authority from the civilian *liangtai* commissioners. He kept the books and never gave account about his revenue which included the local Lijin tax, contributions for transport fees (*juanshou fuma jiaojia* 捐收夫馬腳價) and grain contributions. After the break with Lei

Zhengwan, Zhou Xiancheng became even more ruthless. He used coercive methods to extort grain contributions from the people, allocating quota without regard to actual wealth of the families. In Ningzhou, a county belonging to Qingyang prefecture, he set a grain quota for each of the four sub-county boroughs and announced that he would come with his ten battalions of troops to supervise the delivery.[82] When the troops finally arrived a few days after the announced days, they were quartered in people's houses where they looted, raped and even killed inhabitants who resisted. Several villages in the county were completely pillaged.[83]

Governor-general Yang Yuebin, however, who was a military man, came to the defence of Zhou Xiancheng. Yang appreciated the efficiency of Zhou Xiancheng's work as a supply officer. He had experienced the collapse of the grain market in Gansu himself, which happened just at about the time of the riot, when, as newly appointed governor-general en route to the capital Lanzhou, he could not procure any grain and had to cut the daily rations of his army.[84] Thus, he conceded that given the situation in Gansu, Zhou had no other choice than to use coercion to extort supplies from the people. Recalling to his own experience, he said:

> When I came to Gansu I still had silver to purchase grain, later when war funds did not come in, I sent out upright gentry members to solicit contributions. But as for contribution there was no grain store which opened for our support, and for purchases, [the merchants] hoarded their grain to push up prices. Everybody hid his grain in deep caves and pretended that he did not have a bushel at home. If the said general took away grain by force, it was because he had no other choice.[85]

Instead of being punished, he concluded, Zhou should still be responsible for military supplies. This was despite the fact, that the Ningzhou incident was by no means a singular incident. Zhou was well known in the area for his ruthless methods of requisitioning food supplies for his troops and gained the nickname Zhou 'Hellgod' in the Qingyang area long before Yang Nengge made his accusation.[86] In the final verdict, Zhou Xiancheng was vindicated and remained in charge of military supplies, albeit stripped of his rank. When he died two years later in battle he was remembered as a military hero. Only one of the army's grain procurement agents was actually executed for killing and looting, but he was also known as an adversary of Zhou; others got minor penalties.[87]

Qing armies were small by contemporary European standards, but Gansu could not sustain even these armies in the long run. All the well-tested strategies of living off the land that had been helping the Qing military to survive multiple large-scale rebellions in the south and east of the empire – requisitioning stocks of county granaries, proxy purchase by county administrations to be refunded by the army or later by provincial authorities, purchase from merchants or grain contributions from the population – did not work here.

Even if there was grain to buy, army funds remained unpaid, and the troops relied on contributions to a larger degree than in other corners of the empire. Of course, government troops were not supposed to use coercion to compel people to contribute. Gansu did have a legal basis for rewarding contributions with rank and office, the Gansu Contribution Regulations (Gansu Juanshu Zhangcheng 甘肃捐输章程), according to which grain deliveries to the army camps were commuted into a silver value according to the market price, and this silver value was then used to calculate the price of offices and rank to bestow upon the contributor.[88] Agents of both the Qingyang *liangtai* and the armies also extended their search far into Shaanxi province, as seen in the example in the introduction, mostly asking for silver in these places since grain was considered too bulky to be transported back into Gansu. During the first two or three years, army requests were met with a certain degree of enthusiasm on the side of the Han Chinese populations who greeted the government military as allies in their war against the Muslims. In Qingyang, local gentry volunteered to organize contributions with good success.[89]

However, resources soon wore thin. Shaanxi's provincial judge Zhang Jixin stayed with the almost 10,000-strong army of Cao Kezhong, one of Shaanxi's 'guest armies', outside the walls of the county seat of Jingning in eastern Gansu for a little more than two months from late November 1864 to February 1865 with contributions being their only source of supplies.[90] Zhang found out that the county authorities in Jingning had earlier tried to collect 400,000 catties of wheat flour to support the approaching army. Moreover, every household was made to pay 80 pieces of copper cash to pay for the district's own defence measures and militia organization. When the army finally arrived, it turned out that only over 200,000 catties had been delivered, but were already consumed by the town's own militia while the contributed money had disappeared into the pockets of corrupt officials.[91] The standard daily grain ration of a man was 1.6 catties, so 400,000 catties could have fed Cao Kezhong's troops for almost a month, but now Zhang had to send out agents in search for contributions. The villages and towns mentioned in his diary included a radius of 200 km.[92]

A local expectant prefect (probably a purchased rank) compiled a list of wealthy households in Zhuanglang and offered to collect contributions. When he arrived, the assistant magistrate tried to protect his constituency and closed the gates of the town to obstruct these attempts. Although Zhang Jixin impeached the assistant magistrate in order to give his agent a free rein, the latter soon reported that he hadn't found any wealthy households contradicting his earlier statements. After a month, only 100 taels of silver were collected.[93] Initially the army did not use coercion to solicit contributions. In the cases described by Zhang, one or two agents were sent without armed guards. If there were two agents, one was a civilian official, the other one held a military title. Higher-ranking officials were favoured. If more people appeared in front of the town gates, this could lead to serious irritations.[94] The agents would persuade the local gentry to make pledges, and the gentry was also involved in the process of persuasion and collection. However, soon after Zhang Jixin was unlucky in Zhuanglang, Lei Zhengwan sent his own agents to the place who were said to extort contributions using coercive methods.[95] In the end, efforts to obtain provisions locally in the poor province of Gansu proved unsuccessful. The hungry and demoralized troops started to mutiny leading to military backlash and plunging the province into years of bloody carnage that left 70 per cent of its population dead.[96]

By the end of 1865, it was finally clear that grain could not be procured locally in Gansu and that the independent civilian-staffed *liangtai* was not a sustainable model for fighting a war in this province. The new governor-general of Gansu, Yang Yuebin, impeached director of logistics, Yang Nengge, for his oversized operational budget that left front-line troops hungry, alleging that the only reason why Lei Zhengwan had not blamed Yang for the collapse of his troops was because the two were good friends.[97] Eventually, Yang Yuebin and Shaanxi governor Liu Rong requested the abolition of the Qingyang *liangtai*, arguing that the *liangtai* was a burden for Qingyang prefecture, but of no advantage for the army in eastern Gansu. The request included a proposal for a new institutional structure which would concentrate responsibility for supplying the Gansu army outside the province. Xi'an should establish a rear *liangtai*, which would be headed by provincial treasurer Lin Shoutu 林壽圖, and include a procurement bureau (*caimaiju* 採買局), a manufacturing bureau (*zhizaoju* 製造局) and a transport bureau (*zhuanyunju* 轉運局). Jingzhou 涇州, Qinzhou 秦州 (in southern Gansu where Shaanxi 'guest armies' were deployed and also an alternative route to Lanzhou) and Qingyang 慶陽 would be designated as branches of the latter with only the basic staff necessary to forward funds and, more importantly, grain to the troops.[98]

With the new setup Yang Yuebin not only conceded that grain had to be purchased in Shaanxi and shipped to Gansu, but he also proposed that to unify the war chest in the hands of Shaanxi's provincial authorities would eliminate the unfortunate separation between the Gansu *liangtai* and Shaanxi's 'guest armies'. It would bundle interprovincial assistance in one hand and would entice Shaanxi's authorities to increase their efforts in fighting the war in Gansu.[99] Yang Yuebin also started to enlarge the network of supply and transport bureaus beyond Gansu. He had earlier established a rear *liangtai* in Wuchang in Hubei, mainly to collect interprovincial assistance funds for his own army which accompanied him on his way into Gansu.[100] And while Shaanxi governor Liu Rong as late as in mid-1865 still refused to establish a network of grain procurement bureaus in his province,[101] Yang at the end of the year established two such bureaus close to the Gansu border, one in Hanzhong to supply Lanzhou via Qinzhou and one in Fengxiang to supply the eastern Gansu war theatre with the branch bureau in Jingzhou.[102] Additionally, he requested permission to solicit grain contributions in several counties in Shaanxi as well as in northern Sichuan which had a river connection into southern Gansu.[103] But little came to fruition during Yang's short tenure (he would resign in 1867), and it would be his successor Zuo Zongtang who fully developed an external supply network for Gansu that became a textbook case in Chinese military history.

However, Zuo Zongtang did not enter Gansu before 1869. In late 1866, Shaanxi plunged back into war due to a contemporaneous invasion of a Nian rebel force from the east and starved Muslim rebels pouring back into Shaanxi from the west.[104] Zuo was appointed governor-general of Shaanxi and Gansu in September of that year, but since he was also commander-in-chief of the Gansu–Shaanxi wars and was leading his own army in campaigns against the Nian and Muslim rebels in both provinces, he did not arrive in Lanzhou before November 1869. In the meantime, Mutushan, whose successful reconciliation policy had solved the stalemate with the Muslim force in Ningxia, resided in Lanzhou as acting governor-general.[105]

Yang Yuebin's attempt to unify the war chest in the hands of the rear *liangtai* in Xi'an failed. In early 1867, he ordered the relocation of the rear *liangtai* including all its bureaus to Hanzhong prefecture (actually Lueyang) in south-western Shaanxi. Ostensibly this was because the rebellion blocked the road between Xi'an and eastern Gansu, but the fact that the *liangtai* was to be managed by an expectant circuit intendant (rank 4a) instead of a man of the rank of provincial treasurer as in Xi'an raises the suspicion that Yang as governor-general of Gansu hoped to gain more control over the supply station.[106] This suspicion was soon confirmed by the protracted conflict between Yang's successor Zuo Zongtang and his proxy in Lanzhou, acting governor-general Mutushan. While Zuo Zontang reversed Yang Yuebin's order and instructed the rear *liangtai* to stay in Xi'an, Mutushan, as the de facto superior official of *liangtai* manager and provincial treasurer Lin Shoutu accused the Xi'an *liangtai* of withholding funds from Lanzhou in and favouring the Qingyang war theatre and continued to order the relocation of the *liangtai*. The struggle, which the throne haplessly tried to mitigate, dragged on for a year until it was finally decided that both *liangtai* would continue to coexist. Provinces were ordered to ship half of all interprovincial assistance funds directly to Hanzhong, thus circumventing the Xi'an *liangtai*.[107]

In fact, Qingyang was hardly preferred. By 1868, the situation in Qingyang town had deteriorated so much that the guard army and the staff of the transport bureau left for the countryside, because they could not find food in the town. So exposed, the town was looted alternatively by rebels and equally hungry government troops whose local commander reportedly tortured inhabitants to death to find the last grain hideouts. Only in spring 1869 did a systematic government offensive retake the town. By this time, Qingyang was devastated. Reportedly it took three months until the stench of the rotting corpses had disappeared and a new civil administration could move in.[108] When the new governor-general, Zuo Zongtang, finally entered Gansu in spring 1869, he wondered that it had become an entrenched practice of armies to extort food by force from the people. In one case, a battalion was deployed permanently in a county which was forced to supply it with wheat flour on a regular basis, amounting to annual deliveries that exceeded the regular tax quota of the county by several times. Zuo had been fighting in the south for more than a decade, but claimed to have never seen anything like that.[109]

Establishing an empire-wide supply network

After a series of unsuccessful attempts to quell the uprising in Gansu, Zuo Zongtang was the last leader entering Shaanxi and Gansu with an army of Hubei Army irregulars. He was not only credited with military success, but also with establishing a functioning logistic network which solved the food crisis of the Gansu war. However, Zuo Zongtang was also known for moving his troops only when supplies could be secured. And this meant that success did not come quickly. It took another one-and-a-half years before the Muslim stronghold at Jinjibao was quashed by Zuo's troops, and not before November 1873 was the rebellion suppressed in all of Gansu.[110]

Zuo's efforts to secure Gansu were helped by the suppression of the Nian Rebellion, in which he of course had a part, which allowed an economic recovery of the eastern provinces. Now, altogether 11 provinces were participating in interprovincial assistance for Zuo's campaign, and although payments were still not completely reliable and the war chest, which was styled the Western Expedition Campaign Fund (*xizheng xiangxang* 西征餉項), received only about 65 per cent of the transfers the throne allocated to the provinces, but the total was still over 30 million taels making an annual average over seven years (1867–1873) of slightly over four million per year.[111] Interprovincial assistant funds were supplemented by an empire-wide contribution campaign starting in early 1868, which yielded a reported seven million taels or 17 per cent of the total income of the war chest until 1873.[112]

After he entered Gansu in 1869, Zuo used these funds to establish and improve an empire-wide logistic network that made it possible to transport all the grain and equipment used by the armies into Gansu from outside. The network still included the two rear *liangtai* in Xi'an and Wuchang, which Zuo Zongtang took over from Yang Yuebing, but restaffed with his own confidants. But they were mainly in charge of interprovincial assistance fees and were kept away from disbursement, which was traditionally under the control of the *liangtai*, but now reported directly to Zuo Zongtang.[113] Zuo was very careful to keep the logistic system under his direct control. Shortly after his appointment in 1866, Zuo had commissioned a close confidant, Wang Jiamin, to head the Hubei rear *liangtai*. This *liangtai* collected the interprovincial assistance payments from the south-eastern provinces and shipped them up the Han and Danshui rivers to Jingziguan 荊紫関 and from there overland to Xi'an. Through all the struggles for funds and authority, the Hubei rear *liangtai* faithfully served its purpose and forwarded funds to both Xi'an and Hanzhong, as well directly to the paymaster bureau (*zhiying ju*) of Zuo Zongtang's army, which was not under the Xi'an *liangtai* (which by that time exclusively served Gansu), while it was operating in Shaanxi.[114]

Only in the spring of 1869 did Zuo Zongtang manage to unify the war chest of his own army, with those of Gansu and Shaanxi province under the name of the Western Expedition Fund. In order to give more impetus to his demands for interprovincial assistance, he then replaced the Xi'an rear *liangtai* with the

Western Expedition *Liangtai* under Yuan Baoheng 袁保恒. In both Hubei and Xi'an, the *liangtai* managers were lower ranking officials (Wang was an expectant circuit intendant, rank 4a; Yuan had a metropolitan office as Reader of the Hanlin Academy, rank 4b) devoted full time to the management of the *liangtai*.[115] None of them had anything to do with the disbursement of the interprovincial assistance funds they collected and forwarded. The paymaster general of the campaign was Shen Yingkui 沈應奎, also a circuit intendant by rank and manager of the Central Military Supply Bureau (*zongli junxu ju* 總理軍需局) in Xi'an, which was responsible for the salaries of the troops (including the salaries of the *liangtai* employees) and the procurement of grain and equipment.[116] The Military Supply Bureau also controlled the network of bureaus for procurement and transport which was reaching as far as Shanghai and Tianjin on the eastern coast.[117]

Zuo Zongtang would later justifying his decision to dispose of an independent *liangtai*, commissioned by imperial decree and high enough in rank to communicate with a governor-general at par by referring to the greater efficiency of a clear line of command which he had seen in the coastal settlements of the Western powers in China:

> The commissioners of each bureau are selected by careful inspection and continuously supervised. Abuse and negligence are not tolerated. And according to the bureau statutes, a ten-day-report has to be filed no later than three days after the decade in question, a monthly report no later than ten days after the end of the reported month. There is a succession from the one to the other, so that abuses are easy to detect. In light cases, the commissioner is dismissed, in serious cases there will be an investigation. They are supervised by the circuit intendants of Ganzhou, Liangzhou and Anxi [in western Gansu], but I hold the final control, and the officers and soldiers commissioned do not dare to neglect their duties. The good thing is that their official rank is relatively low so that they are easy to remove, and with a tightly [controlled] bureau and pressing deadlines, it is difficult to commit abuses. This is like in the Western trade cities where there are few official, but affairs thrive. If we investigate previous practice in [Chinese] warfare, mostly were *liangtai* established, with those in charge given huge responsibilities over the income and expenditures. And that too many of these ended in embezzlement and corruption, is a clear proof [of the inefficiency of the *liangtai* system].[118]

By 1869, when Zuo Zongtang had finally managed to set up a unified supply system, large parts of both Gansu and Shaanxi were scorched earth. Western travellers who came to Gansu in the 1880s still found many villages and whole towns in ruins and unoccupied.[119] When Zuo Zongtang's troops entered Qingyang prefecture, there were hardly any survivors to greet them as liberators, and those whom they found fled on the sight of government troops who were hardly distinguishable from robbers and rebels. Only gradually did surviving gentry members came forward with accusations against military commanders. A few minor officers were killed and their heads paraded for the angry locals, but the main perpetrators Lei Zhengwan and his brother Lei Shengyuan would not be touched.[120]

During the ensuing five years none of the problems that had plagued Gansu for such a long time completely ended. Soldiers occasionally still pillaged, there were still guest armies, funds continued to be short and the armies collected local taxes and contributions.[121] However, Zuo Zongtang could avoid much of the haggling between overlapping military and civilian authorities that precluded any effective operation under his predecessors. He moved cautiously, kept his army small but supplied by importing grain from outside the province, settling refugees and founding military colonies. Under Zuo Zongtang, the army, from being a predatory force in conflict with civilian authorities who tried to protect their constituencies, also became a leading force in rehabilitation measures itself, often using military funds to feed refugees and provide them with seeds and tools to recover the fallow lands.[122] This strategy in addition to the exhaustion and devastation of the region helped him to finally win the war and start the re-conquest of Xinjiang.

Conclusion

During the Era of Rebellions, much of the Chinese empire was engulfed in a total war, using all available resources to suppress multiple rebellions. A mixture of interprovincial assistance and local resources provided the material basis for most of the Qing armies, and the conflict between the pressing needs of a violent force and a volatile population was mitigated by a complicated bureaucratic relationship between civil and military administrations as well as the system of contributions that allowed the exaction of resources from the wealthy gentry in exchange for giving them a stake in the government's success by selling them office and rank. This strategy failed in resource poor Gansu province in the north-west to the extent that the armies were overriding civil administrations and indulged in looting and extortion. In order for the province to emerge out of the mayhem, it had not only to be completely supplied by an empire-wide surge in interprovincial commitment, but the complicated network of checks and balances between various provincial and military administrations was also brought under the control of governor-general Zuo Zongtang who unified the war chests of two provinces with that of his own army and streamlined the supply chain under his command. At the same time, given the devastation of the population and administration of Gansu, the military leadership also had to take over the management of civil rehabilitation measures previously assumed by civil administrations.

Notes

1 Jiang Tonggang 姜同岡 and Guo Siwei 郭四維 (eds), *Sanshui xianzhi* 三水縣志, (n.p. 1873), vol. 4, ch. 11, 15a–16b.
2 Cao Shuji 曹樹基, *Zhongguo renkou shi: di wu juan, Qing shiqi* 中國人口史: 第5卷, 清时期 (Population History of China: Vol. 5, Qing Dynasty). (Shanghai, Fudan daxue chubanshe, 2001), pp. 592–593.
3 Jiang Tonggang and Guo Siwei (eds), *Sanshui xianzhi*, vol. 2, ch. 6.
4 Lu Weidong 路韋東, 'Qingdai Shaanxi huimin de renkou biandong 清代陝西回民的人口變動' (Changes in Muslim Population in Shaanxi during the Qing Dynasty), *Huizu Yanjiu*, vol. 52, no. 4 (2003), pp. 71–77.
5 Jiang Tonggang and Guo Siwei (eds), *Sanshui xianzhi*, ch. 11, 19a.
6 Hui Dengjia 惠登甲, *Qingfang Jilue* 慶防紀略 (History of the Defence of Qingyang) (ed. Ma Xiao 馬嘯) (Tianjin: Tianjin guji chubanshe, 2011), p. 82.
7 Cao Shuji, *Zhongguo renkou shi*, p. 618.
8 Zhang Jixin 張集馨, *Dao Xian huanhai jianwen lu* 道咸宦海見聞錄 (An Eyewitness Account of the Officialdom during the Daoguang and Xianfeng Reigns, 1820–1861) (Beijing: Zhonghua shuju, 1981), pp. 365, 371, 380.
9 Jiang Tonggang and Guo Siwei (eds), *Sanshui xianzhi*, vol. 2, chap. 5.
10 For studies of military finance and logistics in the eighteenth century see: Dai Yingcong, 'The Qing State, Merchants, and the Military Labor Force in the Jinchuan campaigns', *Late Imperial China*, vol. 22, no. 2 (December 2001), pp. 35–90; Ulrich Theobald, *War Finance and Logistics in Late Imperial China: A Study of the Second Jinchuan Campaign (1771–1776)* (Boston, MA: Brill, 2013); Lai Fushun 賴福順, *Qianlong zhongyao zhanzheng zhi junxu yanjiu* 乾隆重要戰爭之軍需研究 (Military Supply of Qianlong's Important Campaigns) (Taibei: Guoli gugong bowuyuan, 1984). A summary of military finance until 1840 is: Chen Feng 陳鋒, *Qingdai junfei yanjiu* 清代軍費研究 (Military Finance in the Qing Dynasty) (Wuhan: Wuhan daxue chubanshe, 1992). For late nineteenth-century logistics, see: Chen Chonqiao 陳崇橋 and Zhang Tianyu 張田玉, *Zhongguo jindai houqinshi* 中國近代軍事後勤史 (A history of military logistics in modern China) (Beijing: Jindun chubanshe, 1993); Gao Genghao 郜耿豪, 'Lun Yongying bingzhi xia de jindai liangtai 論勇營兵制下的近代糧臺' (The Modern Liangtai under the Irregular Army System), *Hunan Daxue Xuebao*, vol. 19, no. 1 (January 2005), pp. 89–104; Gao Genghao 郜耿豪, 'Wanqing junshi houqin yanjiu shulun 晚清軍事后勤研究述論' (A Survey of Research on Military Logistics in Late Qing China), *Junshi Lishi Yanjiu*, no. 3, (2002), pp. 175–188; Gao Genghao 郜耿豪, 'Lun jingzhi bingzhi xia de chuantong liangtai 論經制兵制下的傳統糧臺' (The Traditional Liangtai under the Standard Army System), *Junshi Lishi Yanjiu*, no. 4 (2004), pp. 96–104.
11 Franz H. Michael, *The Taiping Rebellion: History and Documents* (Seattle, WA: University of Washington Press, 1966–1971).
12 Ssu-yü Teng, *The Nien Army and their Guerrilla Warfare, 1851–1868* (Paris: Mouton, 1961).
13 Ling Ti'an, *Xian Tong Guizhou junshi shi* (Taibei: Wenhai chubanshe, 1967); Robert Darrah Jenks, *Insurgency and Social Disorder in Guizhou: The 'Miao' Rebellion, 1854–1873* (Honolulu, HI: University of Hawaii Press, 1994).
14 David G. Atwill, *The Chinese Sultanate: Islam, Ethnicity, and the Panthay Rebellion in Southwest China, 1856–1873* (Stanford, CA: Stanford University Press, 2005).
15 Wen-chang Chu, *The Moslem Rebellion in Northwest China, 1862–1878: A Study of Government Minority Policy* (The Hague, Mouton, 1966).
16 Douglas Hurd, *The Arrow War: An Anglo-Chinese Confusion, 1856–1860* (London, Collins, 1967); Jack Beeching, *The Chinese Opium Wars* (London: Hutchinson, 1975).

17 Liu Rong in a memorial dated 27 November 1864 (TZ 3.10.29.), in *Liu Zhongcheng (Xiaxian) zoushu* 劉中丞(霞仙)奏疏 (Memorials by Liu Rong) (Taibei: Wenhai chubanshe, 1974), vol. 1, ch. 6, pp. 429–440.

18 Cao Shuji, *Zhongguo renkou shi*, pp. 690–772. Similar estimates are given by Ge Jianxiong 葛劍雄, Hou Yangfang 侯楊方 and Zhang Genfu 張根福, *Renkou yu Zhongguo xiandaishi: 1850 nian yilai* 人口與中國現代史: 1850年以來 (Population and China's Modern History: Since 1850) (Shanghai: Xuelin chubanshe, 1999), pp. 4–111; Lu Yu 路遇 and Teng Zezhi 騰澤之, *Zhongguo renkou tongshi* 中國人口通史 (Jinan: Shandong renmin chubanshe, 2000), pp. 790–802.

19 Cao Shuji, *Zhongguo renkou shi*, pp. 691–701. Provinces with a population gain include Shandong, Zhili, Hunan, Sichuan, Guangxi, Guangdong and Guizhou.

20 Thomas Francis Wade, 'The Army of the Chinese Empire', *Chinese Repository XX* (1851), pp. 254–255.

21 John King Fairbank, Kwang-Ching Liu and Denis Crispin Twitchett (eds), *Late Ch'ing, 1800–1911*, vol. 11, Pt 2 of the Cambridge History of China Series (illustrated edn) (Cambridge: Cambridge University Press, 1980), pp. 202–203.

22 Philip A. Kuhn, *Rebellion and its Enemies in Late Imperial China: Militarization and Social Structure, 1796–1864* (Cambridge, MA: Harvard University Press, 1970), pp. 37–63.

23 'Tuanlian Juanzi 團練捐資' (Contributions to Militias), in Board of Revenue (ed.), *Qinding hubu haifang zhenggong xinli zhangcheng* 欽定戶部海防鄭工新例章程 (Imperially Approved New Statutes of the Board of Revenue for Maritime Defense Contribution and the Zhengzhou River Building Contribution) (Taibei: Chengwen chubanshe, 1968), vol. 2, pp. 614–615. For *tuanlian* as an 'adjunct to the system of rank purchase' see Kuhn, *Rebellion and its Enemies in Late Imperial China*, p. 205.

24 'Tuanlian Juanzi 團練捐資'.

25 Chen Feng 陳鋒, *Qingdai junfei yanjiu* 清代軍費研究 (Military Finance in the Qing Dynasty) (Wuhan: Wuhan daxue chubanshe, 1992), pp. 8–9.

26 Elisabeth Kaske, 'Fundraising Wars: Office Selling and Interprovincial Finance in Nineteenth Century China', *Harvard Journal of Asiatic Studies*, vol. 71, no. 1 (June 2011), pp. 69–141.

27 During the suppression of the Lin Shuangwen Rebellion on Taiwan in 1787–1788, rice was procured and transported from provinces as far as Sichuan and Huguang (Hunan and Hubei): Paul Lococo, 'The Military Campaign to suppress the Lin Shuangwen Rebellion, 1787–1788' (PhD dissertation, University of Hawaii, 1998), pp. 97–98, 256.

28 Levying the Lijin tax was under the administration of the liangtai, see Qiao Songnian memorial dated November 1860 (XF 10.10.), in *Qiao Qinke gong (Songnian) zouyi* 喬勤恪公 (松年)奏議 (Memorials of Qiao Qinke gong (Songnian)) (ed. Qiao Lianbao 喬聯寶) (Taipei: Wenhai chubanshe, 1971), vol. 1, pp. 241–244.

29 For the eighteenth century see Kent R. Guy, *Qing Governors and their Provinces: The Evolution of Territorial Administration in China, 1644–1796*, (Seattle, WA: University of Washington Press, 2010).

30 Gao Genghao 郜耿豪, 'Lun jingzhi bingzhi xia de chuantong liangtai' 论经制兵制下的传统粮台, *Junshi Lishi Yanjiu*, no. 4 (2004), pp. 96–104.

31 Qiao Songnian, memorial dated 26 January 1860 (XF 10.1.26.), in *Qiao Qinke gong (Songnian) zouyi*, vol. 1, pp. 49–50.

32 Qiao Songnian, memorial dated 2 July 1860 (XF 10.5.14.), in *Qiao Qinke gong (Songnian) zouyi*, vol. 1, pp. 147–154; Qiao Songnian, memorial dated July 1860 (XF 10.6.), in *Qiao Qinke gong (Songnian) zouyi*, vol. 1, p. 185.

33 Li Hongzhang, memorial dated 10 December 1862 (TZ 1.10.19.), in *Li Hongzhang quanji* 李鴻章全集 (Collected Works of Li Hongzhang) (eds Gu Tinglong 顧廷龍 and Dai Yi 戴逸) (Hefei: Anhui jiaoyu chubanshe, 2008), vol. 1, pp. 129–130.

34 Qiao Songnian, memorial dated July 1860 (XF 10.6.), in *Qiao Qinke gong (Songnian) zouyi*, vol. 1, pp. 179–183.

35 The accounts on which these estimates are based are to be found in the Grand Council copies of the palace memorials (*Junjichu lufu zouzhe* 軍機處錄副奏摺, hereafter abbreviated *JLZ*), category 'military supplies' (*junxu* 軍需), microfilm roll 11, as held in the First Historical Archives in Beijing. The document numbers are (in parentheses the dates by the Chinese calendar, XF=Xianfeng reign (1851–1961), TZ=Tongzhi reign (1862–1874, followed by month and day): 4802–31, 32 (XF 7.9.13.–XF 7.10.25.); 4802–33, 34 (XF 7.10.25.–XF 8.3.6.); 4803–18, 19 (XF 9.1.20.–XF 9.6.30.); 4803–17 (XF 9.7.1.–XF 9.12.30.); 4803–16 (XF 10.1.1.–XF 10.8.13.); 4804–90 (XF 10.8.20.–XF 11.6.30.); 4804-91, 92 (XF 11.7.1.–TZ 1.2.5.); 4805-36 (TZ 1.2.6.–TZ 1.6.30.); without number (TZ 1.7.1.–TZ 1.11.21.). Unfortunately, the data in the accounts are cumulated in a way that makes it difficult to determine the exact amounts of interprovincial assistance payments for each province or their exact share in the total.

36 As Note 35.

37 Du Wenlan 杜文瀾, *Jiang nanbei daying jishi* benmo 江南北大營紀事本末 (A Chronological Account of the History of the Jiangnan and Jiangbei Army Groups) (Taibei: Tailian guofeng chubanshe, 1969), ch. 2, pp. 66–74. Wang Jianhua falsely assumes that the Jiangbei *liangtai* supplied the Jiangnan Daying under Hechun 和春 until its collapse in 1860, see: Wang Jianhua 王建華, 'Lun 1856–1860 nian de qingjun jiangnan daying 論 1856–1860 年的清軍江南大營' (On the Jiangnan Army Group of the Qing armies, 1856–1860) *Junshi Lishi Yanjiiu*, no. 3 (1993), pp. 117–124.

38 Qiao Songnian, memorial dated 17 February 1860 (XF 10.1.26.), in *Qiao Qinke gong (Songnian) zouyi*, vol. 1, p. 35.

39 Qiao Songnian, memorial dated 17 February 1860 (XF 10.1.26.), in *Qiao Qinke gong (Songnian) zouyi*, vol. 1, p. 35.

40 Accounts for Yangzhou, 10 March to 1 August 1864 (TZ 3.2.3.–TZ 3.6.29.), in *Li Hongzhang quanji*, vol. 2, pp. 214–215.

41 Qiao Songnian, memorial dated 17 February 1860 (XF 10.1.26.), in *Qiao Qinke gong (Songnian) zouyi*, vol. 1, p. 35.

42 Accounts of the Xuzhou liangtai in *JLZ*, military affairs (junwu 軍務), roll 10: 21 May 1860 to 29 January 1862 (XF 10.4.–XF 11.12.), 4793-72; 22 November 1862 to 13 August 1863 (TZ 1.10.1.–TZ 2.6.29.), 4798-31; 14 August 1863 to 7 February 1864 (TZ 2.7.1.–TZ 2.12.30.), 4798-34; 8 February to 1 August 1864 (TZ 3.1.1.–TZ 3.6.29.), 4798-32; 2 August to 17 September 1864 (TZ 3.7.1.–TZ 3.8.17.), 4798-33. For the practice to use paper note contribution to amortize debt to the soldiers see Elisabeth Kaske, 'Silver, Copper, Rice, and Debt: Monetary Policy and Office Selling in China during the Taiping Rebellion', in Jane Kate Leonard and Ulrich Theobald (eds), *Money in Asia (1200–1900): Small Currencies in Social and Political Contexts* (Leiden: Brill, 2014), pp. 399–460.

43 Wu Tang, memorial dated 15 April 1863 (TZ 2.2.28) [date of edict], *JLZ*, military affairs, roll 10, 4793-56.

44 Qiao Songnian, memorial dated 1861 (XF 11.), in *Qiao Qinke gong (Songnian) zouyi*, vol. 1, pp. 287–293.

45 He followed a long row of predecessors that included Wenyu 文煜, Yang Nengge 楊能格, Liang Zuozhong 梁佐中 and Lianying 联英. See Qiao Songnian, memorial dated 17 February 1860 (XF 10.1.26.), in *Qiao Qinke gong (Songnian) zouyi*, vol. 1, p. 35.

46 Qiao Songnian, memorial dated 7 December 1859 (XF 9.11.14.), in *Qiao Qinke gong (Songnian) zouyi*, vol. 1, p. 33; Qiao Songnian, memorial dated 28 May 1860 (XF 10.4.8.), in *Qiao Qinke gong (Songnian) zouyi*, vol. 1, pp. 89–92.

47 At the beginning of Qiao's tenure, the liangtai supplied 33,000 troops and had monthly expenditures of 160,000–170,000 taels, see Qiao Songnian, memorial dated 17 February 1860 (XF 10.1.26.), in *Qiao Qinke gong (Songnian) zouyi*, vol. 1, p. 35.

48 Qiao Songnian, memorial dated July 1860 (XF 10.6.), in *Qiao Qinke gong (Songnian) zouyi*, vol. 1, pp. 179–183.

49 Qiao Songnian, memorial dated July 1860 (XF 10.6.), in *Qiao Qinke gong (Songnian) zouyi*, vol. 1, pp. 179–183.

50 Qiao Songnian, memorial dated 9 April 1860 (XF 10.3.19.), in *Qiao Qinke gong (Songnian) zouyi*, vol. 1, pp. 73–75.

51 Qiao Songnian, memorial dated 22 March 1860 (XF 10.3.1.), in *Qiao Qinke gong (Songnian) zouyi,* vol. 1, pp. 65–67. Soon thereafter, the office of the governor-general of the southern course of the Grand Canal was abolished (Īppolīt Semenovīch Brunnert (ed.), *Present Day Political Organization of China* (Shanghai: Kelly and Walsh, 1912), p. 400; Qian Shipu 錢實甫 (ed.), *Qingdai zhiguan nianbiao* 清代職官年表 (Chronological Table of the Qing Officialdom) (Beijing: Zhonghua shuju, 1980), vol. 2, 1474–1475).

52 Qiao Songnian, memorial dated 17 February 1860 (XF 10.1.26.), in *Qiao Qinke gong (Songnian) zouyi*, vol. 1, pp. 39–41; Qiao Songnian, memorial dated July 1860 (XF 10.6.), in *Qiao Qinke gong (Songnian) zouyi*, vol. 1, pp. 179–183.

53 Qiao Songnian, memorial dated 1861 (XF 11.), in *Qiao Qinke gong (Songnian) zouyi*, vol. 1, pp. 287–293; Qiao Songnian, memorial dated 1861 (XF 11.), in *Qiao Qinke gong (Songnian) zouyi*, vol. 1, pp. 261–263; Qiao Songnian, memorial dated 2 December 1861 (XF 11.11.1.), in *Qiao Qinke gong (Songnian) zouyi*, vol. 1, pp. 305–306.

54 Yang Yuebin, memorial dated 10 November 1866 (TZ 5.10.4.), in *Yang Yongque gong yiji* 楊勇慤公遺集 (Collected Works of Yang Yuebin) (Taipei: Wenhai chubanshe, 1968), ch. 11, pp. 1215–1220; Wang Hongzhi 王宏志, *Zuo Zongtang ping xibei Hui luan liangxiang zhi chouhua yu zhuanyun yanjiu* 左宗棠平西北回亂糧餉之籌畫與轉運研究 (Military Supply during Zuo Zongyang's Campaign to Pacify the Muslim Uprising) (Taipei: Zhengzhong shuju, 1973), ch. 67, pp. 136–137.

55 Yang Changjun 楊昌濬, *Pingding Guan Long jilue* 平定關隴紀略 (Taibei: Chengwen chubanshe, 1968), vol. 3, pp. 1383–1384.

56 Liu Rong, memorial dated 27 November 1864 (TZ 3.10.29), *Liu Zhongcheng (Xiaxian) zoushu*, vol. 1, ch. 6, 429–440; TZ 3.6.9., Enlin 恩麟, memorial dated 12 July 1864 (TZ 3.6.9.), in *Qinding pingding Shan Gan Xinjiang Huifei Fanglue* 欽定平定陝甘新疆回匪方略 (Imperially Approved History of the Campaign to Pacify the Muslim Rebels in Shaanxi, Gansu and Xinjiang) (ed. Yixin 奕訢) (Taibei: Chengwen, 1968) (hereafter: *Fanglue*), vol. 6, 3235.

57 The exact course of the war is described by Chu Wen-djang, *The Moslem Rebellion in Northwest China 1862–1878* (The Hague: Mouton, 1966).

58 Lei Zhengwan had already arrived in Gansu, see Duolonga 多隆阿, memorial dated 15 December 1863 (TZ 2.11.5.), in *Fanglue*, vol. 5, pp. 2648–2649.

59 Enlin 恩麟, memorial dated 19 March 1864 (TZ 3.2.12.), in *Fanglue*, vol. 6, pp. 2858–2860.

60 He arrived in Lanzhou on 22 March 1863, see Chu Wen-djang, *The Moslem Rebellion in Northwest China*, p. 65; *Zhang Jingyi* 張精義 (ed. Qingyang Xianzhi 慶陽縣志, gazetteer of Qingyang) (Lanzhou: Gansu wenhua chubanshe, 2004), p. 664; Xilin 熙麟, memorial dated 2 March 1864 (TZ 3.1.24.), in *Fanglue*, vol. 6, p. 2787; Enlin 恩麟, memorial dated 19 March 1864 (TZ 3.2.12.), in *Fanglue*, vol. 6, pp. 2858–2860.

61 Wenyu, memorial dated 22 July 1864 (TZ 3.6.19.), in *Fanglue*, vol. 6, pp. 3278–3282; Edict dated 15 January 1865 (TZ 3.12.18.), in, *Muzong Yi huangdi shilu* 穆宗毅皇帝實錄 (Annals of Emperor Muzong) (ed. by Academia Sinica Institute of History and Philology), Hanji dianzi wenxian ziliaoku 漢籍電子文獻資料庫 (Chinese electronic document database), ch. 124, pp. 736a–b; Chu Wen-djang, *The Moslem Rebellion in Northwest China*, pp. 72–73; Yang Nengge 楊能格,

memorial dated 10 September 1865 (TZ 4.7.21.), in *Fanglue*, vol. 10, pp. 5525–5529.

62 The Manchu Wenyu was demoted from his former position as governor-general of Zhili province. He was dismissed and punished with military service (*juntai*) in Henan, but commuted the punishment by paying a penalty (*taifei* 台費). He was reinstated to the rank of Lieutenant-General of the Banner forces and sent to Qingyang, see: Zhao Erxun 趙爾巽, *Qingshi Gao* 清史稿 (Draft History of the Qing), ch. 388, biographies, 11717–11718. Yang Nengge had been out of office since his dismissal from his post as provincial treasurer of Jiangning (in charge of the Jiangbei liangtai), see: Edict date 6 November 1858 (XF 8.10.1.), *Wenzong Xian huangdi shilu* 文宗顯皇帝實錄 (Annals of Emperor Wenzong), ch. 266, pp. 1120–1121; Edict dated 27 November 1864 (TZ 3.10.29.), *Muzong Yi huangdi shilu* 穆宗毅皇帝實錄 (Annals of Emperor Muzong), ch. 119, pp. 643–641.

63 Entry dated 12 January 1865 (TZ 3.12.15.), Zhang Jixin, *Dao Xian huanhai jianwen lu*, p. 394.

64 Accounts of the Qingyang liangtai in *JLZ*, military affairs (*junwu* 軍務), roll 11: 10 May 1864 to 13 March 1865 (TZ 3.4.5.–TZ 4.2.16.), 4798-4; 14 March 1865 (TZ 4.2.17.–TZ 5.3.4.), 4805-66.

65 Lei Zhengwan 雷正綰, memorial dated 11 February 1865 (TZ 4.1.16.), in *Fanglue*, vol. 8, p. 4406.

66 Yang Nengge 楊能格, memorial dated 27 December 1865 (TZ 4.11.10.), in *Fanglue*, vol. 11, pp. 6010–6017.

67 Yang Nengge, memorial dated 15 November 1865 (TZ 4.9.27.), in *Fanglue*, vol. 10, pp. 5822–5824.

68 Zhang Jixin 張集馨, memorial dated 15 May 1864 (TZ 3.4.10.), in *Fanglue*, vol. 6, p. 3013; Zhang Jixin, memorial dated 27 October 1864 (TZ 3.10.29.), in *Fanglue*, vol. 7, pp. 3941–3953; Liu Rong 劉蓉, memorial dated 16 December 1864 (TZ 3.11.18.), in *Fanglue*, vol. 8, pp. 4088–4091; Chu Wen-djang, *The Moslem Rebellion in Northwest China*, pp. 72–73.

69 Wenyu 文煜, memorial dated 19 October 1864 (TZ 3.9.19.), in *Fanglue*, vol. 7, p. 3754; Edict, dated 19 October 1864 (TZ 3.9.19.), *Fanglue*, vol. 7, pp. 3763–3764.

70 Liu Rong, memorial dated 27 November 1864 (TZ 3.10.29.), in *Liu Zhongcheng (Xiaxian) zoushu*, vol. 1, ch. 6, pp. 429–440.

71 Liu Rong 劉蓉, memorial dated 16 December 1864 (TZ 3.11.18.), in *Fanglue*, vol. 8, pp. 4088–4091.

72 Tao Maolin, memorial dated 26 July 1865 (TZ 4.6.4.), in *Fanglue*, vol. 9, p. 5193.

73 Liu Rong, memorial dated 27 November 1864 (TZ 3.10.29.), in *Liu Zhongcheng (Xiaxian) zoushu*, vol. 1, ch. 6, pp. 429–440.

74 Tao Maolin, memorial dated 26 July 1865 (TZ 4.6.4.), in *Fanglue*, vol. 9, p. 5158; Liu Rong, memorial dated 27 November 1864 (TZ 3.10.29.), in *Liu Zhongcheng (Xiaxian) zoushu*, vol. 1, ch. 6, pp. 429–440.

75 Yang Changjun 楊昌濬, 'Pingding Guan Long jilue 平定關隴紀略' (A History of the Pacification of Shaanxi and Gansu), in Bai Shouyi 白壽彝 (ed.), *Huimin Qiyi* 回民起義 (Shanghai: Sheng zhou guo guang she, 1953), vol. 4.2, p. 202.

76 Chu Wen-djang, *The Moslem Rebellion in Northwest China*, pp. 74–75.

77 Yang Yuebin 楊岳斌, memorial dated 29 August 1865 (TZ 4.7.9.), in *Fanglue*, vol. 10, pp. 5444–5445; Yang Nengge 楊能格, memorial dated 15 November 1865 (TZ 4.9.27.), in *Fanglue*, vol. 10, pp. 5814–5818.

78 Yang Yuebing, memorial dated 12 November 1865 (TZ 4.9.24.), in *Fanglue*, vol. 10, pp. 5799–5802; Dexinga 德興阿 and Liu Rong, memorial dated 21 October 1865 (TZ 4.9.2.), in *Fanglue*, vol. 10, pp. 5713–5714; Dexinga 德興阿 and Liu Rong, memorial dated 2 November 1865 (TZ 4.9.14.), in *Fanglue*, vol. 10, p. 5753; Chu Wen-djang, *The Moslem Rebellion in Northwest China*, p. 80.

79 Yang Yuebin, memorial dated 10 December 1865 (TZ 4.10.23.), in *Yang Yongque gong yiji*, vol. 2, pp. 661–667.
80 Yang Nengge 楊能格, memorial dated 15 November 1865 (TZ 4.9.27.), in *Fanglue*, vol. 10, pp. 5814–5818; Lei Zhengwan, memorial dated 6 December 1865 (TZ 4.10.19.), in *Fanglue*, vol. 11, pp. 5959–5962; Yang Nengge 楊能格, memorial dated 27 December 1865 (TZ 4.11.10.), in *Fanglue*, vol. 11, pp. 6010–6017.
81 Yang Nengge 楊能格, memorial dated 15 November 1865 (TZ 4.9.27.), in *Fanglue*, vol. 10, pp. 5822–5824; Yang Yuebin, memorial dated 10 December 1865 (TZ 4.10.23.), in *Yang Yongque gong yiji*, vol. 2, pp. 661–667.
82 Yang Nengge 楊能格, memorial dated 27 December 1865 (TZ 4.11.10.), in *Fanglue*, vol. 11, pp. 6010–6017.
83 Edict dated 18 January 1866 (TZ 4.12.3), in *Muzong Yi Huangdi shilu*, ch. 162, pp. 743b–745a.
84 Yang Yuebin, memorial dated 2 December 1865 (TZ 4.10.15.), in *Yang Yongque gong yiji*, vol. 2, pp. 607–615.
85 Yang Yuebin, memorial dated 23 January 1866 (TZ 4.12.7.), in *Yang Yongque gong yiji*, vol. 2, pp. 755–767.
86 Zhang Jingyi, *Qingyang Xianzhi*, p. 666.
87 Yan Shusen 嚴樹森, memorial dated 30 January 1863 (TZ 1.12.12.), in *Fanglue*, vol. 3, p. 1533; Edict dated 23 April 1866 (TZ 5.3.9.), in *Fanglue*, vol. 11, p. 6440; Edict dated 19 December 1871 (TZ 10.11.8.), in *Muzong Yi Huangdi shilu*, ch. 323, p. 271a.
88 Zeng Guofan, memorial dated 8 November 1856 (XF 6.10.11.), in *Zeng Guofan quanji: Zougao* 曾國藩全集: 奏稿 (Collected Works by Zeng Guofan: Memorials) (ed. Peng Jing 彭靖 *et al.*) (Changsha: Yuelu Shushe, 1987), vol. 2, pp. 772–773; Zuo Zongtang, memorial dated 29 June 1871 (TZ 10.5.12.), in *Zuo Zongtang quanji* 左宗棠全集 (Collected Works of Zuo Zongtang) (ed. Liu Qingbo 劉晴波 *et al.*) (Changsha: Yuelu shushe, 1986), vol. 6, pp. 217–219.
89 Zhang Jingyi, *Qingyang Xianzhi*, p. 664.
90 Zhang Jixin, *Dao Xian huanhai jianwenlu* (Beijing: Zhonghua, 1981), pp. 355–420.
91 Zhang Jixin, *Dao Xian huanhai jianwenlu*, pp. 387, 389; Lei Zhengwan 雷正綰, memorial dated 11 February 1865 (TZ 4.1.16.), in *Fanglue*, vol. 8. pp. 4401–4406.
92 Zhang Jixin, *Dao Xian huanhai jianwenlu*, p. 395. According to Zhang Jixin, 80,000 *jin* (catties) of flour could provide daily rations for six days to feed 12 battalions on the move. According to the Hubei Army standard battalion size of 688 men, this would make 49,536 man/days and would yield an equivalent of 1.6 *jin*/day for one man.
93 Zhang Jixin, *Dao Xian huanhai jianwenlu*, pp. 355, 371, 380, 391.
94 Entry dated 10 December 1864 (TZ 3.11.12.), Zhang Jixin, *Dao Xian huanhai jianwenlu*, p. 368.
95 Zhang Jixin, *Dao Xian huanhai jianwenlu*, pp. 395–396.
96 Chu Wen-djang, *The Moslem Rebellion in Northwest China*, pp. 75–87.
97 Yang Yuebin, memorial dated 18 January 1866 (TZ 4.12.2.), in *Yang Yongque gong yiji*, vol. 2, pp. 749–754; Yang Yuebin, memorial dated 17 May 1866 (TZ 5.4.4.), *JLZ*, military affairs, roll 11, pp. 4806–4843.
98 Yang Yuebin 楊岳斌 and Liu Rong 劉蓉, joint memorial dated 9 April 1866 (TZ 5.2.24.), in *Fanglue*, vol. 11, pp. 6414–6416.
99 Yang Yuebin, memorial dated 2 April 1866 (TZ 5.2.17.), in *Yang Yongque gong yiji*, vol. 2, pp. 879–881.
100 Yang Yuebin, memorial dated 13 January 1867 (TZ 5.12.8.), in *Yang Yongque gong yiji*, vol. 3, pp. 1315–1317.
101 Liu Rong, memorial dated 15 August 1865 (TZ 4.6.24.), in *Liu Zhongcheng (Xiaxian) zoushu*, vol. 2, ch. 10, pp. 833–839.
102 Yang Yuebin, memorial dated 4 February 1866 (TZ 4.12.19.), in *Fanglue*, vol. 11, pp. 6193–6194.

103 Yang Yuebin, memorial dated 2 December 1865 (TZ 4.10.15.), in *Yang Yongque gong yiji*, vol. 2, pp. 607–615.
104 Lin Shoutu, memorial dated 2 March 1867 (TZ 6.1.26.) , in *Fanglue*, vol. 13, p. 7307; Chu Wen-djang, *The Moslem Rebellion in Northwest China*, p. 87.
105 Chu Wen-djang, *The Moslem Rebellion in Northwest China*, pp. 87, 95–98; Qian Shipu (ed.), *Qingdai zhiguan nianbiao*, vol. 2, p. 1486.
106 Lin Shoutu, memorial dated 2 March 1867 (TZ 6.1.26.), in *Fanglue*, vol. 13, pp. 7307–7309.
107 Mutushan 穆圖善, memorial dated 24 June 1867 (TZ 6.5.23.), in *Fanglue*, vol. 14, pp. 7743–7748; Lin Shoutu, memorial dated 6 July 1867 (TZ 6.6.5.), in *Fanglue*, vol. 14, pp. 7773–7776; Zuo Zongtang, memorial dated 6 December 1867 (TZ 6.11.11.), in *Fanglue*, vol. 14, pp. 7999ff.; Mutushan, memorial dated 17 July 1868 (TZ 7.5.28.), in *Fanglue*, vol. 15, pp. 8746–8747; Edict dated 8 May 1868 (TZ 7.4.16.), in *Muzong Yi huangdi shilu*, ch. 229, pp. 157a–158b.
108 Zhang Jingyi, *Qingyang Xianzhi*, pp. 670–680.
109 Zuo Zongtang, memorial dated 28 June 1869 (TZ. 8.5.19.), in *Zuo Zongtang quanji*, vol. 4, pp. 102–105.
110 Chu Wen-djang, *The Moslem Rebellion in Northwest China*, ch. IV.
111 Originally, the annual allocation was 4.5 million, but this did not include assistance transfers to Liu Dian in Shaanxi (600,000 annually) and to Mutushan's Gansu forces (1.2 million annually). In 1869, Zuo Zongtang united the war chests of Shaanxi, Gansu and his own army. With the new allocations by the throne, he was now entitled to interprovincial assistance funds of 7.53 million annually. See Chu Wen-djang, *The Moslem Rebellion in Northwest China*, pp. 113–114; Zuo Zongtang, memorial dated 11 August 1874 (TZ 13.6.29.), in *Zuo Zongtang quanji*, vol. 6, pp. 72–74; Wang Hongzhi, *Zuo Zongtang ping xibei huiluan liangxiang zhi chouhua yu zhuanyun*, pp. 4–9.
112 The contribution campaign has been described in detail in Kaske, 'Fundraising Wars'.
113 Lu Fengge 盧鳳閣, *Zuo Wenxiang gong zheng xi shilue* 左文襄公征西史略 (A short history of Zuo Zongtang's western campaign) (Taipei: Wenhai chubanshe, 1971), pp. 104–112; Wang Hongzhi, *Zuo Zongtang ping xibei Hui luan liangxiang zhi chouhua yu zhuanyun yanjiu*, pp. 125–135.
114 Wang Hongzhi, *Zuo Zongtang ping xibei Hui luan liangxiang zhi chouhua yu zhuanyun yanjiu*, pp. 125–126.
115 Yang Yuebin, memorial dated 18 February 1867 (TZ 6.1.14.), in *Fanglue*, vol. 13, pp. 7251–7254; Edict dated 18 March 1869 (TZ 8.2.6.), in *Muzong Yi Huangdi shilu*, ch. 252, pp. 514–511.
116 Wang Hongzhi, *Zuo Zongtang ping xibei Hui luan liangxiang zhi chouhua yu zhuanyun yanjiu*, pp. 133–134.
117 For an overview of these bureaus see: Lu Fengge, *Zuo Wenxiang gong zheng xi shilue*, table attached to p. 104.
118 *Zuo Zongtang quanji*, vol. 6, 202.
119 Mark Bell, 'The Country of the Dungan Rebellions of 1861 and 1895–1896', *Imperial and Asiatic Quarterly Review*, 3rd Series, vol. II (1896), pp. 23–37; Hermann Michaelis, *Von Hankau nach Su tschou: Reisen im mittlern und westlichen China 1879–81* (Gotha: Perthes, 1888.
120 Hui Dengjia, *Qingfang Jilue*, pp. 95–96.
121 Yang Changjun, *Pingding Guan Long jilue*, pp. 202–205.
122 Yang Changjun, *Pingding Guan Long jilue*, pp. 206–208; Wang Hongzhi, *Zuo Zongtang ping xibei Hui luan liangxiang zhi chouhua yu zhuanyun yanjiu*, pp. 90–95; Chu Wen-djang, *The Moslem Rebellion in Northwest China*, pp. 89–162; Ma Xiao 馬嘯, *Zuo Zongtang zai Gansu, 1969–1880* 左宗棠在甘肅 (Zuo Zongtang in Gansu, 1969–1880) (Lanzhou: Gansu renmin chubanshe, 2005).

17 European military experience in South Asia

The Dutch and British armies in Sri Lanka in the eighteenth century

Channa Wickremesekera

The late eighteenth and early nineteenth centuries were a significant period in European military history. The revolutionary developments brought about by the advent of gunpowder weapons had reached a stage where European armies had become well-drilled bodies of infantry, pouring forth deadly volleys of musket fire supported by mobile artillery. Cavalry still remained a potent element though in an increasingly auxiliary capacity. Fortifications too had assumed an unprecedented complexity with fortresses girthed by thick walls and bristling with and surrounded by elaborate defence works. As a consequence, sieges too became complicated, involving intricate trench systems to approach and storm the sophisticated defences. European military organization was geared to dealing with long sieges and fighting decisive pitched battles.

The period was also one of European expansion outside Europe, particularly in Asia. As they established a measure of control and influence over parts of India, Indo-China and Africa, European powers also faced military resistance from indigenous rulers and peoples. European forces found fighting the armies of indigenous rulers a very different experience from warfare in Europe. Asian rulers fielded armies of tens of thousands of courageous yet undisciplined men, often supported by artillery and firearms that did not follow any of the rules or styles of fighting current in Europe. Despite the numerical superiority of 'native' armies, often this contributed to the success of European forces. However, native armies fighting in their local environment posed their own unique challenges to European forces making European military campaigns against native rulers far from painless.

This chapter will look at European military experience against one of the smaller kingdoms in South Asia, the kingdom of Kandy in the central highlands of Sri Lanka. Kandy (or *Kande Uda Rata*) had been one of several autonomous kingdoms on the island in the sixteenth century when the Europeans – the Portuguese in this instant – first landed in Sri Lanka. From the late sixteenth century, with the kingdoms of Kotte and Jaffna in the lowlands increasingly dominated by the Portuguese, Kandy became the centre of indigenous resistance to European presence in the island. The Portuguese made several unsuccessful attempts to bring Kandy also within their orbit. Between 1594 and 1638, they occupied Kandy and devastated the Kandyan lands several times but were never able to completely conquer the kingdom. In the late eighteenth and early nineteenth centuries the Dutch and the British too tried and failed. Eventually, Kandy was delivered to the British by its nobles who wished to get rid of their king along with their independence.

In military terms, Kandy presented an interesting challenge to the Dutch and the British. The Kandyans were, in many ways, a pedestrian military power which was on par with some of the smaller zamindari armies on the mainland of the Indian subcontinent. Yet, the Europeans found the going tough and the Kandyan peasant hard to defeat in his rugged highlands. This chapter will look at that experience and its dynamics.

Europeans against Kandy: 1761–1818

The Dutch had taken over the Portuguese possessions on the coast between 1638 and 1658 and added to these gains in the decades that followed. By the second half of the eighteenth century, they had come to control a considerable extent of territory around the Sri Lankan coast. Their headquarters were in Colombo where a modern fortress dominated the fledgling town. From Colombo their control extended northward up to the port of Negombo while to the south, Dutch held sway over the coast all the way up to the small harbour at Hambantota. Between Negombo and Hambantota they also controlled a large area inland, extending up to the Kandyan foothills. This territory, called the Colombo Dissawany was controlled and administered by outposts in the interior and forts at the harbours (Kalutara, Galle, Matara and Negombo). In the north, the Dutch ruled the Jaffna peninsula and a swathe of territory on the mainland between Mullaithivu on the east coast and Manna on the west coast. This territory, the Jaffna Dissawany, was administered by the fort in Jaffna with several other lesser forts dotting the coastline. The Dutch also controlled the ports of Trincomalee and Batticaloa where they had forts and even though there was no Dutch outpost between Hambatota in the south and the Batticaloa in the east, the coastline between the two ports was also considered part of the Dutch dominion on the island. The control over these territories served commercial and strategic purposes. The territory under the Colombo Dissawany produced the bulk of the coveted cinnamon crop in the island. The control of the ports helped to monitor the Kandy's interaction with foreign powers that may be inimical to Dutch interests.

The military confrontations between the Dutch and the Kandyans had little to do with any grand expansionist designs on the part of the Dutch. The origins of the confrontations lay in the friction that existed between Kandy and the Dutch since the establishment of Dutch power on the Sri Lankan coast. The Kandyan monarch viewed the Dutch as trespassers. The Kandyan king Rajasinghe II had enlisted the support of the Dutch in the seventeenth century to oust the Portuguese on the understanding that the Dutch were to hand over the Portuguese possessions to the king. The Dutch failure to do so had angered Kandy who now found one set of intruders replacing another. Tensions led to a major flare up in 1675–1676 but the confrontation failed to loosen Dutch grip on the coast and the low country. Thereafter things remained calm even though tensions remained. Instead of confronting the Dutch head on, Kandy now resorted to harassment and intimidation. The king often used the cinnamon peelers in the Dutch territories to exert pressure on the Dutch. The peelers were often incited to rise against their European masters and the Kandyan territories were opened to fugitives from the Dutch territories.

Eventually the king intervened militarily, making use of a widespread rebellion among cinnamon peelers in the Dutch territories in 1760. Kandyan armies swept down the hills in the south-west taking a swathe of territory and laid siege to Matara. The fort fell to the Kandyan after a brief siege but the Dutch hit back with reinforcements from Europe and re-took the lost territory. To put an end to the Kandyan menace once and for all the Dutch now invaded the highlands in 1764. The invasion was abandoned shortly after the main column entered Kandyan territory but a bigger effort made in the following year saw the Dutch forces enter Kandy and set fire to the palace. The invaders, however, found that invasion was easier than occupying Kandy and after weeks of futile campaigning, the Dutch abandoned Kandy.

The British who took over the Dutch territories in 1796 soon ran into trouble with the Kandyans. An independent Kandy in the central highlands was a thorn in the side of the British who wished to have an unimpeded sway over their possessions on the island. They decided to put an end to the irritant by invading Kandy in 1803. Like the Dutch in 1765, they occupied the town but failed to capture the king. Harassed by Kandyan attacks and ravaged by sickness they decided to retreat to Colombo only to be surrounded and forced to surrender. The European soldiers were butchered while their Malay auxiliaries were forced to take service with Kandy. Buoyed by the victory, the Kandyans invaded British territories in the low country only to be beaten back. The British retaliation saw Kandyan lands adjacent to British territories mercilessly ravaged. Another British expedition was also planned for the following year but was later cancelled. However, one detachment of about 300 soldiers under Major Arthur Johnstone that did not receive the order of cancellation in time and marched into Kandy. They again found Kandy abandoned and soon realizing their error they fell back on Trincomalee on the east coast in the face of heavy Kandyan attacks.

When Kandy finally fell to the British in 1815 it was not through military conquest. A band of disgruntled Kandyan nobles conspired with the British to deliver the king to them. This, for the first time, brought the Kandyan highlands under the control of a European power. However, the conquest of Kandy was seriously challenged within three years. In late 1817, a revolt swept through the highlands, placing the British under immense pressure. The British crushed the rebellion, devastating the countryside and summarily executing those found with arms.

Kandyan military

The military challenge posed by the Kandyan kingdom was different from that presented by the major contemporary South Asian powers in many ways. The

Kandyan 'army' was not like the Mughal Army or even like an army of the lesser states on the neighbouring subcontinent. It consisted of peasant levies called up in a time of emergency, rarely numbering more than a few thousand. The military technology available to them was of a very rudimentary kind. Firearms were available and artillery was in use but the quality of these weapons left much to be desired. Yet, the lack of a disciplined, standing army and the backwardness of military technology were compensated for some extent by the ruggedness of the country they inhabited and the guerrilla tactics they employed.

The political economy of the Kandyan state did not favour a permanent standing army. The Kandyan economy was dominated by agriculture with almost the entire population living off the land. The Kandyans practised mainly subsistence agriculture while trade was limited to obtaining essentials such as cloth and salt through barter from itinerant, chiefly Muslim, merchants from the coast. The Kandyan craftsmen – and there were many – produced and repaired simple tools and equipment that were required in an agricultural economy and to meet the needs of Buddhist rituals. The people were tied to their land and the king was, theoretically, the owner of all land. As such, he was able to call upon the services of his people. Military service was one such service the peasantry was obliged to perform.

The peasantry was mobilized through their local and regional chiefs. The local chiefs gathered the men under their jurisdiction and the 'army' thus mobilized were placed under the command of the regional chiefs or Dissavas. Forming the core of this army was a militia which, though far from a regular army, formed the regular retinue of the Dissavas. They often attended the Dissavas' residence and performed work for him, accompanying him on his travels, often with arms, to bolster his standing in the eyes of the populace. In wartime they formed the core of a large pool of men the king could rely on in every province. A large number of villages in the kingdom were also designated 'Gabada' or store villages. The inhabitants of these villages cultivated a portion of the land for the king's exclusive use. In wartime, they also provided manpower for the army. However, when there was a dire emergency the king, as the 'owner' of all land in the kingdom, could mobilize all able-bodied males for military duty.[1]

There was also a small contingent of foreign troops. In the sixteenth and seventeenth centuries, soldiers from the neighbouring subcontinent formed a substantial part of these troops. Some of them were adventurers while others had been sent by sympathetic rulers who shared the Kandyan king's hostility to the Portuguese. In 1739, the Kandyan throne passed over to an Indian dynasty and with this occurred an influx of Tamils into the kingdom, some of whom functioned as bodyguards for the king. With the advent of Europeans, the foreign contingent in the Kandyan Army began to be filled with Europeans as well, some of them deserters, others prisoners. Several Portuguese, Dutch and British served the king of Kandy in this capacity. Several Malays and Africans also deserted the service of their European masters or were persuaded as prisoners to join the Kandyans. The Malays in particular came to play a prominent role in the Kandyan armies in the eighteenth and nineteenth centuries, being in the forefront of many campaigns. However, these troops made up only a very small portion of the Kandyan armies, the majority of whom consisted of peasant levies.[2]

The army thus mobilized was armed with a variety of weapons. Firearms formed only a small proportion of this arsenal. According to a Dutch observer of the eighteenth century, if 3,000 Kandyans were mobilized only 500 would be armed with firearms.[3] Many of these weapons were captured from their European enemies at various times but Kandyans were also adept at producing their own firearms. These were long barrelled matchlock pieces that had curiously curved stocks.[4] Artillery however, was a different proposition. The Kandyan material and technological resources were not sufficient to meet the demands of cannon foundry. The quantity of iron produced in Kandyan territories was neither large enough nor of such quality as to be useful in casting cannon. The Kandyan technology was also not up to the challenge of producing artillery. The simple technology of the Kandyan kingdom was geared to meeting the needs of its agriculture and the rituals of the Buddhist faith. It was sufficient to tackle the demands of producing simple firearms too, but forging cannon was a different matter altogether.

Due to these limitations, the Kandyans only produced cannon of the lightest possible kind, carrying balls of up to a pound in weight.[5] These however, were not widely used. Greater reliance was placed on lighter weapons that could be more easily produced. These were the *kodituwakku* or *gingal*s or *Springaans* as they were called by the British and the Dutch respectively. They were crudely built pieces of light artillery, often looking like enormous pistols or oversized muskets propped up on wooden or iron legs at the front. They carried a ball weighing less than a pound and though hopelessly inaccurate and too light to cause any significant damage to structures, could still inflict serious casualties among massed troops.[6]

For the heavier guns, the Kandyans relied on captured European weapons. Their numerous successes against their European enemies provided the Kandyans with a considerable arsenal of big guns. However, the problem with these was that the Kandyans did not possess gunpowder in sufficient quantities to fire them often and with effect. There are only a very few recorded instances of Kandyans firing artillery at their enemies. More often they simply adorned their batteries, little more than a demonstration to deter European attackers than as weapons with lethal potential.

These human and material resources did not spell significant military power. However, the Kandyans possessed one valuable asset: their country. Much of the kingdom was rugged, mountainous country, covered in jungle. When combined with the terrain the peasantry firing primitive guns could present a considerable challenge, especially to an enemy that was a stranger to the country and the tactics of the Kandyans.

Kandyan guerrilla war

Kandyan military strategy was aimed at countering the challenge posed by their European adversaries. This challenge was twofold: on the one hand there was the threat of a European invasion of Kandyan lands, while on the other hand they also curtailed Kandy's freedom by controlling the ports. Kandyan strategy sought to meet these two threats. Without a disciplined, standing army and a motley array of weapons at their disposal, the Kandyans naturally preferred indirect rather than direct confrontation. What is now called guerrilla warfare became the norm. Accordingly, the Kandyans usually avoided meeting European invasions head-on and withdrew into their mountain fastness. The Europeans were allowed to enter Kandyan territories, sometimes without resistance and often with minimal opposition. Once in the Kandyan territory the enemy was subjected to harassing attacks aimed at weakening him and forcing him to retreat. Outposts and detachments foraging in the countryside came under attack from Kandyans perched on commanding heights and concealed behind thick foliage. Often, the narrow, winding trails were blocked with trees felled across them while Kandyans also fired on the enemy from stockades built at strategic locations along the paths. The baggage handlers and non-European auxiliaries were a prime target, the Kandyans singling them out in order to force them to desert.[7]

The Kandyans also counted on the climate in the mountains, to which the Europeans were not accustomed, to take its toll on the enemy in terms of sickness. The tired and harassed Europeans, ravaged by sickness and deserted by their auxiliaries were considered a prime target for destruction with overwhelming force. In 1765 and 1803, the Dutch and the British were assailed in their main camp in Kandy while in 1765 the Dutch force retreating from Kandy was also attacked heavily.[8]

The Europeans also confronted Kandyan armies when the Kandyans invaded European possessions in the low country. The Kandyans took advantage of any perceived weakness in the Europeans to invade European possessions. In 1761 and 1803, they resorted to this tactic. Once again, guerrilla tactics were employed widely but in the low country the Kandyans also threatened European forts. The European outposts close to the frontier were surrounded and isolated, sometimes forcing them to surrender or evacuate. In 1761, the Dutch lost control over a swathe of territory in the south-west including Matara on the coast. The British were forced to abandon some of their outposts in 1803. This, then, was the challenge posed by the Kandyans to the European armies in the eighteenth century. How did the Europeans fare against this challenge?

European armies and the Kandyan challenge

The numbers of troops deployed by the European powers against Kandy in the eighteenth and nineteenth centuries were small. They never exceeded 5,000 in a campaign. These armies of the Dutch and the British consisted of a small core of European troops and a larger number of non-Europeans. The Dutch used soldiers from Indonesia called 'Easterners' while the British used Malays, Indian sepoys and Africans. The main Dutch force during the invasion of Kandy in 1765 comprised 900 Europeans, 800 Malays and about 100 sepoys.[9] The British also made use of Indian sepoys and Malays. The majority of the troops that occupied Kandy in 1803 were Malays while every detachment and outpost had a proportion of Malay troops and sepoys.[10] The non-European troops were subject to a degree of regular discipline and training. The Malays in the Dutch service had been first organized under their own chiefs with little attempt at training or providing up-to-date weapons. This had changed in the eighteenth century when efforts were made to integrate them into the regular organizational structure of the VOC (Dutch East India Company). These were rudimentary measures but it was far better than employing, raw, untrained 'native' troops. The sepoys employed by the British were even better placed, being trained and armed in the manner of fighting then current in Europe

In a pitched battle, even the smallest army of troops disciplined and armed in the European manner was markedly superior to the Kandyan hordes. Though small in number the European armies in Sri Lanka possessed greater firepower and discipline. Like most European armies of the day they were armed with the flintlock musket and were trained in the disciplined use of these weapons. This included delivering devastating volley fire at close range and manoeuvring into a number of formations in the field. This was much more than the Kandyans possessed. Even a company of European or European-style 'native' troops with their muskets supported by one or two pieces of light artillery could wreak havoc on even large Kandyan armies if the latter was caught in the open.

However, this is not how things always developed in their wars against the Kandyans. To be sure, in open confrontations the Europeans could defeat the Kandyans easily. The few instances when the Kandyans attempted to take on their European adversaries on pitched battle provide ample testimony to this. In 1765, a Kandyan force of 400–500 strong offered battle to the invading Dutch on a slope covered with scrub at a place called Periwille.[11] The Dutch brushed aside the opposition without incurring any casualties. During another campaign in the previous year the Dutch met the Kandyans on the banks of the Kandapola Oya, drawn up in a regular order, awaiting battle. This time the resistance was stiffer but the result was similar: a Dutch victory for the loss of only six killed and 15 wounded.[12]

Behind the walls of their 'forts' too, the Europeans were usually safe from even large Kandyan forces. Without heavy artillery even flimsy walls were often a strong enough barrier against the light *gingal*s and the light artillery the Kandyans could bring to the field. The attack on Hanwella in 1761 was withstood easily, the Dutch losing only one killed and another wounded. The government house in Hambantota was defended by 60 sick Malays and in August 1803 the little fort in Halawatha was held by two young civil servants and 60 sepoys against a force of Kandyans reported at around 3,000.[13] A few European forts did fall to the Kandyans. For instance, Katuwana and Matara succumbed in 1761 and Dambadeniya and Tangalle fell in 1803. This was due largely to the fact that these outposts were far from the main European bases and when surrounded by Kandyans they were hard to supply. Occasionally, when the Kandyans could muster a considerable number of cannon they could exert pressure on even fairly large European bases. Matara was such a place, forced to evacuate in 1761 by a large Kandyan Army armed with a train of light artillery. Still, the Europeans had little to fear in their well-fortified bases like Galle or Colombo.

However, when the Kandyans chose to seek the refuge of their own walls of trees and scrub the Europeans found many of their advantages fast disappearing. Kandy was a natural fortress, ringed by hills with access only through a few narrow passes. No roads linked Kandyan villages, only footpaths and these were hemmed in by often impenetrable jungle and overlooked by forest clad hills and defiles. As Robert Knox who spent 20 years as a captive of the king of Kandy observed:

> The kingdom of Conde Uda is strongly fortified by nature, for which way soever you enter into it, you must ascent vast and high mountains and descend little or nothing. The ways are many but very narrow, so that but one can go abreast. The hills are covered with wood and great rocks so that 'tis scarce possible to get up anywhere, but only in the paths.[14]

Indeed the greater part of the country was thickly forested. Joao Robeiro a Portuguese soldier who campaigned for 18 years in the low country, speaks of jungle too thick to see beyond a few paces.[15] In the late eighteenth century this situation had not improved by much. In the Kandyan territories this situation was exacerbated by the ruggedness of the terrain as well.

While it provided ideal conditions for guerrilla fighting, the rugged terrain negated many of the advantages of European firepower. The narrow paths made it difficult to march in any other order than single file while drawing up to deliver the withering musket fire that was the hallmark of contemporary European warfare was impossible. Deploying artillery was not easy either. The guns had to be of the lightest possible kind and they had little scope for being used in their role of supporting drawn-up infantry. Furthermore, the Kandyans presented at best an elusive target behind the trees and scrubs.

As a consequence, fighting the Kandyans became a case of European soldiers and their sepoy/Malay troops trudging up and down narrow trails winding around rugged mountainous territory and evading desultory fire from Kandyan peasants perched on hilltops. Such campaigns incurred few combat casualties. The abortive Dutch expedition to Kandy in 1765 incurred fewer than 150 deaths from Kandyan bullets and arrows and the British campaigns against the Kandyans during the Great Rebellion in 1817–1818 cost them fewer than 100 battle casualties. In 1804, a British detachment of just over 200 had retreated all the way from Kandy to Trincomalee losing fewer than 50 killed and wounded.

However, the issue was not that simple. Soldiers were not the only men trudging the long winding trails of the Kandyans highlands. Campaigns in Sri Lanka involved serious logistical problems, as was the case for Europeans in many parts of the tropical regions. The Europeans could not rely on the hospitality of the people whose lands they were invading or the bounty of the countryside and, therefore, all supplied had to be carried with them. These included all regular needs of the army such as ammunition, as well as needs specific to European armies. European soldiers required either beef or fish as part of their diet in addition to rice. They also needed substantial amounts of arrack as a stimulant. They needed tents to sleep in and camp kettles to cook the food. The European officers in particular carried much baggage, and often took numerous bulky items such as tables, chairs and other things that provided a measure of comfort in the jungles. The fighting men could not afford to carry all this, not even their own food and ammunition. It was too much to carry in the first place and in hostile country where attacks were numerous and frequent, they had to be free of encumbrances to meet any emergency. The wounded too needed tending to. Last but not least, the need to stick to the narrow paths often required labour to clear them.

As a result, even a small European detachment which expected to be away from their bases for several days carried a substantial baggage train with them. These consisted of porters (coolies), pack animals to carry the burdens and pioneers to clear the roads. Often the baggage train was as large as the army. About half the 7,000 men in the Dutch Army that invaded Kandy in 1765 were coolies and camp followers.[16]

The camp followers were also exposed to all the privations and dangers of campaigning. While the soldiers were hardened to them, the followers were often not. Knowing this 'soft spot' in the European armies, the Kandyans often targeted the baggage train causing the porters to abandon their loads and flee. Sometimes, the fighting itself caused enough panic for the porters to desert. The Dutch campaign in 1764 had to be abandoned after many of the porters deserted.

But, the more serious problem was disease. Many Europeans found the Kandyan climate insalubrious. Nights in Kandy could be cold and wet the damp often penetrating the triple covering of a European field tent.[17] The wet weather and the swampy ground presented an ideal ground for mosquitoes to thrive. This was the perfect setting for the spread of disease, especially fevers that showed symptoms similar to that of yellow fever – 'excessive depression of spirits and strength, weakness, anxiety and aggression'.[18] To campaign in such weather was no picnic. Even if the Kandyan bullets and arrows were few and erratic, the Dutch and the British still had to spend long hours clambering over rocks and creeping through undergrowth, always on the lookout for the enemy lurking in the shadows. It placed considerable strain on the physique and caused immense anxiety. Often troops on the march were exposed to rain and the early morning dews, and were forced to bivouac on wet ground without cover.

The spread of disease was aided by the lack of food. Europeans found it hard to maintain their supply lines in the interior as their lines of communication ran through hostile territory. While in Kandyan territory it was hard to live off the land as the Kandyans had hidden all stocks of food and chased their cattle into the jungle. The rice supply for the Dutch dwindled steadily in 1765 as foraging parties could find little in the abandoned Kandyan villages while the supply lines with the low country were steadily disrupted.[19] The lack of meat was particularly troublesome for the European troops in 1765 and 1803 with the Kandyans having taken particular care to hide or scare off the cattle in the countryside.[20]

The lack of food caused general physical weakness which made troops vulnerable to diseases. The lack of fresh food in particular paved the way for beriberi. The British in 1803 suffered particularly severely from this disease which caused general physical debility and lethargy.[21] All this was compounded by the rigorous campaigning. Marching over a long distance and fighting, even if the latter consisted of little more than dodging bullets and arrows, sapped the strength of many troops, particularly the Europeans who were not used to heavy physical exertion in a tropical climate. As pointed out by a contemporary medical expert in 1803, sickness was mainly caused by the 'excessive fatigue and labour which the men sustained in the service of foraging, during the hottest part of the day, from which they sometimes returned drenched with rain'.[22] A good example of the effects of over-exertion on European health is provided by the aftermath of the capture of the Kandyan outpost at Girihagama in 1803. The British suffered only two European soldiers wounded in the assault but the strain of assaulting the precipitous height in the extreme heat of the day was enough to exhaust the men of the 51st Regiment who consequently became vulnerable to the bouts of fever that gripped them after the occupation of Kandy.[23]

All this contributed to a dramatic increase in the sickness among the troops, particularly the Europeans. Disease caused the largest number of deaths. The Dutch occupying army in 1765 was so ravaged by sickness that the number of guards had to be reduced.[24] The British occupiers suffered even more severely in 1803 – the British commander Major Davy writing that the 19th Regiment was finding it hard to find a single man fit for duty.[25] Almost every man who escaped the Kandyans under Captain Johnstone in 1804 had to enter hospital immediately.[26] During the Kandyan rebellion in 1818, HM 73rd Regiment of Foot suffered death at the shocking rate of 412 out of 1,000.[27] Therefore, although the guns and arrows of Kandyans posed few dangers, the demands of campaigning in the highlands posed serious challenges to the Europeans. The Kandyan guerrillas may have been fairly innocuous but the 'total experience' of fighting them was hardly to be envied.

The shifting balance of power

However, in spite of the daunting challenges posed by the terrain, Kandyan guerrilla tactics and the climate, ultimately it was the Europeans who held the upper hand. Despite the elusiveness of the Kandyans and the difficulty of using their firepower to full advantage, the Europeans possessed sufficient tactical ability to make Kandyan tactics less effective and even to take the fight to them. Even though the Europeans had little chance of using European tactics in the Kandyan highlands they still possessed an advantage over the Kandyans in having superior firearms. The flintlock muskets used by all troops in the British and Dutch service were far more reliable than the assorted badly maintained muskets of the Kandyans. The Kandyan trails did allow only the deployment of the lightest pieces of artillery but the Dutch and the British in the late eighteenth and early nineteenth centuries benefited from the availability of light artillery and mortars. The Dutch carried several light pieces and mortars during their expeditions in

1764–1766 while the British also followed suit. The light mortars of the period were particularly useful in that they took very little space and could be easily manhandled by two people, enabling them to be used in the rugged terrain without much trouble. The British were partial to the Coehorn Mortar which was a particularly small and light weapon, with a base a little over two feet long and a fraction over a foot broad. It could easily be carried by two persons, not a very difficult task even in the Kandyan highlands.[28]

The availability of light artillery and mortars meant the negation of many of the advantages the Kandyans enjoyed as guerrilla fighters. The presence of field artillery enabled the Europeans to tackle the Kandyans from a distance, even in their commanding heights. The light mortars, in particular, were very useful in this regard. Not only could they hit the enemy from afar but their high trajectory could also reach the Kandyans seeking refuge behind natural features and fortifications and their explosive shells could rip through thick foliage. The diarist of the 1765 campaign noted with satisfaction how the mortars could keep the enemy 'at a greater distance than could be done by musket fire'.[29] Nearly four decades later in 1803, the British officer Heber Beaver echoed this view, declaring confidently that whenever he discovered the precise location of an enemy force he was 'sure to rout them with a few shell from the Coehorns'.[30] When that was the case at a distance, any attempt to overwhelm a European force supported with artillery by massed attacks was bound to prove disastrous. This is clearly shown by an attempt to assault the British garrison in Kandy in 1803. A single grapeshot put an end to the ambitious attempt killing 24 Kandyans on the spot.[31]

As a consequence Kandyan guerrilla tactics had limited chance of inflicting losses on the Europeans. Even without mortars and cannon, at close range, a small force of Europeans and sepoys/Malays with their flintlocks had the potential to keep the Kandyans at a safe distance due to their high rate of fire. In 1762–1766 and 1803–1805, we find such detachments often holding their own against large forces of Kandyans in the interior. For instance, in March 1803, Captain Herbert Beaver cleared the province of the Four Korales of Kandyan intruders with only 30 men while in, February 1818, Major Delattre marched with a detachment of just over 100 men from Trincomalee to Kurunegala through hostile territory without suffering any serious harm.[32] The usual Kandyan tactic was to play it safe and engage even tiny enemy detachments from a safe distance even when operating deep in Kandyan territory.

As explained above, the biggest losses to the Europeans were caused by sickness rather than combat. This, however, was not a devastating loss. The Europeans in Sri Lanka had the great advantage of being part of global maritime empires which could supply them with men and material. The depleted manpower could be easily replenished by summoning troops from other European colonies in the region. The Dutch had their bases in the East Indies while the British could always fall back on their massive army on the neighbouring subcontinent.

The inability of the Kandyans to inflict significant loss on the Europeans in combat and the availability of sizeable reserves of troops were decisive factors in the wars between the Kandyans and their European adversaries. The Kandyans may have been relatively safe on their cliffs and behind their trees and scrub but it did not prevent the Europeans from visiting devastation on Kandyan territories causing immense suffering to the peasantry. The destruction of homes and fields forced the peasantry to take refuge in jungles and expose themselves to the ravages of hunger and disease. The Dutch ravages in 1764 and 1765 reduced many parts of Kandyan territory, especially the provinces of Four and Seven Korales into waste-lands, their fruit trees cut down and the fields destroyed.[33]Agriculture was at a standstill while the supply of salt and dried fish was badly disrupted.[34] In 1804, the Four Korales suffered again, and British spies reported that the people in the region were unable to sow their fields and were deficient in salt. The lack of salt in particular was the result of the economic blockade of the Kandyan territories. In late 1800 and early 1805, it was reported that getting even a handful of rice was difficult for the Kandyans and the people of Seven Korales were said to be migrating in large numbers to the less troubled Magul Korale in the modern Puttalam district.[35] The frontier areas of Hanwella and Avisswella were turned into a deserted, desolate land.[36] The British ravages of 1817–1818 were truly devastating, preventing the cultivation of fields for two seasons, forcing the peasantry to sustain themselves on whatever grain they had hidden away.[37]

At the same time, European naval power enabled them to maintain a stranglehold on the Kandyan economy. The European powers were all maritime powers and as such could impose a blockade on Kandyan trade with the coast. A complete blockade was never a possibility but European naval power could ensure that contact with the world outside for Kandy was a difficult exercise. Kandy was trapped behind the very mountains that offered a measure of natural protection, able to harass their enemies but struggling to prevent them from piling misery on their lands. Therefore, despite the stress and strain of hunting for armed peasants in the Kandyan highlands, overall the European military experience was one of dominance and confidence. Kandyan conditions constrained many of the advantages of the European military system but not sufficiently to deter the European forces from gaining the upper hand.

Notes

1 Channa Wickremesekera, *Kandy at War: Indigenous Resistance to European Expansion in Sri Lanka 1594–1818* (New Delhi: Manohar, 2004), pp. 65–66.
2 Channa Wickremesekera, *Kandy at War*, pp. 57–63.
3 'Aanmerkingen ove de Oorlog met de Sinaleesen, Mackenzie Ms. 39', p. 11, Oriental and India Office Collection, British Library, London.
4 See the guns depicted in P.E.P. Deraniyagala, 'Sinhala Weapons and Armour', *Journal of the Royal Asiatic Society (Ceylon Branch)*, vol. 35, no. 95 (1942), pl. IV and p. 121 and also D.H.D.H. de Silva, *A Catalogue of Antiquities and Other Cultural Objects from Sri Lanka Abroad* (Colombo: Government Printers, 1975), p. 150.
5 'Aanmerkingen', pp. 5–6.
6 For a description of the Kandyan *gingal* see Joao de Ribeiro, *The Historic Tragedy of the Island of Ceylon* (tr. P.E. Pieris) (Colombo: Lake House, 1948), p. 146; 'Aanmerkingen', p. 5.
7 Channa Wickremesekera, *Kandy at War*, pp. 136–141.

8 'Aanmerkingen', pp. 20–21; Kandy Diary entry 7 June 1765, in R. Raven Hart (ed.), *The Dutch Wars with Kandy* (Colombo: Govt. Printers, 1964), p. 121; Greeving's Diary, Vimalananda Tennekoon, *British Intrigues in the Kingdom of Kandy* (Colombo: Gunasena, 1973), p. 504.
9 R. Raven-Hart (ed.), *The Dutch Wars with Kandy*, p. 21.
10 James Cordiner, *Description of Ceylon containing an Account of the Country, Inhabitants and Natural Productions with a narrative of the Tour around the Island in 1800, the Campaign in Kandy in 1803 a Journey to Ramisseram in 1804*, 2 vols (1807, reprint, Delhi: Navrang, 1983), vol. 2, p. 197.
11 R. Raven-Hart (ed.), *The Dutch Wars*, pp. 93–4.
12 R. Raven-Hart (ed.), *The Dutch Wars*, p. 74.
13 Geoffery Powell, *The Kandyan Wars* (London: Leo Cooper, 1973), p. 141; James Cordiner, *Description of Ceylon*, vol. 2, pp. 229–230.
14 Robert Knox, *An Historical Relation of the Island of Ceylon* (ed. J.H.O. Paulusz) (1681, reprint, Colombo: Tisara Publishers, 1989), pp. 14–15.
15 Joao de Ribeiro, *The Historic Tragedy*, p. 31.
16 Diary of the 1765 Campaign, in Raven Hart (ed.), *Dutch Wars with Kandy*, p. 93.
17 A. Lennox Mills, *The Advent of the British to Ceylon* (London: Frank Cass, 1964), pp. 4–5.
18 V.M. Methley, 'The Expedition of 1803', *Transactions of the Royal Asiatic Society*, 4th series, vol. 1 (1917), p. 108. For a detailed view of the sickness which affected the British during the campaign in 1803, see 'The Medical Report in April 1803', in James Cordiner, *Description*, vol. 2, pp. 262–286.
19 Kandy Diary, entries for 3, 6 and 28 August 1765, p. 35, Sri Lanka National Archives (SLNA), Colombo 1/4944, pp. 35, 40, 46.
20 Thomas Ajax Anderson, *Journal of the Proceedings of the Trincomalee Detachment* (London: 1809), pp. 181–182.
21 Thomas Ajax Anderson, *Journal of the Proceedings of the Trincomalee Detachment*, pp. 181–182.
22 James Cordiner, *Description of Ceylon*, vol. 2, p. 280.
23 V.M. Methley, 'The Expedition of 1803', p. 104.
24 Kandy Diary, 23 August 1765, SLNA 1/4944, p. 43.
25 Major Davy to Colombo, 10 June 1803, SLNA 7/43, p. 685.
26 Henry Marshall, *Ceylon: A General Description of the Island and its Inhabitants* (1842, reprint, Dehiwela, Srilanka: Tisara Publishers, 1982), p. 190.
27 Geoffery Powell, *The Kandyan Wars*, p. 251.
28 A.F.G. Hogg, *Artillery, its Origin, Heyday and Decline* (London: C. Hurst & Company, 1970), pp. 64–65.
29 Diary of the 1765 Campaign, in R. Raven-Hart (ed.), *Dutch Wars*, p. 101.
30 Geoffery Powell, *The Kandyan Wars*, p. 139.
31 Greeving's Diary, in Vimalananda Tennekoon, *British Intrigues in the Kingdom of Kandy*, p. 504.
32 R.L. Brohier, *The Golden Age of Military Adventure in Ceylon* (Colombo: Plate Ltd., 1933), pp. 4–7.
33 Sri Lanka National Archives (SLNA) 1/4944, pp. 14–15.
34 V. Kanapathipillai, 'Dutch rule in Maritime Ceylon, 1766–1796' (PhD thesis, University of London, 1969), p. 88.
35 The deposition of Talagama Unnanse, SLNA 7/44, pp. 156–161, deposition of Pinnala Unnanse, 1 December 1804; Vimalananda Tennekoon, *British Intrigues in the Kingdom of Kandy*, p. 457; Deposition of Kar Bocus, a Malay Priest, March 1805, Vimalananda Tennekoon, *British Intrigues in the Kingdom of Kandy*, pp. 478–479.
36 John Davy, *An Account of the Interior of Ceylon and of its Inhabitants with Travels in that Island* (Dehiwela, Sri Lanka: Tisra Publishers, 1969, rpt of 1821 edn), p. 262.
37 R.L. Brohier, *The Golden Age of Military Adventure in Ceylon*, p. 34.

18 Military revolution and state formation reconsidered

Mir Qasim, Haider Ali and transition to colonial rule in the 1760s

Douglas M. Peers

Colonel Baillie's defeat in September 1780 at Polilur, vividly captured in the murals painted on Tipu Sultan's summer palace (*Dariya Daulat*) in Seringapatnam, was declared by Thomas Munro to be the biggest defeat suffered by the British in the 75 years following the Battle of Plassey.[1] At Polilur, 3,800 East India Company (EIC) troops under Baillie, which included several hundred Europeans, were cut off and decimated by a military force which contemporaries recognized as one which was well equipped, ably led and well supported by the state of Mysore which had over the previous two decades put into motion a number of key reforms. In her discussion of these murals Janaki Nair astutely observes that neither Haider Ali nor Tipu Sultan is depicted as anything but as a warrior.[2] Yet, tellingly is the prominent place given in these murals to Mir Sadiq, the head of Haider Ali's revenue and finance department.[3] Ironically, these troops were waiting for a relief column commanded by Colonel Hector Munro, the officer who 15 years before had smashed the forces of Mir Qasim and his allies at the Battle of Buxar. Mir Qasim had also sought to revitalize and reform the military and administrative capacity in Bengal, though his successes were more limited than those which took root in Mysore.

The extension of British power from their beachheads at Madras and Calcutta from 1755 onwards was neither inevitable nor was it seamless. In fact, the British encountered stiff resistance in both regions in the 1760s when confronted by newly established, powerful, creative and credible opponents, namely Haider Ali in Mysore and Mir Qasim in Bengal. The former established a regime which would not be crushed until 1799 when his son and successor Tipu Sultan died at the Siege of Seringapatnam. In the first of what would become several wars between Mysore and the EIC, Haider Ali's troops swept through the coastal plains of Madras in 1768–1769 and then dictated terms under the shadow of the walls of Fort St George. This prompted one contemporary to ponder, 'Who ever thought that the intended conquerors of the Mysore country would be reduced to make peace at the gates of their own capital.'[4] Not surprisingly, Haider Ali became a key figure in the nationalist pantheon, leading Jawaharlal Nehru to observe that 'he was a remarkable man and one of the notable figures in Indian history', drawing particular attention to his strategic vision and organizational skill.[5]

In the north, following the Battle of Plassey, British efforts to impose a political order in Bengal favourable to them led them first to attempt to work through Mir Jafar. When Mir Jafar failed to provide the British with an alliance to their liking, they then looked to his son-in-law, Mir Qasim, and installed him as *Nawab* of Bengal. A combination of unreal expectations on the part of the British and Mir Qasim's own quest for greater autonomy quickly eroded their relationship. Mir Qasim had initiated measures to create the fiscal and military preconditions for independence, and the officer who led the forces that eventually toppled him was impressed by their results, observing that,

> whilst he was amassing Treasures, increasing the number of his troops, whom he disciplined and cloathed in the English manner, and filling his magazines with artillery and all kinds of military stores, which he spared no expence to obtain and mount according to some models he had procured from the English.[6]

But, unlike the reforms instigated by Haider Ali, the ones Mir Qasim introduced were insufficient to stave off an attack from the British and arguably triggered a pre-emptive strike by them. His regime was much shorter lived, lasting no more than four years, but at least initially he was thought by some to be capable of driving the British out of their all-important territories in Bengal.

By looking at these two figures and the regimes they forged, collectively as well as comparatively, we can better appreciate the military, political, social and economic dynamics at play in the decades immediately following the mid-eighteenth-century surge in European activity in India. The states which these rulers ushered into being are notable not only because they sought to resist the encroachments of the EIC and its employees (the latter often acting without the EIC's approval),[7] but that they did so by effecting widespread organizational and fiscal transformations. A better appreciation of these transformations checks the tendency to see the British conquest as an inevitable manifestation of Western technical and administrative superiority, or, in other words, proof positive of the so-called Military Revolution hypothesis. Just as importantly, they offer an excellent opportunity through which we can begin to explore the varieties of hybrid military and political formations which flourished in post-1750 India, and seek answers to the question why the transformations in Mysore enabled Mysore to hold out for 30 years yet those in Bengal quickly failed.

A partial clue as to why this was the case is offered in the observation of one contemporary who noted of Mir Qasim that 'had he been as good a general as he was a consummate politician, we should have had by this time little reason to boast for our sway in this country'.[8] Both Haider Ali and Mir Qasim embarked on broadly similar programmes of reform intended to strengthen their armies, improve the flow of resources to sustain those armies, and ensure adequate oversight over both their armies and the financial administration of their territories. But, only Haider Ali was ultimately able to create the winning combination. Mir Qasim, an astute administrator, was not as militarily adept like his counterpart

Haider Ali. Moreover, his scope for action was constrained both by the closer proximity of the British as well as his decision to retain many aspects of Mughal political culture which ultimately frustrated his efforts at political and military integration. Haider, on the other hand, enjoyed greater latitude and was able to distance himself more successfully – militarily, administratively and culturally – from his predecessors as well as his neighbouring rivals.

It is important to recognize that the kinds of military, political and administrative transformations which occurred, no matter how short lived, were not merely intended to mimic Western forms. Instead, they fused together indigenous forms and practices with innovations gleaned from their contacts with Europeans – introduced in some but not all cases by the recruitment of European soldiers of fortune and technical experts. It should also be noted that Europeans in India in general, and the British in particular, were also developing hybrid military cultures.[9] The end results were amalgams of Indian and European institutions and ideas, creative fusions which were the product of ongoing experimentation as Europeans and Indians alike adjusted to the shifting military, political and economic landscape.

British conquests and colonial historiography

British campaigns against Mir Qasim and later Haider Ali and Tipu Sultan were instrumental not only in determining the shape of Britain's empire in India, but were also critical in shaping subsequent efforts at accounting for historical developments there. The popular demonization of Tipu Sultan and Haider Ali is beyond the scope of this study and in any case it has been examined in depth by several scholars.[10] But, it is worth briefly considering such characterizations as they have shaped military historiography. Contemporary accounts of Tipu Sultan almost uniformly cast him in a very negative light, caricaturing him as a violent, intolerant despot, one whose success could be attributed at least in part to his barbarity and cruelty. Such images, while revealing of British attitudes and deeply suggestive of the Orientalist constructions that were already underway, have unfortunately drawn attention away from the less emotive but ultimately more significant military and political developments that enabled the rise and consolidation of his power.

The anxiety caused by Tipu Sultan was largely derived from the very real threat that he posed to the British. His army and his state had effectively and imaginatively incorporated European practices and technologies while also preserving and modifying a number of effective indigenous practices and structures. However, to concede such points would have undermined a number of key imperial assumptions, and consequently he had to be shown to be barbaric to sustain the sense of difference needed to differentiate Indian rulers and their states from what the British were establishing in India. Contemporaries however were more willing to acknowledge the strength and threat posed by Mysore. They were very alive to the challenge posed by Haider Ali. Charles Smith warned the Madras Secret Committee that 'of all the country powers, Hyder Ali is by far the most dangerous and the most powerful – his connections with the French – his having seaports ... his army consisting of cavalry, infantry, and artillery'.[11] Alexander Dow, one of the earliest first-hand chroniclers of British activities in India, wrote:

> Hyder Ali is an extraordinary character, even in a country where men have an ample field for the exertion of great talents.... He is possessed of a surprising memory, deep penetration, quickness of resolution, great perseverance, conduct aided by personal courage.[12]

When examining military developments in eighteenth-century India, we must be mindful of the assumptions and characterizations that have fashioned subsequent historical accounts. The Orientalist tone illustrated by the preceding quote has proven to be remarkably resilient.[13] Orientalism not only yielded stereotypical views of Asiatic rulers, but also relied upon an extensive array of juxtapositions which included pitting Western military science against its Asian counterparts. Military technology and organization have come to be appreciated as more than simply the mechanism by which conquests have been made – generations of historical commentators have turned to them as yardsticks by which civilizations can be examined, and, more importantly, ranked. Both Edmund Burke and James Mill highlighted the rudeness of the military arts in India as demonstrable proof of their state of civilization.[14] By taking the idea of Western technological and military superiority as our starting point, historians have not only ended up misreading the historical record, but have also contributed to the perpetuation of some key Orientalist myths.

Military Orientalism, when combined with other manifestations of Western triumphalism, has encouraged the propensity to exaggerate the gulf between the armies of the EIC and those of their opponents. An earlier commentator on the military situation in India noted of the Company regiments of the 1760s, that 'all details were carried on according to their own ideas, the establishment of the regiments being more Asiatic than European'.[15] The various military forces at work in eighteenth-century India were often more similar than they were different. The idea of a Military Revolution, one which coupled technological developments to organizational reform, and then embedded this within a more general theory of state formation, gained ground in the second half of the twentieth century.

Many of the Military Revolution models in circulation are heirs to this tradition, seen most obviously in the tendency to assume that it is Western military practices, technologies and organizations that are the norm. We must steer clear of what Patrick Porter has aptly termed the 'essentialist' approach when dealing with military cultures and non-European warfare.[16] Common symptoms of Military Orientalism include inflated estimates of the opposing armies, neglecting to list local auxiliaries when tallying up the size of European forces, underappreciating the effectiveness of low-technology inventions, and attributing examples of innovation to European deserters and/or other means by which

European expertize could be directly sampled. The latter line of reasoning proved to be popular throughout the eighteenth and nineteenth centuries, and even explanations for Haider Ali's success occasionally dipped into its reservoir. One contemporary reported:

> According to the best authorities [Haider Ali] was once a private soldier in the French army; in this situation he made such a rapid progress in the military art, that he soon thought himself qualified to fill a more respectable appointment under some of the country princes.[17]

Moreover, we need to consider that not all innovations require a single point of origin, and in fact many can be independently arrived at. One of the essayists in a recent collection pointedly reminds us that 'Understanding the potential of military technology and how to integrate that effectively with existing military organization and institutions was no Western prerogative, and grasping these principles evidently did not require an attentive reading of Vegetius.'[18]

Nevertheless, and notwithstanding a widely shared belief in Western superiority, a number of historians of South Asia have challenged the idea of innate Western military superiority. P.J. Marshall pointedly noted that much of what we assume were the critical advantages enjoyed by the West did not really materialize until the late nineteenth century. By the latter half of the nineteenth century, the combination of a growing technological differential, increased arms production and decreased costs of shipping weapons to India combined to create more favourable conditions for Western arms.[19] In the eighteenth century, technology was readily picked up by many Indian rulers, and given the production capacity then available in India, firearms and cannon of the quality that could be produced in Europe were just as readily available in India. While it is true that access to quality artillery did not necessarily translate into its effective integration into Indian military formations – many Indian rulers were still inclined to favour heavy siege artillery when the increasing mobility of warfare favoured light field pieces – there were certainly many instances of its successful adoption.[20] The Marathas and the Sikhs are noted examples from the first half of the nineteenth century. There are fewer illustrations from the eighteenth century, but notably Haider Ali and Tipu Sultan, as well as Mir Qasim to a lesser extent, showed a willingness to experiment and innovate.

Some historians, having conceded that technology cannot in and of itself account for European success, point instead to the fragmented character of Indian politics. According to them, the most pressing problem facing many Indian rulers was that 'the usual Indian politics of permanent sedition and ever shifting alliances remained completely at odds with the new military principles'.[21] But even here we risk exaggerating the differences between Europe and India by calling into play the notion that politics in India were innately more slippery and more venal than those in Europe. And, by extension, loyalty within Indian institutions including the military became suspect from the outset. Much of the reasons for this became subsumed into the idea of Oriental Despotism, which is but another manifestation of Orientalism and as such frequently has become inscribed into the historical record.

Close scrutiny of political interactions at ground level in India reveals that the differences between British and Indian individuals and institutions were less entrenched than we might otherwise believe. While it is true that Indian rulers often failed to uphold their commitments to their allies, the British record was far from unblemished. In fact, one of the principle reasons for the resumption of fighting between Haider Ali and the British after 1769 was that the British did not render him the assistance to which he was entitled under the peace treaty when he was subsequently attacked by the Marathas.[22] Nor did they live up to their treaty obligations in 1783 when Mysore was again under siege. Definitions of discipline, of loyalty and the importance of personal bonds of patron and client were more similar than dissimilar between India and Britain in the eighteenth century.[23] This can be seen most clearly in the extent to which loyalty was a discretionary commodity for much of this period, and that the reliability of soldiers and officers alike could not be taken for granted. Haider Ali, Mir Qasim and the East India Company all had to contend with serious outbreaks of indiscipline from their troops as well as from their senior officers, usually over issues of pay. Not surprisingly, the Court of Directors of the EIC in 1785 specified that pay and allowances for their European and Indian troops would have the highest priority.[24] In so doing, they paralleled policies introduced by Mir Qasim and Haider Ali.

Mir Qasim and politico-military reform in Bengal

One of the most pressing issues facing the British in the aftermath of the Battle of Plassey was finding a successor to the defeated provincial governor or *Nawab* Siraj-ud-daula. At this point, the British were unwilling and arguably were even incapable of standing forth as rulers in their own right. They lacked the resources, the legitimacy and the detailed local knowledge that was required to introduce direct rule over Bengal. Yet, they were very conscious of their position in Bengal, and of the profitability and power that they had secured.[25] They required a capable yet compliant ruler to ensure stability and profitability – a challenging combination in the very volatile conditions which then prevailed.[26] Initially, they turned to Mir Jafar to serve as *Nawab* of Bengal. Mir Jafar had been one of Siraj-ud-daula's most distinguished generals but had remained on the sidelines during the Battle of Plassey. His position proved to be an untenable one for not only were British demands incessant but the British were themselves riven by faction. Mir Jafar could not count upon the backing of the EIC's officials in India – nor would Mir Qasim when he came into power, for as one contemporary observed, Cossim Ally Khaun (Mir Qasim) was 'Mr Vansittart's Nabob' and Jaffier Ally Khan (Mir Jafar) was 'Mr Clive's Nabob'.[27]

Mir Jafar was in fact driven out by the constant financial pressures of the British – he could not meet their demands or those of his own commanders, and by 1760 he was facing widespread mutinies amongst the forces he had raised. In a letter to the Council, Mir Jafar reported that the sepoys would not move until they had been paid, and that he had little alternative but to melt down his plate to meet their needs.[28] Not only were his troops in arrears, but he had fallen behind in the payments he had promised the British as well as to Indian financiers like the Jagath Seths who had helped to underwrite some of his earlier costs. As he fell further and further into default, factions within the British Council began to look to possible replacements. Mir Qasim emerged as the most likely contender. He had earned a reputation for administrative competence as *faujdar* of Rangpur, and was described in yet another example of Orientalist type-casting as an 'altogether abler man than his drug-addicted father-in-law'.[29] Other commanders and the Jagath Seths then began to rally to Mir Qasim in the hopes that he could meet their needs. The Governor, Henry Vansittart, who was already well disposed to Mir Qasim, came to the conclusion that Qasim could succeed where his father-in-law failed. It is important to note that in the turbulent conditions in post-Plassey Bengal, Mir Qasim was also seeking to position himself better and he had made his own overtures to Vansittart.[30] He had also begun to strike out on his own. He signalled his intent to move his capital to Birbhum, tried to forge an alliance with Shah Alam, the Mughal Emperor, as well as befriend a number of the larger zamindars such as Kamgar Khan, with the intention of driving out Mir Jafar.[31]

Mir Qasim's regime started on a very good footing with the British. He raised sufficient revenues to meet the immediate demands from the British as well as from the troops that had been recruited by his predecessor. Rajat Ray attributes Mir Qasim's initial success to his ability to rebuild relations between warriors, bankers and landlords – relations which had rapidly dissolved in the aftermath of Plassey. Yet, unlike his predecessors who in the absence of the British had enjoyed greater latitude of action, Mir Qasim had more limited means to achieve this. Consequently he would increasingly have to resort to harsh and arbitrary justice to secure the obedience of his followers.[32] Mir Qasim was pressured to provide presents to those who had facilitated his succession. The districts of Burdwan, Chittagong and Midnapur were transferred to the EIC, and John Holwell, Vansittart and other key officials received individual grants of as much as £50,000.[33] The many private deals struck between members of Mir Qasim's courts when coupled with the actions of British officials and traders threatened not only to deprive the *nawab* of much-needed revenues but also signalled a considerable and embarrassing check to his sovereign authority.[34]

Mir Qasim would quickly discover that the tensions that had undermined Mir Jafar's regime would threaten his own rule. It was the pressure of private interests, articulated through deep factions within Calcutta, which hastened the rupture with Mir Qasim. In Rajat Ray's apt words, 'His father-in-law, Mir Jafar, had bought the whole council in 1757. His mistake was that he bought less than half and antagonized the rest.'[35] By early 1763, it was widely rumoured in the *bazaars* (markets) of Bengal and Bihar that Mir Qasim was seeking to reduce the power and limit the presence of the British.[36] In June 1763 it was reported that Mir Qasim's agents were trying to seduce sepoys to leave the EIC's service.[37] Two high-ranking officials in particular, Peter Amyatt and William

Ellis aligned themselves against Mir Qasim. Neither had been well disposed to him at the outset, and over time their hopes for greater commercial freedom came into conflict with Mir Qasim who had assumed that having turned over three districts to the British, he would enjoy unchallenged authority over the remainder of his territories. For his part, 'Mr Ellis appears to have been a man of a very warm temper, and had, unfortunately for himself and those under his command, conceived the greatest dislike to the Nabob Cossim Ally.'[38] Peter Amyatt had expected to succeed Robert Clive as governor and had been embittered by Vansittart's assumption of the office.

Mir Qasim's intention to establish a more secure and durable foundation for his authority became quickly evident following his accession. Within a year, he had not only paid off many of the troops that had threatened to mutiny for want of pay, but he had begun to carve out a more independent power base further inland. According to Robert Clive,

> Soon after Cossim Ally was raised to his new dignity, he was suffered to retire to a very great distance from his capital, that our influence might be felt and dreaded as little as possible by him. He was suffered to dismiss all those old officers who had any connection with or dependence upon us.[39]

He made Monghyr (Mungher in Bihar) his capital to distance himself from the British where

> he occupied himself in attending to his finances, and forming an army on our model. For this purpose he received our seapoys and their officers into his service to teach his troops our discipline, made a vast quantity of firelocks instead of matchlocks, and a good train of field artillery.[40]

When Mir Qasim became *Nawab*, his army amounted to about 90,000 men on paper, and of that no more than a tiny proportion could be considered well trained and loyal. This was all made clear in Mir Qasim's first operation, an attempt to impose authority over the *Raja* of Birbhum. His army was so unreliable and incompetent that he had to secure assistance from the EIC's forces. But, in the meantime he endeavoured by every method to increase his own force, entertaining European and Armenians to discipline his sepoys.[41] More significant reforms began in earnest when Mir Qasim

> discharged by degrees the whole of the ill-armed and disorderly rabble that had hitherto constituted the standing army of his predecessors, and entertained new troops upon a perfectly different footing. The cavalry was formed into regiments ... none but efficient men were admitted, and great attention paid to the manner in which they were mounted.[42]

He enjoyed some success in recruiting select troops from Mir Jafar's army. Many were willing to shift their obedience to Mir Qasim for he had a reputation for paying salary arrears promptly.[43] Sepoys willing to desert the British and join his forces were offered between Rs20 and 25 per month if they brought in their own arms, and Rs15 was paid to those who came in without arms.[44] These wages were well above what the EIC was currently offering.

His infantry was modelled along the lines of the EIC, and comprised two classes, '*Nujeebs* and *Telingas*': the former donned traditional dress and were armed with matchlocks, the latter were uniformed and armed like the sepoys of the EIC. The *Telingas* under their commanding officer Mahmed Taki Khan proved to be a very effective force. Pay for officers was as much as Rs100 a month, sepoys were formed into units of ten which were then grouped into units of 100 and then 1,000, drilled regularly and given opportunities for range practice. They travelled lightly as draught cattle and camels were provided to carry most of their kit.

Other onlookers were impressed by the speed with which Mir Qasim raised such a well-armed and well-funded army, noting that 'he was amassing and manufacturing as many guns and flint-muskets as he could'.[45] According to an early historian of the Bengal Army, the muskets which Mir Qasim brought into use were

> found to be superior to those of English manufacture, particularly in the barrels, the metal of which was of an admirable description; the flints also were of excellent quality, composed of agates found in the Rajmahal Hills, and were much preferred to those imported.[46]

As late as 1814, the Marquess of Hastings spoke admiringly of the workmanship of the gunsmiths of Mungher.[47] Most of Mir Qasim's artillery was locally cast of brass, and the carriages were considered equal to those imported. It was also noted that 'the powder was of excellent quality'. Here, Mir Qasim benefitted from proximity to rich supplies of saltpetre.[48] Bihar was a particularly important provider and Patna was a major distribution centre. It has been estimated that 70 per cent of the world's supply passed through Patna. Fear that this trade would slip out of British hands was yet another factor in their estrangement from Mir Qasim.

The sweeping reforms initiated by Mir Qasim would not have been sustainable had he not also tightened his administrative controls. In particular, he introduced measures to ensure that revenues adequate to his military needs were paid promptly and that his financial pipeline had as few leaks as possible. A contemporary chronicler observed:

> He was hardly established on the mesned [*masnad*] when he instituted a scrutiny in the department of finances, where he was amazed and thunderstruck at the emptiness of the treasury, and at the immediate balances which he had to pay to his own troops, as to those of Mir Jafar, for whose arrears he had pledged himself, and all that, over and above the large sums that were due to the English army and to the English rulers.[49]

The chronicler went on to note: 'It must be remembered that Mir Cassem-Qhan was not only an able accomptant [accountant] and a keen discerner of the intricacies of public accounts, but that he had given himself able ministers who helped him assiduously.'[50] Mir Qasim was careful to recruit some officials of the pre-1750 era who helped to review and re-establish revenue administration.[51] Yet, revenues would fall short of expenditure. He had to borrow heavily from the Jagat Seths and also pursue economies elsewhere. But, in his zeal to raise revenues and check attempts to undermine his authority, Mir Qasim often resorted to harsh measures. He directed his attention particularly towards the zamindars, most of whom he mistrusted. Ghulam Khan would later claim that Mir Qasim was 'in his heart ... an enemy to Zemindars'.[52] Mir Qasim had hoped to restore previous practices whereby districts would be entrusted to officials loyal to him alone.

On balance, however, Ghulam Khan felt that Mir Qasim's strengths outweighed his weaknesses. He particularly credited Mir Qasim with

> unravelling the intricacies of affairs of Government, and especially the knotty mysteries of finance; in examining and determining private difference; in establishing regular payments for his troops, and for his household; in honouring and rewarding men of merit, and men of learning; in conducting his expenditure, exactly between the extremities of parsimony and prodigality.[53]

More recently, Rajat Ray has highlighted Mir Qasim's success in rebuilding relations between warriors, bankers and landlords – relations which had rapidly dissolved in the aftermath of Plassey. But, lacking the resources with which he could reward loyalty, Mir Qasim was forced to rely more on fear than consensus.[54]

The military and financial reforms instituted by Mir Qasim, when coupled with the efforts he was making to ensure greater loyalty from the aristocracy of Bengal and Bihar, could conceivably have produced a state that was not only willing but was also able to withstand British pressures. Certainly, his actions sparked some anxieties amongst the British in Bengal and added to the chorus demanding that he be deposed. Even Governor Vansittart sought to constrain him, urging him to lessen his dependency on local recruits and trust instead to forces which the British could provide for his use.[55]

The underlying issue that ignited the war between Mir Qasim and the British was the flagrant abuse of privileges by British officials and traders. *Dastaks* were passes that enabled the Company to avoid duties and taxes on internal trade. Company officials had taken to using such passes to facilitate their own private trade, and in some cases had given *dastaks* to their Indian agents. Such abuses not only deprived the *nawab* of revenues but in their brazen disregard of his rules, British officials were challenging his authority.[56] A delegation was sent to Mungher but nothing came of it because

> jealousy and distrust seemed to be heightened on both sides from the insolence of several of the Nabob's officers and from the unjust authority

> assumed by some of the English agents and *gomastahs* [EIC's Indian agents] in different parts of the country.[57]

The situation continued to deteriorate. The breaking point came in October 1762 when Mir Qasim directed his officials to intercept boats operating on behalf of European traders and levy on them the taxes he felt were his due. The crisis did not come as a great surprise to the British – conditions were becoming increasingly volatile, and some officials, Ellis and Amyatt in particular, were trying to manoeuvre Mir Qasim into a position where a rupture would become inevitable.[58] Henry Vansittart went to Mungher where he negotiated what he thought was a compromise whereby Europeans would begin to pay customs on inland trade. He was however overruled by his Council following outcries amongst Europeans who stood to lose under the proposed terms. On 23 May 1763, Mir Qasim was presented with a list of demands, including the annulment of Vansittart's short-lived treaty, the release of some prisoners he had taken, payment of reparations and that the British would post a resident in his court.[59] Mir Qasim retaliated by suspending all duties for two years – thus enabling Indian traders to compete on the same terms as British traders.[60]

On 6 June 1763 Ellis with a detachment under Captain Carstairs launched an attack on Patna. Patna fell quickly, but this triggered a popular uprising as the 'whole of the country about Patna were in arms, all complaining of the treacherous manner in which the city had been attacked by the English'.[61] Ray notes that in the subsequent attacks on British factories, participants were drawn from a wide swathe of people: at Dacca, for example, *fakirs* were prominent.[62] The British were soon besieged in Patna, and with well-sited artillery playing upon them, they tried to flee across the Ganges. Short of provisions and ammunition, the British were forced to surrender. Following the battle, in a note heavy with irony, Mir Qasim wrote to the governor that,

> Mr Ellis has turned out to be his sincere friend. The governor would not part with two or three hundred muskets, but Mr Ellis supplied His Excellency with all the muskets and cannons in his possession; for he assaulted the fort like a night robber and plundered the city for hours together, but was at last defeated and captured with all his muskets and cannon.[63]

He concluded his note by requesting that the EIC return Burdwan and other lands which he had ceded to them in payment for their military support, as it had become clear to him that the Company's army no longer served his interests.

The EIC retaliated by reinstating Mir Jafar and despatching a force to capture Mir Qasim. When news reached Mir Qasim that he had been deposed, he responded by ordering the execution of 56 prisoners, an act that led the EIC's first official historian to observe that 'the massacre at Patna is proof that they look upon us in the character of usurpers'.[64] The main force under Major Adams encountered resistance part way between Murshidabad and Rajamahal where Mir Qasim's troops made an 'obstinate stand'.

> They consisted of six battalions of seapoys cloathed and armed like ours, a great number of our Patna seapoys among them. They had twenty guns among which the field pieces we lost at Patna, and 150 of our Europeans were engaged by fair means or foul to serve them.[65]

The threat posed by Mir Qasim persuaded Major Adams that there was 'not a moment to hesitate, lest the enemy flushed with success should first overrun the Company's lands, and effectually prevent our carrying on an offensive war, by cutting off all our supplies'.[66] At Katwa, in early July, the British were fortunate that one of their cannon shot decapitated the local commander.[67] Yet, Adams noted that the attack launched by 2,500 Pathan horse and 1,000 sepoys was very disciplined, and that the quality of their marksmanship was very high. He grudgingly praised their 'obstinate resistance'; 'Their cavalry in particular shew'd uncommon resolution in charging the 84th Regiment when separated from the main body.'[68]

Major Adams was able to capture Mungher without much loss of life owing to the betrayal of Mir Qasim by the local governor: 'Areb-aaly-ghan the Governor, who was naturally a coward ... sent them word that if they would give him a sum of money, he would surrender the fortress.'[69] The seizure of Mungher not only deprived Mir Qasim of his capital, but also many of the grain and ammunition stocks that he had built up. Mir Qasim was forced to flee the province of Bengal, seeking sanctuary from his one-time enemy, Shuja-ud-daula, the *Nawab* of Awadh. The troops which accompanied Mir Qasim sufficiently impressed the Awadh *Nawab* that the latter used them to deal with some refractory zamindars.[70] Yet, Mir Qasim was to all intents and purposes kept as a prisoner, even during the subsequent battle of Buxar.

The British forces under Hector Munro met the hastily and shakily forged alliance of Shuja-ud-daula, Mir Qasim and Shah Alam, the Mughal Emperor at Buxar, where it has been estimated that British deployed 7,000 troops against the 50,000 men fielded by the allies. If the alliance was rather unstable, so too were the troops formed under Colonel Munro. Munro's columns were plagued by acts of disloyalty. Colonel Champion reported that his troops would only march when given promises of additional pay and allowances.[71] He also recorded in his journal that 'CaptLieut Bylong has been confined to quarters – he had put in his resignation even though a battle was imminent, it was refused, and he protested. He was court martialed on the 28th and dismissed from the service.'[72]

At Buxar, the EIC's troops were set out in a line, with 20 field guns protecting their flanks. When Shuja's troops attacked the left flank of the EIC's battle line, the Company's infantry formed into a square and supported by the artillery they were able to repel the cavalry. Shuja's troops were not supported by his artillery, and they were soon beaten back. Mir Qasim's troops were much better disciplined and more capable on the battlefield. Of the *Nawab* of Awadh's troops, it was noted,

> there was so little order and discipline amongst these troops, and so little were the men accustomed to command, that in the very middle of the camp,

> they fought against each other, killed and murdered each other, plundered each other, and went out a plundering and marauding without the least scruple, or the least control.[73]

Had the majority of the troops at Buxar been trained and equipped like those of Mir Qasim, the results could have been very different. Instead, they succumbed to the long-standing Mughal custom of highly individualized warfare which prevented the warrior elite from cohering together in the battlefield. There was no general staff, no period of extended drill training and no well-established and accepted chain of command.[74] Even then, Buxar was a very near run thing; had the British not succeeded, Shuja-ud-daula and Mir Qasim would have been in a situation where there were no troops between them and Calcutta. British supply lines were very tenuous and Munro only had ten days' provisions at hand.[75]

Haji Mustafa, a contemporary chronicler of French origins, believed that had Mir Qasim been able to stave off a rupture with the British for another two years, he could have perhaps bested them in battle as delay would have allowed him not only to build up a larger army but he could have also strengthened his control over the zamindars. Time was not on his side. Nor was the fact that while 'he is a cunning artful politician; [he] will never venture himself in the field'.[76] But, he quite clearly instigated a transformation that posed a serious potential threat to the British, so much so that when Mir Jafar was reinstated, the British took special care to limit his scope for administrative and military reform.

Haider Ali and the revolution in South India

Haider Ali's military record as well as that of his son Tipu Sultan has been much better documented and more widely celebrated. No doubt this is due in large part to the sustained resistance they mounted against British expansion in southern India, the vigour of which has been variously credited to their personal ambitions, French intrigues and religious zeal. More recently, historians have also explored the military and fiscal reforms which underpinned their successful campaigns against the British, and much attention has been directed at Mysore as exemplifying some of the traits associated with the fiscal military state. As Burton Stein observed, 'Thrusting centralization was the signature of regimes of eighteenth-century India, and everywhere this was dictated by military requirements.'[77] The roots of this transformation have been shown to have predated Haider Ali by about a half century and to have taken shape under the previous Wodeyar dynasty.[78] Organizational reform coupled with military innovation and expansion accelerated under Haider Ali and would peak under Tipu Sultan. Haider's military genius and administrative acumen fascinated contemporaries, and while there was often an element of hyperbole in contemporary descriptions, the quote below gives a sense of the kind of regard with which he was held.

> With all the vigour of his councils he can neither read nor write; but these qualifications are not essential to a military genius, nor can learning give

> strength of mind. He is possessed of surprising memory, deep penetration, quickness of resolution, great perseverance, conduct aided by personal courage, and that inexorable temper of mind, where from policy or disposition, that never forgives a crime, in his Durbar, where he smokes his Hookah, he pronounces sentences of death by waving his hand; or, if he uses words upon the occasion, they are only these TAKE HIM AWAY. He thinks no more of the subject, but proceeds coolly to other business. Notwithstanding this summary kind of barbarous justice, his decisions are generally equitable: he is feared, obeyed, and beloved.[79]

Haider Ali's origins lay in northern India; along with many other Muslim soldiers, he was recruited to serve in southern India where there was great demand for experienced cavalrymen. At this point in time, cavalry was more effective than infantry, largely because foot soldiers were then armed with matchlocks, not flintlocks, and consequently were more cumbersome and less reliable. Moreover, cavalry warfare, with its emphasis on individual feats of bravery and one to one combat, resonated better with the cultural values of the age. By the early 1760s, Haider had become one of the principal commanders in Mysore, and when Krishna Raja died in 1766, Haider Ali was instrumental in installing a successor. Soon after, he emerged as the ruler of Mysore where he began to institute sweeping reforms, though it should be noted that he was careful to maintain a number of the customs and symbols valued in the region.[80]

Haider Ali had first-hand experience of the Anglo-French struggles in India during the Seven Years War, and was persuaded early of the value of drilled flintlock equipped infantry.[81] He reconfigured much of his infantry so as to produce tightly drilled formations, and encouraged the production of up to date weapons. But, his approach to military reform was not simply a slavish imitation of European practices; instead, he melded together infantry technology, drill and tactics from Europe with indigenous innovations, such as rockets, and retained elements of local military culture that had proven its value, particularly light cavalry and bullock trains. This enabled him to combine firepower with mobility in a form with which the British could not easily keep pace. Another distinguishing feature of Haider Ali's reforms was his drive towards standardization. He replaced the bewildering array of variously sized and trained bodies of troops with units of fixed strength, of which the most important was the *risala* of 1,000 troops. He even laid the foundations for a navy – by 1766 he had two ships, seven smaller vessels and 40 *galivats*.[82]

The changes he introduced to the tactics, technology and organization of his army would be buttressed by sophisticated logistical and administrative innovations. Haider Ali closely supervised the different departments of the army. It was reported that he met each morning with the 'majors of the army' who the author describes as analogous to the adjutant generals in European armies. Significantly, we are told that these officers had risen to these positions on the basis of merit. 'They are not persons of distinction, but men of approved diligence and fidelity, chosen out of the subaltern officers of cavalry and infantry.'[83] He also took care to ensure that communications were carefully tracked. When messages were sent, they were accompanied by a sheet wherein the time of its arrival at each staging post was recorded. Personnel in the army were also closely monitored. Each soldier in the army was issued a *bati*, a small form which listed the name of the individual, his father and grandfather, the date he entered service, his station and his rate of pay, and these were published in three languages: Persian, Marathi and Kanarese. Moreover, Haider took personal responsibility for signing the monthly statement of accounts. According to a French officer who served under him, 'no payment is made without the signature of Hyder, or, in his absence, of the general commandant'.[84]

One of the defining characteristics of military fiscalism, and a critical enabler of the military transformations initiated by Haider Ali and carried on by his son, was the establishment of a centrally controlled and accountable group of officials responsible for the collection of revenues. Without them, it was both difficult to calculate with precision the revenues available and ensure that these revenues reached the state's treasury. Haider Ali initiated the practice of shifting responsibility for revenue collection from *jagirdars*, individuals who were assigned lands in return for military service, to paid officials accountable to the state. This was a major transformation, and while the shift would largely occur under Tipu Sultan, the outlines of a new system emerged under Haider Ali. In the meantime, Haider Ali distributed much of the lands taken during the wars to warriors through auctions though he was careful to impose restrictions on the authority they enjoyed over these lands. Further checks included replacing tax farmers with officials, *amildars*, who reported to the local governor.[85] By starting to curb the power of intermediaries such as the *jagirdars*, the authority of the centre was strengthened. Not only did the state secure more direct control over revenues, but it could then use those revenues to recruit and reward its troops rather than rely upon military entrepreneurs like the *jagirdars.* Haider Ali also took special care to solicit the support of merchants and suppliers of military material. He was particularly keen to remain on very good terms with horse dealers.[86] His careful attention to merchants and traders led General Eyre Coote to ruefully note that Haider Ali was more popular amongst the local population than was their erstwhile ruler (and British ally), the *Nawab* of the Carnatic.[87]

Perhaps the most intriguing illustration of Haider's commitment to military innovation and systematization came shortly after his death with the commissioning of a manual on the art and science of war, which has usually been attributed to Tipu Sultan. Its purpose was to make available, 'this noble science and graceful art, which is not to be met with or found in the cities of Hind'.[88] The book appears to have been written in 1783 or 1784, just after the death of Haider Ali. It goes into tremendous detail on a whole range of issues related to raising and maintaining an army, including advice on how to establish up to date musters, provide proper physical drill and effectively maintain the army's equipment. Little is known of its author. According to one source, Mir Zainul Abedeen Shushtari was asked by Tipu Sultan to produce a compilation of the regulations and procedures which had been instituted by his father.[89]

In 1766, the Marathas launched an attack on Haider Ali, and while there was no formal treaty between the Marathas and the Nizam of Hyderabad, the latter joined in, and the British followed, in part because Haider Ali had laid claim to some of the lands of their ally, the *Nawab* of Carnatic. A British force under Colonel Joseph Smith was dispatched to join the Nizam of Hyderabad and jointly it was intended that they would besiege Bangalore while the Marathas attacked the northern parts of Haider's territories. But, the situation was a very fluid one, and the British came to regret that the alliance which they joined had not been firmed up.[90] Haider managed to placate the Marathas who then withdrew, and followed this by reaching a separate arrangement with the Nizam of Hyderabad, thereby isolating the British. In a classic example of understatement, one officer noted that, 'the terms of which portended no good to the English'.[91] The Nizam's troops were deeply in arrears and the promise of monies from Hyder Ali helped to engineer the switch.[92] The combined forces of Haider and the Nizam poured into the Carnatic in August 1767 and by September Haider's cavalry was harassing the outskirts of Madras. The alliance between Haider Ali and the Nizam of Hyderabad proved to be short lived, and the Nizam separately sued for peace in February 1768 following attacks launched against Hyderabad by the EIC's detachments.

In September 1767, the army under Joseph Smith consisted of two regiments of Europeans which with the artillery, amounted to about 1,400 rank and file, 11 battalions or about 9,000 sepoys 'and a few horse, not worth mentioning, being too few to be of any consequence in a battle, surrounded by such superior numbers'.[93] Colonel Smith's force was little more than a flying column, but one with limited mobility, for it was short of cavalry. It also lacked sufficient artillery to undertake sieges effectively.[94]

The real edge enjoyed by Haider Ali was his mobility – his forces could move more quickly than those of the British, and he could count upon them being better and more quickly supplied. Robert Clive complained that,

> With regards to the operation of the army, they have consisted in little else but marching and countermarching without a possibility of bringing Hyder to an action as he has it in his power to avoid it whenever he pleases.[95]

One British officer complained that 'what Hyder marched in two days, we did with difficulty in six'.[96] He went on to report that Haider Ali used to refer to them disparagingly as the 'English cripples'.[97] The British observers were also impressed by Haider's preparations for crossing the rivers.

> Hyder had a great number in his army; two men carry the skeleton of one (made of bamboo), and two more the skin: in a quarter of an hour they are ready for use; and one boat will carry twenty-five men or a piece of cannon.[98]

Another noted that in one campaign, Haider's men stripped down, covered their knapsacks with waxed cloth, so as to be more mobile during the monsoon season.[99]

This allowed Haider Ali to avoid easily what the British wanted most – a general action where the British advantage in regular infantry would count most. As one officer lamented, 'We want to bring him to an engagement, but he is so artful a warrior he won't let us.'[100] Instead of manoeuvring for the decisive battle, Haider Ali chose to strike at the most vulnerable point of the British – their communications and supply systems. He looked to his cavalry to pressure the British from multiple directions, using scorched earth tactics to deprive the British of the resources needed to keep their large and ponderous army in the field. At one point, Smith had to dedicate a third of his available troops to protecting his baggage from Haider's cavalry.[101] Observers also commented on Haider's troops using 'platoon firing' very effectively and that this had been well coordinated with his artillery.[102]

The bleak situation facing the British was aptly summed up in the following summary:

> You will hear from all hands how truly unprofitable the war has proved to us, and that indeed we were never before so handled. Excepting Cuddalore, Trichenopoly, Madura and Pallamcotah, we have not a foot to the southward of Pondicherry that we can call our own: all the country to the southward entirely laid waste; not an hut or inhabitant to be seen for sixty miles together, so terrible have been, and still are, the devastations of the enemy's horse. The Coimbatoor country we once called our own, but alas, every fort surrendered again on the first summons, and we are in a much worse situation than ever we were.[103]

In exasperation, Colonel Wood, the second-in-command, complained to Haider Ali about his refusal to do battle. To this Haider Ali replied:

> Give me the same sort of troops that you command and your wishes shall be granted. Shall I risk my cavalry, which cost a thousand rupees each horse against your cannon balls which cost two pice? No, I shall march your troops until their feet shall meet their bodies. You shall not have a blade of grass, nor a drop of water.[104]

In early 1769 negotiations were opened between Haider Ali and the British; Haider was concerned that the Marathas could strike again and the British were being ground down by the costs of the war. A deal eluded them, however, and Haider Ali launched another wave of attacks on the Carnatic in February 1769. Smith tried to engage Haider in a pitched battle, but unlike Mir Qasim, Haider astutely avoided set-piece battles when he could, and instead relied upon the fleetness of his forces to maintain a constant pressure on the British. Smith, lacking cavalry and hampered by a cumbersome logistics network, could not keep pace with Haider Ali. Haider made it as far as St Thomas's Mount, eight miles from Madras, where peace was agreed upon, the terms of which included restoration of properties seized by each side, an exchange of prisoners and a defensive alliance between the two powers.

Conclusion

In the end, the biggest advantage that the British had when compared to the many indigenous states vying for authority in eighteenth-century India was a relatively robust and centralized financial system which enabled them to provide more consistent war funding which was so vital in securing the loyalty of the troops they had raised. Most of the Mughal successor states, as well as the petty kingdoms beyond the frontiers of the old Mughal Empire, while often receptive to new technologies and eager to implement them, failed to initiate appropriate administrative reforms, particularly of their tax systems, which would have enabled them to support new technologies and their attendant military formations in a sustainable fashion. In the two cases studied above, we can see that Mir Qasim and Haider Ali sought to reconcile indigenous military practices with new technologies and the tactics they made possible, and underpinned these efforts with financial and administrative reforms. Haider Ali was particularly successful in this transition, prompting one British officer to grudgingly concede that 'Hyder has so gallantly and with so much scientific knowledge in his possession, changed the nature of the war.'[105] One of the reasons why Haider Ali enjoyed more success than Mir Qasim was that he was better able to neutralize the major advantage enjoyed by the British, the number and quality of disciplined infantry at their disposal, by playing on the major advantage then available to Indian rulers – access to light cavalry.[106]

Mir Qasim and Haider Ali both pursued with varying degrees of success administrative reforms intended to put their militaries on sounder financial footings. They also proved to be receptive to new technologies, though they carefully selected the technological and organizational developments that best suited their needs. For example, Haider Ali has been praised for his 'brilliant combination of mobile cavalry organized on the Mughal pattern with his increasingly disciplined musket-using infantry'.[107] In the south, where the British did not have access to reliable and sufficiently large forces of cavalry, this would greatly hamper their ability to check his advances.[108] One key further difference between Mir Qasim and Haider Ali is that Mir Qasim had to operate within a much more fluid world of loyalties. Haider Ali had made more of the administrative and political adjustments necessary to ensure that his new style military could function effectively. Mir Qasim's failure to secure the loyalty of his subordinates meant that the British were not so disadvantaged when it came to raising the cavalry they needed to protect their supply lines, collect valuable reconnaissance and engage in hot pursuit. The situation for the British in the south was not so favourable.

Notes

1 G.R. Gleig, *The Life of Major-General Sir T. Munro, Bart … Late Governor of Madras: With Extracts from his Correspondence and Private Papers* (London: Colburn & Bentley, 1830), vol. 1, p. 25.

2 Janaki Nair, *Mysore Modern: Rethinking the Region under Princely Rule* (Minneapolis, MN: University of Minnesota Press, 2011), p. 37.

3 Janaki Nair, *Mysore Modern*, p. 38.
4 William Martin Goodlad to Robert Palk, 30 June 1769, Historical Manuscripts Commission; *Report on the Palk Manuscripts in the Possession of Mrs Bannatyne of Haldon, Devon* (London: HMSO, 1922), p. 109.
5 Irfan Habib (ed.), *Confronting Colonialism: Resistance and Modernization under Haidar Ali and Tipu Sultan* (Delhi: Tulika, 1999), p. xvii.
6 Major Thomas Adams to the Earl of Egremont, 15 October 1763, H/MISC/98/8, p. 100, Oriental and India Office Collections (hereafter OIOC), British Library, London.
7 P.J. Marshall, *East Indian Fortunes: The British in Bengal in the Eighteenth Century* (Oxford: Oxford University Press, 1976).
8 Thomas Rumbold to Richard Smith, 3 February 1764, MSS EUR OV 21/25, OIOC.
9 Douglas M. Peers, 'Army Discipline, Military Cultures, and State-Formation in Colonial India, *c.*1780–1860', in H. Bowen, V.E. Mancke and J.G. Reid (eds), *Britain's Oceanic Empire: Atlantic and Indian Ocean Worlds,* c.*1550–1850* (Cambridge: Cambridge University Press, 2012), pp. 282–306.
10 Kate Brittlebank, *Tipu Sultan's Search for Legitimacy: Islam and Kingship in a Hindu Domain* (Delhi: Oxford University Press, 1997); Kate Teltscher, *India Inscribed: European and British Writing on India, 1600–1800* (Oxford: Oxford University Press, 1995).
11 Charles Smith to Madras Secret Committee, 10 November 1779, R/C/67, p. 130, OIOC.
12 Alexander Dow, *The History of Hindostan* (1770, reprint, Delhi: Today and Tomorrow Publishers, 1973), p. 223.
13 See Patrick Porter, *Military Orientalism: Eastern War through Western Eyes* (New York: Columbia University Press, 2009). For India in particular, see Douglas M. Peers, ' "There is Nothing More Poetical than War": Romanticism, Orientalism, and Militarism in J.W. Kaye's Narratives of the Conquest of India', in J.F. Codell (ed.), *Imperial Co-Histories: National Identities and the British and Colonial Press* (Madison, NJ: Fairleigh Dickinson University Press, 2003), pp. 273–299; Douglas M. Peers, 'Conquest Narratives: Romanticism, Orientalism and Intertextuality in the Indian Writings of Sir Walter Scott and Robert Orme', in M. Franklin (ed.), *Romantic Representations of British India* (London: Routledge, 2006), pp. 238–258.
14 Douglas M. Peers, 'Revolution, Evolution, or Devolution: The Military and the Making of Colonial India', in W.E. Lee (ed.), *Empire and Indigenes: Intercultural Alliances, Imperial Expansion, and Warfare in the Early Modern World* (New York: New York University Press, 2011), p. 87.
15 Colonel G. Fitzclarence, 'Observations on the Employment of Mahommedan Mercenaries in Christian Armies', *Naval and Military Magazine* vols 2, 3, 4 (1827, 1828), p. 571.
16 Patrick Porter, *Military Orientalism*. See also James Belich whose work foreshadowed and anticipated the writings of Porter: James Belich, *The New Zealand Wars and the Victorian Interpretation of Racial Conflict* (Montreal-Kingston: McGill-Queens University Press, 1989).
17 John Le Couteur. *Letters, Chiefly from India: Containing an Account of the Military Transactions on the Coast of Malabar during the Late War* (London: John Murray, 1790), p. 111.
18 David Parrott, 'Had a Distinct Template for a "Western Way of Warfare" been Established before 1800?', in H. Strachan and S. Scheipers (eds), *The Changing Character of War* (Oxford: Oxford University Press, 2011), p. 49.
19 P.J. Marshall, 'Western Arms in Maritime Asia in the Early Phases of Expansion', *Modern Asian Studies*, vol. 14, no. 1 (1980), pp. 13–28.
20 G.J. Bryant, 'Asymmetric Warfare: The British Experience in Eighteenth-Century India', *Journal of Military History*, vol. 68, no. 2 (2004), p. 458. See also Kaushik

Roy's, *War, Culture and Society in Early Modern South Asia, 1740–1849* (London: Routledge, 2011) for a comprehensive overview of the eighteenth century.

21 Jos Gommans, 'Indian Warfare and Afghan Innovation during the Eighteenth Century', *Studies in History*, vol. 11, no. 2 (1995), p. 268.

22 Meer Hussein Ali Khan Kirmani, *The History of Hydur Naik, otherwise styled Shums ul Moolk, Ameer ud Dowla, Nawaub Hydur Ali Khan Bahadoor, Hydur Jung* (London: Oriental Translation Fund, 1862), pp. 285–287. See also John Clunes, *Historical Sketch of the Princes of India, Stipendiary, Subsidiary, Protected, Tributary, and Feudatory* (London: Smith, Elder & Co., 1833), pp. 58–61.

23 Douglas M. Peers, 'Army Discipline', pp. 282–306.

24 William Hough, *Precedents in Military Law: Including the Practice of Courts Martial* (London: W.H. Allen, 1855), p. 38.

25 The notion that the British blundered, almost unthinkingly, from being merchants to being territorial sovereigns has been convincingly discredited by Philip Stern's fine study: Philip J. Stern, *The Company-State: Corporate Sovereignty and the Early Modern Foundations of the British Empire in India* (Oxford: Oxford University Press, 2011).

26 For a more detailed background to these events, see P.J. Marshall, *Bengal: The British Bridgehead, Eastern India, 1740–1828* (Cambridge: Cambridge University Press, 1988) and Rajat Kanta Ray, *The Felt Community: Commonality and Mentality before the Emergence of Indian Nationalism* (New Delhi: Oxford University Press, 2002).

27 *An Address to the Proprietors of East-India Stock* (London: C. Henderson, 1764), p. 16.

28 Mir Jafar to Bengal Council, 25 March 1759, in India, Imperial Record Department, *Calendar of Persian Correspondence: Being Letters, Referring Mainly to Affairs in Bengal, which Passed Between Some of the Company's Servants and Indian Rulers and Notables* (New Delhi: Government Publishers, 1914), vol. 10 (hereafter *Calendar of Persian Correspondence*).

29 Rajat Kanta Ray, *The Felt Community*, p. 251.

30 Mir Qasim to Vansittart, 15 September 1760, *Calendar of Persian Correspondence.*

31 Rajat Kanta Ray, *The Felt Community*, pp. 251–252.

32 Rajat Kanta Ray, *The Felt Community*, pp. 252–253.

33 Nicholas Dirks, *The Scandal of Empire: India and the Creation of Imperial Britain* (Cambridge, MA: Harvard University Press, 2006), pp. 50–51.

34 P.J. Marshall, *East Indian Fortunes*, p. 123.

35 Rajat Kanta Ray, *The Felt Community*, p. 263.

36 Nandalal Chatterji, 'The Genesis of Mir Qasim's War with the English', *Calcutta Review*, vol. 47, no. 2 (1933), p. 319.

37 Nandalal Chatterji, 'Genesis of Mir Qasim's War', p. 321.

38 Captain Williams, *An Historical Account of the Rise and Progress of the Bengal Native Infantry, from its first formation in 1757, to 1796* (London: John Murray, 1817), p. 120.

39 Robert Clive to Court of Directors, 27 April 1764, MSS EUR OV 21/6, pp. 198–199, OIOC.

40 *Reflections on the Present Commotions in Bengal* (London: G. Kearsley, 1764), pp. 9–10.

41 Major Adams, 'Journal of Military Operations, September 1761 to January 1764', MSS EUR OV 4, p. 5, OIOC.

42 Arthur N. Broome, *History of the Rise and Progress of the Bengal Army* (Calcutta: Thacker, 1851), p. 351.

43 Seid Gholam Hossein-Khan, *The Seir Mutaqherin; or View of Modern Times: Being a History of India from the Year 1118 to the Year 1194 of the Hedjrah … Volume the Second* (Calcutta: n.p., 1789), vol. 2, p. 145 (hereafter *Seir Mutaqherin*).

44 Translation of Documents from Mir Kasim, MSS EUR OV 21/27 p. 217, OIOC.
45 *Seir Mutaqherin*, vol. 2, p. 185.
46 Arthur N. Broome, *History of the Rise and Progress of the Bengal Army*, p. 351.
47 Douglas M. Peers, 'Revolution, Evolution, or Devolution', pp. 93–94. This would seem to bolster Daniel Headrick's idea that the real breakthrough came with what he has termed the 'breechloader revolution'. Daniel R. Headrick, *The Tools of Empire: Technology and European Imperialism in the 19th Century* (Oxford: Oxford University Press, 1981), pp. 96–102.
48 James W. Frey, 'The Indian Saltpeter Trade, the Military Revolution, and the Rise of Britain as a Global Superpower', *Historian*, vol. 71, no. 3 (2009), pp. 507–554.
49 *SeirMutaqherin*, vol. 2, p. 153.
50 *SeirMutaqherin*, vol. 2, p. 181.
51 *Seir Mutaqherin*, vol. 2, p. 151.
52 *Seir Mutaqherin*, vol. 2, p. 156.
53 *Seir Mutaqherin*, vol. 2, p. 197.
54 Rajat Kanta Ray, *The Felt Community*, pp. 252–253.
55 Vansittart to Mir Qasim, 26 May 1761, *Calendar of Persian Correspondence*, pp. 101–102.
56 J.Z. Holwell, *A Defence of Mr Vansittart's Conduct, in concluding the Treaty of Commerce with Mir Cossim Aly Chawn at Mongheer* (London: T. Beckett, 1764).
57 Robert Orme, *A Short Sketch of the Troubles in Bengal to the 28th July 1763*, MSS EUR, OV 21/15, ff. 137–138, OIOC.
58 Summary notes relating to disputes over trading rights, September 1763, MSS EUR C463, OIOC.
59 Dispatch to Mir Qasim, 23 May 1763, *Calendar of Persian Correspondence*, pp. 218–219.
60 Rajat Kanta Ray, *The Felt Community*, pp. 274–275.
61 Captain Williams, *An Historical Account of the Rise and Progress of the Bengal Native Infantry*, p. 125.
62 Rajat Kanta Ray, *The Felt Community*, p. 289.
63 Mir Qasim to Vansittart, 17 July 1764, *Calendar of Persian Correspondence*, p. 227.
64 E. Stephenson to Richard Smith, December 1763, MSS EUR OV 21/24, OIOC.
65 Robert Orme, 'A Short Sketch of the Present Troubles in Bengal, 1763', MSS EUR, OV 21/16, p. 145, OIOC. It has been estimated that Major Adams pocketed Rs3,00,000 from this short campaign: P.J. Marshall, *East Indian Fortunes*, p. 211.
66 Major Thomas Adams to the Earl of Egremont, 15 October 1763, H/MISC/98/8, p. 102, OIOC.
67 Major Thomas Adams to Major-General Stringer Lawrence, 6 August 1763, H/MISC/98/8, p. 148, OIOC.
68 Major Thomas Adams to Major-General Stringer Lawrence, 6 August 1763, H/MISC/98/8, p. 150.
69 *Seir Mutaqherin*, vol. 2, p. 285.
70 *Seir Mutaqherin*, vol. 2, p. 303.
71 Colonel Champion's Diary, 8 March 1764, H/MISC/198, OIOC.
72 Colonel Champion's Diary, 26 March 1764, H/MISC/198, OIOC.
73 *Seir Mutaqherin*, vol. 2, p. 306.
74 Rajat Kanta Ray, *The Felt Community*, p. 292.
75 C.E. Oldham, 'The Battle of Buxar', *Journal of the Bihar and Orissa Research Society*, vol. 12, no. 1 (1926), pp. 1–38.
76 Major Thomas Adams to the Earl of Egremont, 15 October 1763, H/MISC/98/8, p. 109, OIOC.
77 Burton Stein, 'State Formation and Economy Reconsidered', *Modern Asia Studies*, vol. 19, no. 3 (1985), p. 391.

78 Burton Stein, 'State Formation', pp. 387–413; Sanjay Subrahmanyam, 'Warfare and State Finance in Wodeyar Mysore, 1724–25: A Missionary Perspective', *Indian Economic and Social History Review*, vol. 26, no. 2 (1989), pp. 203–233.
79 Muhammad Qasim Hindu Shah Astarabadi Firishtah and Alexander Dow, *The History of Hindostan* (London: T. Beckett and P.A. De Hondt, 1770), vol. 2, p. 239.
80 Kate Brittlebank, *Tipu Sultan's Search for Legitimacy.*
81 Irfan Habib, *Confronting Colonialism*, p. xx.
82 A *galivat* is a small armed vessel with oars and sails. Irfan Habib, *Confronting Colonialism*, p. xxii.
83 M.D.L.T. S.f.M (Maitre de la Tour), *The History of Hyder Shah, alias Hyder Ali Khan Bahadur: Or, New Memoirs concerning the East Indies with Historical Notes* (London: J. Johnson, 1784), p. 16.
84 M.D.L.T., *The History of Hyder Shah*, p. 90.
85 Burton Stein, 'State Formation', p. 402.
86 Meer Hussein Ali Khan Kirmani, *The History of Hydur Naik*, p. 502.
87 Eyre Coote to Bengal Secret Committee, 10 November 1780, H/MISC/156/9, p. 53, OIOC.
88 Mahmud Husain (ed.), *Fath-Ul-Mujahideen: A Treatise on the Rules and Regulations of Tipu Sultan's Army and his Principles of Strategy, compiled by Mir Zainul Abedeen Shushtari* (Karachi: Urdu Academy Sind, 1950), p. ix.
89 Mahmud Husain (ed.), *Fath-ul-Mujahideen*, pp. ix–x.
90 Colonel John Call to Robert Palk, 19 March 1767, *Palk Manuscripts*, pp. 24–25.
91 Colonel Robert Kelly, Memorandum, *c.*1785, MSS EUR D1146/6, OIOC.
92 Robert Palk to Thomas Saunders, *c.*1767, *Palk Manuscripts*, p. 63.
93 Colonel Robert Kelly, Memorandum, *c.*1785, MSS EUR D1146/6, OIOC.
94 Joseph Smith's Diary, annotated by Robert Orme, 2 April 1769, MSS EUR OV 215 p. 82, OIOC.
95 Clive to Verelst, 12 December 1768, MSS EUR G/56, OIOC.
96 Colonel Robert Kelly, Memorandum, *c.*1785, MSS EUR D1146/6, OIOC.
97 Colonel Robert Kelly, Memorandum, *c.*1785.
98 M.D.L.T., *The History of Hyder Shah*, p. 74.
99 M.D.L.T., *The History of Hyder Shah*, pp. 75–76.
100 Thomas Palk to Robert Palk, 30 September 1768, *Palk Manuscripts*, p. 81.
101 Colonel Robert Kelly, Memorandum, *c.*1785, MSS EUR D1146/6, OIOC.
102 Colonel Robert Kelly, Memorandum, *c.*1785.
103 William Martin Goodlad to Robert Park, 11 March 1769, *Palk Manuscripts*, p. 95.
104 J.D. Forbes, *Oriental Memoirs* (London: R. Bentley, 1813), vol. 3, p. 287.
105 Colonel Robert Kelly, Memorandum, *c.*1785, MS EUR D1146/6, OIOC.
106 G.J. Bryant, 'Asymmetric Warfare: The British Experience in Eighteenth-Century India', *Journal of Military History*, vol. 68, no. 2 (2004), pp. 431–469.
107 Irfan Habib, *Confronting Colonialism*, p. xx.
108 G.J. Bryant, 'The Cavalry Problem in the Early British Indian Army, 1750–1785', *War in History*, vol. 2, no. 1 (1995), pp. 1–21.

19 Contrasting Indian and British concepts of race and authority in the East India Company armies

Michael H. Fisher

'Race' and the East India Company's colonial armies

The contested and contingent concept of 'race' (variously defined as we will discuss) stands central to the recruitment and disposition of soldiers in many armies. Military commanders, in constituting military contingents and recruiting people for them, often use race to ascribe specific physical and moral characteristics to men (and women) of particular communities, regions, ethnicities or other identities. They then try to equip, organize and deploy these soldiers accordingly. The physical bodies of recruits matter in virtually all armies – including appearance, height, weight, the presence and functioning of all limbs, age, sex and gender – but many cultures have presupposed that body types also correlate with particular races. Further, military commanders have historically associated specific moral features – like courage, steadfastness, initiative and discipline (especially under the stress of battle) – with particular races. Such racial differentiation proved especially fundamental in colonial armies, including those of the English East India Company (EIC).

The EIC was uniquely a European joint-stock commercial corporation which conquered and ruled as the colonial power over diverse peoples and cultures in South Asia, ambiguously recognizing both British and Mughal sovereignty. As the EIC entered the contested world of South Asia, it creatively synthesized its military forces, incorporating and adapting pre-existing military cultures, practices and manpower from both India and Europe. As the EIC's armies grew in complexity and size (peaking in 1857 at 271,000 regulars plus many militia/local forces), they used shifting definitions of race to categorize and organize them.

Throughout the EIC's existence, its British policy makers in London, Calcutta, Madras and Bombay – and factions at each site – often came to contradictory conclusions about the ways that their various and changing conceptualizations of racial difference should or could be implemented. Over this period, British society as a whole was redefining race, including which men made the best soldiers, in what functions and at what rank. Especially from the early nineteenth century onward, Britons increasingly based their definitions of race on biological rather than cultural criteria: ancestry rather than comportment.[1] This led to increasing racial discrimination by the EIC's Directors in

London, who eventually reserved their officer corps in regular regiments for 'white' Europeans, largely relegated 'mixed race' European officers to irregular units and depended on 'black' Indian soldiers (sepoys) for most of their manpower while generally restricting them to 'Native' regiments.[2] However, the EIC at various times also distinguished different military aptitudes among white Europeans, for example, between Protestants and Catholics, among English, Irish, Scottish, French and Portuguese, and among different social classes within each.

In India, the EIC's Bengal, Bombay and Madras armies also each gradually developed and inconsistently institutionalized its own forms of racial discrimination for Indians and among the Indians. Each army's commanders developed ideologies that they thought would produce the best configuration of their forces given their circumstances. Thus, the EIC armies differentially classified and assessed the martial abilities of Indians of different races, by which Britons meant variously different castes (*jati*s), religions, regional origins and/or types of military deportment.

One crucial and recurrent issue faced by each EIC army was whether to recruit Indian soldiers mainly from one race or to balance recruitment among several. If the latter, then whether to have similar Indians serve together in 'single class' companies, battalions and/or regiments or else mix them in integrated units. If men from the same race served together, this might enhance their corporate solidarity and create units which specialized in particular martial roles (e.g. artillery, light infantry or irregular cavalry) ascribed to that race. But, such solidarity might come at the cost of collective resistance to their EIC commanders.

Many individuals fit uneasily into such racial categories. Indeed, the various men – of European, Asian or mixed ancestry or culture – who sought to serve in the EIC's armies as officers or soldiers had their own interests and concepts of race, which often did not accord with those of the EIC prevailing at that moment. Due to asymmetries in power between the EIC as employer and those seeking employment, aspirants found it difficult to challenge the EIC's entire current system of racial definition. However, an individual could seek to conform to or manipulate the EIC's racial classification of him in particular by modifying his own self-representation through clothing, religion, education, claimed mother language or putative ancestry.

Thus, while the EIC used race to classify men in its armies, the men who served in these armies also held or developed their own conceptions of race, for themselves and for those serving around, below or above them. These changing definitions varied according to the often disparate factors of contrasting British and India cultural values, available European and Indian manpower, and the EIC's self-perceived military needs. Thus, race remained central to the EIC armies, from their earliest formation in the early seventeenth century until 1858, as they gradually conquered (directly or indirectly) India's four million square kilometres and 200 million people.

Precursors to the EIC's armies

South Asia had a strikingly diverse military history from long before the EIC began to construct its armies. The subcontinent contains a wide range of physical environments and social communities, each with implications for the composition and deployment of armies. Perennially or seasonally humid regions alternate with arid ones, each conducive to different kinds of military campaigns and equipage, for example favouring either infantry or cavalry. Different diets, based on that region's predominant crop (broadly dry crops like wheat or wet crops like rice), affected the physical growth of people living there. Bodily features like complexion vary widely between and within regions. Local environment shaped people's identities as well, for example as forest dwellers, mountaineers, or coastal or inland peoples. For instance, many *adivasi* communities (traditionally forest dwellers) also valued characteristic martial arts, especially archery. Further, regions were often defined by language (e.g. Bengali, Telugu or Marathi speakers). Specific communities (often identified by *jati* 'birth group', sometimes translated 'caste') acclimatized to and identified with each region. Each *jati* had a specific *dharma* ('code for conduct') including claimed ancestry (from a deity or heroic human progenitor), diet, deportment and, for some *jati*s, supposed affinity for specific types of weaponry and military service; from youth, men of these *jati*s customarily drilled in their community's particular martial traditions. Some *jati*s celebrated their skills as horsemen, others as foot soldiers, some as swordsmen, others as musket men, yet others as *lathi* (metal-tipped staff) fighters, each with long and heroic histories of exceptional martial prowess.

While warrior *jati*s were putatively hereditary and endogamous, in practice they were open to incoming groups and individuals who conformed to that *jati*'s *dharma*. Immigrants who conquered, settled and then intermarried with already established clans could become over generations members of that local martial *jati*. For example, *jati*s of Rajputs (literally 'son of a king') were composed of intermarrying clans which each boasted a distinguished patrilineal descent, typically from the sun, moon or fire sacrifice. Elite Rajputs stereotypically fought as cavalry; most Rajputs, however, served as infantry. Particular comportment, including facial hair, clothing, weaponry and battlefield behaviour customarily typified a Rajput. Nevertheless, many conquering immigrant peoples and also local *jati*s who were not historically identified as Rajput found that model attractive and adopted it, modifying their deportment and genealogies to conform.[3] Conversely, under the influence of Muslim rulers from the thirteenth century onward, some Hindu Rajput clans converted to Islam, yet retained their Rajput identity.[4]

Further, men from various different *jati*s might choose to conform to a regionally based military type.[5] For example, as the EIC began recruiting in north India from the late eighteenth century, they favoured Rajput and Bhumihar (landholding) Brahmin *jati*s and other 'Easterners' (*Purbiyas*) from the lower Ganges plain identified as *buxaria* – warriors expert in sword, shield and musket, and identifiable by particular clothing and comportment. Such *buxaria* had customarily been recruited into seasonal war bands near the town of Buxar (Baksar, in Bihar).[6] Pindaris in west-central India were another martial type, composed of men of diverse *jati*s who specialized as irregular light cavalry and infantry and who assembled into war bands around charismatic leaders; they were often paid by what they looted.[7]

Some religious traditions that recognize inclusion by conversion also developed a strongly martial orientation, for example Muslim *ghazi*s (warriors for Islam). Diverse Hindus joined orders of *sannyasi*s ('renouncers'); some orders formed militant bands which trained religiously with particular weapons (for example devotees of Shiva might use the *trishul*, 'trident', associated with that deity) and which wore distinctive (or even virtually no) clothing. Large troops of such *sannyasi*s served as shock troops hired by various rival regional rulers of the late eighteenth century.[8] Sikh generals mobilized much of their religious community into powerful armies in the Punjab during the eighteenth and early nineteenth centuries.[9] When the Portuguese and later other Europeans encountered these *jati*s, military types and religious communities, they indiscriminately used the terms caste (meaning 'breed') and 'race' for each of these groups, failing to appreciate how fluid, permeable and distinctive each historically was.[10]

Traditionally, Indian rulers used a patrimonial model for their military forces. They built upon the organizational core of their own permanent household retainers by temporarily hiring military labour gangs in order to constitute as needed the bulk of their army's infantry and also to add contingents of specialists like light cavalry to harass the enemy's supply lines. As Peter Lorge explains 'One did not buy arms and marry them to one's soldiers; one bought men who used certain weapons.'[11] Standing armies were expensive and their commanders might pose a threat to the ruler. Thus, rulers customarily sought to expand or shrink their forces rapidly, hiring bodies of men but demobilizing them thereafter.

Rulers customarily hired a military labour jobber, often called a *jemadar*, who assembled and commanded blocks of men from that region's diverse military labour market, each man with presupposed martial characteristics based on *jati*, military type and/or religious community. Such *jemadar*s advanced money to soldiers in their employ in order to purchase weapons, horses, other equipage and/or provisions. The rulers and *jemadar*s often left soldiers long unpaid to discourage desertion and they often expected post-battle (or pre-battle) looting. Typically, battles were preceded by heated bargaining between rulers and *jemadar*s.

The Mughal Empire (1526–1858) had been the dominant military and political power across most of South Asia. This empire had its own ways of classifying peoples and their martial qualifications. At the empire's core was a dynasty that drew upon the Central Asian martial traditions (mounted archers) and the authority of descent from Mongol Changiz Khan and Turkoman Timur. This Timurid clan (called Mughals by Persians) began conquering north India in

1526. In their first decades in India, they used cannon and wagons to secure their centre while their light cavalry archers outflanked the enemy. Over the next century, the Mughals built a patrimonial-bureaucratic empire, assembling vast armies; over time, the Mughals shifted to heavy armoured cavalry, more gentlemanly but less mobile.[12]

The Mughals recruited and deployed officers and men with their *qaum* or *ulus* ('race', 'ethnicity' or 'tribe') as a major consideration.[13] One mid-seventeenth century Mughal account detailed the diverse imperial *mansabdar*s (rank holding officers):

> From different races, Arabs, Persians, Turks, Tajeks, Kurds, Lars, Tatars, Russians, Abyssinians, Circassians, etc. and from the country of Rum (Turkey), Egypt, Syria, Iraq, Arabia, Persia, Gilan, Mazandran, Khurasan, Sistan, Trans-Oxiana, Khwarazm, the Qipchaq Steppes, Turkistan, Gharijstan, Kurdistan, various classes and groups of persons from every race and people have sought asylum in the [Mughal] Imperial Court, and various groups of Indians, men possessed of knowledge and skill as well as men of the sword, viz., Bukhari and Bhakkari, Saiyids of correct lineage, Shaikhzadas of noble ancestry, tribes (*ulusat*) of the Afghans, such as Lodis, Rohilas, Khwaishgi, Yusufzai, etc. and clans of Rajputs ... such as Rathors, Sisodias, Kachhwahas, Haras, Gaurs, Chauhans, Jhalas, Chandravats, Jadauns, Tonwars, Baghelas, Bais, Badgujar, Panwars, Bhadurias, Solankis, Bundalas, Sekhawats, and all other people of India, such as the races of Ghakkars, Langahs, Khokars, Baluch and other races ... of the deserts and mountains, from the country of Karnatik, Bengal, Assam, Udaipur, Srinagar, Kamaun, Bandhu, Tibet, and Kishtwar, etc. ... have obtained the privilege of kissing the threshold of the Imperial court.[14]

Each of these races had, in Mughal eyes, distinctive physical and moral qualities, revealed by their dress, language, physiognomy, religion and inherent martial characteristics.

Prudent and potent Mughal emperors like Akbar (r. 1556–1605) specified that *mansabdar*s (with a few exceptions) must recruit a substantial proportion of their subordinates from races other than their own. This policy hoped to preclude ethnic solidarity ranged against personal loyalty to the emperor. In particular, Afghans, the most frequent opponents of Mughal emperors, had the infamous reputation as being especially unreliable to whomever they nominally served and instead remaining loyal to fellow Afghans of their own clan, even those in an enemy army. The Iranis had the collective reputation for sophisticated courtly manners but being less brutal on the battlefield than Turani Turks, for example. Hindu Rajputs, Sikhs and Indians who had converted to Islam also received characteristic martial identities. In each instance, a group's self-perception might differ from views held of them by outsiders. Occasionally, factions in the Mughal court and armies, often designated as predominately Irani or Turani (but rarely composed solely of that ethnicity), formed around prominent men (or imperial wives). While many Mughal *mansabdar*s were recent immigrants, the vast bulk of Mughal armies consisted of people long settled in India.

A persistent problem for Mughals remained how to bind the vast, diverse and long-established military labour markets of each region exclusively into its imperial service.[15] One authoritative Persian language source from the late sixteenth century listed for the core Mughal provinces 342,696 horsemen and 4,039,097 infantry as the available military manpower base, some 10 per cent of the male population.[16] The Mughal *mansabdar*s hired blocks of soldiers, including by enticing them away from a rival chieftain or an enemy army. Two of the many factors causing the fragmentation of the Mughal Empire from the late seventeenth century onward were its inability to persuade regionally and ethnically based rulers to identify with the Mughals, and to pay the salaries of enough martially inclined men to fight for the Mughals instead of against them.

Various Europeans affected South Asia's military history significantly from the sixteenth century onward.[17] Over that century, the Portuguese Navy dominated the Indian Ocean but Portuguese rule never penetrated far inland from its coastal enclaves. However, individual Portuguese and other European mercenaries, and their descendants from Indian women, joined men from the Ottoman Empire and elsewhere in West Asia and East Africa by entering the Indian military labour market. Europeans and West Asians, whatever their personal military experience, often specialized in the artillery since these races had the reputation of expertise in that art. The Venetian immigrant Niccolao Manucci (*c.*1639 to *c.*1717) described how any European could secure lucrative and easy employment as a supposed expert cannoneer: 'European artillerymen who took service in that branch had only to take aim; as for all the rest – the fatigue of raising, lowering, loading, and firing – this was the business of [Indian] artificers or labourers.'[18] European mercenaries who remained Christian also profited from running – or fronting for – alcohol distilleries, since Mughal officials accepted that all Europeans were addicted to liquor.

The early infusion of European people, cultures and technologies produced new military types. For instance, the *topasi* was defined (in the words of an eighteenth-century Indian soldier) as 'a tawney race of foot soldiers, descended from the Portuguese marrying natives, and called Topasses, because they wear hats' (from either the Hindi *topi* or Portuguese *topo* meaning European-style hats).[19] Alternate etymologies of *topasi* conveying similar Western and Indian combinations include: Turkish *top-chi*, 'cannoneer' and Hindi *do-bhashiya*, 'someone of two-languages'. This new *topasi* military type included specialists in Western derived firearms including men with either a Western biological ancestor or Indian converts to Roman Catholicism. The *Telinga* was another new hybrid type: men armed with a musket and a partially European style military uniform, including not only the Telegu-speaking people from central India (particularly Naiks) but also Muslims from the Carnatic and Tamil speakers from south-east India.[20]

Especially over the eighteenth century, the English and French East India Companies recruited Indian, European and mixed ancestry mercenaries in order to defend their respective commercial outposts, defeat each other and extend their rival political influences inland. The 'sepoy' then emerged as the most significant new Indian military type.[21] The Persian term *sipahi* had been long current in India to mean a cavalryman. From the 1740s, however, the French and English EICs adapted a new model sepoy for the bulk of their armies: an Indian infantryman, trained, uniformed and armed in a semi-European manner, commanded by European officers.[22]

EIC armies as constructed in India by Britons and Indians[23]

After some London merchants secured in 1600 Queen Elizabeth's royal charter for the monopoly on English trade in the East Indies, the EIC began sending armed merchant ships to India's major ports. Eventually, the EIC established fortified warehouses ('factories') for the security of the specie and English goods they had imported and of the Indian goods they had purchased for export. Three factories developed into the coastal port colonies of Madras (acquired 1640, now Chennai), Bombay (acquired by King Charles II from the Portuguese in 1661 and rented to the EIC from 1668, now Mumbai) and Calcutta (acquired 1690, now Kolkata); each eventually became designated a presidency (since a president headed each colony's Council). Defending these unevenly expanding presidencies and advancing the EIC's interests beyond them required considerable and ever more military manpower. But, European troops were extremely expensive to recruit, bring to India and maintain there. Many died or deserted for more lucrative employment as military experts under Indian rulers. Hence, each presidency rapidly and increasingly also hired a variety of Indians from the extant local military labour market.[24] Gradually, EIC policy makers in Britain and India inconsistently sought to distinguish between white and black troops. But, this dichotomy failed to define the diversity of groups and individuals who comprised the officers and men in these armies as they developed over time.

In India, each presidency army emerged in a particular political context and local military labour market. For example, Bombay's hinterland was controlled by expanding Maratha powers which initially kept it more confined territorially than the Bengal and Madras presidencies.[25] But, Bombay's location also provided easier access to Arab, East African and West Asian soldiers, initially called 'peons'. To expand their forces, the presidencies all initially followed the extant military tradition that an Indian jobber would be employed to provide and command a contingent of soldiers. Further, British cultural presuppositions about the military value of particular groups also shaped recruitment. Thus, in 1653, the Bombay Council, believing that Rajputs from the north made superior soldiers, contracted with two Indian jobbers each to supply and command one company (100 men) of such Rajputs.[26] By 1762, the Bombay army included 2,500 black (i.e. Indian) soldiers.[27]

Similarly, the Madras Presidency army known as the Madras Army recruited from its own local military labour market, particularly *Telinga*s (Telegus from south India, i.e. Madras region) and also men who had moved south to serve in

Mughal successor states' armies.[28] In the early 1750s, the Madras Army briefly experimented with military formations that integrated Indian sepoys and European troops; European soldiers were encouraged to marry Indians so as to produce mixed race recruits.[29]

Madras initially relied on established Indian jobbers. For instance, in 1709, this presidency offered: 'such as would bring fifty recruits the rank of jemadar in the service, the rank of havildar [sergeant] to such as brought thirty, and to those that brought twenty the rank of a naig [corporal]'.[30] The most prominent military jobber-commander was Muhammad Yusuf Khan (*c.*1725–1764), an experienced commander who alternately served the French EIC against the English EIC and vice versa. Around 1751, the Madras Presidency employed him as the *subedar* (captain) of the block of soldiers he supplied. Three years later, Madras made him 'commandant' over all Indian soldiers in its employ. He campaigned sometimes as an independent commander, sometimes under British EIC or British Army officers. He still, however, received payment from Madras for all Indian soldiers under him. For example, in 1759, he received Rs300,000 for the past five years of service by the 1,996 cavalry and 1,425 infantry under his command.[31] His independence and occasional service to the French EIC, however, led the Madras Presidency to use other parts of the Madras Army to attack, capture and then execute him in 1764. The Madras Army at that time contained about 9,000 Indian soldiers.[32]

Based partly on the EIC's experience with Muhammad Yusuf Khan, it thereafter in principle opposed this system of jobber-commanders. Nonetheless, the exigencies of its manpower needs meant that this practice occasionally recurred as late as 1781 – the larger the body of recruits, the higher the military rank their Indian jobber received.[33] But, the EIC refused to allow autonomy to its Indian officers. Thus, although Indian subaltern officers still commanded companies composed of other Indians, they did so under a British officer.[34]

The EIC only gradually evolved its own formal officer corps, however. Until the mid-eighteenth century, the EIC did not recruit any European employees specifically as military cadets. Rather, the EIC began sending out retired or on-leave ('half-pay') British Army officers to provide martial expertise.[35] As the scale of European military involvement in India continued to escalate, the British and French Royal Armies occasionally deputed their own regiments to fight against each other and rival Indian rulers. Further, some EIC commercial employees already in India, based on their personal inclination or the pressing need of the moment, commanded and trained the EIC's Indian soldiers. Most famously, Robert Clive (1725–1774) went to Madras as a civilian 'writer' but, after Madras fell to a French EIC assault in 1746, he volunteered for its military wing as an ensign.

The EIC's Bengal Presidency army had the least promising beginning but, from the mid-eighteenth century it expanded more extensively than the Bombay or Madras armies. After Calcutta fell in 1756 to the surrounding Mughal successor state of *Nawab* Siraj al-Daula (r. 1756–1757), Robert Clive led contingents from the Madras and Bombay armies, supported by the Royal Navy, to recover it. Having recaptured Calcutta, he then expanded his forces by rapidly hiring from the local military labour market of the lower Ganges plain, especially Brahmins and Rajputs, many of whom had traditions of service as *buxaria*.[36] This composite Bengal Army of 2,100 Indian and 900 European men then ousted Siraj al-Daula at Plassey in 1757, thus taking effective control over rich territories three times the size of England.[37] To hold this vast conquest, to defeat the French and to seize more land revenues, Clive wanted many more troops. But, as Clive explained,

> A small force from home [Britain] will be sufficient, as we always make sure of any number we please of black troops, who, being both much better paid and treated by us than by the country powers, will very readily enter into our service.[38]

The *Telinga*s and *topasi*s (often the offspring of Portuguese men and Indian females) whom Clive had brought with him to Bengal either returned to the Madras Army or left the EIC's service.[39]

In the Bengal Army's subsequent rapid expansion it recruited individuals and also entire units of *buxaria*s and other martial types in the Ganges plain. The Bengal Army rose in numbers from its initial 300–400 sepoys in 1757 to 8,289 by 1762. By 1765, Clive had further expanded the Bengal Army and reorganized it into three brigades (each consisting of seven battalions of sepoys plus one regiment of European infantry, one company of European artillery and one troop of cavalry) totalling 14,000–15,000 Indians and 3,000 Europeans.[40] Then, each year from 1766 to 1768, the Bengal Army expanded on average by 5,000 Indians until, in 1770 the Bengal Army totalled 27,277 serving Indian officers and men (in addition to 522 European officers and 2,722 European soldiers).[41] By 1782, 52,500 Indians were serving in the Bengal Army, and over 115,000 in the three presidency armies combined. Race divided these three armies: white units usually held the designation 'European' (for example the 'Bengal European Regiment') in contrast to more numerous black regiments of 'Native Infantry'.

European military science during the seventeenth and eighteenth centuries developed its much discussed European 'military revolution(s)'.[42] Asian societies also had their own 'military revolutions'.[43] When the French and British EICs brought to India key European innovations – especially standardized equipage, supply, discipline, organization and salaries for infantry – and combined them with Indian traditions and innovations, this produced a 'military synthesis' that would prove critical in the British EIC's conquest of India.[44] From the 1750s, the EIC armies not only made Indian sepoys wear uniforms that combined Indian and European features and bore emblems that signified their newly standardized ranks, they also equipped them with British made standard issue weapons. The quality of European armaments was not then superior to the best that India could produce by hand. Nevertheless, Britain's system of mass manufacture meant that large numbers of identical weapons of reasonable quality could be supplied at relatively low cost. British-manufactured flintlock muskets performed better and more reliably in battle under a range of climatic conditions than Indian matchlocks and also enabled precisely drilled volleys, expeditious reloading and standardized ammunition. Socket bayonets enabled protection against cavalry and a rapid shift to close-order fighting. To a degree, a sepoy's military career professionalized, with regular pay and a pension on injury or retirement, and continuity of service to a lasting institution. This provided far more coherent and shared identity throughout a sepoy regiment than the traditional Indian model of the heroic, individually distinguished and motley clothed soldier bearing a non-standard weapon requiring custom made ammunition, recently hired as part of a military labour gang by an Indian jobber-commander, and temporarily rentable by the highest bidding ruler.

European military scientists had discovered empirically that rigorous close order drill of such a standing professionalized army enabled trained officers to control and reposition orderly bodies of infantry even in battle. In India, this meant sepoys with European style trained European and Indian officers could stand up to – and manoeuvre while under attack by – the heavy cavalry and artillery that formed the core of many Indian armies. Further, the larger groups of less drilled foot soldiers that filled out the forces of many Indian rulers usually gave way before the sometimes fewer but frequently more disciplined and uniformly armed EIC sepoys. As a late-eighteenth-century commentator wrote in Persian: so long as the British-commanded sepoys 'maintain their formations, which they call lines', they are like an immovable volcano spewing artillery and rifle fire like unrelenting hail on the enemy, and they are seldom defeated'.[45] The sepoy thus formed the dominant Indian martial type in the new model colonial armies, making the military difference in India and elsewhere in Asia and in Africa.[46]

Many British officers discriminated among various races of Indians and sought to implement their stereotyped preconceptions about the martial characteristics of each. For example, in 1768, Clive wrote he wanted the Bengal Army to have 'each Battalion composed of an equal number of Gentoos [Hindus] and Mussulmen [Muslims], and to encourage a rivalship of disc[ipline] between them' and to prevent 'their holding Cabals'.[47] The Bengal Army throughout this period preferred to recruit high caste Purbiya men, whom it designated 'the fighting Tribes of the Hindoos and Musselmen, and as many of the latter as can be procured' as opposed to Bengalis from the lower Ganges plain whom it discounted as soldiers.[48] Indeed, in the Bengal Army during the late eighteenth century, Muslims consistently composed nearly half of the higher Indian officer corps, about two-fifths of the lower ranked Indian officers and about one-third of the sepoys, far in excess of their one-quarter proportion of the general population.[49]

In addition, for policing and revenue enforcement, the EIC recruited many units outside its regular Bengal Army. The EIC created such units ad hoc for specific local needs and they often consisted of races of men whom the British did not consider worthy of inclusion in its Native Infantry. After having proven themselves in service, however, some of these units evolved into regular regiments.

While European troops continued in short supply, of great cost and often of brief lifespan in India, the EIC's directors in London and many of its European officers believed them the heart of its armies (although they only comprised 12–15 per cent of the EIC's forces). Until the final defeat of Napoleon in 1815, the long European wars made able-bodied males hard to recruit there (even for the British Army which, by law, had precedence in recruiting over the EIC). Consequently, the EIC engaged contractors ('crimps') to supply Europeans (of any nationality, including French, German and Swiss prisoners of war) at a rate of 1–5 guineas per individual.[50] Indeed, London newspapers reported that the EIC's directors had arranged for European men to be kidnapped and illegally impressed into its armies, a charge the directors vehemently denied.[51] Only after 1778, did the British Government allow the EIC to recruit in Ireland, and the Irish recruits quickly rose to about half those Europeans entering the EIC's armies.[52] The EIC even tried the ultimately expensive and unsuccessful experiment of hiring two battalions of German troops (2,000 men) under the auspices of King George III (as Elector of Hanover).[53] On average over the decade 1762–1772, each of the 17,082 (mostly unsatisfactory) European soldiers cost the EIC £7.4 to recruit and £13.8 to transport to India (whether they arrived alive or not).[54]

During this period, EIC officers in India continually complained, and London regularly made excuses, about the inadequate quantity and low quality of European recruits. Mortality rates on the voyage to India sometimes reached 50 per cent. Additionally, the high rates of deaths in India from disease, and occasionally from wounds, meant a constant, and largely unmet, demand for European soldiers. In 1770, for example, only 452 European soldiers survived to reach India or could be recruited there, ten men short of the casualties and desertions that year. This left the Bengal Army with only half the European soldiers active that it wanted. On average over the decade 1762–1772, about 70 per cent of the European troops that the EIC sent each year to India went to replace soldiers who had died, deserted or retired from the army, making any expansion in their number extremely slow.

Further, despite nominal requirements for age, size and health, many European recruits who actually reached India proved unfit. The commander-in-chief of the Bengal Army wrote in 1768 of the latest recruits:

> they are exceedingly bad ... the refuse of our metropolis ... The Company are at a great expence to send abroad annually a number of soldiers when in fact, instead of recruiting our army, they only serve to increase our Hospitals ... [A]t present our European Regiments compared to a Battalion of Sepoys appear like a Regiment of Dwarfs.[55]

Further, once in India, each European soldier demanded pay and generated expenses many times that of a sepoy.

European officers and European troops 'mutinied' about as frequently as Indian troops. Yet, based on racial prejudices, the EIC tried, as much as possible, to rely on European troops. In 1766, the Bengal Council wrote to London:

> we cannot avoid being alarmed at the extraordinary disproportion of numbers between your European and Black Infantry. The very signal instance of fidelity and attachment exhibited by the Sepoys during the late Mutiny of the [British] Officers, might indeed be sufficient to quiet our apprehensions and remove all suspicion of their conduct, if experience the preceding year of their dangerous insolence and turbulent spirit had not evinced the necessity of keeping the Black troops in awe and subjection.[56]

Many Britons believed sepoys would only stand firm in battle if European regiments provided 'stiffening'. Thus, the EIC made great efforts, and went to much expense, not only to recruit but further to guard these European regiments from unnecessary danger, tending to reserve them for the most crucial operations. At the same time, the attitude and treatment by EIC officers towards European soldiers remained far from humane by today's standards.

Not until 1764 did the Bengal Army formally institutionalize its European-type organization and discipline over its Europeans and Indians. Major Sir Hector Munro of the British Army (recently appointed the Bengal Army's commander-in-chief), concluded the army's existing relatively informal organization and discipline had led to mutinies by its European officers, European soldiers and Indian sepoys. In 1764, Munro established a regular code of orders and manoeuvres, based on the Duke of Cumberland's standing orders for the British Army.[57] This body of standing orders sought to bring further uniformity to the military but also specified that Indians and Europeans served under different terms and were subject to different military regulations and law.

An ongoing debate within the EIC remained whether Indians were racially able to conform to European standards. Munro declared in 1764 his 'firm resolution to treat the seapoys in every respect as soldiers, to give them all their just rights when they behave well, and their just punishment when they behave scandalously'.[58] Other British officers doubted Indians were capable of learning or sustaining European style discipline. For instance, Lieutenant-Colonel Carnac of the Bengal Army wrote in 1765:

> I would recommend the shortest and most simple method of exercise, especially for the Sepoys; it is needless to teach the latter the forming of the Square, or any of those Manoeuvres to which recourse is had when ... being obliged to retreat, as they can never be taught sufficient firmness under such an emergency to reduce their lessons to practise.[59]

Even within the Native Infantry regiments, EIC officers strictly enforced their current understanding of racial boundaries. As one European officer's wife explained in the late eighteenth century:

> There is no instance (I have heard of) of an European soldier being under a black man, for the [white] sergeants are superior in command to even the first black officer in the army. Therefore although in the bulk of the army

> the natives are most numerous, the power and command is vested entirely in the Europeans.[60]

The EIC believed European artillery to be virtually a secret weapon, giving its armies vital advantage over its Indian enemies. India produced expensive brass and less serviceable cast or wrought iron cannon but Europe had developed the technology for casting iron cannon at relatively low cost and weight. The directors demanded its artillery, and its technological secrets, be kept a secure monopoly. Even European Catholics (who might have pro-French sympathies) or men married to Catholics were for a time officially barred by the EIC from its artillery.[61] The directors repeatedly enjoined that Indians could only serve as manual labourers (*lascar*s):

> As it is very essential that the natives should be kept as ignorant as possible both of the theory and practice of the artillery branch of the art of war, we esteem it a very pernicious practice to employ the people of the country in working the guns, and therefore direct that in future … no native be trusted with any part of this important service, unless necessity should require it.[62]

In fact, given Indian conditions, this exclusion proved unfeasible. Thus, in 1778, the EIC still had three battalions of 'native artillery' (*gol-andaz*, 'ball throwers') totalling some 2,962 Indian officers and men serving under 35 European officers.[63]

The cavalry remained one of the weakest arms of the EIC's armies. The supply and maintenance of horses had always been costly in India where few regions are suitable for horse breeding and many are quite unhealthy for them. Since horses do not ship well, they mostly had to be acquired by land from Central Asia. While the EIC did have a few elite European cavalry units, it often relied on 'irregular cavalry', often designated as 'country horse', which could be employed and dismissed relatively quickly.[64] Thus, troops of Mughal style cavalry frequently entered and then left the employ of the Bengal Army as collective bodies of men.[65]

Officers of mixed European and Indian ancestry, like James Skinner (1778–1841), found employment in the Bengal Army only through recruiting and commanding irregular cavalry: he raised Skinner's Horse in 1803. He fruitlessly protested the EIC's racial policies:

> [while serving in a Maratha army (1796–1803)] My merits were rewarded, my birth not questioned. When I entered the British service … I thought I had now served a nation that had no prejudices against caste or colour, but alas I was mistaken … For if they will allow me to be a native, reward me as such … Although my birth has been the cause of my not gaining what I deserved (that is rank) … I served my king and country as loyally as any Briton in India.[66]

Only after 1828 could 'Eurasian' officers like him get regular EIC commissions or military honours like Bath.

On their part, many sepoys continued to regard service with the EIC's army as a temporary situation, to be entered into or left at their pleasure. In 1781, for example, the EIC complained that whenever a sepoy regiment relocated to a different region, locally recruited sepoys regularly deserted it and re-enlisted in the new regiment transferring in, rather than accept relocation.[67] The EIC sought to reorient sepoys until they saw themselves as bound professionally by 'honour' to its army as a whole and to their regiment in particular.[68]

Some Indian regional rulers, seeing the power of the EIC's sepoy-based armies, sought to emulate them.[69] Several hired individual European officers who claimed expertise in training sepoy units. But, these European officers and their European model weapons and training of sepoy units became a constant drain on their employer's treasury. In the end, most Indian regional rulers could not command the resources necessary to sustain the continuous employment of tens of thousands of European-trained European and Indian officers and men as did the EIC.[70]

Instead, many regional rulers made subsidiary alliances with the EIC, hence gaining access to the services of its sepoy armies. But, this proved a superficially attractive but ultimately even more costly strategy. These troops went on to the Indian ruler's payroll but remained under EIC command. Hence, the EIC subsidized large portions of its armies by essentially renting its troops to its Indian allies. Military dependence on the EIC, however, meant these rulers gradually lost much of their treasuries, sources of revenue and, finally, independence. By the mid-nineteenth century, all the Indian rulers had succumbed either to annexation or to indirect control by the EIC.

As the EIC encountered new peoples who appeared particularly 'martial', British officers incorporated stereotypes about them into their evolving racial categories of military service. For example, when the EIC armies struggled to defeat the advancing Gurkha kingdom in the Himalayan foothills (1814–1816), Britons came to admire the war-like character of the Gurkha race. Consequently, the Bengal Army recruited men from this region, equipping them as if they were Gurkhas – most notably with the *khukri* (broad curved knife) – although most of these recruits were not actually from the Gurkha community but rather different peoples the Gurkhas had recently conquered.[71] Some initially irregular Gurkha units became regular regiments in the Bengal Army (and continue in the British Army today). Similarly, after two hard fought wars in the Punjab (1845–1846, 1849) against predominantly Sikh armies, the EIC created the Sikh regiments of the Punjab Irregular Frontier Force. Many Hindu Punjabis wishing to enlist had to adopt a Sikh identity to qualify.

Over their last 50 years, each of the EIC's three armies continued to develop distinct racial ideologies and practices. The Bengal Army largely monopolized recruitment from the central Ganges plain of high caste Brahmins and Rajputs and also Muslim Pathans/Pakhtuns and Rohilla Afghans. Hence, this army tended to institutionalize these separate community identities through 'single class' regiments. The Bombay Army attempted to keep a 'bipolar balance between Hindustanis [from north India] ... and Konkanis [from the west coast]'.[72] Further, it recruited more widely from the surrounding society and thus tended to mix Indians of different communities within its more integrated units. The Madras Army tried to balance recruitment from three major south Indian groups: Tamils, Telugus and Muslims from Karnataka, classifying them as racially different but seeking to keep a degree of balance among them.[73] By 1857, these armies collectively contained 228,000 Indian soldiers (in addition to 43,000 Europeans).

British racial attitudes in the EIC headquarters in London

While the EIC's armies were developing in India, its directors in London were following their own attitudes, policies and practices concerning race. These reflected British popular opinion but also their own pragmatic interests. While the directors largely agreed on the principle of appointing only appropriate white British officer gentlemen and excluding the racially unworthy, their racial definitions and actual practices were inconsistent and varied over time.

Only from the mid-eighteenth century did the directors started appointing Britons as cadets in a professional military career track. This marked recognition of the EIC's permanent military engagement in Indian politics. Since nomination for these appointments were at the patronage of individual directors, each used his own criteria. Further, this patronage had a market value so some directors sold their nominations to be cadets (and also civilian officials); although illegal, such sales were common. For example, the London newspapers in 1783 advertised cadetships for 50 guineas (civilian appointments went for 1,000 guineas).[74] Further, men who held such nominations occasionally sold them to third parties. As a point of comparison, the British Army officers legally purchased commissions until 1871. But, such patronage recruitment had little basis in merit or accomplishment. Once cadet nominee, William Hickey, described the pro-forma interview he underwent in 1768, completely unprepared as he was, before obtaining the cadetship which his despairing father had purchased as the last resort for a wayward son.[75]

From the beginning the directors were determined to keep its officer corps purely European and composed of gentlemen, and specifically to exclude people it deemed 'natives of India'. But, well into the nineteenth century, the directors struggled with how they could define race and how to fit into this definition the various men who sought to appointments in its service. Until the late eighteenth century, the conception of race in Britain tended to be based more on culture than on biology. With rare exceptions, at first only white, Protestant, gentlemen, born in Britain could receive commissioned rank in the EIC's armies.[76] Catholics and non-Europeans, by regulation, could not hold such commissions except under special dispensation.[77] Men who seemed European gentlemen, not only by birth and upbringing but also even if they had Anglicized themselves, also occasionally appeared appropriate to the directors as officers in their nascent armies. Not all Europeans were, however, classed as white, the Irish and Portuguese being prime examples during the eighteenth century. Given the difficulties of recruitment of European soldiers and non-commissioned officers, the criteria for these had much more latitude. From the perspective of candidates, men had every motivation to prove themselves 'British', or, conversely, not a 'Native of India'. But, there were no objective criteria for these categories or means of judicial appeal, which left the directors with much arbitrary power.

In 1770, a particularly anomalous case arose which illustrated the complexities of race and their effects on a man's career. Emin Joseph Emin (1726–1809), a Christian Armenian, who had followed his merchant family from his birthplace in Hamadan, Iran, to Calcutta and then had worked his way as a seaman to London in 1751. There, he secured the patronage of various notables, including Edmund Burke, the Duke of Northumberland, and the royal Duke of Cumberland. This distinguished patronage gained him admission to the royal military academy at Woolwich. He then served about five years as a gentleman officer with the English and Prussian royal armies, periodically fighting on the continent against the French during the Seven Years War. Describing the Battle of Hastenbeck, Germany (26 July 1757) in which he fought, Emin identified the French with Indians: 'The [French] Enemy begun to advance with their Musick, and Drums, making a very great noise, more like Indians than Europeans … [The French] are vastly like black Indians, [both] fire at a great distance and run away'.[78] After Emin's return to Calcutta, he sought officer rank in the EIC's Bengal Army but he ran up against stronger social and racial barriers in Bengal than those he had faced in London. Unable to get himself 'naturalized' as British, he could gain entry into the Bengal Army only at the most marginal officer's rank: brevet ensign in the irregular cavalry (*Turksowar*s).[79] After two years of thankless military employment, Emin lost active service status in 1772, when this cavalry demobilized. Emin then was compelled to retire. But, he regained European status as a member of the 3rd European Invalid unit in Calcutta.[80] Even then, due to his anomalous status, he had to petition the EIC for 25 years before he received his back pay and pension.[81] His case illustrates that racial distinctions in the EIC's colonial army in India tended to be more strictly defined than for the British Army in Europe, although somewhat arbitrarily in both places.

Indeed, not until 1791, when the EIC's directors were faced with another anomaly between their implicit racial qualification for their officer corps and an individual's apparent racial features, did they articulate a formal rule about race. John Turing was the highly Anglicized son of a British former EIC official and an Indian woman. Through his father's influence, he had secured nomination as an EIC cadet officer in the Madras Army. However, when he presented himself in person before the EIC's Shipping Committee in London, that Committee's Chairman visually noted that Turing 'appears to be a Native of India'.[82] On 19 April 1791, the directors as a whole called Turing in for their own ocular scrutiny, and then immediately, having ordered him to withdraw, created a policy by unanimously resolving: 'That no Person, the son of a Native Indian, shall henceforth be appointed by this Court to employment in the Civil, Military, or Marine Service of the Company.' The next day, Turing successfully applied for and purchased legal permission from the directors to go to Bengal as 'a Native'.[83]

The timing of this formative 1791 resolution reflected both British desires and fears of the time, but also the immediate need to decide an individual case. The regulation thus articulated the directors' hitherto implicit beliefs and formally established the EIC's official policy for the next two decades. Yet, this was not always a clear policy to enforce since no definition of 'Native' was specified.

Faced with another anomalous case in February 1792, the directors resolved: 'That in future no foreigners shall be admitted into the Company's service, as Writer or Cadet.'[84] They did not, however, define 'foreigner', leading to future ambiguity in this definition as well. Further, the directors occasionally voted by ballot to suspend their new rule to appoint an acknowledged 'Native of India' when that candidate had a sufficiently powerful patron.[85]

The directors continued to struggle to determine whom they wanted to exclude from EIC employment as officers (and civilian officials) based on visual and biological criteria. They, or a subcommittee of the directors, scrutinized nominees and, occasionally based on 'complexion', rejected men. To cope with new anomalies as they arose, the directors tinkered with (usually to expand) the 1791 resolution, for instance by voting to exclude 'Persons born in the West India Islands, whose Parents are not Natives of Great Britain or Ireland.'[86] This issue, however, would persist, especially as the directors' racial definitions and desires conflicted with those prevalent in Britain, and especially Parliament.

In October 1816, the directors again moved to expand their 1791 resolution to include both ancestry and skin complexion:

> no person, the grandson of a Native Indian shall henceforward be appointed by this Court to employment in the Civil, Military, or Marine Service of the Company; and that no person whether a Native of Asia, the West Indies, America, or Africa, shall be eligible to employment, whose colour shall appear exceptionable.[87]

The complexity and implications of this resolution, however, meant that, before voting, the directors first referred it to their Committee of Correspondence (the subcommittee which handled particularly vital issues). That committee was completely divided over this issue.[88] Finally, after more than seven months of deliberation, the committee resolved not to record any opinion at all.[89] This avoidance of the issue, however, did not make it disappear.

For much of the early nineteenth century, the directors designated their Addiscombe Military Seminary Committee to inspect candidates for EIC employment as officers or officials. Yet, this Committee increasingly doubted what they were doing. In November 1821, they protested: 'your Committee having frequently been called upon to decide upon the nomination of young Men … the circumstances of whose birth, combined with their complexion, left [the Committee] in doubt as to the precise extent of the Court's intentions'.[90] Consequently, the committee pleaded for 'more determinate rules', requesting the Directors as a whole

> to establish ... whether the term 'Son of a Native Indian' shall be construed to limit the exception to young men actually the sons of Natives, or of Native women by European fathers; or to extend the same to descendants of the offspring of such connexions if the Colour of the parties shall exhibit clear signs of such origin, although in a second or third generation.[91]

In short, should complexion and other visible physiological features override distant ancestry?

The directors themselves remained divided. They took up consideration of this issue four times in December 1821 and January 1822, but each time postponed coming to any decision. Finally, they moved a series of detailed specifying resolutions that combined issues of gender, 'blood', legitimacy and education and also specified the precise type of documentary evidence about ancestry and education requisite to prove the candidate's identity. Apparently because of the exacting but not always consistent detail in these motions about the ways to classify a 'native of India', the directors debated extensively and then voted to adjourn without deciding any of them. They therefore left their scrutinizing subcommittees in the same dilemma as before.

Over the early nineteenth century, British public opinion increasingly stressed biological 'purity', which official regulations haltingly reflected. The Military Seminary Committee, after struggling for nearly seven years, suddenly came to the conclusion in 1827 that the definition of 'native Indian' was 'obvious':

> to remove all doubts as to the meaning of the Term 'Native Indian' ... in accordance with its obvious meaning, the exclusion of Persons from the Service under that [1791] Regulation be limited to the offspring of aborigines, that is so say, of Parents of whom either one, or both, are of pure unmixed Native Extraction.[92]

Thus, this committee excluded from appointments all men it considered tainted by full Indian ancestry on either side. Yet, this biological definition for officials and officers contradicted the geographical criteria fixed by the current law set by Parliament to define 'Indians': everyone, regardless of ancestry or appearance, who was born or domiciled in the EIC's geographic arena, which extended from the tip of Africa to the Philippines.

In 1833, Parliament effectively closed the matter for the directors, however. As a by-product of its 20 year (1834–1854) suspension of the EIC's Charter to trade in Asia, Parliament specified that no 'native' of the EIC's territories or 'natural born subject of His Majesty' would be excluded from any appointment in the army or civil service on the basis of religion, place or birth, descent or colour. Nevertheless, in practice, racial identities did continue to matter in a man's career in the British Indian Army.

Sequel: Indian armies under the British Raj (1858–1947)

Following the traumatic events of 1857–1858, when much of the Bengal Army rose up against its British officers and nearly drove the EIC out of north India, many British and Indian racial attitudes changed. During the bloody fighting, tens of thousands of people lived or died according to their ascribed racial identity. Thereafter, the Bengal Army in particular substantially reorganized, disbanding many of its established native regiments and reconstituting others. Further, the British Parliament took the governance of India from the EIC, made its armies into the British Indian Army and transferred its European regiments to the British Army. But, race continued as a major factor in the Indian Army's composition and organization despite the assurance of Queen Victoria's 1858 Proclamation that 'our subjects, of whatever race or creed, be freely and impartially admitted to offices in our service, the duties of which they may be qualified, by their education, ability, and integrity, duly to discharge'.[93]

Indeed, over the late nineteenth century, the Indian Army revised its recruitment policies largely around the idea of 'effeminate races' and 'martial races'. The British included in the former many of the high *jati*s which had provided much of the manpower of the Bengal Army in its century of martial expansion across north India. But, because many men from these groups had risen up against the British in 1857, they appeared to be no longer suitable for military service. Thereafter, the British sought ideologically to demobilize and disempower them with the gendered epithet of 'effeminate races'. In contrast, those soldiers who had fought on the British side in 1857–1858, especially Gurkhas and Punjabis, appeared in British eyes and policies as 'martial races'. They thereafter received highly disproportionate preference for military recruitment. Many scholars have analysed the ways that authorities in the British *Raj* reconceptualized and applied race, especially in terms of recruiting and organizing their military forces.[94]

Tracing as we have how the EIC constructed its armies over more than two centuries, we have seen how these armies arose out of synthesis of European and Indian patterns of military service. Both Indian and British conceptions of race and what that should mean for military service proved highly contingent and inconsistent. Yet, these conflicted and contested categorizations of race affected the lives of individuals profoundly.

Notes

1 See Kathleen Wilson, 'The Island Race', in Tony Claydon and Ian McBride (eds), *Protestantism and National Identity* (Cambridge: Cambridge University Press, 1998), pp. 265–290; Roxann Wheeler, *Complexion of Race* (Philadelphia, PA: University of Pennsylvania Press, 2000); Mark Harrison, *Climates and Constitutions* (Delhi: Oxford University Press, 2000).

2 The EIC armies variously organized companies around regiments, battalions and/or brigades. For consistency, I use regiment as the generic unit.

3 John Hitchcock, 'Idea of the Martial Rajput', in Milton Singer (ed.), *Traditional India* (Philadelphia, PA: American Folklore Society, 1959), pp. 216–223.

4 For example, William H. Sleeman, *Journey* (ed. P.D. Reeves) (Cambridge: Cambridge University Press, 1971), p. 240.
5 Dirk H.A. Kolff, *Naukar, Rajput and Sepoy: The Ethnohistory of the Military Labour Market in Hindustan, 1450–1850* (Cambridge: Cambridge University Press, 1990), pp. 169–181.
6 Shiv Das Lakhnawi, *Shahnama Munawwar Kalam* (tr. Syed Hasan Askari) (Patna: Janaki Prakashan, 1980).
7 B.K. Sinha, *The Pindaris: 1778–1818* (Calcutta: Bookland, 1971).
8 See William R. Pinch, *Warrior Ascetics and Indian Empires* (Cambridge: Cambridge University Press, 2006); Gerald Bryant, 'Pacification in the Early British Raj, 1755–1785', *Journal of Imperial and Commonwealth History*, vol. 14 (1985), pp. 3–19.
9 Fauja Singh Bajwa, *Military System of the Sikhs* (Delhi: Motilal Banarsidass, 1964).
10 Dirk H.A. Kolff, *Naukar*, pp. 169–181.
11 Peter A. Lorge, *The Asian Military Revolution: From Gunpowder to the Bomb* (New York: Cambridge University Press, 2008), p. 135.
12 William Irvine, *The Army of the Indian Moghuls: Its Organization and Administration* (London: Luzac, 1903); Stephen P. Blake, 'The Patrimonial-Bureaucratic Empire of the Mughals', *Journal of Asian Studies*, vol. 39, no. 1 (November 1979), pp. 77–94.
13 Stephen F. Dale, *The Garden of the Eight Paradises: Babur and the Culture of Empire in Central Asia, Afghanistan and India (1483–1530)* (Leiden: E.J. Brill, 2004), pp. 101, 158; Iqtidar Alam Khan, 'Nobility under Akbar', *Journal of the Royal Asiatic Society*, vols 1–2 (1967), p. 35.
14 Shandrabhan Barhaman, *Guldasta*, cited in M. Athar Ali, *The Mughal Nobility under Aurangzeb* (Bombay: Asia Publishing, 1966), p. 15.
15 Dirk H.A. Kolff, *Naukar*, p. 3.
16 Abul Fazl Allami, *Ain-i Akbari* (tr. H.S. Jarrett), 2 vols (Delhi: Motilal Banarsidass, 1989), vol. 2, pp. 141–368; Dirk H.A. Kolff, *Naukar*, p. 3.
17 Sanjay Subrahmanyam, *Explorations in Connected History* (Delhi: Oxford University Press, 2005).
18 Niccolao Manucci, *Storia do Mogor* (tr. William Irvine, ed. Michael Edwardes), 4 vols (London: John Murray, 1913), vol. 1, pp. 54–55.
19 Henry Yule and A.C. Burnell, *Hobson-Jobson* (Delhi: Munshiram Manoharlal, 1968), s.v. 'Topaz'.
20 Sake Dean Mahomed, *First Indian Author in English* (ed. Michael H. Fisher) (Delhi: Oxford University Press, 1996), Letter XVI.
21 Henry Yule and A.C. Burnell, *Hobson-Jobson*, s.v. 'Sepoy'.
22 Henry Dodwell, *Sepoy Recruitment in the Old Madras Army* (Calcutta: Superintendent Government Printing, 1922), pp. 1–12.
23 Studies of the EIC Armies include: Seema Alavi, *The Sepoys and the Company: Tradition and Transition in Northern India, 1770–1830* (Delhi: Oxford University Press, 1996); Premansukumar Bandyopadhyay, *Tulsi Leaves and the Ganges Water* (Calcutta: K.P. Bagchi, 2003); Pradeep P. Barua, *Gentlemen of the Raj: The Indian Army Officer Corps, 1817–1949* (Westport, CT: Praeger, 2003) and *The State at War in South Asia* (Lincoln, NE: University of Nebraska, 2005); Nile Green, *Islam and the Army in Colonial India* (Cambridge: Cambridge University Press, 2009); T.A. Heathcote, *The Military in British India: The Development of British Land Forces in South Asia, 1600–1947* (Manchester: Manchester University Press, 1995); Jeffrey Greenhunt, 'Sahib and Sepoy: An Inquiry into the Relationship between the British Officers and Native Soldiers', *Military Affairs*, vol. 48, no. 1 (1984), pp. 15–18; Dirk H.A. Kolff, *Naukar*; Douglas M. Peers, *Between Mars and Mammon: Colonial Armies and the Garrison State in Early Nineteenth Century India* (London: I.B. Tauris, 1995); Kaushik Roy (ed.), *War and Society in Colonial India: 1807–1945* (Delhi: Oxford

University Press, 2006); Kaushik Roy, 'Armed Expansion of the English East India Company: 1740s–1849', in Daniel P. Marston and Chandar S. Sundaram (eds), *A Military History of India and South Asia: From the East India Company to the Nuclear Era* (Westport, CT: Praeger, 2007), pp. 1–15; Kaushik Roy, 'Recruitment Doctrines of the Colonial Indian Army', *Indian Economic and Social History Review*, vol. 34, no. 3 (1997), pp. 321–354; Channa Wickremesekera, *'Best Black Troops in the World': British Perceptions and the Making of the Sepoy, 1746–1805* (Delhi: Manohar, 2002).

24 S.C. Hill, *Yusuf Khan, The Rebel Commandant* (London: Longmans, Green, 1914); G.J. Bryant, 'Indigenous Mercenaries in the Service of European Imperialists', *War in History*, vol. 7, no. 1 (2000), pp. 2–28.

25 Randolf G.S. Cooper, *The Anglo-Maratha Campaigns and the Contest for India: The Struggle for Control of the South Asian Military Economy* (Cambridge: Cambridge University Press, 2005).

26 Edward Rivett-Carnac, *Presidential Armies of India* (London: W.H. Allen, 1890), pp. 77, 84.

27 Arthur Broome, *History of the Rise and Progress of the Bengal Army* (London: W. Thacker, 1850), appendices, p. 6.

28 Henry Davison Love, *Vestiges of Old Madras, 1640–1800*, 4 vols (London: J. Murray, 1913).

29 Neema Cherian, 'Spaces for Races', *Social Scientist*, vol. 32 (2004), pp. 32–50.

30 Henry Dodwell, *Sepoy Recruitment*, pp. 20–21.

31 Madras Council (3 March 1759) cited in S.C. Hill, *Yusuf Khan*, pp. 88–89.

32 Arthur Broome, *History*, appendices, p. 6.

33 Madras Military Consultations (21 November 1768), fol. 1709–1716; (15 March 1781), fol. 668–669; (13 November 1798), fol. 6853–6855; (19 February 1799), fol. 1018–1019; Bengal Public Consultation (9 July 1782), British Library (hereinafter BL), London; Henry Dodwell, *Sepoy Recruitment*, pp. 19–21, 29, 119.

34 Henry Dodwell, *Sepoy Recruitment*, pp. 20–21.

35 See Gerald Bryant, 'Officers, of the East India Company's Army', *Journal of Imperial and Commonwealth History*, vol. 4 (1977–1978), pp. 203–227; Suresh Chandra Ghosh, *Social Condition of the British Community in Bengal, 1757–1800* (Leiden: Brill, 1970), p. 31; P.E. Razzell, 'Social Origins of Officers in the Indian and British Home Army', *British Journal of Sociology*, vol. 14, no. 3 (September 1963), pp. 248–260.

36 Arthur Broome, *History*, pp. 92–93; John Williams, *Historical Account of the Rise and Progress of the Bengal Native Infantry* (London: John Murray, 1817), p. 4.

37 George William Forrest, *Life of Lord Clive*, 2 vols (London: Cassell, 1918), vol. 1, p. 437.

38 Robert Clive to William Pitt (7 January 1759), in Michael H. Fisher (ed.), *Politics of the British Annexation of India* (Delhi: Oxford University Press, 1993), p. 63.

39 Bruce Lenman, 'The Transition to European Military Ascendancy in India', in John Lynn (ed.), *Tools of War* (Bloomington, IN: University of Indiana, 1990), p. 121.

40 Arthur Broome, *History*, pp. 92–93; John Williams, *Historical Account*, p. 3.

41 General Return of the Troops (31 July 1770), National Archives of India. See also Bengal Public Consultation (16 December 1769) in *Fort William–India House Correspondence* (Delhi: National Archives of India, 1949), vol. 5, pp. 99–100.

42 See Jeremy Black, *A Military Revolution? Military Change and European Society, 1550–1800* (Basingstoke: Macmillan, 1991); Clifford J. Rogers (ed.), *The Military Revolution Debate: Readings on the Military Transformation of Early Modern Europe* (Boulder, CO: Westview, 1995); Geoffrey Parker, *The Military Revolution: Military Innovation and the Rise of the West* (New York: Cambridge University Press, 1996).

43 Peter A. Lorge, *The Asian Military Revolution.*

44 Kaushik Roy, 'Military Synthesis in South Asia', *Journal of Military History*, vol. 69, no. 3 (July 2005), pp. 651–690 and *War, Culture and Society in Early Modern South*

Asia, 1740–1849 (London: Routledge, 2011); Stewart Gordon, 'Symbolic and Structural Constraints', in Richard Barnett (ed.), *Rethinking Early Modern India* (Delhi: Manohar, 2002), pp. 155–178.

45 Mir Abdul-Latif Khan, *Tuhfat al-alam va zayl al-tuhfah*, cited in Juan Cole, 'Invisible Occidentalism', *Iranian Studies*, vol. 25, nos 3–4 (1992), p. 10.

46 See Letter to Court (3 September 1766) in J. Long (ed.), *Selections from Unpublished Records of Government* (Calcutta: Office of the Superintendent of Government Printing, 1869), vol. 1, p. 427; Bruce P. Lenman, 'The Weapons of War in Eighteenth-Century India', *Journal of the Society for Army Historical Research*, vol. 36 (1968), pp. 33–43; Gayl D. Ness and William Stahl, 'Western Imperialist Armies in Asia', *Comparative Studies in Society and History*, vol. 19 (1977), pp. 2–29.

47 Court of Directors to Fort William (16 March 1768), *Fort William–India House Correspondence*, vol. 5, p. 100.

48 Minutes of Bengal Council (22 April 1782); General Orders of the Commander-in-Chief (11 May 1782), National Archives of India.

49 I calculated these from 'Infantry Native Officers and Sepoys', Bengal Military Consultations 1778–1784, BL. The Madras Army had similar proportions, see Henry Dodwell, *Sepoy Recruitment*, pp. 11, 40–49.

50 For example, Court of Directors to Fort William (27 May 1779), *Fort William–India House Correspondence*, vol. 8, p. 242. See also Home Miscellaneous Series, 24:113, BL; Arthur N. Gilbert, 'Recruitment and Reform', *Journal of British Studies*, vol. 15 (1975), pp. 89–111.

51 Minutes of Court of Directors (16 September 1785), BL.

52 This high rate did not persist. See Arthur N. Gilbert, 'Recruitment and Reform', p. 100, n. 47.

53 Court of Directors to Fort William (25 January 1782), *Fort William–India House Correspondence*, vol. 9, pp. 10–15.

54 I have calculated these from Great Britain, Parliament, 'Account of the Number of Non-Commissioned Officers and Private Men sent to the East Indies from September 1762 to September 1772', in *Eighth Report from the Committee of Secrecy* (25 May 1773), BL.

55 Richard Smith letter (2 November 1768), Bengal Secret and Military Consultation (17 November 1768), BL.

56 J. Long (ed.), *Selections*, p. 467.

57 Arthur Broome, *History*, vol. 1, pp. 455–458.

58 Charles Caraccioli, *Life of Robert, Lord Clive*, 4 vols (London: T. Bell, 1775–1777), vol. 2, pp. 46–47.

59 Carnac to Fletcher (6 November 1765), MS Eur. 128/4, BL.

60 Jemima Kindersley, *Letters from the Island of Teneriffe, Brazil, the Cape of Good Hope, and the East Indies* (London: Norse, 1777), p. 209.

61 Arthur Broome, *History*, vol. 1, pp. 42–43.

62 Court of Directors to Fort William (23 March 1770), *Fort William–India House Correspondence*, vol. 6, p. 35.

63 The Company periodically disbanded and then reconstituted these battalions. See E. Buckle, *Memoir of the Services of the Bengal Artillery* (ed. J.W. Kaye) (London: William H. Allen, 1852), pp. 51–61.

64 Seema Alavi, 'Makings of Company Power', *Indian Economic and Social History Review*, vol. 30, no. 4 (October 1993), pp. 437–459.

65 Court of Directors to Fort William (25 March 1772), *Fort William–India House Correspondence*, vol. 6, p. 159.

66 James Skinner, *Military Memoir* (ed. James Baillie Fraser) (London: Smith, Elder, 1851), p. 128.

67 Court of Directors to Fort William (11 April 1781), *Fort William–India House Correspondence*, vol. 8, p. 295.

68 See Philip Mason, *A Matter of Honour: An Account of the Indian Army, its Officers and Men* (New York: Holt, Rinehart & Winston, 1974).
69 G.J. Bryant, 'Asymmetric Warfare', *Journal of Military History*, vol. 68, no. 2 (2004), pp. 431–469.
70 C.A. Bayly, *Indian Society and the Making of the British Empire* (Cambridge: Cambridge University Press, 1988); Douglas M. Peers, 'War and Public Finance', *International History Review*, vol. 11, no. 4 (November 1989), pp. 628–647.
71 Seema Alavi, *The Sepoys and the Company*, pp. 264–291.
72 Kaushik Roy, *Brown Warriors of the Raj: Recruitment and the Mechanics of Command in the Sepoy Army, 1859–1913* (Delhi: Manohar, 2008), pp. 20–29, 36–37.
73 James Hoover, *Men without Hats* (Delhi: Manohar, 2007).
74 See Edward Blagdon, *Cadetship in the Honourable East India Company's Service* (Oxford: Oxford University Press, 1931); James Mayer Holzman, *Nabobs in England* (New York: Author, 1926), p. 22n.
75 William Hickey, *Memoirs* (ed. Alfred Spencer), 4 vols (London: Hurst & Blackett, 1919–1925), vol. 1, pp. 124–125.
76 Gerald Bryant, 'Officers, of the East India Company's Army', pp. 203–227.
77 French and men of mixed ancestry faced similar barriers to officer rank. See Rosie Llewlyn-Jones, *Very Ingenious Man, Claude Martin in Early Colonial India* (Delhi: Oxford University Press, 1992); Antoine-Louis Henri Polier, *European Experience of the Mughal Orient* (tr. Muzaffar Alam and Seema Alavi) (Delhi: Oxford University Press, 2001).
78 Joseph Emin, *Life and Adventures* (ed. A. Apcar) (Calcutta: Asiatic Society of Bengal, 1918), pp. 78–81.
79 V.P.C. Hodson, *List of the Officers*, 4 vols (London: Constable and Phillimore, 1927–1947), vol. 2, p. 137; Joseph Emin, *Life and Adventures*, pp. 439, 483.
80 Emin to Hastings (3 March 1774) Warren Hastings, General Correspondence, Addl 29134 fol. 322–323, BL; Joseph Emin, *Life and Adventures*, pp. 441–455.
81 Joseph Emin, *Life and Adventures*, pp. 483, 513.
82 Court Minutes (19 April 1791), BL.
83 Court Minutes (20 April 1791), BL.
84 Court Minutes (2 February 1792), BL.
85 Court Minutes (1 March 1792), BL.
86 Court Minutes (19 February 1800), BL.
87 Court Minutes (23 October 1816), BL.
88 Correspondence Committee Minutes (24 January 1817), BL.
89 Correspondence Committee Minutes (4 June 1818), BL.
90 Seminary Committee Reports (28 November 1821), BL.
91 Court Minutes (5 December 1821, 2–16 January 1822), BL.
92 Correspondence Committee Minutes (30 October 1827); Court Minutes (31 October 1827, 7 November 1827), BL.
93 'Proclamation by the Queen in Council, to the Princes, Chiefs, and People of India' (1 November 1858); East India (Awadh Proclamation), Great Britain, Parliament, Accounts and Papers, *Sessionals* (3 February to 19 April 1859), vol. 18, pp. 296–297.
94 See Pradeep Barua, 'Inventing Race: The British and India's Martial Races', *Historian*, vol. 58, no. 1 (1995), pp. 107–116; Stephen P. Cohen, *The Indian Army: Its Contribution to the Development of a Nation* (New York: Oxford University Press, 1990), pp. 36–49; Kaushik Roy, 'Beyond the Martial Race Theory: A Historiographical Assessment of Recruitment in the British-Indian Army', *Calcutta Historical Journal*, vols 21–22 combined (1999–2000), pp. 139–154 and 'Construction of the Regiments in the Indian Army, 1859–1913', *War in History*, vol. 8, no. 2 (2001), pp. 127–148.

20 The East India Company, the Indian Army and the China Wars, 1839–1860

Rob Johnson

Outside the walls of Canton on 28 January 1841, on the Hill of the Five-Storeyed Pagoda during the First Anglo-Chinese War (1839–1842), a force of 700 British and Indian regular troops were confronted by a Chinese militia 10,000 strong. At the height of a torrential downpour, the militia had poured out of the city. A British officer of the 37th Madras Native Infantry recorded:

> At this time, not a single musket would go off, and little resistance could be offered against the enemy's long spears. [Our] men ... were soon surrounded by thousands of the enemy, who, had they possessed the slightest determination, could have at once annihilated them.[1]

He continued:

> the rain, ceasing to fall for a time, enabled a few of the men to discharge their muskets. The enemy was not removed above fifteen yards and every shot told as a matter of course ... as each shot told with great effect upon the crowd, the enemy was forced to retire.

The small British Indian force then extracted itself, and, once threatened with a naval bombardment, the Chinese agreed to withdraw their militia while a peace settlement was arranged. To the Chinese, it seemed that irregular forces had been able to prevent the capture of the city, and, in contrast to their power at sea, the British and their Indian allies were not thought to be militarily effective on land. Yukien, the Governor of Jiangsu province, told the Emperor the British 'are stiff and their legs straight. The latter, further bound with cloth, can scarcely stretch at will. Once fallen down, they cannot again stand up. It is fatal to fighting on land.'[2] The British, however, believed that, with a few exceptions, the Chinese lacked courage and were given to ruses that breached accepted norms of war. This was a seminal moment in the mutual misunderstanding that had developed from the end of the eighteenth century, but led to an almost continuous conflict for 20 years.

In 1841, Indian troops were not only on the march against China. Under British officers they had already been involved in the brief subjugation of

Afghanistan, assisted in the capture of Aden, and had participated in a number of operations across India and Burma. In 1856–1857, they contributed to the defeat of the Persian Empire.[3] By the time of the Second China or Arrow War (1858–1860), the old Mughal Empire had ceased to exist and India was firmly under British control. Yet, only the year before, Indian troops of the Bengal Presidency had been in action against their European rulers and the crisis had brought down not only the last vestiges of the 'ancien regime' in Delhi, but also the commercial enterprise that had governed a significant part of India for decades, namely the East India Company (henceforth EIC). By contrast, Indian and European commercial agents clamoured to open up China and their influence appeared to increase throughout the nineteenth century. The paradox is that China's own empire, which was defeated twice by the European and Indian military–commercial combination, would outlive the Mughal system by over 50 years.

These shifts and dynamics demand a much closer comparative analysis than they have received in the past, not least because, as the editors of this volume point out, India and China seemed poised to recover from their subordinate positions in history and play an important military and commercial role in the future. This chapter seeks to examine war between India and China in a period of significant transformation in the nineteenth century, examining the changes in the Army of the EIC, the relative importance of technological developments and naval power, the state of the Chinese forces in the period, the combat operations and their impact, and the military-strategic thinking that underpinned the key decisions in these campaigns.

The nexus of commerce and war

The EIC had raised local military forces to augment its European security contingents from the beginnings of its operations on the coasts of the subcontinent, at Madras, at Bombay and at the mouth of the Hugli in Bengal. The presidencies, or councils of merchants, operating under a Royal Charter, sought to make their 'factories', or warehouses, the focal point of their commercial success. However, rivalry with the French *Compagnie des Indes* led to open warfare and expansion of the EIC's armed forces. The succession of wars with France (1743–1815) had drawn the British into closer alliances, or conflict, with the various successor states of the Mughal Empire. The desire to acquire land holdings for their revenue potential had also spurred Company officers on into the interior. By the 1840s, much of the subcontinent was under EIC rule or allied to it. The Bombay Presidency had recruited a large number of local sepoys so that by 1805 it could muster over 26,000 men, while Madras and Bengal could each field a further 64,000 each.[4]

In 1807, the government-appointed Board of Control in London established a Military Department to assist the governor-general and the commander-in-chief in India, itself a coordinating appointment set up in 1786. However, the EIC army suffered from a parsimonious attitude from its directors, and there were no senior officer appointments until Lord Cornwallis, the first commander-in-chief, reorganized the staffing of regiments. Crown officers were encouraged to serve in the Company forces but existing EIC officers managed to prevent their full subordination.[5] However, Indian personnel found that an influx of Europeans deprived them of their former responsibilities. Indian officers were consulted only with regard to cultural matters and welfare and ceased to enjoy operational command. Nevertheless, many senior officer appointments were not filled even by the mid-nineteenth century. Moreover, the lack of supervision and complacency meant many European junior officers neglected their troops. The Europeanizing of Company administration added to the gulf emerging between the sepoys and the British officer corps and led, ultimately, to the Mutiny of 1857.[6]

The EIC had profited directly from the First War with China in 1839–1842. The war indemnity imposed by Britain amounted to £5.75 million in silver bullion and the Chinese were forced to cede the island of Hong Kong as an entrepôt for British goods. Compensation for loss of trade during the conflict was established at £21 million, payable in instalments. Fixed tariffs on imports and exports were imposed. The treaty between the powers that ended the war stipulated that the ports of Canton, Foochow, Ningpo and Shanghai were all to be opened up to British merchants, but years of wrangling followed.

It is perhaps erroneous to label the first war as the 'Opium War'. While the fighting had broken out over the smuggling of opium, it was really confiscation of property and what was perceived as a haughty Chinese attitude towards Europeans and free trade that turned a commercial dispute into a full-scale war. There was no mention of the sale of opium in the peace treaty, and Indian-grown opium was banned briefly after the war. It was Chinese officials who ignored the ban and encouraged sales in China in the hope of gaining lucrative tax revenues for themselves. Some British merchants even campaigned against the continued sale of opium on the grounds that it deprived Chinese consumers of cash that could be spent on legitimate goods.

However, before the war, British and Parsi merchants had supplied an increased demand for opium in China, exchanging their 30,000 chests offshore with Chinese intermediaries. One-seventh of the EIC's revenue was derived from opium in the 1820s, but the commodity was licensed not only for its financial value. The Chinese frequently refused to buy Western goods except silver or opium, and the demand for tea and other Chinese goods in Britain was so great that the Company accepted the trade as a vital part of its enterprise. Although illegal, the Chinese also levied an unofficial tax which was supplied directly to the Emperor, while the Shanxi banks supported the opium merchants for their own gain. Triad secret societies handled internal distribution for profit, while the Hongs (from *gonghang* or *cohong*: those using the court trading system) emerged as the pre-eminent traders in Canton. These organizations were blamed for malpractice by British and Indian merchants.

The Chinese authorities were nevertheless alarmed by the degradation of their troops through opium addiction in the 1832 campaign against the Yao, and the drain on silver bullion, used as currency, by the increased volume of trade.[7] A ban in 1836 was supported by the British, but opium merchants upgraded the armaments on board their ships, landed their goods anyway and engaged Chinese junks in a series of skirmishes. Commissioner Lin Zexu reinforced the ban, made arrests, seized £3 million worth of opium and held the British and Indian merchants hostage on Lintin Island under threat of capital punishment.[8] It was this internment that precipitated the conflict in September 1839. In November, Chinese demands to surrender another British national led to a naval battle between two British vessels and 28 Chinese war junks. Three Chinese ships were sunk and others were badly damaged, for the loss of one British sailor slightly wounded. By June 1840, a further 16 British warships, four steamers and 28 transports had arrived near Canton carrying 4,000 troops to enforce the British demand for compensation.

The Chinese were taken by surprise by the Westerners' willingness to go to war for the sake of merchants, and they themselves did not differentiate between private traders and their national government.[9] The Chinese believed that all nations were subordinate to the Emperor and their significance diminished with their distance from the centre of the Celestial Empire.[10] To Chinese officials, British actions represented gross impertinence.

The Second Anglo-Chinese War (1856–1860) occurred at the supreme moment of crisis for the Manchu/Qing dynasty. The proposed reforms of the new Empreror Hien-Feng precipitated unrest, and Hung Sin-Ts'uan led a revolt of Tai-ping from Kwangsi province that led to the seizure of the old capital of Nanking. Hung Sin Ts'uan propelled his followers with a political and religious zeal, promising a realm of heavenly peace to replace the moribund and corrupt old order. However, at Canton, the high-handedness of Commssioner Yeh Ming-ch'en led to renewed antagonism with European merchants. The seizure of the Hong Kong registered ship *Arrow* and the arrest of its crew was interpreted as an insult to the British that would inevitably lead to greater obstruction. The internment of the crew was used as the pretext for naval operations against Canton in October 1856. However, in December, local unrest in the city led to the destruction of European property and it was decided that further reprisals would be necessary. Ostensibly, however, the real motive was to further open up China to European commerce. Canton was bombarded a second time and assaulted by a joint Anglo-French landing party, and then placed under European jurisdiction. The Emperor's refusal to negotiate led to further operations on the Peiho.

The Europeans believed the Chinese to be a people who had largely turned their back on warfare because of Confucianism. John Crawfurd, writing for the Royal United Services Institute in 1858, argued the Chinese regarded war as 'an idle and profitless employment'.[11] Yet, Chinese inefficiency against Indian and European forces owed more to their experience of maintaining order within China, and a pragmatic response to overwhelming firepower, than any lack of courage or belligerence. Chinese troops that appeared to break and run were capable of reforming readily once they were out of immediate danger. What the Chinese lacked was the training, discipline, leadership and unit motivation to wage war against Europeans and Indians effectively.

In confronting internal rebels, traditionally Chinese forces aimed to bring greater mass to the battlefield than any of their opponents. This display of overwhelming force was sometimes sufficient to quell unrest. If, however, an enemy was prepared to field a force for combat, the army would invariably attempt to overawe the rebels by psychological means. The array of banners, bright uniforms, yelling, wildly charging and discharging firearms in a noisy fusillade from a distance could startle or terrify an enemy into flight. The emphasis on display meant that Chinese troops deployed in full view of their enemy. In the 1860s, Dr J. Lamprey observed that if the enemy made an advance, the Chinese would fire off every firearm, even though the adversary was entirely out of range. If on the defensive, he noted the Chinese might briefly break and then reform further away in order to repeat the manoeuvre.[12]

Chinese authorities were also used to offering financial reward for the destruction of their enemies. The British were shocked when the Emperor issued a set of bounty rates for the capture of a Man of War, a foreign prisoner or for killing a foreigner, for which some grisly evidence had to be produced.

Most Chinese senior officers hoped to achieve victory without fighting, but if there was to be an engagement, they preferred to try and induce the enemy to attack the strongest part of their line. They were frequently dismayed when the British adopted turning movements and Robert Fortune noted in 1857 that 'Having prepared their works, and posted their men so as to slaughter an enemy in attack from the front, they regard an attack in flank as low-minded, if not cowardly.'[13] In attack against a weaker enemy, the Chinese Army ordinarily attempted to envelop both flanks while pinning the centre of the line. The envelopment was carried out by cavalry who would deliver a charge about 200 yards from the enemy line. However, these attacks, when repulsed or held in check by a disciplined line of sepoys or Europeans, would break up in confusion. Some of those from the steppes of the north displayed excellent horsemanship skills and eyewitnesses noted that some horsemen were capable of stretching alongside the flanks of the horse to avoid a sword thrust or pistol shot. Cavalrymen carried a variety of weapons, including heavy matchlock gingalls which would be fired when dismounted. Mounted men appeared to be able to fire lighter matchlocks from the saddle and reload while riding away at speed.

Chinese forces marched in five divisions in the shape of a cross which permitted them to move off in any direction led by a vanguard and protected by a rearguard. Local irregular troops were used to screen the main force as it manoeuvred. In closer detail, infantry formed deep columns where the leading man in each file discharged his firearm and then withdrew to reload to be replaced by the next man. This continuous rotation could also be used in defensive positions but with selected men acting as the single firer and others passing loaded weapons up to him from behind. These formations, if well drilled, could be extended into more linear arrangements in order to work around the flanks of an enemy, all the time maintaining a steady stream of fire from those so armed. If an infantry attack was considered, the advance would be preceded by skirmishing archers, a few gingalls and swordsmen protected with shields. Entire formations could adopt an extended skirmishing order if checked or forced to withdraw.[14] Captain Grenville Loch recorded:

> If drilled under English officers, they would prove equal, if not superior, to the Sepoys. The matchlock man carries the charge for his piece in bamboo tubes, contained in a cotton belt fastened about his waist. He loads without a ramrod, by striking the butt against the ground after inserting the ball; the consequence is that he can charge and fire faster than one of us with a common musket.[15]

Regardless of the manoeuvre, Chinese soldiers would accompany their actions by 'a terrible display of huge flags, roaring gongs, horridly painted bamboo shields and the most extravagant waste of gunpowder'.[16] A mixture of weapons in units prevented orderliness. Against the Europeans, Chinese tactics rarely had a chance to work. The fact that so few were trained was evident in the First Anglo-Chinese War where 'after the first discharge of their firearms, all order ceased, and every individual did as he liked, abandoning his rank, and either running away or fighting single-handed, sometimes with great determination'.[17] However, most preferred to avoid hand-to-hand fighting. Episodes of mêlée generally occurred where escape routes had been cut off, or where more elite units were involved in the defence of a specific position. Defensive squares hedged with pikes to resist cavalry were possible amongst some experienced units but most attempts collapsed in confusion, caused by lack of training or the pressure of British and Indian firepower. Mutual support in defensive positions was also rather theoretical once any single position had been taken. Units tended to operate individually allowing the Indian and British troops to tackle them piecemeal.

The principal failing of the Chinese forces was not in their soldiers. Many commented on the willingness of the rank and file to endure terrible privations, to obey and endure strict discipline. It was absence of good leadership in their officers that mattered most. In the First Anglo-Chinese War, eyewitnesses commented on the absence of officers from the point of danger. Senior officers were content to arrange the men, guns and horses but barely directed any actions. This absence meant that opportunities were lost; there was no exploitation or initiative and morale of the troops often failed at the critical moment.

The Chinese Army and Navy

The Chinese Army of the mid-1800s was largely unchanged in its organization from the seventeenth century when the Manchus had overthrown the Ming Dynasty. The Manchu organization was the Eight Banners, where soldiers were given hereditary rights to rations and equipment in military service. In addition, the older indigenous Chinese Army survived as the Troops of the Green Standard but in the 1800s these men were mostly employed in garrison duties. In 1842, the combined army was thought to number between 800,000 and 1.8 million men. Soldiers of the Eight Banners were designed as being in one of three ethnic classes or *ku-sai*, with irregulars added as necessary. The first class was Manchus arranged in 678 separate companies of 100 men. On occasions, the *ku-sai* was organized as five regiments. The second class was Mongolians in just 211 companies (or two regiments) and the third class was 270 Chinese companies (again sometimes referred to consisting of five regiments). Most of the latter were mounted, although the Chinese companies also made up the entire complement of artillerymen in the army. The British referred to these as 'Tartars'.

The fourth class was the men of the Green Standard, estimated to be 500,000 strong, but, unlike the Bannermen, they did not receive rations of rice, arms, horses and pay. Instead the Green Standard men received land grants from which they were supposed to provide for themselves. Many of the soldiers were virtually indistinguishable in dress from the local peasantry, and when battles went against them, they would quickly return to civilian settlements for concealment. Pay rates fell gradually through time but by the 1860s it was estimated that Bannermen and Green Standard troops received as little as one to four taels a month, when even Chinese labourers in British service could earn six taels a month. Unsurprisingly many soldiers earned extra incomes through work as porters, nightwatchmen and cleaners. Periodic musters for drill degenerated into rituals rather than effective training.[18] Cadets as young as seven were brought on by their companies and soldiers could remain in service until the age of 60.

On tactics, a report in the *United Services Magazine* noted:

> The principal drill in which the troops, both Mandshoo [Manchu] and Chinese, are exercised, is shooting with the bow, both when mounted and dismounted: a portion of them are practised in discharging firearms without locks or ramrods; and a minor portion in loading and firing cannon.... The horseman is rapid in his movements, and advances to the attack with much impetuosity – at least when no enemy is before him. But his little, slender horse, with his short, quick step, wants the qualities of a charger.[19]

The report also noted that the appearance of the Chinese soldiers failed to impress: 'The feminineness of his apparel, his silk boots and the fan with which he refreshes himself in hot weather, make him cut a very effeminate and somewhat laughable figure and ... renders his appearance still more anti-martial.' It was not unknown for Chinese troops to carry umbrellas and fight beneath their canopies in wet weather.

The vast majority of the army was stationed around the capital as a security cordon, but some of the best units were stationed in Mongolia and Manchuria and took no part in the early stages of the First Anglo-Chinese War. Nevertheless, the garrison of the capital was considered the elite of the army. Consisting of six units including the Imperial Bodyguard, they provided internal security as well as the defence force for the walls and fortresses of the city. Specialist roles were not unknown for the army too. The *lu-chiao ping* (deer's horns men) were theoretically deployed ahead of the battle line to build *chevaux de frise* as an obstacle against cavalry. The *Chi'en-jui* (light division) was used for scaling walls, indicating the importance of seizing and retaining cities in Chinese warfare, while marine units were expected to move seamlessly from land to sea operations in order to take ports and enemy vessels.

A vast militia was created from clustering manpower in villages and rural districts under the direction of a local leader loyal to the regime. However, in the mid-nineteenth century, some militia commanders were religious or even secret society leaders. Mustered to deal with the threat of bandits or rebels at the local level, they were disbanded as soon as any emergency had passed in order to deal with agricultural demands. City militias were also created and tended to be better armed although the majority still carried bladed weapons and spears.

The *Yung-ying* ('braves battalions') forces were raised and disbanded like the militia but unlike the obligations on the militiamen, all braves were volunteers, recruited from across the Empire, and paid and supplied by the Imperial authorities. Pay rates were up to three times that of the Green Standard men and so the appeal was financial rather than patriotic. Rebel groups could sometimes be purchased and formed into *Yung* units as the means to settle unrest. However, the quality of the units was generally low, and groups might break off to indulge in looting as much as fighting an enemy. Training was only gradually introduced and in the 1850s, in response to revolts, Hunan province raised 136,000 *Yung-ying* men who could rival the Green Standard troops in efficiency and cohesion. Subunits varied in size from 50 to 100 and loyalty to the commander was based on his personal example, a stark contrast to the regular army units where contact between officers and men was actually discouraged. The irregulars would furnish 175,000 men from within China and a further 500,000 Mongolian light cavalry.

The standard firearm for the Chinese was the matchlock although a few more modern weapons appeared later. The rest of the infantry and cavalry were armed with bows, while the militia carried a mixture of spears, halberds and swords. The heavy matchlock, or Gingall, would be operated by a small, trained crew. Selected 'Tigermen' were sent ahead of the army to act as skirmishers and to delay and disrupt enemy formations, especially cavalry. Their bright uniforms, sudden rushes, yelling or flourishing of coloured shields were supposed to startle and disorientate horses and men. Firecrackers and other fireworks were designed to have the same psychological effect but could also cover a withdrawal, their continuous detonation mimicking the sound of sustained volley fire.

The navy or 'Water Force' was an extension of the army and followed the pattern of Bannermen and Green Standard troops. Levels of pay, training and efficiency were low, and corruption flourished. Some vessels, designed for riverine operations against piracy, were actually hired out to pirates. The size of the fleets had gradually declined over time and following the destruction of many junks in the First Anglo-Chinese War, there were too few even for coastal policing. However, *Yung-ying* forces meant that China could deploy as many as 570 vessels with trained crews by the outbreak of the second conflict. Formerly war junks had often been no more than converted merchantmen with a number of small calibre brass pieces aboard. However, war junks and smaller craft in the

1850s were divided into five specialist categories. By 1858, this fleet was armed with 2,000 guns of which 40 per cent were purchased from foreign manufacturers and ranged from 4-pdrs to 24-pdrs. River forces were divided into *ying* and *shao* formations, complete with flagships and supply vessels, and they were instrumental in suppressing the Tai'ping. In the first Anglo-Chinese War, fire ships had also been used. Loaded with gunpowder, rockets, hooks to snare rigging or masts, and grapnels trailed below the surface to catch cables, they nevertheless proved ineffective. Equally, Europeans tended to dismiss Chinese efforts at psychological warfare at sea. The brandishing of tridents, banging of gongs and even pulling faces or hurling stink pots and fireworks failed to impress the Royal Navy which could engage and destroy the Chinese long before they drew close to British ships.

The army of the East India Company, technological developments and naval power

By the time of the China war, the British had transformed their factory guards of the early eighteenth century into regular regiments of Indian infantry and cavalry, drilled and trained in the European style. In 1808, there were some 155,000 Indian sowars, sepoys, gunners, miners and pioneers in the army of the EIC, making it one of the world's largest standing armies. The use of close order linear tactics, to maximize the firepower of musketry, combined with mobile artillery and light cavalry, had proven an effective combination against the irregular cavalry armies of South Asia, and they were to demonstrate a similar effectiveness against Chinese regular forces and light cavalry between 1839 and 1860.

Indian units that saw service in China were drawn from the Madras Presidency army. There were a number of changes in the first decades of the nineteenth century to the Madras forces, but by the outbreak of the First Anglo-Chinese War there were 50 regiments of infantry, and eight regiments of Madras Light Cavalry.[20] There were troops of both foot and horse artillery, three of which were composed of Indian personnel, and a corps of pioneers and Sappers and Miners. The 2nd, 6th, 14th, 37th and 41st Madras Native Infantry regiments and Madras Sappers and Miners all received battle honours for their operational duties in China.

What was distinctive about the Madras regiments, compared to the units raised in Bengal, was their wide social base of recruiting. There was little consideration given to caste sensibilities, with men serving together in composite battalions. By contrast, the Bengal units were dominated by high caste Brahmins who guarded their status jealously and rose in rebellion when they were offended by a series of reforms and changes in the 1850s.[21] Overseas service had been a cause of complaint, even though Indian units had served in Egypt in 1801, Burma in 1824–1825, China and Persia. Madras units, which had provided some of those contingents, were unaffected by the Mutiny of the north. After the Mutiny, following the Peel Commission (1858), the Indian Army was subject to further changes, with larger numbers of Punjabi Muslims and Sikhs being recruited, but the army that went to war in China in 1858–1860 was still largely organized along its traditional lines. Despite proving its worth in China, the Madras Army was the victim of a preference for northerners, mountain-dwellers and 'martial races'. Madrassi units were not deployed on operations again until 1885 and sepoys drawn from poor rural communities tended to stay in service for long periods. British officers often tried to avoid service in Madras in favour of better promotion prospects elsewhere and the Madras Army was gradually reduced in size and strength until 1895.

Indian infantry units were organized in the same manner as the Europeans, with battalions subdivided into eight companies. Until 1858, two companies were designated as Grenadier and Light, the latter being deployed as a skirmishing force. However, in the Second Anglo-Chinese War, all companies were expected to be able to perform the tasks of light infantry. These deployed ahead of the main battle line in order to skirmish against the Chinese.

Indian sepoys in the First Anglo-Chinese War were armed with flintlock muskets and bayonets, while officers and cavalry carried the *arme blanche*. Although the effective range of the muskets was barely 100 yards, used en masse and fired in volleys against the densely packed ranks of Chinese horse and foot, musketeers could inflict significant casualties. Loading and firing the one ounce ball from its paper cartridges took time but a trained soldier could expect to get off three rounds a minute. Having shocked and depleted the Chinese ranks by volley fire, the expectation was that sepoys would press home the attack with the bayonet. Junior European officers were expected to lead these assaults from the front and great store was placed on their exhibition of courage and sangfroid. The combination of short range weapons, the desire to close with the bayonet and the aggressive leadership of the European officers meant that the British Indian forces were invariably on the offensive and retained the initiative at a tactical level. The British also possessed a significant advantage in artillery in that their complement of horse-drawn field artillery and howitzers was augmented by the firepower of naval guns. Horse troops and field batteries were divided into four-gun or two-howitzer teams or divisions. The bulk of gun ammunition was round shot, with the rest divided between shrapnel and canister. Howitzers carried some specific incendiary rounds but mainly shrapnel, devastating against infantry inside enclosed spaces who had no overhead protection. Gun and howitzer teams could expect to fire between two and three rounds a minute, and a 6-pdr piece could engage targets effectively at 600 yards and the 12-pdr howitzer at 1,000 yards. And the guns and crews outclassed their Chinese equivalents.

In the Second Anglo-Chinese War, the British infantry were armed with Enfield rifled muskets which gave them greater range and accuracy. The use of Minié bullets (introduced in 1853) ensured that the round had a tighter fit in the grooved barrel and a more aerodynamic shape which greatly improved its efficiency and flight. Enfields could engage targets from 400–900 yards allowing British troops to inflict casualties long before the Chinese could reply. The introduction of percussion caps (1842) made the discharge of the weapon more reliable and ensured a higher rate of fire, and consequently trained riflemen were able to maintain a rate of ten rounds a minute. However, Indian units retained the older smooth-bore musket for some time and were only gradually upgraded.

In assaulting Chinese forts, artillery was used to soften the defences before infantry advanced in skirmish order, firing as they advanced to keep those of the ramparts pinned down. In theory, storming parties carrying ladders and those acting as supports would follow 200 yards behind the skirmish line. On a number of occasions, however, the British and Indian troops avoided a frontal assault and manoeuvred around the flanks of defensive positions, or established a firing point that overlooked fortress walls. Only when the defenders' fire began to slacken would the order be given to enter a position and engage in hand-to-hand fighting.

Indian and British cavalry was organized and fought in traditional ways throughout both wars. The smallest subunit was the troop, nominally 62 men with officers, NCO, trumpeters and farriers attached. Although used in reconnaissance, skirmishing, patrols or guarding flanks and baggage, the objective was always to keep squadrons and regiments concentrated for the shock effect of a charge or pursuit at the opportune moment. In China, the King's Dragoon Guards delivered a charge in the traditional style and routed a formation of 'Tartar' horse, despite an attempt by Chinese archers to deliver a Parthian shot as they withdrew.

Rifled Armstrong breech loader guns of 6-pdr and 12-pdr were used in China in 1860 by the field artillery. Their bullet shaped projectiles had a range of two miles, greatly outdistancing the old smooth bore cannons of the first war. The 12-pdrs were also lighter, requiring fewer horses in the limber teams to draw them, and both field and howitzer troops and batteries adopted a six gun arrangement. Not all experiments with breech loaders were a success and for a time the army reverted to muzzle loading rifled guns, while in India smooth bore cannons were not replaced for years, but, in China they appear to have proven their worth.

The Indian Navy, known as the Bombay Marine, was the world's first all steam force, having converted entirely to paddle steamers in 1837.[22] This small but formidable fleet was instrumental in determining the outcome of the conflict in 1842. The paddle steamer *Nemesis* was a 700 ton, iron-hulled private ship crewed by Indians and by commanded by officers of the Royal Navy which alone destroyed 11 Chinese war junks.[23] At Amoy, just two sailing ships of the line inflicted similar levels of destruction. A three hour bombardment smashed 26 war junks and silenced all the Chinese shore batteries without loss. Of the fleet that made its way up the Yangtze in 1842, 11 were under sail and ten were steamers, with a number of transport ships and vessels to carry coal. Steam had the advantage of allowing the British to operate against the tides and winds if necessary and deploy landing parties with impunity along the Chinese coasts, but their firepower made perhaps the greatest contribution of all.

Naval guns were also evolving slowly. Paddle wheel houses restricted the number of guns that could be deployed along the decks, and so the Royal Navy had experimented with the use of fewer but heavier guns. In 1826, 32-pdrs were adopted as the standard weapon for its ships of the line, but some paddle steamers were armed with 68-pdrs and by the time of the First Anglo-Chinese War both steamers and ships of the line mounted a few of these heavier guns.[24] In 1838, the British also introduced shell alongside the standard round shot, but the cannon ball from a 32-pdr remained the ammunition of choice because its range was 1,300 yards while an eight-inch shell was effective only at 800 yards. Although the Indian Navy was wound up in 1863, the use of steam propulsion, ironclads and improved gunnery had demonstrated its value beyond all doubt in the First Anglo-Chinese War.

In the Second Anglo-Chinese War, a combination of sailing ships and paddle steamers took part in the initial bombardment of Canton in November 1856, but in 1857–1858 both the British and French made use of screw-propelled steam ships in the Pearl River and at the Peiho. Nevertheless, smaller gunboats that attempted to operate inshore were vulnerable, and in the second assault on the Taku Forts in June 1859, three were lost to Chinese shore guns. These operations were taking place at a time of significant change in naval technology. It was during the war that the British laid down HMS *Warrior*, its first iron-hulled warship with a speed of 14 knots and bristling with modern guns. France had also launched its own *Gloire* and within years, wooden walls had been entirely replaced with ironclads and iron hulls. Crucially, naval power was at its most effective when it could project into the littoral and literally steam to the gates of the Chinese capital.

Operations and their impact: the first Anglo-Chinese 'Opium' war

In the early stages of the British operations against China, a fleet was assembled off Macao on 21 June 1840 where it was subjected to a night attack by 18 fire rafts made up of fishing boats and cargo flats lashed together and set ablaze. The attack failed to damage any British vessels but the subsequent decision by the navy commander to send most of the fleet towards the mouth of the Peiho to the north gave the appearance to the Chinese that their tactics had succeeded. There was therefore surprise when the British landed troops at Chusan Island in Hangzhou Bay on 5 July and bombarded the port of Ting-hai. British and Indian troops were landed and made their way across waterlogged ground to the walls of the port, which was surrounded by a wide canal. From an adjacent hill, musket fire raked the ramparts while field artillery hurled round-shot into the city. The following morning the Chinese capitulated. The 3,000 strong British and Indian force was then gradually depleted by a combination of malaria, bad rations and disease. By December, 450 personnel had died.

The Chinese authorities initially believed that the British would make no further moves inland since they were only able to fight well at sea, but when the British fleet began to move up the Peiho towards Nanking, the strategy was changed: negotiators were to keep the Europeans on the coast and exhaust them until a force could be assembled to destroy them. The British agreed to withdraw their sickly force from Chusan and continued talks off Canton for over a month. However, the lack of progress towards a settlement prompted a renewal of action in January 1841. A landing force was sent ashore to seize the Bogue Forts that guarded the mouth of the Pearl river estuary.[25] In the engagement, the British suffered only 38 wounded, but the Chinese suffered an estimated 300–400 killed. The British had marched around the flanks of the position which they noted was deeply entrenched, and mutually supported by other fortifications and field guns. The ramparts were densely packed with Chinese troops who shouted, beat gongs and fired fusillades the moment the British Indian force came into view. There was an exchange of fire for about 20 minutes before the assaulting British and Indian troops drove the defenders off. Fighting continued across the woods and hills adjacent to the forts. However, the last defenders were at a distinct disadvantage when high ground that overlooked the forts fell into British hands, which suggests a major design fault in the forts' construction.

In the estuary, ten war junks mounted their own attack, but were swiftly sunk, crippled or captured. About 200 Chinese sailors lost their lives. Despite the overwhelming defeat, the Chinese authorities reported that they had achieved a victory, using the absence of any pursuit or the fall of Canton as their evidence. The British therefore pressed their advantage and assaulted the Inner Bogue Forts and the Wantong shore batteries. Once again Chinese losses were severe, but only five British Marines were wounded. On 27 January, yet more batteries were destroyed and over 2,000 defenders were driven back into Canton with a loss of a further 300. Just five miles from the city, 13 war junks were sent against the British ships in the river, but ten were captured and the admiral's own vessel exploded when a rocket fired by the steamer *Nemesis* entered the magazine. A fire raft attack was launched at night, but, once again, it proved ineffective.

Major General Sir Hugh Gough deployed his land forces to make the final attack on Canton, and his Indian troops captured the two easternmost forts after a significant cannonade lasting half an hour. The two nearest bastions were then assaulted leaving only the walls in Chinese hands. Hoping to negotiate a surrender and avoid considerable loss of life, Captain Charles Eliot, the Superintendent of Trade for Canton and the British government's representative, ordered a ceasefire. The Chinese militia who made up much of the vast garrison interpreted the pause in fighting as the turn of the tide of battle and swept out of the city during a storm that rendered the British and Indian muskets temporarily inoperative. Indian troops used their turbans to dry their firelocks as soon as the rain began to lessen, and their ability to resume their fire at close range drove the militia back. Canton remained in Chinese hands as negotiations were finally concluded, creating the myth that irregular Chinese forces had checked or even defeated the British and their Indian allies.

The Emperor repudiated the negotiations in Canton and sent reinforcements to the city, stating that any idea of ceding Hong Kong to the 'barbarians' was 'repugnant'. In January he had appointed his cousin, I-Chan, to lead an Army of Extermination. Meanwhile, at Canton, General Yang-Fang, a 70-year-old deaf officer, recruited a local 'Patriotic Militia', rebuilt the redoubts and rerouted canals. Elliot ordered the new defences to be destroyed before they were complete and Yang-Fang was forced to accept a truce, which was, in turn, rejected by I-Chan. In May, a great fleet of war junks were sent against the British but 71 were destroyed or sunk and in the follow-up operations, 60 shore batteries were destroyed or overrun. Once again, Canton was at the mercy of a British assault. Eliot hesitated and accepted negotiations. His terms were that the military garrison was to evacuate the city and the province, but their departure led to wholesale looting and disorder by local gangs. The Imperial reaction was to dismiss the affair and the advice that the army should wait until the British ships left, then simply retake and refortify the entire area.

Sir Henry Pottinger replaced Elliot off the Chinese coast in August 1842 with instructions to punish the Chinese and extract compensation for the loss of trade and the costs of naval and military operations.[26] The expeditionary force allocated to the mission consisted of 25 warships, 14 steamers and 10,000 troops in transports. The first objective was the port of Amoy which was protected by a half mile line of batteries, forts protected with granite walls and a flotilla of 50 war junks. British ships sailed close to the forts, but their shot was deflected by the stonework. Chinese guns could make little impression on the British ships either and an artillery duel therefore produced little result. However, the Chinese garrison, some 9,000 strong, could not prevent a landing party overrunning the forts. Overcome by the dishonour, the Chinese second-in-command ran into the sea to drown himself and another officer was seen to slit his own throat as the British entered the embrasures. The Chinese troops were forced to evacuate their defences and the city fell to the British. The Chinese suffered heavy casualties while the Europeans lost just two killed and 15 wounded. A great quantity of arms and military stores was taken, including a gun casting foundry. At Tinghai the Chinese managed to hold out for three days but this too was taken and the garrison fled in panic. At Ningpo, there were further signs that the Chinese forces were at their limit of endurance. When the river-mouth fort of Chen-Hai was taken, the city forces fled. The British dismantled the barricades that barred the city gates with the assistance of local Chinese and there was no fighting. Ningpo proved to be an ideal base for the winter and the British set about garrisoning the points they had taken. The Imperial Chinese forces nevertheless regarded the lull in operations as the opportunity to regroup and prepare a spring counter-offensive.

I-Ching, the Emperor's Director of the Hunting Parks and Gardens, as well as Commander of the Forbidden City's police, was appointed the commander of the area's forces. He assembled 12,000 soldiers with 33,000 militiamen, believing that the small British and Indian forces would be overwhelmed by the sheer mass of Chinese. So confident was I-Ching of victory that tea parties, celebrations and poetry readings were organized to commemorate the impending victory. However, there was no clear chain of command, provincial officers were ignored by regulars and the entire logistics system collapsed. By the time the force reached Ningpo in the rain, many of the troops were starving. I-Ching's plan had been to send the main force to retake Ningpo while simultaneously having a marine force of 10,000 retake Chusan and a division of 15,000 soldiers attack Chen-hai. This would divide and weaken the Europeans who could not be strong everywhere. However, almost two-thirds of the army acted as bodyguards for their commanders and only 4,000 men were tasked with the assault on Ningpo. The timing of the attack was dictated by tradition: 0400 on 10 March 1842 equated to the Tiger Hour of the Tiger Day or the Tiger Month of the Tiger Year.

The first Chinese assault was by men from Szechuan, but owing to a misunderstanding about the need to withhold their fire until the last minute for surprise, they believed this to mean they should not carry firearms at all. Their attack was destroyed by gunfire before any of the 700-strong regiment could reach the city defences. Fresh waves were ordered forward but these were also cut down. One party was permitted to scale the walls and enter a pre-designated killing area. They were shot down with volley fire, and, as they tried to escape, canister was fired into them at close range from small field guns. The attack at Chen-hai was also unsuccessful. Orders to commit the reserve were never enacted because their commander was in a stupor induced by smoking opium while he waited. The marine force was embarked as planned but never landed at Chusan: their commander chose to sail along the coast sending in reports of imaginary victories over the Royal Navy. The only success the Chinese enjoyed was floating boxes into the river which they knew the British would recover. On being opened, they exploded. The operations ended with a British counter-attack at Tse-kee. The Chinese base there was defended by Imperial bodyguards and fighters from Chinese Turkestan. The Chinese regarded them as invincible, and they put up a robust defence with gingalls and matchlocks, but they were attacked from three directions, broke and fled with heavy losses. An estimated 900 were slain. For the rest of the spring, the Chinese engaged in a more protracted resistance. Individual Indian and British soldiers were captured and executed, sometimes after torture. The British in turn inflicted reprisals.

In May 1842, Pottinger ordered the advance towards the capital. Chapu was taken after an escalade, but the 300-strong Tartar Manchu garrison was unable to escape and defended a centrally located house. Subjected to fire on every side, the building was eventually set alight. Only 53 were taken alive. In the defence, a total of 1,200 Chinese troops had been killed. However, far more shocking was the discovery that many Manchus in the city had believed they would be massacred by the British and therefore they had killed their own families. Some had been slashed across the throat while others had been drowned in ponds and wells. Chinese camp followers accompanying the British force completed the looting and destruction of the city, and the local Chinese fled as refugees from the advancing Europeans.[27]

The gradual collapse of order was reinforced by a growing sense that the foreigners were being assisted by *nei-ying*, 'fifth columnists'. Manchu commanders executed suspects and began to target Han Chinese and smugglers as the most likely culprits. Punishments became so random that many Chinese feared the regime more than the foreigners. Rumours began to spread, exaggerating still further the sense of fear that pervaded. Shanghai fell without a shot being fired but sustained resistance was offered at Jinjiang. In the street fighting, British troops noted that Chinese soldiers started to commit suicide to avoid capture or when their defeat was imminent.

As the British ships entered the Yangtze they were confronted with more shore batteries. Of these, 376 guns were captured. It was found that many had been fitted with a pivot to swivel the barrel, with gunsights made of bamboo, suggesting efforts to adapt to the new threat. All the British ships had been hit by round shot from these guns: HMS *Blonde* was pierced 14 times by cannon balls but was still seaworthy and operational. At Chinkiang on 21 July, the pattern of previous engagements was repeated: storming parties scaling walls or blowing in gates; sustained hand-to-hand fighting taking place until the Chinese were overwhelmed; isolated elements of the garrison that were cut off from any escape fighting to the last man; and the systematic suicide or murder of Manchu families by their own kinsmen. Hai-lin, the commander of the garrison, ordered his house to be set alight and he died in the blaze. Once again, looters flooded in from the countryside and organized gangs stripped the city of its wealth. Bodies that were left unburied became the vectors of disease while the intense summer heat caused further fatalities on both sides.

However, the approach of British ships to the gates of Nanking and the inadequate nature of the garrison compelled the Emperor to negotiate. To the end, the Emperor had been misinformed about the progress of the Europeans. After the fall of Amoy in 1841, he had been told by his mandarins that the city had fallen only after his forces had burned and sunk five of the barbarians' warships and a steamer. They claimed that, as a result, 'The south wind blew the smoke in our soldiers' eyes, and Amoy was lost.' The Court attempted to keep the worse news from the Emperor in a conspiracy of silence and wishful thinking, but the hubristic nature of the regime simply could not contemplate the idea that its own forces could be defeated by 'barbarians'.

Operations and their impact: the second Anglo-Chinese 'Arrow' war

Having bombarded Canton's defences on 27 September 1856, British seamen and Marines had stormed the city two days later but the Chinese had maintained a steady sniping during the brief British occupation. A subsequent naval bombardment defeated the Chinese war junks at anchor in the Pearl river and a small fort was taken. However, the Chinese made a more determined defence of the Bogue Forts than in 1841. Some 200 guns were mounted on the ramparts and Chinese gunners remained serving them until the British were actually inside the defences. At that point, boats which had been prepared for the evacuation of the garrison were launched mainly to save the Chinese officers. The remainder fled into the water where they were rescued and then released by the British sailors. The British nevertheless lacked the manpower to hold the positions they had taken, or the city, and operations were largely suspended pending negotiations. During this lull, one Chinese party disguised themselves as passengers and massacred the crew of a mail steamer. The attack, thought to be outside the norms of war, led to reprisals from the British: they burnt down the western suburbs of Canton. The Chinese in turn responded by offering bounties for the capture or death of Europeans, threatening any ethnic Chinese who assisted the foreigners with beheading. In January 1857, Chinese agents attempted to poison the bakeries of Hong Kong with arsenic.

That May, a British flotilla destroyed the Chinese vessels in the Canton estuaries and when shallow water prevented pursuit, British Marines and sailors rowed upriver, surprised a fleet at anchor and destroyed it. At Fatshan Creek, the Chinese had armed their forts with more artillery in the hope that this superior firepower would deter further British raids, but the British Marines and Bluejackets stormed the forts and blew up any junks they could reach. When the Chinese tried to retire further upstream they were often compelled to abandon the vessels that ran aground on mud flats. It was at this moment that news arrived of the outbreak of the Mutiny in India and ships and men were diverted to the subcontinent, which gave the Chinese a crucial respite.

Operations were resumed in December 1858 after fruitless negotiations. Canton's forts were again assaulted and seized, and the city itself fell on the 28th. 'Tartar' troops made one last attempt at a sortie as the city was being overrun, but they were destroyed. A new round of negotiations failed to produce any settlement, and a fresh expedition was mounted against the capital.

On 19 May, the Allied forces encountered the Taku Forts which extended for one mile on either side of the Peiho river mouth. Some 87 heavy guns were mounted on the ramparts with a further 49 in supporting batteries beyond. Earthen walls had been constructed to absorb European gunfire, while ditches and abbattis protected the glacis against land assault. The Chinese had demonstrated their ability to adapt to the Europeans' advantages in firepower. However, in a bombardment lasting 90 minutes, the Chinese guns were silenced, and the fire ships and embrasures upriver appeared to be abandoned. No further resistance was offered so the Allies could request terms at Tien-tsin, a treaty being concluded on 26 June 1858. Nevertheless when new ambassadors were despatched, the accompanying squadrons found the Taku Forts rebuilt and the river blocked with stakes, chains and wrecks, the offer of negotiations having been used only to buy time. When a party blew up some of the booms, the Chinese replaced them overnight. An attempt to force a passage led to the loss of three gunboats to Chinese artillery fire, although most of the defences at Taku were badly damaged and two forts were silenced. When a landing party got caught in mud, it was subjected to a sustained barrage of arrows, musket and cannon fire. As the British tried to evacuate at night, rockets were used to pinpoint and illuminate the rescue parties.[28]

This setback led to the deployment of a much larger force, 11,000 strong, including Sikh volunteers. The French added a further 6,000 and the whole Allied army was supported by a fleet of 200 ships. In August, the force was able to move from its beachhead on the Peiho. Two columns set out but one was again caught in waterlogged ground and had to drive off Chinese cavalry that had tried to seize the opportunity for another tactical victory. Artillery fire proved effective and Sikh horsemen were instrumental in driving off the Chinese as they withdrew. Moving closer to the defences of Tangku, the British guns, including their new Armstrong breech-loaders, soon silenced the Chinese batteries. As the British manhandled their artillery to within 350 yards of the earthen walls they were subjected to a great fusillade of gingalls and musket fire, but this proved no match for the gunfire of the British and Indian troops and the positions were all carried by assaulting infantry. At the Taku Forts themselves, the pattern was repeated with Madras Sappers making a significant contribution in laying tracks across the numerous ditches and boggy ground. However, flooded canals with sharpened bamboo stakes and thicker earth walls proved a significant obstacle to assaulting infantry. The British troops were forced to wade and in some cases swim across them while under a rain of grenades, stink pots, vases of lime, matchlock balls and rockets. When the Allies fought their way into the defences, resistance collapsed and all the other forts were quickly abandoned. Some 400 Chinese guns were taken.

The seizure of a British negotiating party and ambush at Tang-Chow on 18 September was construed as a breach of the rules of war, but the incident probably occurred as the result of a misunderstanding about the state of hostilities between the two sides. The Chinese mustered 20,000 troops rapidly and greatly outnumbered the Allied force. The British quickly formed a battle line to receive the onrush of Chinese horsemen, and a single volley drove them off, whereupon they were pursued by the British cavalry. Hundreds of Chinese were killed and wounded for the loss of one dead and 17 wounded. At Pa-le Chiao, the Chinese army had reformed. Their cavalry attempted a charge but was driven off by British horsemen. Another charge against the British line was cut down with heavy loss when the horse artillery fired canister into them at short range. Indian Cavalry, Fane's Horse and the 1st Sikh Cavalry, joined into the battle but had to make a charge across fields of maize and millet stubble which cut the horses legs and bellies.[29] Armstrong guns pitched shells right amongst the Chinese cavalry and they maintained a distance of 1,000 yards between themselves and the Indians, gradually withdrawing towards Peking. By October, Allied guns were within range of the walls of the city, but the Emperor fled and the capital was consequently surrendered without further resistance.[30] Subsequent negotiations concluded the war.

Conclusion

The Anglo-Chinese wars of 1842–1860 brought the two Asian powers of China and India into direct confrontation. Yet, whilst India's armed forces had been transformed by contact with the British, China's military remained relatively unchanged until the late nineteenth century. The British had succeeded in creating an expeditionary strike arm made up of British and Indian troops, and a powerful navy, with which to destroy the Chinese defence system, although it would be wrong to attribute this success entirely to a technological advantage. Leadership and organization were crucial to the outcome. The impact of the military defeats on China was profound and traumatic for an Empire that had believed it was the centre and apogee of the world. Yet, for India too, the Anglo-Chinese wars occurred in a time when British rule was undergoing considerable changes. The defeat of the Bengal Army's Mutiny and its associated rural rebellion (1857–1858), styled as the First War for Independence by the nationalists, was no less traumatic. The British had taken the internally divided subcontinent of India piecemeal over a period of more than 100 years, while China, although not annexed or partitioned, was subjugated in a far shorter period through military and naval asymmetry and at a time of great internal stress.

The British objectives in China were not unlike those expressed about India in the early 1700s, namely to avoid the acquisition of territory (which would necessitate expensive formal administration) in favour of opening up commerce. However, the British did not try to acquire land, as they had in India, beyond controlling the entrepôts of Hong Kong, Shanghai and other treaty ports. The key factor common to both conquests was naval power. While China had traditionally looked towards its long landward frontiers as the direction of its main threat, it had underestimated the amphibious power of the Western nations.

Wang Gungwu attributes the failure to recognize this greater threat or the technological advance of rival powers as complacency, but there was arrogance in the Court's thinking, and a willingness to discount any information that did not comply with the orthodoxy that China was a vast and superior empire.[31] Wei Yuan drew up a study in the wake of the First Anglo-Chinese War arguing for the urgent reorganization of the Chinese navy, the retraining of its sailors in Western methods and the adoption of modern technology. Yet, the Imperial Court remained inert and under pressure fell back on trusted methods of war. Although the Chinese fleet expanded in size and armament, it remained a riverine force effective only against internal enemies.

In the historiography, nationalist historians of the twentieth century criticized the Manchu dynasty for failing to prepare China against the inevitable Western onslaught. The chief problems, they argued, were the corruption and treachery of the Court officials and the only exception was Commissioner Lin, a figure held up as a paragon of national resistance. Some military officers who held on to the coastal forts despite the odds against them were also singled out for praise. Not surprisingly, many Chinese historians wanted to find examples of popular resistance that aligned with their communist interpretations. The villagers of Sanyaunli near Guangzhou were celebrated for their defiance of British troops.[32] However, in Chinese decision making, the British military intervention was a minor threat compared with the great Tai'ping uprising from 1851. This was followed by a succession of other risings in the north, north-west and south-west that lasted for 30 years.[33] From the Chinese Court's perspective, all these great rebellions on land were suppressed, while local Chinese forces assisted in the defeat of revolts in Yunnan and Sinkiang. The European threat was coastal and could, they reasoned, be contained.

It was not until Mao Haijian published *Tiancho de Bengkui* in 1995 that the entire Imperial system, including Lin, were criticized for failing to study the Westerners and their methods, and for underestimating their military and naval power.[34] It was not a question of merely adopting Western technology but of completely re-gearing the political and security systems of the Empire to ensure its defence. Ironically, after the Second China War, the British were eager to preserve China and ensure it could defeat its internal enemies, and they trained and equipped the Bannermen battalions in the use of modern firearms. Nevertheless, all of the Empire's resources were absorbed in defeating rebellions and in the retaking of Xinjiang so there was little effort in building naval power beyond the establishment of a merchant fleet at Foochou (Fuzhou) and the training of a handful of personnel in Britain. Chinese military reforms were never truly effective, and, beyond outward appearances of modernization, the Chinese system was vulnerable and at risk of failure.

This chapter examined the encounter between the British-Indian Army, the Royal Navy and Indian Navy (as they transferred from sail to steam), and the Imperial forces of China. What emerges is a stark contrast between rapidly modernizing British-Indian forces and the traditional organization, training, doctrine and tactics of the Chinese. There are scores of works on the Anglo-Chinese wars and a number of secondary works dealing specifically with the Chinese political perspectives, but, there are few works on the Chinese Army in the period and fewer still that attempt to make comparative judgements with India in the same era. Making those comparisons, analysing the navies as well as the armies, and contextualizing the development of those forces is crucial if we are to understand the character of war in Asia in the mid-nineteenth century.

Notes

1 'An Officer of the Force', *United Services Magazine*, vol. I, (1843), pp. 242–243.

2 Pin-Chia Kuo, *A Critical Study of the First Anglo-Chinese War* (Westport, CT: Hyperion Press, 1973), p. 261; Peter Fay, *The Opium War, 1840–42* (Chapel Hill, NC: University of North Carolina Press, 1975), p. 230.

3 George Henry Hunt and George Townsend, *Outram and Havelock's Persian Campaign, to which is prefixed a Summary of Persian History, and Account of the Various Differences between England and Persia and an Inquiry into the Origin of the Late War* (London: Routledge, 1858).

4 Philip Mason, *A Matter of Honour: An Account of the Indian Army, its Officers and Men* (London: Jonathan Cape, 1974), p. 140; H. Wilson, *A History of the Madras Army from 1746 to 1826* (Madras: Government Press, 1882).

5 Raymond Callahan, *The East India Company and Army Reform 1783–1798* (Cambridge, MA: Harvard University Press, 1972); Peter Stanley, *The White Mutiny: British Military Culture in India, 1825–75* (London: Hurst & Company, 1998); Byron Farwell, *Queen Victoria's Little Wars* (London: Allen Lane, 1973), p. 135.

6 S.L. Menezes, *Fidelity and Honour: The Indian Army from the Seventeenth Century to the Twenty-first Century* (New Delhi: Penguin, 1993).

7 'Tabular View of the Quantity of Opium Exported from Bengal and Bombay, with Profits Derived therefrom by the East India Company', enclosure no. 4 in Despatch no. 26 from Sir John Bowring to the Earl of Clarendon, Hong Kong, 8 January 1856, in *Parliamentary Papers: Papers Relative to the Opium Trade in China, 1842–56*, p. 50.

8 Hsin-pao Chang, *Commissioner Lin and the Opium War* (1964, reprint, New York: Norton Company, 1970), pp. 92–98.
9 F.W. Mote, *Imperial China, 900–1800* (Cambridge, MA: Harvard University Press, 1999), pp. 949–956.
10 Wang Gungwu, *Anglo-Chinese Encounters since 1800: War, Trade, Science and Governance* (Cambridge: Cambridge University Press, 2003), p. 50.
11 John Crawfurd, 'China and the Chinese', *Journal of the Royal United Services Institution*, vol. I (1858), pp. 51–60.
12 Dr J. Lamprey, 'The Economy of the Chinese Army', *Journal of the Royal United Services Institution*, vol. XI (1867), pp. 403–433.
13 Robert Fortune, *A Residence Amongst the Chinese* (London: John Murray, 1857).
14 Arthur Fisher, *Personal Narrative of Three Years' Service in China* (London: Richard Bentley, 1863).
15 Cited in Ian Hernon, *The Savage Empire: Forgotten Wars of the Nineteenth Century* (Thrupp, Stroud: Sutton, 2000), pp. 74–75.
16 Augustus Lindley, *Ti-ping Tien Kwoh: The History of the Tai'Ping Rebellion* (London: Day & Son, 1866), vol. I.
17 Ian Heath, *Armies of the Nineteenth Century: Asia, II, China* (St Peter's Port: Foundry, 1998), p. 32.
18 T.F. Wade, 'The Army of the Chinese Empire', *Chinese Repository*, vol. XX (1851), pp. 250–280.
19 *United Services Magazine*, vol. II, (1842), pp. 415–416.
20 W.J. Wilson, *Historical Record of the Fourth Prince of Wales' Own Regiment Madras Light Cavalry* (Madras: Government Office, 1877).
21 David Omissi, *The Sepoy and The Raj: The Indian Army, 1860–1940* (London: Macmillan, 1994), p. 4.
22 Andrew Lambert, 'The Introduction of Steam', in Robert Gardiner (ed.), *Steam, Steel and Shellfire: The Steam Warship, 1815–1905* (London: Conway Maritime Press Ltd, 1992), p. 28.
23 Lawrence Sondhaus, *Naval Warfare 1815–1914* (London and New York: Routledge, 2001), p. 36.
24 Lawrence Sondhaus, *Naval Warfare 1815–1914*, pp. 22–23.
25 Bogue was a corruption of the Portuguese *Boca Tigre* (Tiger's Gate), itself an interpretation of the *Ha-mun*, or river narrows.
26 Secret Department, China Correspondence (1841), IOR/L/PS/9/194-5, India Office Records, British Library, London.
27 Robert Rait, *The Life and Campaigns of Hugh First Viscount Gough, Field Marshal* (London: Archibald Constable, 1903).
28 *Illustrated London News*, 17 September 1859; *North China Herald* cited in Robert Giddings, *Imperial Echoes* (London: Leo Cooper, 1996), pp. 78–80; Field Marshal Sir Garnet Wolseley, *The Story of a Soldier's Life* (London: A. Constable and Co. Ltd, 1903).
29 Papers of Major-General Philip Durham Henderson, Madras Cavalry 1857–1895, Mss Eur D/1205/5, BL.
30 Byron Farwell, *Queen Victoria's Little Wars*, p. 140.
31 Wang Gungwu, *Anglo-Chinese Encounters*, p. 15.
32 Guandong Sheng Wenshi Yanjiuguan (ed.), *Sanyuanli renmin kang Ying douzheng shiliao* (Documents on the Anti-British Struggle by the People of Sanyuanli) (Beijing: Zhonghua Publishing, 1978).
33 John K. Fairbank, *Rebellion and its Enemies in Late Imperial China, Militarisation and Social Structure, 1796–1864* (Cambridge, MA: Harvard University Press, 1964).
34 Mao Haijian, *Tiancho de Bengkui: Yapian zhanzheng zai yangyui* (The Collapse of the Heavenly Kingdom) (Beijing: Sanlian, 1995). This also appeared in *China Review International*, vol. 15, no. 4 (2008), pp. 533–538.

Index

Page numbers in *italics* denote tables, those in **bold** denote figures.

Abdali, Ahmad Shah 94–5, 96, 97
Afghanistan 6, 91–120; Afghans, origin of 91–2; Ali Masjid fort 110; Baluchis 112; British policy towards tribes 92–3; counter-insurgency (COIN) 103, 114; First Anglo-Afghan War (1839–1842) 101–10; geographical information 91; Gilzai tribes 107–8; Great Game 100, 113; guerrilla warfare 97, 99; invasion of Persia 93–4; iron ore 111–12; Islam 92; Kabul insurrection, November 1841 108–9; Kuhurd-Kabul pass 108; logistics 105, *106*, 107; Mughal decline and Afghan revival 1707–1830 93–100; Pathan tribes 111, 113; prelude to British intervention 100–1; raids on India 95–7; raids on India against the British 98–9; small war along the Indus frontier (1843–1860) 111–13; subsidies 114; Waziri tribes 111–12, 113
Afif, Shams-i Siraj 78
Ahmad Shah 95, 96–7
Ain-i Akbari (Fazl) 80–2, 83, 84–5
Ajatasatru 35
Akbar 80–2, 83, 85, 87, 328
Akbarnama 148
Alamgirnama (Kazim) 84
Alexander 35, 75
Ali, Haider 302, 303–4, 304–5, 306; administrative and revenue reforms 315–16; confrontations with the British 317–18; contemporaries' opinion of 314–15; military reforms 315–16; and the revolution in South India 314–18; success of 319; war manual 316
Ali, Rustam 94
Allahabad *Prasasti* 41, 42
Amatya, Ramchandrapant 183
amphibious warfare 146–65; boats 154–5, 157, 159–60; command of fortified locations 153–4; early amphibious battles 147; geography and geopolitics of Mughal amphibious warfare 149–51; historical sources 148; logistics of river warfare 155–62; manpower and war animals 160–2; military aspects of Mughal amphibious engagements 152–5; Mughal campaigns in Indus and Ganga-Brahmaputra Deltas, outline 151–2; post-nomadic empires 146–7
Amyatt, Peter 308–9
Analects, The (Confucius) 216, 219
Ancient Greek Way of Warfare 1, 2
Angre, Kanhoji 189–92, 193
Anguttara Nikaya 34
archery/archers 2, 5, 6, 7, 62, 147, 327, 328
armies: civil domination culture 12–13; court control 10; and external security 47–8; military culture, control and construction of armies 8–13; mutinies 13; race and ethnicity, use in construction 13
Army of Indus 102–3
Art of War (Sunzi) 5
Arthasastra (Kautilya) 4–5, 34, 37–8, 39, 42, 44, 45, 46, 47; on elephants 75–6
artillery 7, 78–9, 94, 112, 131, 174, 176–7, 293, 306; casting cannon 136, 310, 336; China Wars (1839–1860) 356, 357; Chinese use of cannons against Mongols 122–7; exclusion of Indians from knowledge of 336; *folangji* 136; *hongyi pao* 136; *kodituwakku* 293; Ming-

Manchu war 134–45; mortars 298–9; water transport 152–3
Aryans 74–5, 91–2
Asian Elephant, The: Ecology and Management (Sukumar) 74
Asoka Maurya 35–7
Ataide, Pêro de 169
Atharvaveda 75
Atkinson, James 105, 107
Aurangzeb 85, 93, 186, 188

Babur 78–9, 155, 159
Baburnama 159, 162
Badakshan 5
Baharistan-i Ghaybi (Nathan) 148, 154, 158–9, 160–1, 162
Bai Juyi 208
Baillie, Colonel 302
Balkh 5
bannerman painters 224–33; demand for Zhu Lunhan's paintings 227; finger painting 227–31, **230**; hybridity discourse in the Qing social structure 231; influence of Gao Qipei on Zhu Lunhan 228; military skills and career of Zhu Lunhan 225–6; *Mountain Landscape with Architecture* (Zhu Lunhan) 229–31, **230**; praise for Zhu Lunhan's poetry 225–6; as a symbol of *wu* 231–2; tiger paintings 228, **229**; Zhu's physical appearance 224–5, **225**
Baqi hua lu 224
Baqi wenjing 225
Barakzai 97–8
Barani, Ziya ud-din 77
Barfield, Thomas 146
Barr, Lieutenant William 103
beasts of burden 105, *106*, 107
Beaver, Heber 299
Bellew, H. W. 91–2
Bengal 148, 149, 151–2; army 10, 113, 331–2, 333, 337–8, 339, 342, 348; Mir Qasim and politico-military reform in Bengal 307–14
Bérrio (ship) 168
bheda (dissension) 45
Bhimbetka 74
Bian Lingcheng 205, 209
Bimbisara 34–5, 39
Black, Jeremy 1, 2
boats 154–5, 157, 159–60
Bombay 186, 187, 188, 190, 330, 348
Book of Songs 220
Boone, Charles 190–2
border defence 121–2
border garrisons, Liaodong 57–73; anxieties about Liaodong as a border region 62; horse markets 62–4; integration of Mongol, Jurchen and Korean populations 65–6; intelligence 66–8; Koryŏ and Ming conquest 58–61; military capacity 61; military labour 64–5; Mongol transformation and legacy 57–8; paradoxes of a border region 61–8; private trade and bonuses 64; state investment 61–2
Brahmanism 2, 39
bribery 5, 19, 45, 204, 206, 234, 240
bridges 155–9
Brihadratha 10
Brihat Samhita 76
Britain: conquests and colonial historiography, India 304–7; First Anglo-Afghan War (1839–1842) 101–10; military confrontations with Kandyans 291, 294–8; naval attempts to dislodge Khanoji Angre 190–2; power in India 303–4; prelude to intervention in Afghanistan 100–1; tactical advantages in Sri Lanka 298–300
British-Indian Army 9; First Anglo-Afghan War (1839–1842) 101–10; logistics 105, *106*, 107; officers 107, 111; *see also* China Wars (1839–1860); East India Company (EIC)
Burke, Edmund 305
Burnes, Alexander 101
buxaria 326–7, 332

C3I (Command, Control, Communications and Intelligence) 9
Cabral, Pedro Álvares 168–9, 177
Caldwell, Earnest xii, 17–31
camels 95, 104, 105, *106*, 107
camp followers 297, 361
Cao Mie's Battle Array 5, 17–31; assessing the strengths and weaknesses of enemy leadership 21–2; deployment of the forces, opportune moments 21–3; discovery and description 17–18; doubt 26–7; failure by rulers to invest in generals 22–3; *ji* and the question of perspective 19–21; morale 27, 29; opportune moments after battle 27–9; opportune moments during battle 25–7; opportune moments when the enemy is scattered or congregating 23–4; rewards and punishments 27–9; translation of opportune moments 18–19

Cao Shuji 260
Caroe, Olaf 92
Carre, Abbe 183
cavalry 9, 95, 147, 289, 315, 336, 351, 357
cave paintings, Bhimbetka 74
Cen Shen 215, 218
Chakraborty, Dabraj xii, 7, 74–90
Chakravarti, P.C. 33
Chakravarti, Ranabir 42, 43
Chandra Gupta I 41
Chandragupta II Vikramaditya 42
Chandragupta Maurya 35
Changzhuo, battle of 25–6, 30
chariots 35, 48, 75, 126
Charney, Michael 148
Chase, Kenneth 6
Chen Jian 19, 20
chiefdoms 32
China: cannons, impact on the border war with Mongols 122–7; development of gunpowder weaponry 135–6; fortifications in the war against the Mongols 127–30; Great Wall of China 12, 122, 126, 127–8; and India: juxtaposing and interlinking military history 2–4; opium 349–50; problems of comparing Chinese and Indian military histories 4; strategies to fight the Mongols 121–2
China Wars (1839–1860) 347–67; army drill and discipline 353; British objectives 365; cavalry 357; Chinese army classes 353; Chinese army marching formations 351–2; Chinese navy 354–5; East India Company (EIC) army technological developments and naval power 355–8; failure of Chinese forces 352; First Anglo-Chinese War (1839–42) 10, 349–50; First Anglo-Chinese War (1839–42), description of operations 358–62; intimidating tactics of Chinese forces 350–1, 354; militia 354; nationalist historians' interpretations 365–6; Second Anglo-Chinese War (1856–1860) 350–2; Second Anglo-Chinese War (1856–1860), description of operations 362–4; specialist army roles 353–4; storming of forts 357; weapons 354, 356–7; *Yung-ying* ('braves battalions') forces 354
Chinggisid polity 57–8, 59
Chosŏn court 66–8
Chouhai tubian (Illustrated Naval Strategy and Tactics) (Zheng Ruozeng) 126
Chu 24
Clausewitz, Carl von 5
Clive, Robert 308, 331–2, 333
Complete Collection of Tang Poetry, The (*Quan Tang shi*) 214
Confucianism 2, 61, 221, 248, 251, 350
Confucius 216, 219
corruption 234–45; confiscation of private assets 239; definitions 234; incorrect use of corruption as a term 242–3; military failure and the corruption case 236–40; possible embezzlement of funds by Zhang 239–40; reinterpretation of Zhang's case in a novel 240–1; waste of material, money and manpower by Zhang 238–9; Zhang accused of poor military methods 237; Zhang as a commander 235–6; Zhang executed 238, 240, 241, 242; Zhang involved in private intrigues, not corruption 239; Zhang's betrayal reported to the emperor 238
'Corruption in Eighteenth Century China' (article) 235
Cosmo, Nicola Di 135, 138
Cotton, Sir Sydney 86
counter-insurgency (COIN) 12, 103, 114
Cura, Nixi xii, 11, 224–33

Da Ming huidian (Collected Statutes of the Great Ming) 125, 130
Dai Shulun 221
Dai Xiuting 215
dana (bribery) 45
danda (force) 37, 45, 46–7
Delhi Sultanate 76–7, 83
dhamma 36–7
dhammavijaya 36
Digby, Simon 77
Dikshitar, V.R. Ramachandra 33, 48
disease 297, 298
Dost Muhammad Khan 100–1, 102, 107, 112
doubt 26–7
Dow, Alexander 305
Du Song 137
Duff, James Grant 183
Durand, Henry 101, 105
Duxinga 267, 270

East India Company (EIC) 99, 112, 188, 303, 305; Addiscombe Military Seminary Committee 340–1; army drill and discipline 333; army officers 331;

army technological developments and naval power 355–8; army uniforms 332; Bengal army 331–2, 333, 337–8, 339, 342, 348; Bombay army 330, 348; British naval attempts to dislodge Khanoji Angre 190–2; British preconceptions of martial characteristics of different races 333, 337; British racial attitudes in the EIC headquarters in London 338–41; capability of Indian soldiers to conform to European standards 335; cavalry 336; and the China Wars (1839–1860) 347–67; commerce and war 348–52; confrontations with Haider Ali 317–18; confrontations with Mir Qasim 312–14; contrasting Indian and British concepts of race and authority in EIC armies 324–46; EIC armies as constructed in India by Britons and Indians 330–8; Emin case 339; European soldiers 334–5; exclusion of Indians from knowledge of artillery 336; First Anglo-Chinese War (1839–42) 349–50, 358–62; gulf between European and Indian army officers 348–9; Indian armies under the British Raj (1858–1947) 342; Madras Army 330–1, 338, 348, 355–6; mixed race officers 336, 339–40; officer commissions 338–9; patronage 338, 339; Polilur 302; precursors to the EIC's armies 326–30; 'race' and the EIC's colonial armies 324–5; racial boundaries 335–6; Second Anglo-Chinese War (1856–1860) 350–2, 362–4; Turing case 339–40; visual and biological criteria for officer/civilian official selection 339–41

Eight Banner Army 261, 266, 352–3

elephants 7, 74–90, 105, *106*, 156, 158, 161–2, 168–9; African elephants 83; care of 76; classification of 81; decline in population 86–7; domestication and use 74–5; elephant forests 75–6; fierceness 81–2; fighting 82; military exercises 86; in Mughal India 78–86; Mughal rulers fondness for 79–80; numbers of 75, 77, 83–4; in pre-Mughal India 74–8; price of 84; sport hunting 86–7; supply sources 77–8; training 82–3, 85; trapping 78, 82, 84–5, 162; war elephants 75, 76, 77, 80; worship of 76

Elephants by the River (painting) 79–80

Elliott, Mark C. 224

Ellis, William 308–9, 312

Emin, Emin Joseph 339

Eminent Chinese of the Ch'ing Period (Hummel) 235

eunuchs 10, 203–13; appointment as commander of a major imperial army 208; army supervisors, introduction and role of 205–6; blame for failure of campaigns 207–8; commandant-protectors 203; criticism of eunuch supervisors 204, 206, 208–9; denunciations of military governors 206–7; direct, operational control of armies 208; Flying Dragon Palace Army 204; An Lushan rebellion 205; palace censorate 204; rarity of punishment 210; role in advising generals 205; successful army supervisors 209–10; usurping of military governors' command authority 207; venality 206

Fan Zuyu 206, 209

Farooque, Abdul Kair Muhammad 158

Fazl, Abul 80–2, 83, 84–5, 151–2, 159, 160

Filipiak, Kai xii, 6–7, 121–33

firearms 79, 94, 102, 103, 293, 295, 298, 356–7; impact on the Chinese/Mongol border war 122–7

First Anglo-Afghan War (1839–1842) 101–10

First Anglo-Chinese War (1839–42) 10, 349–50; description of operations 358–62

First Jinchuan War (1747–1749) 11, 234–45

Firuz Shah (Sultan) 78

Fisher, Michael H. xii, 13, 324–46

Fitch, Ralph 151

fortifications 6–7, 8, 47, 125, 131, 289; *bao*, definition of 249; *bao* fortification and the Miao revolt 249–51; border revolts and local defence initiatives 247–8; Chinese/Mongol border wars 127–30; command of 153–4; fortresses 128–9; local defence and imperial objectives 248–9; mud forts 153–4, 161; and social ordering 246–56; Sri Lanka 295–6; towers 129–30, *130*; walls 12, 122, 126, 127–8, 246–56; White Lotus revolt and *bao* fortification 251–3

Fu Nai 250–1

Fujian tongzhi (Local Records of Fujian) 124

Gajasastra 75
Gama, Vasco da 166, 167, 168, 170
Ganga-Bhamaputra Delta *see* amphibious warfare
Gansu 12–13, 270–8, 279–81
Gao Bing 228
Gao Genghao 263
Gao Qipei 228
Gao Shi 215, 216, 217, 220, 221
Gao Xianzhi 205
Gao Youren 20
Gautamiputra Satakarni 40
gavishti 33
Ghaznavid Sultanate 77
Ghazni Fort 104–5
Ghoshal, U.N. 39, 40
Gilzai tribes 107–8
Giyasuddin Balban 77
Golnabad, Battle of 93
Gommans, Jos 62, 92, 146–7
Gong Jinghan 251, 252, 253
Graff, David A. xiii, 10, 203–13
Great Wall of China 12, 122, 126, 127–8
Green Standard 261, 266, 352, 353
Griffith, Samuel B. 203
guerrillas 9, 97, 99, 292, 293–4, 299–300
guest armies 12–13, 265, 271, 272–3, 276
gunpowder empires 134
Guochao qixian leizheng 235
Gurkhas 337

Habib, Irfan 76, 84
Haider Ali 8
Han shu 17
Hanson, Victor Davies 1, 2
Historiography Institute 235
Holland: military confrontations between Dutch and Kandyans 290–1, 294–8; and Sri Lanka 290; tactical advantages in Sri Lanka 298–300
Hong (battle 637 BCE) 24
Hongwu 59–60, 61, 62, 63, 65
horses 2, 6, 11, 100, 146, 336; horse markets, Liaodong 62–4
Hu Zhenheng 215, 216
Huang Taiji 7, 140, 141
Humayun 80
Hummel, W. 235
Hydaspes, Battle of 35

India: Afghanistan raids on India 95–7; and China: juxtaposing and interlinking military history 2–4; elephants in pre-modern India 74–90; military revolution and state formation 302–23; problems of comparing Chinese and Indian military histories 4; *yuddha* and *vijaya* concepts of war and conquest 32–54; *see also* naval warfare
Indian Navy 357–8
intelligence, border garrisons, Liaodong 66–8
Islam 2, 92

Jafar, Mir 307–8
Jahangir 83, 85
Jai Singh 185
Japan 134
jatis (birth groups) 326–7, 342
jemadars (jobbers) 327, 330, 331
Jesuits 135, 138
Jian, Ming xiii, 11
Jiang, Ming 214–23
Jiang Shiquan 226
Jiang Tonggang 257–8
Jiaqing 247–8, 249, 251
Jiu Tang shu (Old Tang History) 206, 209
Jixiao xinshu (Qi Jiguang) 124–5, 126, 136
Johnson, Rob xiii, 9, 347–67
Jones, Susan Mann 253
Jouher 80
Journal of Global History 146
Jurchens 11, 65–6, 67–8, 139–40

Kalidasa 76
Kalinga 36
Kam Louie 218
Kamandaka 4, 42–4, 45–6, 47
Kamran Shah 101
Kandahar 94
Kandy 9; *see also* Sri Lanka
Karnal, Battle of 94
Kaske, Elizabeth xiii, 12, 257–88
Kathasaritsagara (Ocean of Stories) 45
Kautilya 4–5, 34, 36, 37–8, 39; on elephants 75–6; legacies of: armies and external security 47–8; legacies of: spies and internal security 44–7
Kazim, Muhammad 84
Keigwin, Richard 187
Khan, Ahmad 95
Khan Chishti 152
Khan, Ghulam 311
Khan, Khafi 186
Khan, Muhammad Yusuf 331
Khan, Zain 80
Khanderi 186, 191
Khazanov, Anatoly 146

Ki Sayin-Temür 59
Kirtivarman II 43
Knox, Robert 296
Kongmin 57–8, 58–61
Korea 66–7, 134
Koryŏ 57–8, 63; and Liaodong 58–61
Kuhn, Philip 253, 261
Kulkarni, V.B. 184
Kumar, Amarendra xiii, 8, 181–99
Kushanas 40

Lahore 95–6
Lee, Wayne E. 1
Lei Zhengwan 259, 264, 272, 273, 274–5, 281
Levenson, Joseph 221
Li Bai 220–1
Li Xizhong 221
Li Zhongxi 222
Lianbing shiji (Practical Handbook on Military Training) (Qi Jiguang) 125, 126, 129
liangtai (grain station) 263, 279–80; failure of the independent *liangtai* 270–8; independent *liangtai* 265–70
Liaodong Gazetteer 62
Liaodong (southern Manchuria) *see* border garrisons, Liaodong
Ling, Li 19–20, 21, 23
Liu Tao 22–3
Liu Yi 60
local defence 246; *bao* fortification and the Miao revolt 249–51; border revolts and local defence initiatives 247–8; and imperial objectives 248–9; and social ordering 246–56; White Lotus revolt and *bao* fortification 251–3
Loch, Captain Grenville 352
logistics 105, *106*, 107, 263, 264–5; boats 157, 159–60; bridges 155–9; food supply problems 297–8; manpower and war animals 160–2; riverine warfare 155–62; Sri Lanka 296–7
Lorge, Peter xiii, 1–14, 327
Lower Indus Basin *see* amphibious warfare
loyalty 307, 319, 328

Machiavelli, Niccolo 5
McMahon, Daniel xiii, 12, 246–56
Macrory, Patrick 102
Madras 330–1, 338, 348
Magadha 34–6
Mahabharata 40
mahajanapadas 34
Mahapadmananda 35
Mahavastu 40
Mahmud 99–100
Malacca 167
Man Singh 152
Manasollasa (Delight of the Mind) 45, 76
Manava-Dharmasastra 34, 39–41, 46, 47
Manchu (Qing) Dynasty 6, 146; *see also* Ming-Manchu war
Mannu, Mir 95, 96
mansabdars 328–9
Manu 39–41
Manucci, Niccolao 82, 85–6, 329
Manuel I, Dom 169, 181
Manzhou shilu 138
Manzhou yuanliu kao (Researches on Manchu Origins) 231
Mao Haijian 366
Marathas 96–7, 182–6, 317
Marshall, P.J. 306
Masum, Mir 148
Mathew, K.S. xiii, 8, 166–80
Maurya Empire 10
Mehrab Khan 103–4
Memorial on the Production of Firearms (Weng Wanda) 126
Mencius 219, 220
mercenaries 12–13
Meri (ship) 170, 177
Miao revolt 246, 247, 249–51, 254, 260
military control politics on the Indian west coast 1650–1730 181–99, 187; actions against the Portuguese 187–8, 188–9; attack on Bombay 188; British naval attempts to dislodge Khanoji Angre 190–2; control of the sea notion 181, 182; *Estado da India* 181, 182, 184; European traders 184; Janjira, importance of 185; Maratha-*Sidi* peace 190; Marathas 182–6, 187; Portuguese deceit 184–5; Portuguese dominance of Indian waters 181–2; *Sidis* of Janjira 185, 186, 187, 188, 190
military culture 246
military history: China and India: juxtaposing and interlinking military history 2–4; as a Eurocentric enterprise 1; focus on modern times 1; Military Revolution thesis 1–2, 303, 305; problems of comparing Chinese and Indian military histories 4; West versus the rest paradigm 3; Western Way in Warfare thesis 1–2
military labour 64–5

military revolution 302–23; abuse of privileges by the British 311–12; administrative and revenue reforms of Haider Ali 315–16; Ali Haider and the revolution in South India 314–18; British confrontations with Haider Ali 317–18, 319; British confrontations with Mir Qasim 312–14; British conquests and colonial historiography, India 304–7; challenges to notions of Western military superiority 306–7; exaggeration of the gulf between EIC armies and their opponents 305–6; fragmented nature of Indian politics 306–7; Haider Ali's military reforms 315; Mir Qasim and politico-military reform in Bengal 307–14; Mir Qasim's army reforms 309–10; orientalism 305–6; tightening of administrative controls by Mir Qasim 310–11
Military Revolution, The (Parker) 1
Military Revolution thesis 1–2, 303, 305
military supply and civilian resources in China 257–88; 'braves' (volunteer irregular forces) 261; Central Military Supply Bureau 280; contributions from wealthy landowners 259, 276–7; devastation of towns and villages 258; establishing an empire-wide supply network 279–81; failure of the independent *liangtai* 270–8; funding of total war 261; guest armies 12–13, 265, 271, 273–4, 276; independent *liangtai* 263, 265–70, 279–80; interprovincial assistance 264–5; Jiangbei LHQ 267, 268–9; Jiangnan *liangtai* 266–7; local militia 261; management of army supplies 263–5; military mobilization 261; Ningzhou incident 275; population losses of total war 260; purchase of supplies 257–8; rebellions of 1850s and 1860s in China 259–60; revenue sources 264, 266, 267–8; salary tickets 268; taxation 261–2, 264; total war 259–63; Western Expedition Campaign Fund 279–80; Xuzhou *liangtai* 267–8
Mill, James 305
Ming dynasty (1348–1644) 6; border garrisons, Liaodong 57–73; *see also* Ming-Manchu war
Ming huidan 136
Ming-Manchu war 134–45; creation of first Latter Jin firearms division 141; Dalinghe 140–1; firepower and Jurchen defeats 139–40; Jesuits 135, 138; Jinzhou 140; Ming use of overwhelming firepower 137–8; Ming use of religious contacts to procure weapons 138–9; Ningyuan, Battle of 139–40; Sarhu, Battle of 137; Song-Jin, Battle of 141
Mirat-i Ahmadi 84
Mohan Lal 104, 108
Mongols 57–8, 65, 121–33; cannons, impact on the border war with China 122–7; Chinese strategies to fight the Mongols 121–2; fortifications, use by Chinese in the war against the Mongols 127–30
Monserrate, Father 148, 156, 157, 158
Moosvi, Shireen 84
morale 27, 29
Morillo, Stephen 1
Moriyoshi, Hasumi 66–7
Mountain Landscape with Architecture (Zhu Lunhan) 229–31, **230**
mud 8, 153–4, 161
Mudraraksa 41
Mughal Empire 93, 94, 96, 182; amphibious warfare 146–65; military forces and racial differentiation 327–9; as a post-nomadic empire 146–7; *see also* elephants
Mukherjee, B.N. 36, 40, 48
mules 105, *106*
Munro, Major Sir Hector 302, 313, 335
Muntakhabut-Tawarikh 148
murals 302
Muslim Rebellion of Shaanxi and Gansu 258, 260, 270–8, 279–81
Mustafa, Haji 314
mutiny 10, 13, 206, 209, 211, 274, 277, 307–8, 335, 349, 365

Nadir Shah 93–4, 95
Nair, Janaki 302
Nath, Pratyay xiii, 7–8, 146–65
Nathan, Mirza 148, 159, 162
naval warfare 7–8, 166–80, 300; Angediva 173; Cannanore confrontation 173; capture of ship and elephants 168–9; caravels 175, 176, 179n23; Chaul confrontation 174–5, 176, 177; Chinese navy 354–5, 365; Chittagong 176; Cochin, attack on 172; Cumbalam confrontations 171; Diu 175, 177–8; encounter between Vasco da Gama and the Zamorin of Calicut (1498) 168; fustas 176; gales 175–6, 179n24; Goa

175; Honovar encounter 172; Indian merchant ships 166–7; Indian navigation 167–8, 177; Indian Navy 357–8; Indo-Portuguese naval engagement 31 December 1501 169–70; loss of Portuguese prestige 173–4; naval engagement in Arabian Sea 1502 170–1; naval guns 177, 357–8; Palliport encounter 172; Pantalayani encounter 172; Ponnani 174; Portuguese naval superiority 168–9, 170; shipbuilding techniques 178; steam ships 357; weakness of Indian vessels 177
Nayapaladeva 46
Needham, Joseph 124
Nehru, Jawaharlal 302
Nemesis (ship) 357–8, 359
New Book of Effective Discipline (Qi Jiguang) 123
Nian Rebellion 260, 279
Ningyuan, Battle of 139–40
niti (moral action) 44
Nitisara 42–4
Nitivakyamrta 44
Nocin 235, 237, 238–9, 240, 241
Nova, João de 169–70, 177
Nurhaci 7

Official History of the Koryŏ Dynasty, The 58–9, 60
opium 349–50
opportune moments 17, 18–29; after battle 27–9; assessing the strengths and weaknesses of enemy leadership 21–2; during battle 25–7; for deployment of the forces 21–3; doubt 26–7; failure by rulers to invest in generals 22–3; *ji* and the question of perspective 19–21; morale 27, 29; rewards and punishments 27–9; translation of section in *Cao Mie's Battle Array* 18–19; when the enemy is scattered or congregating 23–4
orientalism 305–6
Ottoman Empire 134, 146
Ovington, A. 188

Pacheco, Duarte 171, 172
Pancatantra 41
Panipat, First Battle of 78–9, 80
Panipat, Third Battle of 97
Park, Nancy 242
Parker, Geoffrey 1
patronage 338, 339
peace 38
Peers, Douglas M. xiii–xiv, 8, 302–23
Perdue, Peter 246
Persia 93–4
Peshawar 100–1
Peterson, Charles 210
Pingding Jinchuan fanglüe 235
pirates 9, 168
Pires, Tomé 167
poetry 11, 214–23, 257–8; access of poets and literati to public office via military service 216–17; Civil Service Examinations 215, 216; constructing the meaning of war 218–22; distinction between Chinese and non-Chinese 220; growing partiality towards *wu* 215; high value of *wu* 216; Hua-Yi distinction 220; just war concept 219–21; landscapes, description of 218; masculine and vigorous tone of poems 217–18; poets' view that war added value to their lives 221; shifting *wen–wu* dyad in Tang frontier poetry 214–18; strengthening of *wen* 215; visits to frontier areas by poets 215, 222n5
Polilur 302
polity, science of 43, 44
Porter, Patrick 305
Portugal 8; actions against, in India 187–8, 188–9; deceit 184–5; dominance of Indian waters 181–2; naval warfare 166–80; and Sri Lanka 289–90
post-nomadic empires 146–7
Powindah Khan 98
Principles for the Construction of Fighting Towers (Qi Jiguang) 129–30
punishments 27–9
Purbiyas 13
Puru 35

Qasim, Mir 8, 303–4, 306; army reforms 308–9; and politico-military reform in Bengal 307–14; relations with British 308–9; tightening of administrative controls 310–11
Qi Jiguang 123, 124–5, 126, 127, 128, 129–30, 136
Qiao Songnian 268–9
Qing (Manchu) Dynasty 11–12; bannerman painters 224–33; fortified walls and social ordering 246–56; *see also* Ming-Manchu war
Qingguoshi 235
Qingshi liezhuan 235
Qingshilu 235

racial differentiation, EIC armies 13, 324–46; Addiscombe Military Seminary Committee 340–1; Bengal army 331–2, 333, 337–8, 339, 342; Bombay army 330; British preconceptions of martial characteristics of different races 333, 337; British racial attitudes in the EIC headquarters in London 338–41; *buxaria* 326–7, 332; capability of Indian soldiers to conform to European standards 335; caste 327; cavalry 336; 'effeminate races' and 'martial races' 342; EIC armies as constructed in India by Britons and Indians 330–8; Emin case 339; European cultures and technologies 329; European soldiers 334–5; exclusion of Indians from knowledge of artillery 336; Indian armies under the British Raj (1858–1947) 342; *jatis* (birth groups) 326–7, 342; Madras Army 330–1, 338; mixed race officers 336, 339–40; Mughal forces 327–9; officer commissions 338–9; patronage 338, 339; racial boundaries 335–6; Rajputs 326, 330; religious traditions 327; renting troops to Indian allies 337; sepoys 329, 332, 333, 337; South Asia 326–7; *Telinga* 329, 330–1, 332; *topasi* 329, 332; Turing case 339–40; visual and biological criteria for officer/civilian official selection 339–41
Raghuvansham (Kalidasa) 76
raja (king) 32
Rajniti (Amatya) 183, 184
Rajputs 326, 330
Ray, Krishnendu xiv, 32–54
Ray, Rajat 308, 311, 312
Raychaudhuri, H.C. 42
Ren Wenjing 215
rewards 27–9
Rigveda 33, 34, 44, 45, 75
risk assessment 22
riverine warfare *see* amphibious warfare
Roberts, Colonel A. 105
Roberts, Michael 1
Robinson, David M. xiv, 6, 57–73
Roy, Kaushik xiv, 6, 91–120
Roy, Nirodbhusan 92
Rudradman 48
Russia 100

Sadiki, Muhammad Muhsin 94
Sadiq, Mir 302
Safavids 146
Sale, Robert 107–8, 109
sama (conciliation) 39, 40, 45
Sambhaji 186–8
Samudra Gupta 41–2
Sanderson, G.P. 81
sandhi 38
Sanshui 257–8
Sarhu, Battle of 137
Sarkar, Jadunath 75
Sarwani, Abbas Khan 80
Saundarananda 41
Second Anglo-Chinese War (1856–1860) 10, 350–2; description of operations 362–4
Second Anglo-Sikh War (1848–1849) 100
security: external security 47–8; internal security 44–7
Seductress of Jinchuan (Xu Zhiyan) 241
Sejo 67
Sen, S.N. 189–90
sepoys 113, 294–5, 308, 310, 313, 329, 330, 332, 333, 337, 348
Serruys, Henry 129–30
Shivaji 182–6, 185–6
Shuja 99, 100, 102, 104, 107, 109
sieges 289
Sima Guang 206
Sind 148, 149
Singh, Ranjit 101
Sirhindi, Abdullah 78
Skinner, James 336
small war 103, 104, 111–13
Smith, Charles 304–5
social ordering *see* local defence
Sodré, Vicente 170–1
Soma 33, 75
Somadeva Bhatta 44, 45
Somadeva III 76
Somesvara 45
Song-Jin, Battle of 141
spies 44–7
Spring and Autumn Annals of Mister Lü, The (*Lüshi chunqiu*) 219
Sri Lanka 289–301; camp followers 297; devastation to peasantry 300; disease 297, 298; European armies and the Kandyan challenge 294–8; European forts 295–6; European naval power 300; European tactical advantages 298–300; Europeans against Kandy 1761–1818 290–1; food supply problems 297–8; foreign troops 292–3; Kandyan economy 292; Kandyan guerrilla

warfare 293–4, 299–300; Kandyan military 291–3; logistical problems for Europeans 296–7; Malay and sepoy troops 294–5; peasant levies 292; Portuguese influence 289–90; rugged terrain 296; weapons 293, 295, 298
state formation 302–23; abuse of privileges by the British 311–12; administrative and revenue reforms of Haider Ali 315–16; Ali Haider and the revolution in South India 314–18; British confrontations with Haider Ali 317–18; British confrontations with Mir Qasim 312–14; British conquests and colonial historiography, India 304–7; challenges to notions of Western military superiority 306–7; exaggeration of the gulf between EIC armies and their opponents 305–6; fragmented nature of Indian politics 306–7; Haider Ali's military reforms 315; Mir Qasim and politico-military reform in Bengal 307–14; Mir Qasim's army reforms 309–10; orientalism 305–6; tightening of administrative controls by Mir Qasim 310–11
steam ships 357
Stein, Burton 314
subsidies 5, 6, 114
Sukumar, Raman 74
Sun Bin bingfa 22
Sun Chengzong 139
Sun Yuanhua 139
Sunga, Pushyamitra 10
Sunzi bingfa 17
Sunzi (Sun Tzu) 4, 5, 28
Swope, Kenneth M. xiv, 7, 134–45

Tabaqat-i-Akbari 148
Tabaqat-i Baburi (Khan) 80
Taiping Rebellion 259–60
Tang jian (Mirror of Tang) 206
Tang period 10, 11; cultural construction of war in frontier poetry 214–23; eunuchs as army supervisors 203–13
Tarikh-i Firuzshahi (Afif) 78
Tarikh-i Mubarakshahi (Sirhindi) 78
Tarikh-i-Sher Shahi (Sarwani) 80
Tarikh-i Sind (Mir Masum) 148
technology 5; metal technology 32; *see also* weapons
Telinga 329, 330–1, 332
Ten Kings, Battle of 33–4
Teng Wen Gong I 220
Tezkereh-al Vakiat (Jouher) 80
Thapliyal, U.P. 33, 48
Theobald, Ulrich xiv, 11, 234–45
Third Anglo-Maratha War (1817–1818) 100
Tian Ji 28
Tiancho de Bengkui (Mao Haijian) 366
Tibet 236, 237–8
Tiebao 224
Timoja 168
Timur Shah 97–8
Tipu (Sultan of Mysore) 8, 302, 304, 306
Tong Yangxing 140, 141
topasi 329, 332
towers 129–30, *130*
treaties 38
Treatise on Fingerpainting (Gao Bing) 228
Tripodi, Christian 92–3, 111
Turing, John 339–40
tusnidanda 44
Tutu Chengcui 208
Tuzuk-i Jahangiri 85, 148

udasina 37
al-Umari, Ibn Fazl Allah 75
Upanishads 75

Vakpati, Guadavaho 76
Vansittart, Henry 308, 311, 312
Veritable Records of Chosŏn Dynasty 67
Veritable Records of King Sejo 67
Veritable Records of Taizu 65
Veritable Records of Yingzong 67
vijaya concept (conquest) 32–54, 33; armies and external security 47–8; Asoka's *dhamma* 35–7; Kamandaka's *Nitisara* 42–4, 47; Kautilya *Arthasastra* 5, 34, 37–8, 42, 44, 45, 46, 47; Manu's *Manava-Dharmasastra* 39–41, 46, 47; politics, warfare and theories of warfare in the Gupta age 41–2; spies and internal security 44–7; Vedic era 33–5
vijigisuraja (king desirous of conquest) 37–8
Visakhadatta 41
Visnusarma 41
vyuhas (battle arrays) 5

Wade, Thomas 261
Waley-Cohen, Joanna 254
walls 12, 122, 126, 127–8, 246–56
Wang Gungwu 365
Wang Hong 125

Wang Wei 215, 218, 219, 221
Wang Zhaochun 123, 124
warfare: armed forces, technology and warfare 5–7; military culture, control and construction of armies 8–13; theories 4–5
Warring States period (*c.*475–221 BCE) 17
Warrior, HMS 358
weapons 48, 260, 289, 292; archery/archers 2, 5, 6, 7, 62, 147, 327, 328; artillery 7, 78–9, 94, 112, 122–7, 131, 152–3, 174, 176–7, 293, 298–9, 306, 310, 336, 356, 357; Chinese army 354; East India Company (EIC) armies 332–3; firearms 79, 94, 102, 103, 293, 295, 298, 356–7; *folangji* 136; *hongyi pao* 136; *kodituwakku* 293; naval guns 357–8; quality 310; water transport 152–3; *zamburaks* 95
Wei Liaozi 20
Wei Yuan 249, 365
wen (civil) 10–11, 228, 231; relationship with *wu* 224; shifting *wen–wu* dyad in Tang frontier poetry 214–18
Weng Wanda 123–4, 126, 127
Western Way in Warfare thesis 1–2
White Lotus rebellion 246, 247, 248, 251–3, 254
Wickremesekera, Channa xiv, 9, 289–301
Wink, Andre 146
wu (military) 10–11, 231; and bannerman painters 231–2; relationship with *wen* 224; shifting *wen–wu* dyad in Tang frontier poetry 214–18
Wu Qi 24
Wuzi 4, 20, 21, 22, 24

Xian-zong 208
Xiang of Song, Duke 24
Xiangyue (Community Contract) (Yin Geng) 128–9
Xianqingtang shiji (Zhu Lunhan) 224–5, **225**, 226
Xiao Fu 208–9
Xichao yasong yi (Tiebao) 224, 225–6, 227
Xin Tang shu (New Tang History) 206
Xu Guangqi 138
Xu Jian 226
Xu Zhiyan 241
Xue Xingzhen 206–7

Yajnavalkyasmriti 34, 39, 43, 46
Yajurveda 75
Yan Ruyi 249, 250, 253
Yang Fuguang 209–10
Yang Jiong 215
Yang Nengge 272–3, 274, 277
Yang Yiqing 128
Yang Yuebin 275, 277, 278
Yao Nai 226
Yao Nanzhong 206–7
Yar Muhammad, Khan 101
Yin Geng 128–9
Yoshihiro, Kawachi 66–7
Yu Qian 68
Yuan court 59–61
Yuan Qianyao 221
yuddha concept (war) 32–54, 33; armies and external security 47–8; Asoka's *dhamma* 35–7; Kamandaka's *Nitisara* 42–4, 47; Kautilya *Arthasastra* 5, 34, 37–8, 42, 44, 45, 46, 47; Manu's *Manava-Dharmasastra* 39–41, 46, 47; politics, warfare and theories of warfare in the Gupta age 41–2; spies and internal security 44–7; Vedic era 33–5
Yue Zhongqi 235, 237–8, 241, 242–3
Yunnan Muslim Rebellion 260

Zaman Shah 98–9
Zamorin of Calicut 168–9, 170–2, 173
Zhang Bin 218
Zhang Geng 231
Zhang Guangsi 11, 234; accusations of using poor military methods 237; betrayal reported to the emperor 238; as a commander 235–6; execution of 238, 240, 241, 242; incorrect use of corruption as a term 242–3; involvement in private intrigues not corruption 239; military failure and the corruption case 236–40; possible embezzlement of funds 239–40; private intrigues rather than corruption 239; reinterpretation of corruption case in a novel 240–1; waste of material, money and manpower 238–9
Zhao Yi 209
Zheng Ruozeng 126
Zhou, Duke of 220
Zhou Xiancheng 274–5
Zhou Yangpu 249
Zhu Di 121–2
Zhu Lunhan 11; demand for paintings 227; finger painting 227–9; as a hybridized symbol 231; influenced by Gao Qipei 228; military skills 225–6; *Mountain Landscape with Architecture* 229–31, **230**; physical appearance 224–5, **225**;

poetry themes 226; praise of poetry and artistic achievements 226, 227; tiger paintings 228, **229**
Zhu Yuanzhang 121
Zhuang of Lu, Duke 18–19, 28, 30
Zi, Etienne 231
Žižek, Slavoj 225
Zizhi tongjian (Comprehensive Mirror for Aid in Government) 206, 209
Zou Yigui 229
Zuo Zongtang 277–8, 279, 280, 281
Zuozhuan 24, 25

Printed by PGSTL